Fundamentals of
Electrochemistry

Fundamentals of Electrochemistry

V. S. Bagotzky

A. N. Frumkin Institute of Electrochemistry
Russian Academy of Sciences
Moscow, Russia

Translated from Russian by
Klaus Müller

Plenum Press • New York and London

Library of Congress Cataloging-in-Publication Data

Bagotskiĭ, V. S. (Vladimir Sergeevich)
    Fundamentals of electrochemistry / V.S. Bagotzky ; translated from
Russian by Klaus Müller.
       p.   cm.
    Includes bibliographical references and index.
    ISBN 0-306-44338-4
    1. Electrochemistry.   I. Title.
QD553.B23   1993
541.3'7--dc20   93-12087
                                                                     CIP

ISBN 0-306-44338-4

© 1993 Plenum Press, New York
A Division of Plenum Publishing Corporation
233 Spring Street, New York, N.Y. 10013

Printed in the United States of America

FOREWORD TO THE ENGLISH EDITION

This book was first published in Russian in 1988 by Khimiya Press, Moscow. In the English edition three supplementary chapters have been added: *Photoelectrochemistry*, *Electrokinetic Processes*, and *Bioelectrochemistry*. Errors in the Russian edition have also been corrected in the English edition.

Dr. Klaus Müller from Geneva had the enormous task of translating this book into English. I am very much indebted to him for his excellent work, for valuable discussions, and for many helpful comments and remarks.

Vladimir Sergeevich Bagotzky
Moscow

FOREWORD

This book provides a rigorous yet lucid and comprehensible outline of the basic concepts (phenomena, processes, and laws) forming the subject matter of modern theoretical and applied electrochemistry.

Particular attention is given to electrochemical problems which are of fundamental significance and yet are often treated in an obscure or even incorrect way in monographs and texts. Among these problems are some which, at a first glance, appear elementary, such as the mechanism of current flow in electrolyte solutions, the nature of electrode potentials, and the values of the transport numbers in diffusion layers.

By considering the theoretical and applied aspects of electrochemistry jointly one can more readily apprehend their intimate correlation and gain a fuller insight into this science as a whole. The applied part of the book outlines the principles of some processes and illustrates their practical significance but does not describe technical or engineering details, nor the design of specific equipment, as these can be found in specialized treatises on applied electrochemistry.

The mathematical tools used in electrochemistry as a rule are simple. However, in books on electrochemistry one often finds equations and relations which are quite unwieldy and not transparent enough. The author's prime aim is that of elucidating the physical ideas behind the laws and relations, and of presenting all equations in the simplest possible, though still rigorous and general, form.

There is a great deal of diversity in the terminology and names used in the literature for electrochemical concepts. It is the author's aim to introduce a uniform terminology in accordance with accepted standards and recommendations. Most of the recommendations of the IUPAC Commission are used, including its suggestion to regard the cathodic current as negative; but other symbols are sometimes used when multiple meanings exist. Throughout the book, SI units are used.

For a profitable reading of the book and understanding of the material presented, the reader should know certain parts of physics (e.g., electrostatics), the basics of higher mathematics (differentiation and integration), and the basics of physical chemistry, particularly chemical thermodynamics.

The author is deeply grateful to Edvard V. Kasatkin, Candidate of Chemical Sciences, who as a lector of this book made many useful comments.

Vladimir Sergeevich Bagotzky
Moscow, October 1987

PREFACE

> Of all electrical phenomena electrolysis appears the
> most likely to furnish us with a real insight into the
> true nature of the electric current, because we find
> currents of ordinary matter and currents of electricity
> forming essential parts of the same phenomenon.
>
> James Clerk Maxwell
> *A Treatise on Electricity and Magnetism*
> Vol. 1, Oxford (1873)

Two very important fields of natural science—chemistry and the science of electricity—matured and grew vigorously during the first half of the 19th century. Electrochemistry developed simultaneously. From the very beginning, electrochemistry was not a mere peripheral field but evolved with an important degree of independence, and it also left very significant marks in the development of chemistry and of the theory of electricity.

The first electrochemical device was the Voltaic pile built in 1800. For the first time scientists now had a sufficiently stable and reliable source of electric current. Research into the properties of this current provided the basis for progress in electrodynamics and electromagnetism. The laws of interaction between electric currents (André-Marie Ampère, 1820), of proportionality between current and voltage (Georg Simon Ohm, 1827), of electromagnetic induction (Michael Faraday, 1831), of heat evolution during current flow (James Prescott Joule, 1843), and others were discovered.

Work involving the electrolysis of aqueous solutions of salts and salt melts which was performed at the same time led to the discovery and preparation of a number of new chemical elements, such as potassium and sodium (Sir Humphry Davy, 1807). Studies of current flow in solutions (Theodor von Grotthuss, 1805) formed the starting point for the concept that the molecular structure of water and other substances is polar, and also led to the so-called electrochemical theory of the structure of matter formulated by Jöns Jakob Berzelius (1820). The laws of electrolysis discovered in 1833 by Faraday had an even greater significance for knowledge concerning the structure of matter. During the second half of the 19th century the development of chemical thermodynamics was greatly facilitated by the analysis of phenomena occurring in electrochemical cells at equilibrium.

Today electrochemistry is a rigorous science concerned with the quantitative relations between the chemical, surface, and electrical properties of systems. Electrochemistry has strong links to many other fields of science. Electrochemical concepts proved particularly fruitful for studying and interpreting a number of very important biological processes.

Modern electrochemistry has vast applications. Electrochemical processes form the basis of large-scale chemical and metallurgical production of a number of materials. Electrochemical phenomena are responsible for metallic corrosion, which causes untold losses in the economy. Modern electrochemical power sources (primary and secondary batteries) are used in many fields of engineering, and their production figures are measured in billions of units. There are other electrochemical processes and devices which are widely used.

Different definitions exist for electrochemistry as a subject. Thus, electrochemistry can be defined as the science concerned with the mutual transformation of chemical and electrical energy. According to another definition, electrochemistry deals with the structure of electrolyte solutions as well as with the phenomena occurring at the interfaces between metallic electrodes and electrolyte solutions. These and other, similar definitions are incomplete and do not cover all subject areas treated in electrochemistry.

By the very general definition adopted today by most research workers, electrochemistry is the science concerned with the physical and chemical properties of ionic conductors, as well as with phenomena occurring at the interfaces between ionic conductors, on the one hand, and electronic conductors or semiconductors, other ionic conductors, even insulators (including gases and vacuum), on the other hand. All these properties and phenomena are studied both under equilibrium conditions when there is no current flow and under nonequilibrium when there is electric current flow in the system. In a certain sense, electrochemistry can be contrasted to electronics and solid state theory, where the properties of electronic conductors and electronic or hole-type semiconductors as well as the phenomena occurring at the interfaces between these materials or between the materials and vacuum are examined.

This definition of electrochemistry disregards systems where nonequilibrium charged species are produced by external action in insulators, e.g., by electric discharge in the gas phase (electrochemistry of gases) or upon irradiation of liquid and solid dielectrics (radiation chemistry). At the same time, electrochemistry deals with certain problems often associated with other fields of science, such as the structure and properties of solid electrolytes and the kinetics of ionic reactions in solutions.

CONTENTS

ABBREVIATIONS

ac	alternating current
AE	auxiliary electrode
cd	current density
dc	direct current
DME	dropping mercury electrode
DSA^R	dimensionally stable anode
ecc	electrocapillary curve
edl	electric double layer
emf	electromotive force
ese	excess surface energy
hap	high anodic potentials
lps	linear potential scan
ocp	open-circuit potential
ocv	open-circuit voltage
Ox, ox	oxidized form
pd	potential difference
pzc	point (or potential) of zero charge
RDE	rotating disk electrode
rds	rate-determining step
RE	reference electrode
Red, red	reduced form
rhe	reversible hydrogen electrode
RRDE	rotating ring-disk electrode
SCE	saturated calomel electrode
SHE	standard hydrogen electrode
WE	working electrode

Part 1
BASIC CONCEPTS OF THEORETICAL ELECTROCHEMISTRY

Chapter 1
ELECTRIC CURRENT IN GALVANIC CELLS. THE ELECTRODES

1.1. FREE CHARGES IN CONDUCTORS

By the nature of conduction and values of conductivity, materials can be divided into conductors, semiconductors, and insulators (dielectrics). It is a special attribute of conductors that free electric charges are present in them which when moving represent the electric current.

Real charge is always associated with well-defined physical carriers such as electrons and ions; this is not so for the idealized "physical" charge considered in electrostatics. Each conductor can be characterized by stating the nature and concentration of the free charges. In the present section we consider free charged particles of atomic (or molecular) size, not the larger, aggregated entities such as colloidal particles.

The concentration of free charged particles of type j can be stated in terms of the number of moles, n_j, of these particles per unit of the volume V: $c_j = n_j/V$.

The electric charge Q_j of a particle can be described as

$$Q_j = Q^0 z_j,$$

where $Q^0 = 1.62 \cdot 10^{-19}$ C is the elementary charge (charge of the proton), and z_j is the charge number (an integer), i.e., the number of elementary charges associated with one particle. The value of z_j is negative for negatively charged particles. The charge of one mole of particles is given by $z_j F$, where $F \equiv N_A Q^0 = 96485$ C/mol (or roughly 96500 C/mol), which is the Faraday constant, and N_A is the Avogadro constant.

The volume density of charge of a given type, $Q_{V,j}$, is defined as

$$Q_{V,j} = z_j F c_j. \qquad (1.1)$$

A conductor is always electroneutral, i.e., in any part of it the combined density of all charges (free and localized), given by $\sum_j Q_{V,j}$, is zero and hence

$$\sum_j z_j c_j = 0 \quad \text{or} \quad \sum_{(+)} z_j c_j = -\sum_{(-)} z_j c_j, \qquad (1.2)$$

where $\sum_{(+)}$ and $\sum_{(-)}$ denote summation over all species of positive and negative charge, respectively.

The electroneutrality condition is disturbed only within thin layers (a few atoms in thickness) directly at the interfaces formed by the conductor with other conductors or insulators, where excess charge of a particular sign can exist in the form of monolayers or thin space-charge layers.

All forms of electrostatic ("coulombic") interaction of the particles with each other and with their environment are determined by the magnitude and sign of the charge and by the concentration of the charged particles. However, in contrast to "physical" charges, the real charged particles experience interactions other than electrostatic. Without discussing these additional interaction forces in depth, we shall designate them as chemical forces. It is because of these forces that each type of real charge has its own chemical individuality. Electrochemistry, in contrast to electrostatics, deals with both the chemical and the electrostatic properties of the free charged particles.

Electric currents in conductors are the directed motions of free charge under the influence of an applied electric field. The conduction can be electronic or ionic, depending on the kind of charge involved. Substances exist where conduction is mixed, i.e., where both ions and electrons are moving.

Electronic conduction is found in all metals and also in certain other substances: carbon materials (graphite, carbon black), some oxides and other inorganic compounds (e.g., tungsten carbide), and a number of organic substances.

Among ionic conductors, different groups can be distinguished, the largest being liquids. Aqueous solutions of salts, acids, and bases, as well as salt melts, are the ionic conductors most widely found and studied. Solid ionic conductors have attracted more and more interest recently.

Ionic conductors are known also as *electrolytes* or *conductors of the second kind*. The term "electrolyte" is used, not only in the sense of an ionic conductor, but also in speaking of substances which ordinarily are not conducting but produce ionic conduction after being dissolved in water or in another solvent (e.g., in expressions such as "electrolyte solution" and "weak electrolyte").

1.2. ELECTRIC CURRENT IN CONDUCTORS

The positively and negatively charged free particles will move in opposite directions when an electric field is applied. Therefore, outwardly the effect of motion of positive charges is exactly the same as that of the motion of negative charges, and the partial currents due to transport of each kind of charge must be summed. The direction of motion of the positive charge conventionally is adopted as the direction of electric current in conductors.

The strength of the electric current I in conductors is measured in amperes, and depends on the conductor, on the electrostatic field strength $\mathbf{E}$ in the conductor, and on the conductor's cross section S perpendicular to the direction of current flow. As a convenient parameter which is independent of conductor dimensions, the current density i is used, which is the fraction of current associated with unit area of the conductor's cross section: $i \equiv I/S$ (units: A/m^2).

The current density (cd) is proportional to field strength:

$$i = \sigma \mathbf{E} \tag{1.3}$$

(a differential form of *Ohm's law*). The proportionality factor σ is the (electrical) conductivity (units: S/m); it quantitatively characterizes the ability of a material to conduct electric current, and for any given material it depends on the temperature but not on the size and geometry of the sample. The reciprocal $\rho \equiv 1/\sigma$ is the resistivity (units: $\Omega \cdot$ m); numerically it is the resistance of a conductor sample 1 m long and 1 m^2 in cross section.

1.2.1. General Considerations Concerning the Flux of Matter

Flux of matter is the motion of matter in space (or transport) occurring under the effect of an external force. The flux density J_j of a substance j

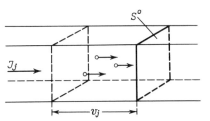

Fig. 1.1. Derivation of Eq. (1.4) for the mass flux density.

is measured in terms of the number of moles crossing unit cross-sectional area ($S^0 = 1$ m^2) perpendicular to the direction of the flux in unit time (its units are mol/m$^2 \cdot$ s).

The flux density depends on the volume concentration c_j and linear velocity v_j of the substance being transported. In unit time, unit cross-sectional area S^0 will be crossed by all particles no farther than v_j from S^0, i.e., by all those residing in a volume numerically equal to v_j (Fig. 1.1). Thus,

$$J_j = c_j v_j. \tag{1.4}$$

The mean velocity of motion v_j depends on the external driving force f_{dr} and on the resistance to motion set up by the medium's viscosity. This retarding force as a rule is proportional to the velocity. Under the influence of the external force, the velocity will increase until it attains the value v_j where the retarding force $\theta_j v_j$ (θ_j is the drag coefficient) becomes equal to the external force. Hence

$$v_j = f_{dr}/\theta_j \tag{1.5}$$

(here and in the following, the driving forces and the retardation factor refer to one mole of the substance). It follows from these equations that

$$J_j = (c_j/\theta_j) f_{dr}. \tag{1.6}$$

In the steady state, the total flux is constant along the entire path. This condition, i.e., that of flux continuity, is a reflection of mass balance; nowhere in a steady flux will substances accumulate or vanish, i.e., their local concentrations are time-invariant.

The condition of continuity of the steady flux is disturbed in those places where substances are consumed (sinks) or produced (sources) by chemical reactions. It is necessary in order to preserve the balance that any excess of substance supplied corresponds to the amount of substance reacting, and that any excess of substance eliminated corresponds to the amount of substance formed in the reaction.

1.2.2. Migration of Free Charges

Free charged particles in conductors are in a state of continuous kinetic molecular (thermal) motion. This motion is chaotic when an electrostatic field is not present, i.e., the particles do not preferentially move in any particular direction, and there is no current flow.

When an electrostatic field of field strength $\mathbf{E}$ is applied, then each kind j of free particle carrying charge $z_j F$ (per mole) finds itself under the effect of an electric force $z_j F\mathbf{E}$ causing the particles to move in a direction given by the field. This motion under the effect of a field is called the *migration* of free charges. Migration is superimposed upon the thermal motion of the particles.

According to Eq. (1.5), the mean velocity of migration is given by

$$v_{m,j} = (1/\theta_j)z_j F\mathbf{E} \qquad (1.7)$$

(the direction of the field is assumed to be that of the motion of positive charges; the velocity of motion in the opposite direction is counted as negative). The absolute value of the velocity of migration corresponding to unit field strength (1 V/m) is called the *mobility* of a species (units: $m^2/V \cdot s$):

$$u_j = (1/\theta_j)|z_j|F. \qquad (1.8)$$

Using this parameter we can write for the velocity of migration:

$$v_{m,j} = (\text{sign } z_j)u_j\mathbf{E},$$

where sign z_j is a factor of $+1$ for positive and -1 for negative values of z_j (sign $z_j \equiv z_j/|z_j| \equiv |z_j|/z_j$). This coefficient characterizes the direction of migration. The expression for migration flux density becomes

$$J_{m,j} = c_j v_{m,j} = (\text{sign } z_j)c_j u_j\mathbf{E}. \qquad (1.9)$$

By definition, the partial current density i_j is the number of charges which in unit time cross unit cross-sectional area on account of the migration of species j, i.e.,

$$i_j = z_j F J_{m,j} = |z_j| F c_j u_j \mathbf{E} \qquad (1.10)$$

(for negative charges the values of z_j and $J_{m,j}$ are negative, i.e., the partial currents are always positive).

In a conductor with different kinds of free charges, the total current density is determined by the fluxes of all kinds of charges:

$$i = \sum i_j = F \sum z_j J_{m,j} = F\mathbf{E} \sum |z_j| c_j u_j. \qquad (1.11)$$

We can see when comparing Eqs. (1.3) and (1.11) that a conductor's conductivity depends on the concentrations and mobilities of all kinds of free charges:

$$\sigma = F \sum |z_j| c_j u_j. \qquad (1.12)$$

The fraction of current transported by species of a given kind is called the *transport number* t_j of these species:

$$t_j \equiv i_j/i = |z_j| c_j u_j \Big/ \sum |z_j| c_j u_j . \qquad (1.13)$$

It is obvious that $0 \leq t_j \leq 1$ and $\sum t_j = 1$. For conductors with a single kind of free charged particles the transport number of these species is unity. For conductors with different kinds of such particles the individual transport numbers of the species depend on their concentrations and mobilities.

Conductivity is a very important parameter for any conductor. It is intimately related to other physical properties of the conductor, such as thermal conductivity (in the case of metals) and viscosity (in the case of liquid solutions).

1.3. BRIEF CHARACTERIZATION OF THE DIFFERENT KINDS OF CONDUCTOR

1.3.1. Metals and Semiconductors

Metals have high values of electronic conductivity, viz., $> 10^7$ S/m. The conductivity decreases somewhat with increasing temperature, i.e., its temperature coefficient is negative.

In isolated metal atoms (in the vapor phase), the valence electrons reside within the bounds of the atom. Each valence electron has a certain kinetic energy, i.e., exists at a certain energy level which is the same for all such atoms. When the atoms approach and form a condensed phase the weakly bound valence electrons are collectivized and can freely move within the entire space between the ions, thus forming the so-called electron gas. In it, the original energy levels of the electrons are split up, since according to the Pauli principle each level can be taken up by no more than two electrons. Thus, the kinetic energy W of the valence electrons is distributed over a large set of different levels very closely spaced. The set of energy levels able to accommodate valence electrons is called the valence band.

For metals, the number of levels in the valence band is larger than 1/2 the number of valence electrons, and part of the levels remain vacant. At a temperature of 0 K all the low-lying levels between $W = 0$ and some level W_F called the Fermi energy or Fermi level are filled up consecutively. According to the theory of metals, the Fermi level is determined by the expression

$$W_F = (h^2/2m_e)(3N_e/8\pi)^{2/3}, \qquad (1.14)$$

where h is the Planck constant ($6.63 \cdot 10^{-34}$ J·s), m_e is the electron mass ($9.1 \cdot 10^{-31}$ kg), and N_e is the number of free electrons per unit volume (for metals, it is 10^{28}–10^{29} m^{-3}, i.e., of the same order of magnitude as the number of ions in the same volume). The Fermi energy is between 3 and 8 eV for different metals.

At higher temperatures part of the electrons are at higher energy levels, because they now have higher kinetic energies.

Under the effect of an electric field acting on the metal, the kinetic energy of the electrons also increases on account of the additional velocity of migration, and the electrons should move to higher energy levels. Since a sufficient number of free levels are available, this transition is possible even at low field strengths where the additional energy is low. Hence migration of the electrons is unimpeded, and they can actually be regarded as free species.

Semiconductors differ from metals by a number of special features; they have lower conductivity values (roughly between 10^{-8} and 10^6 S/m), a positive temperature coefficient of conductivity, and the possibility of so-called hole conduction.

Hole conduction develops when only those charges are free which at a given time find themselves next to a free site, viz., a hole or vacancy

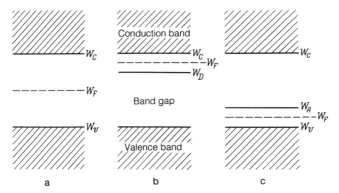

Fig. 1.2. Energy bands in semiconductors with intrinsic conduction (a) and with impurity-type conduction: n-type (b) and p-type (c).

in the crystal lattice. After transfer of the charged particle to this vacancy a new vacancy is formed at the site previously occupied by the particle, and the next-following charged particle can move into this site in turn. Each individual free carrier moves just a short distance, viz., into the next vacancy. Yet the vacancy existing initially gradually moves over a greater distance in the opposite direction. Vacancies of charged particles are equivalent to an equally large excess charge of the opposite sign. Hence formally a mobile hole, which is a missing negative charge, can be regarded as a free positive charge.

Semiconductors in which the electrons are mobile are called n-type semiconductors (from the word "negative"), and those where positive holes are mobile are called p-type semiconductors (from the word "positive").

At 0 K pure semiconductors, in contrast to metals, have the valence band completely filled with electrons to the top level W_v. No free levels are left, and hence under these conditions there is no electronic conduction. However, at a certain level W_c above the valence band another, vacant band of available energy levels starts, which is the so-called conduction band (Fig. 1.2a). The energy interval between the top of the valence band and the bottom of the conduction band is called the *band gap* ("forbidden band"). This is not very wide for semiconductors, viz., no more than 2 eV (metals have no band gap at all). In the case of insulators it is more than 3 eV wide, and practically no free electrons can be transferred to the conduction band.

Intrinsic conduction is observed in semiconductors at higher temperatures, where the thermal energy of the electrons is sufficient for part of the electrons to be transferred from the upper part of the valence band to the conduction band. This leaves vacancies (holes) in the valence band, and gives rise to free electrons in the conduction band. Thus, free carriers of both types, electrons and holes, appear simultaneously, and in equal numbers (i.e., in pairs). Hence intrinsic semiconductors have mixed electronic and hole conduction. At a temperature of 300 K, the number of carriers per m^3 is $2.5 \cdot 10^{19}$ for germanium, and $6.5 \cdot 10^{16}$ for silicon, which is nine to twelve orders of magnitude below the number of free electrons in metals. Intrinsic conduction is seen only in very pure and perfect semiconductors. In the case of intrinsic conduction the Fermi level is situated in the middle of the band gap.

Impurity conduction is caused by impurities present in the semiconductor material. When present in the silicon lattice, an arsenic atom (or atom of another group-V element) can release its superfluous fifth electron, and thus act as an electron donor. If in its original state this electron was at an energy level within the semiconductor's band gap, but not too far away from the bottom of the conduction band (level W_D in Fig. 1.2b), it can readily enough be transferred to the conduction band, and as a result free electrons appear in this band (n-type conduction). A single arsenic atom introduced into germanium per 10^7 germanium atoms will raise the number of free electrons to 10^{21} m^{-3}. When boron (or another group-III element) is the impurity it can accept a fourth electron in its outer orbital. Here the transition is possible when this electron is lifted to a level not very far from the top of the valence band (level W_A in Fig. 1.2c). In this case the impurity atom acts as an electron acceptor, a vacancy appears in the valence band, and the semiconductor acquires p-type conduction.

The number of free carriers (electrons or holes) in a semiconductor can also be raised by the effect of light.

1.3.2. Aqueous Electrolyte Solutions

Acids, bases, and salts (i.e., electrolytes in the second sense of the word) dissociate into ions when dissolved in water. This dissociation can be complete or partial. The fraction of the original molecules which have dissociated is known as the *degree of dissociation*, α. Substances which exhibit a low degree of dissociation in solution are called weak electrolytes, while when the value of α comes close to unity we speak of strong electrolytes.

In general we can write the dissociation equation

$$M_{\tau_+}A_{\tau_-} \rightleftarrows \tau_+ M^{z+} + \tau_- A^{z-}. \tag{1.15}$$

It is evident that

$$z_+ \tau_+ = |z_-|\tau_- = z_k, \tag{1.16}$$

where z_k is the electrolyte's charge number (the number of elementary charges of each sign appearing upon dissociation of one molecule of the electrolyte).

Let c_k be the original concentration of substance k (e.g., of the compound $M_{\tau_+}A_{\tau_-}$) without dissociation. Then the concentrations c_+ and c_- of the ions, the concentration c_n of the undissociated molecules, and the total concentration c_σ of all species dissolved in the solution can be written as

$$c_+ = \alpha\tau_+ c_k, \qquad c_- = \alpha\tau_- c_k, \qquad c_n = (1-\alpha)c_k,$$
$$c_\sigma \equiv c_+ + c_- + c_n = [1 + \alpha(\tau_k - 1)]c_k, \tag{1.17}$$

where $\tau_k \equiv \tau_+ + \tau_-$ is the total number of ions into which one original molecule can dissociate. In the limit of $\alpha = 1$ we have $c_\sigma = \tau_k c_k$.

Thus, electrolyte solutions contain several kinds of particles, but their concentrations are interrelated, and only one of the concentration values is independent, e.g., that of the original compound k (which is c_k). A subscript k is used instead of j to point out, in the following, that this independent component is considered, rather than its dissociation products.

A substance such as $ZnSO_4$ where $z_+ = |z_-|$ ($\tau_+ = \tau_- = 1$) is called a symmetric or $z:z$ electrolyte; a particular case of the $z:z$ electrolytes are the 1:1 electrolytes for which KCl is an example.

In the past, the parameter $1/|z_j|$ mol of a given ion j had been called the ion's chemical dissociation equivalent, and the corresponding mass $M_j/|z_j|$ (where M_j is the molar mass) was called the ion's equivalent mass. The ion's equivalent concentration $c_{(eq)j}$ is related to its mole concentration c_j as $c_{(eq)j} = |z_j|c_j$. However, since the chemical equivalents of any given substance in dissociation reactions may differ from the equivalents relevant in electrochemical reactions (see Section 1.5.3), this term should be used with caution.

Binary electrolyte solutions contain just one solute in addition to the solvent, i.e., two components in all. Multicomponent solutions contain several original solutes and the corresponding number of ions. Sometimes in multicomponent solutions the behavior of just one of the components

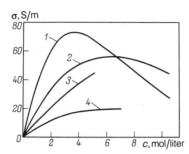

Fig. 1.3. Conductivities (at 25°C) as functions of concentration in aqueous solutions of: 1) H_2SO_4; 2) KOH; 3) NH_4Cl; 4) $NaNO_3$.

is of interest; in this case the term "base electrolyte" is used for the set of remaining solution components. Often a base electrolyte is actually added to the solutions to raise their conductivity.

For the conductivity of a binary electrolyte solution with the degree of dissociation α, we have, according to Eq. (1.12),

$$\sigma = \alpha c_k F(\tau_+ z_+ u_+ + \tau_- |z_-| u_-) = \alpha z_k c_k F(u_+ + u_-). \qquad (1.18)$$

The mobilities, u_j, of ions in solutions are concentration-dependent. They are highest in dilute solutions (the limiting mobilities u_j^0), and gradually decrease with increasing concentration. Hence in dilute solutions of strong electrolytes ($\alpha = 1$) the conductivity is proportional to the total concentration c_k. Because of decreasing mobility, the conductivity rise becomes slower as the concentration increases. In solutions of weak electrolytes this slowdown is more pronounced, since the degree of dissociation decreases in addition to the mobilities. In certain cases the plots of conductivity against concentration go through a maximum (Fig. 1.3).

For binary electrolyte solutions, Eq. (1.13) for the transport numbers becomes

$$t_j = u_j/(u_+ + u_-). \qquad (1.19)$$

The parameters of *molar conductivity of the electrolyte*, $\Lambda \equiv \sigma/c_k$, and *molar conductivity of ions*, $\lambda_j \equiv |z_j| F u_j$ (units: $S \cdot m^2/mol$) are also used to describe the properties of electrolyte solutions (Λ is used only in the case of binary solutions). With Eq. (1.18), we can write for a binary solution

$$\Lambda = \alpha(\tau_+ \lambda_+ + \tau_- \lambda_-). \qquad (1.20)$$

For the change of molar conductivity of the ions which occurs with increasing concentration, only the mobility decrease is responsible; in dilute solutions a limiting value of $\lambda_j^0 \equiv |z_j| F u_j^0$ is attained. A limiting value of molar conductivity Λ^0 implies limiting values λ_j^0 as well as complete dissociation, i.e.,

$$\Lambda^0 = \tau_+ \lambda_+^0 + \tau_- \lambda_-^0. \tag{1.21}$$

In the past, equivalent concentrations had often been used as a basis for the values of Λ and λ_j. In this case the equivalent conductance was defined as $\Lambda_{eq} \equiv \sigma/c_{eq} = \sigma/z_k c_k$, and the ionic equivalent conductance (or equivalent mobility) was defined as $\lambda_{(eq)j} \equiv F u_j$; in this notation,

$$\Lambda_{eq} = \alpha[\lambda_{(eq)^+} + \lambda_{(eq)-}] \quad \text{and} \quad \Lambda_{eq}^0 = \lambda_{(eq)^+}^0 + \lambda_{(eq)-}^0. \tag{1.22}$$

Today these terms are no longer recommended. Instead we consider the molar conductivities of electrolytes and ions as defined above, and where necessary indicate the electrolyte units to which the concentrations refer, e.g., $\Lambda(MgCl_2)$ or $\Lambda(\frac{1}{2}MgCl_2)$, $\lambda(Ca^{2+})$ or $\lambda(\frac{1}{2}Ca^{2+})$. We evidently have $\Lambda(\frac{1}{2}MgCl_2) = \frac{1}{2}\Lambda(MgCl_2)$.

It is a typical feature of aqueous electrolyte solutions that one can change within wide limits the solute concentrations, and hence the conductivities themselves. Pure water has a very low value of σ; it is about 5 μS/m at room temperature after careful purification of the water. In the most highly conducting solutions, viz., concentrated solutions of acids and bases, values of 80 S/m can be attained at the same temperature, i.e., values seven orders of magnitude higher than those found for pure water.

1.3.3. Molten and Solid Electrolytes

Free ions of both signs are formed upon melting of solid salts with ionic lattices. In contrast to solutions, an additional "inert" component (the solvent) is not present here, so that the total concentrations of free ions in the different melts are similar and have values of 10 to 50 mol/liter, which is approximately the same as the concentration of free electrons in metals. High values are found, accordingly, for conductivities in melts; they are between 10^2 and 10^4 S/m. They vary within narrower limits than those of electrolyte solutions, and depend only on the melt type and temperature.

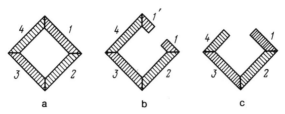

a   b   c

Fig. 1.4. Electric circuits: (a) closed; (b) open, complete (or "properly open," i.e., properly terminating at both ends with identical conductors); (c) open, but incomplete (or "improperly open," i.e., having different terminal members).

The ionic conduction found in certain crystalline solids (solid electrolytes) is related to special structural features of their lattice and, in some cases, to foreign ions present in the lattice. A typical feature of solid electrolytes is the unipolar character of ionic conduction, viz., mobility of only one of the types of ion present in the ionic lattice. Sometimes (particularly at high temperatures) some electronic conduction is also seen, i.e., conduction then is "mixed."

A special kind of solid electrolyte are the solid polymer electrolytes. They contain a macromolecular backbone of organic (or more rarely inorganic) material with fixed ionogenic groups of one sign, e.g., negative sulfonic acid groups. The ions of opposite sign (the counterions) are mobile rather than localized. This situation again gives rise to unipolar conduction of the electrolyte.

1.4. CIRCUITS INVOLVING GALVANIC CELLS

Electric circuits, as a rule, consist not of a single conductor but of several different conductors connected with each other, and forming a sequence of conductors. This circuit can be closed or open (Fig. 1.4). An open circuit is complete (or "properly open") only when terminating at both ends with the same kind of conductor.

In electrical engineering and electronics, circuits are employed which consist of electronic conductors exclusively. Circuits which in addition include at least one ionic conductor are called *galvanic cells*.[1] These have a number of characteristic features not present in purely electronic circuits.

---

[1] Strictly speaking, the terms "galvanic circuit" or "galvanic cell circuit" should be used.

It is one of the more important tasks of electrochemistry to consider these features.

The sequence of conductors constituting a galvanic cell can be described schematically, e.g., as

$$Cu|Zn|ZnCl_2, aq|graphite|Cu \qquad (1.23)$$

($ZnCl_2, aq$ denotes an aqueous solution of $ZnCl_2$). Vertical lines in the scheme denote the areas of contact (interfaces) between two adjacent conductors.

Other galvanic cells contain two or more ionic conductors, which may be in direct contact with each other, or alternate with electronic conductors. An example of the former case is that of the cell

$$Cu|Zn|ZnSO_4, aq|CuSO_4, aq|Cu. \qquad (1.24)$$

The broken vertical line denotes an area of contact between any two ionic conductors, particularly between liquid ionic conductors (electrolyte/electrolyte interface or *liquid junction*). Ions can transfer between phases by diffusion across such a boundary, hence cells containing such an interface are often called cells with transference.

In a galvanic cell, electronic conductors are in contact with ionic conductors in at least two places. An electronic conductor in contact with an ionic conductor is called an *electrode*.

An open galvanic cell without transference properly terminating at both ends with identical conductors can generally be formulated as

$$M_1|E|M_2|M_1, \qquad (1.25)$$

while a similar cell with transference can be formulated as

$$M_1E_1|E_2|M_2|M_1, \qquad (1.26)$$

where M is a metal (or other electronic conductor), and E is an electrolyte.

The technical realization of a galvanic cell as formulated above is also called *galvanic cell* (or cell, electrochemical cell, electrolysis cell).

When an electric current is made to flow in a galvanic cell the current will pass from electrode to electrolyte ("enter" the electrolyte) at one of the electrodes, and it will pass from electrolyte to electrode ("leave" the electrolyte) at the other electrode. The first of these electrodes has been named the *anode* (from Greek ἀνά "up"), the second has been named the *cathode* (from Greek κατά "down"). It follows from this definition that the designations of "anode" and "cathode" depend on the direction of current flow in the galvanic cell. An anode becomes cathode, and vice versa, when the direction of current flow is inverted. Within the electrolyte, the current flow is always from the anode to the cathode. Therefore, the positively charged electrolyte ions migrating toward the cathode have been named *cations*, and the negatively charged ions migrating toward the anode have been named *anions*. In the external parts of the closed circuit ("external" relative to the electrolyte) the current flow is from the cathode to the anode.

1.5. ELECTRODES AND ELECTRODE REACTIONS

1.5.1. Passage of Current through Electrodes

The area of contact between two different kinds of conductor is a special place in any circuit. The character of current flow in this region depends on the phases in contact.

The simplest case is that of contact between two metals. In both conductors the conduction is due to the same species, viz., electrons. When current crosses the interface the flow of electrons is not arrested; all electrons which come from one of the phases freely cross over to the other phase upon their arrival at the interface. No accumulation or depletion of electrons is observed. In addition, current flow at such a junction will not produce any chemical change.

More complex phenomena occur when current crosses interfaces between semiconductors. The most typical example is the rectification produced at interfaces between p- and n-type semiconductors. Electric current freely flows from the former into the latter semiconductor, but an electric field "repelling" the free carriers from the junction arises when the attempt is made to pass current in the opposite direction: holes are sent back into the p-phase, and electrons are sent back into the n-phase. As a result, the layers adjoining the interface are depleted of free charges, their conductivities drop drastically, and current flow ceases ("blocking" of the interface).

When the current is carried by different species in the two adjacent phases, the continuous flow of carriers is interrupted. Charges of one kind come up to (or depart from) the area of contact on one side, and carriers of a different kind come up to (or depart from) this area on the other side.

In order to sustain steady current flow, one needs a steady sink for the particles arriving, and a steady source for those departing.

In galvanic cells the carriers are ions and electrons. In this case, chemical reactions occurring at the interface—at the electrode surface—and involving carriers from both phases (including electrons) are the sink and source for the corresponding particles. Reactions involving electrons are called *electrochemical* or *electrode reactions*. Reactions at anodes are also called anodic, and reactions at cathodes are called cathodic. At an anode, electrons go away from the junction into the metal, hence an anodic reaction must generate electrons. Similarly, at a cathode, electrons supplied by the circuit must react (and thus are eliminated from the reaction zone).

For instance, when current flow is from the right to the left in galvanic cell (1.23), the zinc electrode will be the cathode, and its surface is the site of the cathodic reaction involving the deposition of metallic zinc by discharge of zinc ions from the solution:

$$Zn^{2+} + 2e^- \rightarrow Zn. \tag{1.27}$$

This reaction satisfies the requirements listed above; the zinc ions and electrons arriving at the surface from different sides disappear from the reaction zone. The anodic reaction

$$2Cl^- \rightarrow Cl_2 + 2e^- \tag{1.28}$$

occurs at the surface of the graphite electrode (the anode); it generates electrons while Cl^- ions disappear.

Electron withdrawal from a material is equivalent to its oxidation, while electron addition is equivalent to its reduction. In the anodic reaction, electrons are generated and a reactant (in our example the chloride ions) is oxidized. In the cathodic reaction the reactant (the zinc ions) is reduced. Thus, anodic reactions are always oxidation reactions, and cathodic reactions are reduction reactions for the initial reactants.

In all cases the electrode reaction secures continuity of current flow across the interface, or a "relay"-type transfer of charges (current) from the carriers in one phase to the carriers in the other phase. In the reaction, the interface as a rule is crossed by species of one kind, electrons [in reaction (1.28), for instance] *or* ions [in reaction (1.27), for instance].

In complete galvanic cells, electrochemical reactions occur simultaneously at the anode and cathode. Since the current is of equal strength at the two electrodes, the corresponding electrode reactions are interrelated, in that the number of electrons set free in unit time at the anode is equal to the number of electrons reacting during the same time at the cathode. Electrode reactions subject to such a condition are called *coupled reactions*.

Current flow in cells is attended by an overall chemical reaction, more particularly a *current-producing* (or *current-consuming*) *reaction* in which electrons do not appear explicitly. In the example reported above, decomposition of dissolved zinc chloride,

$$ZnCl_2 \ (\rightleftarrows Zn^{2+} + 2Cl^-) \rightarrow Zn + Cl_2, \qquad (1.29)$$

is the reaction which results when the cathodic and anodic reaction are combined.

In symmetric galvanic cells, i.e., cells consisting of two identical electrodes (e.g., zinc electrodes), current flow does not produce a net chemical reaction in the cell as a whole, and only a transfer of individual components occurs in the cell (in our example, metallic zinc is transferred from the anode to the cathode).

1.5.2. Classification of Electrodes and Electrode Reactions

The type of electrode reaction that will occur depends on the electrode and electrolyte and also on external conditions: the temperature, impurities that may be present, etc. Possible reactants and products in these reactions are: (a) the electrode material, (b) components of the electrolyte, (c) other substances (gases, liquids, or solids) which are not themselves component parts of an electrode or the electrolyte but can reach or leave the electrode surface. Therefore, when discussing the properties or behavior of any electrode we must indicate not only the electrode material but the full electrode system comprising electrode and electrolyte, as well as additional substances which may be involved in the reaction, e.g., $ZnCl_2$, $aq|(Cl_2)$, graphite [the right-hand electrode in (1.23)] or generally $E|(Y), M$, where Y are additional reactants.

Among all the substances that may be involved in an electrode reaction, substances for which the oxidation state does not change (such as complexing agents) must be distinguished from the principal reactants and products, for which there is a change in oxidation state in the reaction.

Using specific examples, we shall consider different types of electrodes and electrode reactions. The examples are cases involving aqueous solutions, but the features pointed out are found as well in other electrolytes. In the examples we first indicate the electrode system, then the cathodic reaction equation. The corresponding anodic reaction follows the same equation, but from the right to the left. Special features of anodic reactions are stated in square brackets in the text that follows.

A unique terminology for the different types of electrodes and reactions unfortunately has not been established so far. Electrodes can be classified according to different distinguishing features.

Reacting and nonconsumable electrodes:

$$\text{AgNO}_3, \ aq|\text{Ag} \qquad \text{Ag}^+ + e^- \rightleftarrows \text{Ag}, \qquad (1.30)$$

$$\text{FeCl}_2, \ \text{FeCl}_3, \ aq|\text{Pt} \qquad \text{Fe}^{3+} + e^- \rightleftarrows \text{Fe}^{2+}, \qquad (1.31)$$

$$\text{HClO}_3, \ \text{HCl}, \ aq|\text{Pt} \qquad \text{ClO}_3^- + 6\text{H}^+ + 6e^- \rightleftarrows \text{Cl}^- + 3\text{H}_2\text{O}, (1.32)$$

$$\text{H}_2\text{SO}_4, \ aq|(\text{H}_2), \ \text{Pt} \qquad 2\text{H}^+ + 2e^- \rightleftarrows \text{H}_2. \qquad (1.33)$$

In the first of the four examples, the electrode material (metallic silver) is chemically involved in the electrode reaction; hence its amount increases [decreases] with time. Such electrodes are called reacting (or consumable) electrodes.

In the other examples, the electrode materials are not involved in the reactions but constitute the source [sink] of electrons. Such electrodes are called nonconsumable. The term "inert electrodes" sometimes used is unfortunate insofar as the electrode itself is by no means inert; rather, it has a strong catalytic effect on the electrode reaction. For reactions occurring at such electrodes, the terms "oxidation–reduction reaction" or "redox reaction" are widely used, but even these terms are not very fortunate, since reactions occurring at reacting electrodes are also reducing and oxidizing in character. Reactions of the type of (1.31) where just a single electron is transferred (or more rarely, two electrons at the same time) will in the following be called simple redox reactions, while reactions of the type of (1.32) where other solution components also are involved will be called complex (demanding) redox reactions.

Electrodes at which gases are evolved or consumed [e.g., by reaction (1.33)] can be called gas electrodes. In the conventional formulation of the electrode system, the reacting gas is indicated in parentheses.

Specific kinds of consumable electrode are designated in terms of the constituent material, e.g., as "silver electrode." Nonconsumable electrodes are designated either in terms of the electrode material or in terms of the chief component in the electrode reaction; for instance, the terms "platinum electrode" and "hydrogen electrode" are used for the electrode (1.33). Neither of these names completely describes the special features of this electrode.

Reacting electrodes with soluble and insoluble reaction products (reactants):

$$KCl,\ aq|Ag \qquad AgCl + e^- \rightleftarrows Ag + Cl^-. \qquad (1.34)$$

Depending on electrolyte composition, the metal will either dissolve in the anodic reaction, i.e., form soluble ions [reaction (1.30)], or will form insoluble or poorly soluble salts or oxides precipitating as a new solid phase next to the electrode surface [reaction (1.34)].

Reacting metal electrodes forming soluble products are also known as *electrodes of the first kind*, and those forming solid products are known as *electrodes of the second kind*.

Electrodes with invertible and noninvertible electrode reactions. Most electrode reactions are invertible in the sense that they will occur in the opposite direction when the direction of current is inverted.[2]

Two kinds of reactions exist which are noninvertible in this sense.

(i) Reactions noninvertible in principle or, more probably, reactions for which no conditions have been found so far under which they will proceed in the opposite direction. An example of such a reaction is the cathodic reduction of hydrogen peroxide

$$H_2O_2 + 2H^+ + 2e^- \rightarrow 2H_2O. \qquad (1.35)$$

The formation of hydrogen peroxide by anodic oxidation of water has so far not been realized.

(ii) Reactions which under existing conditions cannot be inverted because of lack of reactants. Thus, metallic zinc readily dissolves anodically

---

[2]The concepts of invertibility and reversibility must be distinguished. *Invertibility* is the term proposed to be used for reactions which can be made to occur in both directions, regardless of the departure from thermodynamic equilibrium that is necessary to achieve this. *Reversibility* of a reaction means that it occurs with a minimum of departure from the thermodynamic equilibrium state. For further details see Section 3.3.

in sulfuric acid solution [reaction (1.27) from the right to the left], but when this solution contains no zinc salt, the reverse reaction in which zinc is deposited cathodically cannot occur.

Simple and polyfunctional electrodes. At simple electrodes, one sole electrode reaction occurs under the conditions specified when current flows. At polyfunctional electrodes two or more reactions occur simultaneously; an example is the zinc electrode in acidic zinc sulfate solution. When the current is cathodic, metallic zinc is deposited at the electrode [reaction (1.27)] and hydrogen is evolved at the same time [reaction (1.33)]. The relative strengths of the partial currents corresponding to these two reactions depend on the conditions, e.g., the temperature, pH, solution purity, etc. Conditions may change so that a simple electrode becomes polyfunctional, and vice versa.

1.5.3. General Form of the Reaction Equations

Equations for electrode reactions can generally be written as

$$\sum\nolimits_{ox} \nu_j X_j + ne^- \rightleftarrows \sum\nolimits_{red} \nu_j X_j, \tag{1.36}$$

where X_j are the species involved in the reaction and ν_j are their stoichiometric coefficients. Summation index "ox" implies that the sum is taken over the oxidized form of the principal reaction component and the substances reacting together with it, while index "red" implies that the sum is taken over the reactant's reduced form and substances associated with it.

According to Eq. (1.36), the cathodic reaction occurs from the left to the right, and the anodic reaction occurs from the right to the left. For the cathodic reaction, all substances on the right-hand side of the equation are reaction products (index "p"), and those on the left-hand side are the reactants (index "r").

In the following we shall use two further symbolic ways of writing reaction equations, in order to simplify them:

$$\sum \tilde{\nu}_j X_j + ne^- = 0 \tag{1.37}$$

and

$$\sum \tilde{\nu}_j X_j \pm ne^- = 0 \tag{1.38}$$

(the plus sign is for anodic reactions, and the minus sign for cathodic reactions). In both equations the sum is taken over all species involved in the reaction. Stoichiometric coefficients $\tilde{\nu}_j$ in Eq. (1.37) take on positive values for the oxidized component forms and for the substances entering the reaction together with them $[\tilde{\nu}_j = \nu_{(ox)}]$, and negative values for the reduced forms $[\tilde{\nu}_j = -\nu_{(red)}]$. Coefficients $\tilde{\nu}_j$ in Eq. (1.38) are positive for the products $[\tilde{\nu}_j = \nu_p]$ and negative for the reactants $[\tilde{\nu}_j = -\nu_r]$. The value of n is always positive.

When writing an equation for an electrode reaction we must observe balance of the product, reactant, and electronic charges:

$$\sum \tilde{\nu}_j z_j = n \quad \text{or} \quad \sum \bar{\nu}_j z_j = \pm n \qquad (1.39)$$

(it is taken into account here that the electron's charge is negative).

Sometimes a general conventional formulation,

$$\text{Ox} + ne^- \rightleftarrows \text{Red}, \qquad (1.40)$$

can be used for relatively simple redox reactions; here Ox and Red are the oxidized and reduced form of the principal reaction component.

The overall current-producing reaction can be obtained by combining the cathodic reaction occurring at one electrode (index "1") with the anodic reaction occurring at the other electrode (index "2"), while the equations for these reactions must be written so that the values of n in these equations are identical (the reactions must be coupled):

$$\sum\nolimits_{1,ox} \nu_j X_j + \sum\nolimits_{2,red} \nu_j X_j \rightleftarrows \sum\nolimits_{1,red} \nu_j X_j + \sum\nolimits_{2,ox} \nu_j X_j. \qquad (1.41)$$

Symbolic formulations of the overall reaction equation are as follows:

$$\sum\nolimits_1 \tilde{\nu}_j X_j = \sum\nolimits_2 \tilde{\nu}_j X_j \quad \text{or} \quad \sum \bar{\nu}_j X_j = 0. \qquad (1.42)$$

For the overall current-producing reaction, the condition of charge balance of reactants and products can be written as

$$\sum\nolimits_1 \tilde{\nu}_j z_j = \sum\nolimits_2 \tilde{\nu}_j z_j \quad \text{or} \quad \sum \bar{\nu}_j z_j = 0. \qquad (1.43)$$

The parameter ν_j/n mol which can be written for each component of the electrode reaction is sometimes called the chemical equivalent of the component in the reaction named, and the value of $(\nu_j/n)M_j$ is called the equivalent mass (see "chemical equivalent" in Section 1.3.2).

1.6. FARADAY'S LAWS

The number of reactant molecules involved in an electrode reaction is related stoichiometrically to the number of charges (electrons) flowing in the circuit. This is the basic argument of the laws formulated by Michael Faraday in 1833/34.

Faraday's first law reads: "In electrolysis, the quantities of substances involved in the chemical change are proportional to the quantity of electricity which passes through the electrolyte."

Faraday's second law reads: "The masses of different substances set free or dissolved by a given amount of electricity are proportional to their chemical equivalents."

In honor of the discoverer of these laws, the amount of charge which corresponds to the conversion of one chemical equivalent of substance has been named the *Faraday constant*.

An amount of charge nF/ν_j is required to convert one mole of substance j. When an amount of charge Q has been consumed at the electrode, the number of moles Δn_j of substance which have formed or reacted is given by

$$\Delta n_j = \bar{\nu}_j Q/nF. \tag{1.44}$$

This formula unites the two laws of Faraday.

Since the total amount of substance being converted is proportional to the amount of charge, the specific reaction rate v_j, which is the amount of substance j converted in unit time per unit surface area of the electrode, is proportional to the current density i:

$$v_j \equiv (1/S)\,dn_j/dt = \bar{\nu}_j i/nF. \tag{1.45}$$

For this reason the specific rates of electrode reactions are often stated in the electrical units of current density.

Faraday's laws are absolutely rigorous for steady currents. They are the basis for a highly accurate method of measuring the amount of charge, viz., in terms of the mass or volume of substance reacting or set free (coulometry). Faraday's laws have served in the past for defining the unit of electric current: The international ampere was that strength of invariant current which when sent through aqueous silver nitrate solution would deposit 1.111800 mg silver per second at the cathode (another definition is now provided for the ampere as an SI unit).

Deviations from the Faraday laws can be observed in the case of transient currents, when charge aside from being involved in the electrode reactions accumulates in certain parts of the circuit (near interfaces). Such transient currents are also known as "*nonfaradaic.*"

An apparent departure from the laws of Faraday can be observed at polyfunctional electrodes when the set of reactions taking place is not fully accounted for.

1.7. MASS TRANSPORT IN ELECTROLYTES

When a current flows in a galvanic cell, balance should exist, not only with respect to the charges but also with respect to the reactants. Hence, these materials should be brought up to (or carried away from) the electrode surface with the rates at which they are consumed (or produced) by the reaction.

Allowing for Eq. (1.45) we can write the condition for mass balance of the component j as

$$J_j = \bar{\nu}_j i / nF. \tag{1.46}$$

The value of J_j defined by this equation is the flux density of the substance j in the electrolyte stoichiometrically required when the electrode reaction proceeds under steady-state conditions.

With Eqs. (1.9), (1.11), and (1.13) we can at the same time write the migration flux density as

$$J_{m,j} = t_j i / z_j F. \tag{1.47}$$

The values of the fluxes that can be calculated from these two equations differ substantially. Therefore, an apparent contradiction exists between the balance requirements with respect to charges and substances. This contradiction is particularly obvious in two cases.

(i) According to Eq. (1.46), ions not involved in the reaction need not be transported, since $\bar{\nu}_j = 0$, while according to Eq. (1.47) they are transported.

(ii) Reacting neutral molecules should be transported according to Eq. (1.46), but they are not, according to Eq. (1.47), since for them $t_j/z_j = 0$ [cf. Eqs. (1.13) and (1.8)].

Under realistic conditions a balance is secured during current flow on account of additional mechanisms of mass transport in the electrolyte:

diffusion and *convection*. The initial imbalance between the rates of migration and reaction brings about a change in component concentrations next to the electrode surfaces, and thus gives rise to concentration gradients. As a result a diffusion flux $J_{d,j}$ develops for each component. Moreover, in liquid electrolytes hydrodynamic flows bringing about convective fluxes $J_{kv,j}$ of the dissolved reaction components will almost always arise.

Uncharged reaction components are also transported by diffusion and convection, even though their migration fluxes are zero. The total flux density J_j of species j is the algebraic sum of densities of all flux types:

$$J_j = J_{m,j} + J_{d,j} + J_{kv,j}. \tag{1.48}$$

The contradiction mentioned above is an apparent one, since the overall flux density is relevant for the stoichiometrically required flux contained in Eq. (1.46), while only the migrational component is contained in Eq. (1.47). In the steady state, the diffusion and convection fluxes are always set up in such a way as to secure mass balance.

These questions will be considered in more detail in Chapter 4. Any description of current flow in galvanic cells will be incomplete when these additional phenomena are disregarded.

1.8. THE SIGN CONVENTION FOR THE CURRENT AND THE FLUXES

A difficulty arises when the total flux is calculated via Eq. (1.48), since the different types of flux can have different directions. The migration fluxes of the cations are always toward the cathode, and those of the anions are always toward the anode. But the total flux of any reactant is always in the direction of the corresponding electrode, while the product fluxes are always away from the electrode surface.

A particular sign convention is used in electrochemistry to allow for these phenomena. For each of the electrodes (anode and cathode) one introduces the proper system of coordinates x in the electrolyte, which are reckoned from the electrode surface ($x = 0$) in the direction of the other electrode (Fig. 1.5). A flux away from any electrode in the direction of increasing x values (e.g., the total flux of reaction products) is regarded as positive, and a flux toward the electrode surface is regarded as negative. In precisely the same way, the anodic current is regarded as positive and the cathodic current is regarded as negative (we point out that in the

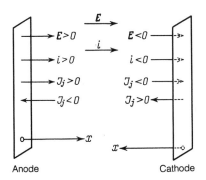

Fig. 1.5. The signs of current and of the fluxes
to and away from the electrode surfaces.

bulk of any conductor the current is always regarded as positive; see
Section 1.2.2). The electrostatic field strength at the anode is positive,
and that at the cathode is negative.

When using this convention most of the equations derived previously
remain valid, both for the anode and for the cathode. The form of the
equation for mass balance becomes slightly more exact, viz., instead of
Eq. (1.46) we must now use the equation

$$J_j = \bar{\nu}_j |i| / nF. \tag{1.49}$$

Here it has been taken into account that the overall flux of reaction products
(for which $\bar{\nu}_j > 0$) is positive for both the anode and the cathode. Hence,
the absolute value of current density appears in the equation instead of the
current density itself, which for the cathode is negative. In many other
equations which result from the equation of balance and which will be
discussed in detail in Chapter 4, values $|i|$ are used instead of i so that
identical equations can be used for the anode and cathode.

In the adopted system of notations and signs, the identity

$$\bar{\nu}_j |i| \equiv \tilde{\nu}_j i \tag{1.50}$$

is valid. Thus, for a cathodic reaction ($i < 0$) an oxidizing agent ($\tilde{\nu}_j < 0$)
is the reactant ($\tilde{\nu}_j > 0$). This interchange is possible, in particular, in the
balance equation (1.49).

Chapter 2
ELECTRODE POTENTIALS

2.1. ELECTROSTATIC FIELDS AND POTENTIALS

Classical electrostatics deals with the interactions of idealized electric charges. Electrochemistry deals with real charged particles having both electrostatic and chemical properties. For a clearer distinction of these properties, let us briefly recall some of the principles of electrostatics.

An electrostatic field can be described with the aid of electrostatic potential ψ or of field strength $\mathbf{E}$, a vectorial parameter that is equal to the negative potential gradient:

$$\mathbf{E} = -\operatorname{grad} \psi. \tag{2.1}$$

The directions of the vector at different points are often pictured as the lines of force. The potential gradient will be $d\psi/dx$ in linear fields, where the lines of force are parallel along the x-axis (the one-dimensional problem). In nonlinear fields the gradient can be written as $d\psi/dx + d\psi/dy + d\psi/dz$, but other coordinate systems which more conveniently describe a given system can also be used, for instance spherical or cylindrical.

The potential difference (pd) $\Delta\psi^{(B,A)} \equiv \psi^{(B)} - \psi^{(A)}$ between points A and B is defined as the work w_e performed by external forces when moving an electric test charge Q_t from A to B, divided by the magnitude of this charge:

$$\Delta\psi^{(B,A)} = w_e/Q_t. \tag{2.2}$$

It is assumed here that the test charge is small so that it will not distort the field (or relative positions of other charges), and that the work performed in moving the charge is only that necessary to overcome electrostatic forces and not any others such as chemical forces, i.e., the charge is ideal.

The potential difference between two points is defined with the aid of Eq. (2.2). The concept of potential of an individual (isolated) point is undefined, and becomes meaningful only when this potential is referred to the potential of another point chosen as the point of reference.

Often the concept of (two-dimensional) surface or (three-dimensional) space charge is employed. Here it is assumed that the charge is distributed in a continuous fashion ("smeared out") over the surface (with area S) or volume (V). Surface and space charge can be described in terms of surface-charge density $Q_S = dQ/dS$ or space-charge density $Q_V = dQ/dV$, which may either be constant or vary between points.

The relation between the spatial potential distribution and the spatial distribution of space-charge density can be stated, generally, in terms of Poisson's differential equation

$$\nabla^2 \psi = -Q_V/\varepsilon_0\varepsilon \qquad (2.3)$$

where ∇^2 is the Laplace operator (in Cartesian coordinates, $\nabla^2 \equiv d^2/dx^2 + d^2/dy^2 + d^2/dz^2$), $\varepsilon_0 = 8.85 \cdot 10^{-12}$ F/m is the permittivity of vacuum, and ε is the relative (dielectric) permittivity of the medium considered.

Boundary conditions must be supplied in addition to the functional relation between charge density and the coordinates in order to integrate this equation.

In some particular cases a simple solution is available for this equation. Figure 2.1 schematically shows the spatial distribution of potential ψ and of the modulus of field strength $\mathbf{E}$ for a number of typical configurations of the (positive) electric charges.

A point charge Q. According to *Coulomb's law*,

$$|\mathbf{E}| = Q/4\pi\varepsilon_0\varepsilon r^2, \qquad \psi = \psi^0 + Q/4\pi\varepsilon_0\varepsilon r, \qquad (2.4)$$

where r is the distance from the point charge, and ψ^0 is the potential of a reference point located at $r \to \infty$. It is assumed as a convention that $\psi^0 = 0$.

A uniformly charged plane with the charge density Q_S. At distances x from the plane which are short as compared to its size (i.e., for a plane of "infinite" size) the field strength is constant and independent of this distance x. The point of reference ($\psi^0 = 0$) can conveniently be

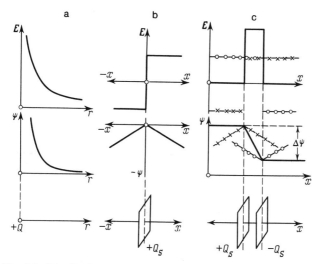

Fig. 2.1. Distributions of the modulus of field strength and of potential in electrostatic fields: (a) point charge (positive), (b) uniformly (positively) charged plane, and (c) electric double layer (× for the positively charged side, o for the negatively charged side, — for the edl as a whole).

placed in the plane itself ($x = 0$). Then

$$|\mathbf{E}| = Q_S/2\varepsilon_0\varepsilon, \qquad \psi = -(Q_S/2\varepsilon_0\varepsilon)x. \qquad (2.5)$$

An electric double layer (edl) consists of two parallel planes a distance δ apart which carry charge densities Q_S equal in magnitude but opposite in sign. The electric fields of the two planes are superimposed. The resulting field strength in the gap between the planes is given by $Q_S/\varepsilon_0\varepsilon$. Outside the edl the two field strength components compensate each other and the resulting field strength is zero. Thus, an edl does not give rise to electric fields in the surrounding space but it divides the space into two regions. The potential within each of these two regions is constant, but there is a potential change in the transit from one region to the other which is accomplished across the edl; between these two zones the potential difference (or "drop")

$$|\Delta\psi| = |\mathbf{E}|\delta = Q_S\delta/\varepsilon_0\varepsilon = Q_S/C \qquad (2.6)$$

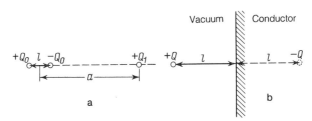

Fig. 2.2. Interaction between a point charge and a dipole (a) and the image of a point charge situated near the surface of a conductor (b).

exists [relation (2.6) is the relation for electrostatic capacitors]. Here parameter $C \equiv \varepsilon_0\varepsilon/\delta$ is the specific electric capacitance of the edl per unit area or "edl capacitance" (units: F/m^2).

Electric fields in conductors. At zero current the field strength within a conductor is zero everywhere, and the potential is constant. During current flow an electrostatic ("ohmic") field is set up along the conductor which secures the required velocity of migration of the free charges. When an isolated conductor as a whole is charged, the excess charges are displaced to the conductor's surface by forces of mutual repulsion.

Charge–dipole interaction. The interaction between a point charge Q_1 and an electric dipole $\mu_0 = Q_0 l$ (which is uncharged as a whole) depends on their distance apart (a) and on their relative orientation. When they are located along a straight line and the dipole is oriented so that the point charge is closer to the dipole's opposite charge (Fig. 2.2a), an attractive force is operative between them. From Coulomb's law we have, for $a > l$,

$$f \approx Q_1\mu_0/2\pi\varepsilon_0\varepsilon a^3 \qquad (2.7)$$

Image forces. An external charge placed close to the surface of an uncharged conductor will act upon the conductor's free charges; it will repel the charges of like sign and (or) attract the charges of opposite sign.

Close to the point where the external charge is located, induced charge of opposite sign will appear, and in the system as a whole an attractive force arises. The magnitude of this force is equal to the force that would operate between the external charge and its "mirror image," i.e., a charge of equal size but opposite sign located equally far from the surface on the side of the conductor, though without the conductor being present (Fig. 2.2b). This is why forces of this type came to be called image forces. For an elementary charge Q^0 placed 1 μm away from the surface, the image force corresponds to a potential difference of about 1 mV.

2.2. INTERFACIAL POTENTIAL DIFFERENCES (GALVANI POTENTIALS)

2.2.1. Metal/Metal Contact

An arbitrary potential difference usually exists between two pieces of metal which are insulated relative to each other; its value depends on excess charges accidentally accumulated on the metal surfaces. When the two pieces are brought in contact, the charges will then undergo a redistribution and the potential difference will become well defined. When identical metals are involved the potential difference will vanish completely, but when different metals are involved a certain potential difference will be set up across the junction (interface) which depends on the conductors. This potential difference, $\varphi_G^{(2,1)} = \psi^{(2)} - \psi^{(1)}$, between arbitrary points within the first and second metal is called the Galvani potential of this junction; $\psi^{(1)}$ and $\psi^{(2)}$ are the inner potentials of the two phases.[1] The sign of the Galvani potential depends on the relative positions stipulated for the phases, and $\varphi_G^{(2,1)} = -\varphi_G^{(1,2)}$.

Galvani potentials are produced by the difference in chemical forces exerted on the electrons within the surface layers by each of the two metals (Fig. 2.3). The unidirectional resultant f_{ch} of these forces causes transition of electrons from one metal to the other. As a result, if initially the two metals are uncharged, one of them will charge up negatively, and the other (owing to its electron deficit) will charge up positively. The excess charges accumulate near the interface and form an edl. The field which arises within this layer stops a further transition of electrons. In

---

[1]Instead of the symbol φ_G preferred here, IUPAC recommends: $\Delta\varphi = \varphi^2 - \varphi^1$, and uses the term of Galvani potential difference. In this book we use the symbol ψ for potentials at selected points within a phase, φ for potential differences between phases (e.g., the Galvani and Volta potential), and χ for potential differences near or across interfaces.

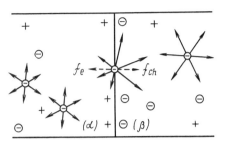

Fig. 2.3. Forces acting on charged particles near the interface between phases (α) and (β).

the end an equilibrium state is established in which the electric force f_e in the edl completely balances the effect of the chemical forces.

The chemical interaction between carriers j (here electrons) and the surrounding medium can also be described in terms of a chemical potential μ_j, which is the potential energy of these particles due to said interaction. By analogy to the electrostatic potential difference, one can define the chemical potential difference for particles in two media as the work w_{ch} performed against chemical forces in transferring these particles, divided by the number n_j of particles transferred:

$$\Delta\mu_j = w_{ch}/n_j \qquad (2.8)$$

(the chemical potential refers to one mole of the substance, hence the values of n_j are stated in moles; a more rigorous definition of chemical potentials will be given in Chapter 3). The total potential energy $\tilde{\mu}_j$ of the charged particles (again per mole, and with the charge $z_j F$) is the sum of a chemical and an electrostatic component:

$$\tilde{\mu}_j = \mu_j + z_j F\psi; \qquad (2.9)$$

it is called the electrochemical potential of the species involved.

The condition of equilibrium of the charged particles at the interface between two conductors can be formulated as the state where their electrochemical potentials are the same in the two phases:

$$\tilde{\mu}_j^{(1)} = \tilde{\mu}_j^{(2)} \qquad (2.10)$$

(the combined work of transfer, $w_e + w_{ch}$, is then zero). From this equality, and allowing for Eq. (2.9), the value of the Galvani potential established at equilibrium will be

$$\varphi_G \equiv \Delta\psi = -\Delta\mu_j / z_j F. \qquad (2.11)$$

It follows from Eq. (2.11) that the equilibrium Galvani potential only depends on the nature of the two phases (their bulk properties, which are decisive for the values of μ_j) but not on the state of the interphase (i.e., its size, any contamination present, etc.).

2.2.2. Metal/Electrolyte Contact

Galvani potentials also arise at interfaces between other types of conductors. For the interface between a metal electrode and an electrolyte, the Galvani potential can be written as $\varphi_G^{(M,E)} = \psi^{(M)} - \psi^{(E)}$, i.e., as the inner potential of the electrode (metal) relative to that of the electrolyte. The statement: "there is a shift of electrode potential in the positive direction" means that the potential of the electrode becomes more positive (or less negative) as compared to that of the electrolyte.

The general way in which a Galvani potential is established is similar in all cases, but special features are observed at the metal/electrolyte interface. The transition of charged species (electrons or ions) across the interface is possible only in connection with an electrode reaction in which other species may also be involved. Therefore, the equilibrium for the particles crossing the interface [Eq. (2.10)] can also be written as an equilibrium for the overall reaction involving all other reaction components. In this case the chemical potentials of all reaction components appear in Eq. (2.11) (for further details see Chapter 3).

If, depending on the external conditions imposed, there are different electrode reactions that can occur, and different equilibria that can be established at a given interface, then the Galvani potentials will differ accordingly; in each case they are determined by the nature of the equilibrium that is established. For instance, at a platinum electrode in sulfuric acid solution through which hydrogen is bubbled, the equilibrium of the hydrogen oxidation–reduction reaction [reaction (1.33)] will be established, but when Fe^{2+} and Fe^{3+} ions are added to the solution the oxidation–reduction equilibrium of these ions [reaction (1.31)] will be established. The values of Galvani potential established between the platinum and the solution will be different in these two cases. Thus, the Galvani potential between metal and electrolyte is determined by the nature of the electrode reaction occurring at the interface between them.

2.2.3. Galvani Potentials Cannot Be Determined

Galvani potentials between two conductors of different kinds cannot be measured by any means. Methods in which the force acting on a test charge is measured cannot actually be used here, since any values that could be measured would be distorted by the chemical forces. The same holds true for determinations of the work of transfer. At least one more interface is formed when a measuring device such as a voltmeter, potentiometer, etc. is connected, and the Galvani potential of that interface will be contained in the quantity being measured. Galvani potentials also cannot be calculated from indirect experimental data in any rigorous thermodynamic way.

Thus, potential differences can only be measured between points located within phases of the same nature.

It might be possible to attempt a theoretical calculation of Galvani potentials on the basis of certain molecular model concepts [e.g., with the aid of Eq. (2.11)]. But at the present level of scientific development such calculations are still impossible, since the full set of chemical forces acting upon charged particles and also the chemical potentials of the corresponding species cannot as yet be accounted for quantitatively. In the case of metals, the chemical potential of the electrons corresponds to the value of the Fermi energy (relative to the ground state). However, theoretical calculations of this value by Eq. (1.14) are highly inaccurate and cannot be used as a basis for calculating Galvani potentials.

Even if Galvani potentials for individual interfaces between phases of different kinds cannot be determined, their existence and the physical reasons why they develop cannot be doubted. The combined values of Galvani potentials for certain sets of interfaces which can be measured or calculated are very important in electrochemistry (see Section 2.3).

2.2.4. Exchange Currents

Equilibria at interfaces between conducting phases are dynamic; every second a certain number of charges cross the interface in one direction, and an equal number of charges cross over in the other direction. Thus, even though the overall current is zero, partial currents constantly cross the interface in both directions, and we observe an exchange of charged particles between the two phases.

At junctions between electronic conductors and electrolytes, the exchange is associated with continuing anodic and cathodic partial reactions.

It therefore follows that equilibrium can only be established for an electrode reaction when this reaction is invertible, i.e., can be made to occur in the opposite direction.

The rate of exchange when stated in electrical units is called the *exchange current* I^0 or (when referring to unit area of the interface) *exchange current density* i^0. The partial current densities in the anodic and cathodic direction are designated as $\vec{\imath}$ and $\overleftarrow{\imath}$ (in accordance with the sign convention described in Section 1.8, the value of $\overleftarrow{\imath}$ is negative).[2] The condition for equilibrium can be written as

$$\vec{\imath} + \overleftarrow{\imath} = 0 \quad \text{or} \quad \vec{\imath} = |\overleftarrow{\imath}| = i^0. \tag{2.12}$$

The values of exchange current density observed for different electrodes (or reactions) vary within wide limits. The higher they are (or the more readily charges cross the interface) the more readily will the equilibrium Galvani potential be established, and the higher will be the stability of this potential against external effects. Electrode reactions (electrodes) for which equilibrium is readily established are called thermodynamically reversible reactions (electrodes). But low values of the exchange current indicate that the electrode reaction is slow (kinetically limited).

2.2.5. Electric Double Layers at Interfaces

It has been mentioned earlier that it is owing to the formation of an electric double layer at the phase boundary that Galvani potentials are established between different phases. It is a special feature of such edl that the two layers forming the double layer are a very small (molecular) distance apart, between 0.1 and 0.4 nm. For this reason edl capacitances are very high, viz., tenths of F/m^2.

2.3. OPEN-CIRCUIT VOLTAGES

2.3.1. Metal Circuits

At zero current, when the potential within each conductor is constant, the potential difference between the terminal members of a sequence of

---

[2] Symbol i is preferred over the recommended symbol j, since I is used for currents and J is used for fluxes.

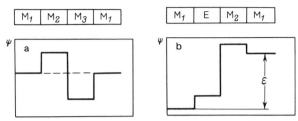

Fig. 2.4. Potential distribution in complete open circuits (with identical metals as the terminals; see Fig. 1.4): (a) circuit of different metals; (b) galvanic cell.

conductors joined together as an open circuit is the algebraic sum of all Galvani potentials at the individual interfaces, e.g.,

$$\psi^{(3)} - \psi^{(1)} = \varphi_G^{(3,2)} + \varphi_G^{(2,1)}. \tag{2.13}$$

As in the case of an individual Galvani potential, this parameter can neither be measured nor calculated for an open circuit such as that pictured in Fig. 1.4c, which is incomplete. But for a sequence of conductors with the same metal at either end (Fig. 1.4b) the obstacle pointed out in Section 2.2 vanishes, because potential differences between the identical terminal members can be measured. This parameter,

$$\mathcal{E} \equiv \psi^{(1\prime)} - \psi^{(1)} = \varphi_G^{(1,3)} + \varphi_G^{(3,2)} + \varphi_G^{(2,1)}, \tag{2.14}$$

is called the *open-circuit voltage* (ocv). When equilibrium exists at all interfaces the term *emf* (from the obsolete concept of an "electromotive force") is also used. The individual components written on the right-hand side of this equation remain unmeasurable.

The ocv of any circuit consisting only of metals or other electronic conductors which are all at the same temperature and not subjected to any external force fields is always zero (*Volta's law*, Fig. 2.4a). In fact, at any interface the Galvani potential is defined as the chemical potential difference of the electrons: $\varphi_G = \Delta\mu_e/F$ (for electrons $z_e = -1$). When these expressions are put into Eq. (2.14) the values of μ_e in the intermediate phases cancel and we obtain the expression $\mathcal{E} = [\mu_e^{(1\prime)} - \mu_e^{(1)}]/F$. Since the terminal members are identical in nature, the chemical potentials in them are the same and the ocv is zero.

It follows from Volta's law that

$$\varphi_G^{(n,1)} = -\varphi_G^{(1,n)} = \varphi_G^{(n,n-1)} + \cdots + \varphi_G^{(3,2)} + \varphi_G^{(2,1)}. \tag{2.15}$$

Therefore, the potential difference between metallic conductors 1 and n will remain unchanged when metallic conductors 2, 3, ..., $n-1$ are interposed between them.

2.3.2. Galvanic Cells

For galvanic cells the ocv generally are not zero. In contrast to metal circuits, where electrons are the sole carriers, in galvanic cells the current is transported by different carriers in the different circuit parts, viz., by electrons and by ions. Hence when substituted into Eq. (2.14) the chemical potentials of the carriers in the intermediate circuit parts will not cancel.

The concept of ocv in the case of galvanic cells always refers to a complete open-circuit arrangement with terminals of the same conductor. In the case of cells involving a zinc electrode, a graphite electrode, and a $ZnCl_2$ solution [scheme (1.23)], it refers to one of the following possibilities:

$$Cu|Zn|ZnCl_2,\ aq\,|graphite|Cu, \qquad (2.16a)$$

$$Zn|ZnCl_2,\ aq\,|graphite|Zn, \qquad (2.16b)$$

$$graphite|Zn|ZnCl_2,\ aq\,|graphite,\ etc. \qquad (2.16c)$$

It follows from Volta's law that the nature of the terminal conductor has no effect on the ocv, so that identical ocv values are obtained for all of the possibilities listed. Hence in the conventional written formulation of cells one usually does not include any terminal phases not in contact with the solution [the phases omitted may normally be taken as the terminals of the measuring instrument, like Cu in cell (2.16a)].

A nonzero ocv of a galvanic cell implies that the potential of one of the electrodes is more positive than that of the other (there is a "positive" and a "negative" electrode). By convention, the potential difference is measured as the potential of the right-hand electrode minus that of the left-hand electrode. The sign of the potential difference depends on the nature of the cell.

For the galvanic cell without transference (1.25), the ocv can be written as (Fig. 2.4b)

$$\mathcal{E} = \varphi_G^{(M_2,E)} - \varphi_G^{(M_1,E)} + \varphi_G^{(M_1,M_2)}. \qquad (2.17)$$

When the cell is symmetric, i.e., consists of two identical electrodes, its ocv will be zero.

In the case of the cell with transference (1.26), the ocv includes an additional *liquid-junction potential* (a potential difference between electrolytes), and

$$\mathcal{E} = \varphi_G^{(M_2,E_2)} - \varphi_G^{(M_1,E_1)} + \varphi_G^{(M_1,M_2)} + \varphi_G^{(E_2,E_1)}. \tag{2.18}$$

True equilibrium cannot be established at the interface between two different electrolytes, since ions can be transferred by diffusion. Hence, in thermodynamic calculations concerning such cells one often uses the corrected ocv, $\mathcal{E}^*$:

$$\mathcal{E}^* = \mathcal{E} - \varphi_G^{(E_2,E_1)}, \tag{2.19}$$

which correspond to the ocv of the same cell without the liquid-junction potential. For two electrolytes with similar chemical properties (e.g., two different solutions having the same solvent) one can calculate somewhat approximately this liquid-junction potential (see Chapter 5). Experimental means are also available for reducing the value of this potential difference. Thus, an approximate value of $\mathcal{E}^*$ can be either calculated or directly measured. However, as a rule this parameter cannot be determined with high accuracy.

In the case being discussed, the cell with transference can be formulated as

$$M_1|E_1\;\vdots\;E_2|M_2|M_1, \tag{2.20}$$

where the dotted vertical double line symbolizes the boundary between two electrolytes where the liquid-junction potential has been eliminated (by calculation or experiment).

Values of the liquid-junction potential and hence the corrected ocv can be neither calculated nor measured when the two electrolytes in contact differ chemically (e.g., when they are solutions in different solvents).

2.3.3. Two Directions of Current Flow in Galvanic Cells

Two directions of current flow in galvanic cells are possible: a spontaneous and an imposed one. When the cell circuit is closed with the aid of electronic conductors, the current will flow from the cell's positive electrode to its negative electrode in the external part of the circuit, and from the negative to the positive electrode within the cell (Fig. 2.5a). In

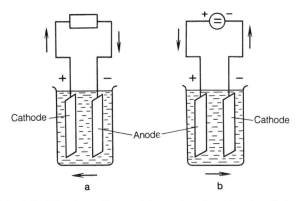

Fig. 2.5. Directions of current flow when the galvanic cell functions as a chemical power source (a) and as an electrolyzer (b).

this case the current arises on account of the cell's own voltage, and the cell acts as a chemical source of electric current or *chemical power source*. But when a power source of higher voltage, connected so as to oppose the cell, is present in the external circuit, it will cause current to flow in the opposite direction (Fig. 2.5b), and the cell works as an *electrolyzer*.

It follows that in chemical power sources the negative electrode is the anode, and the positive electrode is the cathode. In an electrolyzer, to the contrary, the negative electrode is the cathode and the positive electrode is the anode. Therefore, attention must be paid to the fact that the concepts of "anode" and "cathode" are related only to the direction of current flow and not to the polarity of the electrodes in galvanic cells.

2.4. DEFINING THE CONCEPT OF "ELECTRODE POTENTIAL"

A galvanic cell's ocv is the algebraic sum of at least three Galvani potentials, two at the interfaces between the electrodes and the electrolyte, and one at the interface between the two electrodes. Since in the cell two arbitrary electrodes are combined, it will be desirable to state the ocv as the difference between two parameters, each of which being characteristic of only one of the electrodes. In the past, the relation $\mathcal{E} = \varphi_G^{(M_2,E)} - \varphi_G^{(M_1,E)}$, involving the individual Galvani potentials between the electrodes and the electrolyte, had been examined under this aspect. However, this relation disregards the Galvani potential between the metals; moreover, it is not useful insofar as it contains parameters that cannot be determined.

A parameter that is convenient for said purpose is the "*electrode potential*"; it must not be confused with the concept of a potential difference between the electrode and the electrolyte. The electrode potential E is the ocv of a galvanic cell that consists of the given electrode (the one which is studied) and a reference electrode selected by convention.[3] Thus, the potential of this electrode is compared with that of a reference electrode that is identical for all electrodes being studied. In accordance with this definition, the electrode potential of the reference electrode itself is (conventionally) regarded as zero. Any electrode system for which the equilibrium Galvani potential is established sufficiently rapidly and reproducibly can be used as a reference electrode. We shall write the electrode system to be used as the reference electrode, generally, as M_R/E_R.

The same reference electrode can be used to characterize electrodes in contact with different electrolytes; therefore, the cell used to determine the electrode potential often includes a liquid junction (electrolyte/electrolyte interface). In this case the electrode potential is understood as being the corrected ocv value, $\mathcal{E}^*$, which is the value for this cell after elimination of the liquid-junction potential.

In the case of cell (2.16a), the electrode potentials of the electrodes correspond to the ocv values of cells

$$Cu|M_R|E_R|\!|ZnCl_2, aq|graphite|Cu \tag{2.21a}$$

and

$$Cu|M_R|E_R|\!|ZnCl_2, aq|Zn|Cu, \tag{2.21b}$$

and are given by the algebraic sum of Galvani potentials of these cells. For instance, for the zinc electrode (and allowing for Volta's law),

$$
\begin{aligned}
E &= \varphi_G^{(Zn,E)} - \varphi_G^{(M_R,E_R)} + \varphi_G^{(Cu,Zn)} + \varphi_G^{(M_R,Cu)} \\
&= \varphi_G^{(Zn,E)} - \varphi_G^{(M_R,E_R)} + \varphi_G^{(M_R,Zn)}.
\end{aligned}
\tag{2.22}
$$

---

[3]IUPAC uses the same symbol E for the quantities $\mathcal{E}$ and E as defined here.

It follows from this equation that for a given electrode, the value of electrode potential corresponds to the Galvani potential of the electrode/electrolyte interface, up to a constant term: $E = \varphi_G^{(Zn,E)} + \text{const.}$ If for any reason the value of the Galvani potential changes by a certain amount, the electrode potential will change by the same amount: $\Delta E = \Delta \varphi_G^{(Zn,E)}$. Hence the value of electrode potential yields a rather good description of the properties of this electrode. In what follows, the term "electrode potential" will always be understood as being an electrode potential relative to a defined reference electrode.

We can readily show that for cells without transference, the ocv value is equal to the difference in electrode potentials of the two electrodes, i.e., it can be written in terms of two parameters which are measurable, and each of them refers to just one of the electrodes. In the expression for ocv, the φ_G values for the reference electrode cancel, so that the reference electrode itself has no effect on the results (provided all potentials refer to the same reference electrode).

In the case of cells with transference, the difference in electrode potentials is equal not to the total but to the corrected ocv value, $\mathcal{E}^*$.

The following sign convention applies to ocv and electrode potentials. The ocv value is conventionally defined as the potential difference between the right-hand and left-hand electrode (in the written representation of the cell):

$$\mathcal{E} = E_{(\text{rhs electrode})} - E_{(\text{lhs electrode})}. \tag{2.23}$$

Since ocv values $\mathcal{E}$ are commonly reported as positive values, this would imply that in any written representation of cells, the negative electrode is on the left and the positive electrode is on the right. However, this is not necessarily true. Thus, when discussing electrode potentials we always place the reference electrode on the left [see scheme (2.21)]; the value of electrode potential will then be positive or negative depending on the polarity of this electrode relative to the reference electrode.

In practice different kinds of reference electrode are used (for more details see Section 8.3), which gives rise to different scales. The values of the electrode potentials in these scales differ by amounts given by the potential differences between the reference electrodes themselves. Most common is the scale of the standard hydrogen electrode (SHE), and values of potential measured with the aid of other reference electrodes are often converted to this scale.

2.5. NONEQUILIBRIUM ELECTRODE POTENTIALS

Electrochemical equilibrium is not always established at electrode sur-faces. When there is no equilibrium the value of the Galvani potential will not be the same as the equilibrium value defined by Eq. (2.11); the value of electrode potential will also differ accordingly.

Two cases where equilibrium is lacking must be distinguished: that which is unrelated to current flow, and can be observed even for a non-working electrode, and that which is related exclusively to the passage of current through the electrode. The former is the case of nonequilib-rium open-circuit potentials (nonequilibrium ocp), and the latter is that of electrode polarization.

2.5.1. Nonequilibrium Open-Circuit Potentials

Different reasons for lack of equilibrium at electrode/electrolyte inter-faces exist.

Charge transfer is effectively impossible (exchange currents are very small). If, under given conditions, not even a single electrode reaction is possible, any accidental accumulation of charge in the edl will be stable, as it cannot be removed by charge transfer (or "leakage" of charges). One can supply charge from outside, as in the case of ordinary electrostatic capacitors, and vary the electrode potential arbitrarily. Hence the potential has no unique equilibrium value but depends on an accidental or deliberate accumulation of charge in the edl. An electrode having these attributes is called ideally polarizable.

Ideal polarizability usually is observed only within a certain interval of potentials, while outside this interval charge transfer across the edl (current flow) is possible.

Low exchange currents. When an electrode reaction is possible but its exchange current is low, the equilibrium potential is readily disturbed by external effects. Such an influence is exerted in particular by foreign components (contaminants), the reactions of which are superimposed upon the basic reaction.

Several electrode reactions occurring simultaneously. The follow-ing reactions can occur at an iron electrode in HCl solution containing

FeCl$_2$ while a hydrogen atmosphere is maintained:

$$Fe^{2+} + 2e^- \leftrightarrows Fe, \tag{2.24}$$

$$2H^+ + 2e^- \leftrightarrows H_2. \tag{2.25}$$

Each of these reactions has its own exchange current density and its own equilibrium potential. The condition of overall balance at this electrode is not determined by Eq. (2.12) but by the equation

$$\overrightarrow{\imath}_1 + \overrightarrow{\imath}_2 + \overleftarrow{\imath}_1 + \overleftarrow{\imath}_2 = 0, \tag{2.26}$$

where subscripts 1 and 2 refer to reactions (2.24) and (2.25), respectively. Generally $\overrightarrow{\imath}_1 \neq |\overleftarrow{\imath}_1|$, i.e., reaction (2.24) will occur in one direction and reaction (2.25) will occur in the other direction. In our example, iron will dissolve anodically ($\overrightarrow{\imath}_1 > |\overleftarrow{\imath}_1|$) while hydrogen will be evolved at the cathode ($\overrightarrow{\imath}_2 < |\overleftarrow{\imath}_2|$). The nonequilibrium potential being established will be intermediate between the individual equilibrium potentials corresponding to reactions (2.24) and (2.25), and is called a *mixed potential* or, when it is well reproducible and almost time-invariant, a steady-state or *rest potential*.

Equilibrium potentials can be calculated thermodynamically (for more details see Chapter 3) when the corresponding electrode reaction is precisely known, even when they cannot be reached experimentally (i.e., when the electrode potential is nonequilibrium despite the fact that the current is practically zero).

The open-circuit voltage of any galvanic cell where at least one of the two electrodes has a nonequilibrium open-circuit potential will also be nonequilibrium.

Particularly in thermodynamic calculations, the term *emf* is often used for measured or calculated equilibrium ocv values.

2.5.2. Electrode Potentials during the Passage of Current

At nonzero current, the flow of charges crossing the interface in one direction is larger than that crossing it in the other direction:

$$i = \overrightarrow{\imath} + \overleftarrow{\imath}. \tag{2.27}$$

The partial current ($\overrightarrow{\imath}$) is larger than the (absolute value of the) partial current $|\overleftarrow{\imath}|$ in the case of anodic currents ($i > 0$), and vice versa in the case of cathodic currents. Thus, when there is a net current flow, the partial currents are not the same as in equilibrium [when condition (2.12) is observed]. For this change in partial currents a unidirectional force is needed, i.e., the electric force f_e should be stronger (or weaker) than the chemical force f_{ch} (see Fig. 2.3). The electrode potentials change accordingly. Anodic current flow is associated with a positive potential shift (easier transfer of positive charges from the electrode into the electrolyte or of negative charges in the opposite direction); cathodic current flow is associated with a negative potential shift.

The shift of electrode potential caused by current flow,

$$\Delta E \equiv E_i - E_0 \qquad (2.28)$$

(where E_i and E_0 are the potentials at a current density i and at zero current, respectively), has been termed "*electrode polarization*" or "*overpotential*." The quantity ΔE is positive in the case of anodic currents, and negative in the case of cathodic currents. For a given electrode, the absolute value of polarization or departure from equilibrium will be larger at higher currents. The exchange current density is used here as the reference quantity. When the net current is low as compared to the exchange current the departure from equilibrium will be minor and polarization low (for further details see Chapter 6).

2.5.3. The Cell Voltage at Nonzero Current

The way in which the voltage of a galvanic cell changes when a current flows depends on the direction of the current. In a cell working as a chemical power source, the positive electrode is the cathode and the negative electrode is the anode. Owing to polarization the potential of the former moves in the negative direction, and that of the latter moves in the positive direction, i.e., the potentials of the two electrodes move closer together, and the voltage decreases (Fig. 2.6a). In addition, an ohmic voltage drop φ_{ohm} develops in the cell (chiefly in the electrolyte); the potential of the electrolyte will be more negative at the cathode, to which the cations migrate. This leads to an additional voltage decrease. Thus, in this case the voltage under current flow, $\mathcal{E}_i$, will always be lower than the ocv, $\mathcal{E}_0$:

$$\mathcal{E}_i = \mathcal{E}_0 - \Delta E_a - |\Delta E_c| - \varphi_{ohm} = \mathcal{E}_0 - \eta_{cell} \qquad (2.29)$$

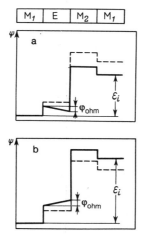

Fig. 2.6. Potential distributions in galvanic cells functioning as a chemical power source (a) and as an electrolyzer (b); the dashed lines are for the zero-current situation.

(subscript a stands for anode, and subscript c stands for cathode).

In cells working as electrolyzers, the positive electrode is the anode and the negative electrode is the cathode. All the factors listed above (for the case of a power source) will now produce an increase in voltage relative to the ocv (Fig. 2.6b):

$$\mathcal{E}_i = \mathcal{E}_0 + \Delta E_a + |\Delta E_c| + \varphi_{ohm} = \mathcal{E}_0 + \eta_{cell}. \qquad (2.30)$$

The modulus of voltage change, $\eta_{cell} \equiv |\mathcal{E}_i - \mathcal{E}_0| = \Delta E_a + |\Delta E_c| + \varphi_{ohm}$, is called the total cell overvoltage.

2.6. POTENTIALS AT THE CONDUCTOR/INSULATOR (OR VACUUM) INTERPHASE

2.6.1. The Surface Potential of a Phase

Consider a conductor α in contact with an insulator (or vacuum) not containing charges (Fig. 2.7). Let x be the distance of a point inside the insulator measured from the interface. The potential will be referred to the conventional point of reference at $x \to \infty$ within the insulator.

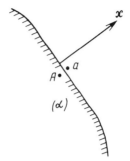

Fig. 2.7. The conductor/insulator interface.

When the conductor as a whole is charged (i.e., has excess charge of one sign in its surface layer), an electrostatic field and a potential gradient will develop in the insulator region adjacent to it. The name of outer potential, $\psi_{ex}^{(\alpha)}$, of the conductor is used for the potential at a point a located in the insulator just outside the conductor. Since point a and the point of reference are located in the same phase, this potential can be measured.

The concept of "just outside" must be more closely defined. When the test charge is moved from the point of reference toward the surface, work is performed due to the (primary) electrostatic field being discussed. However, very close to the surface, image forces start to act upon the test charge; they give rise to an additional work term and distort the primary field. Hence the distance of point a from the surface is selected large enough so that the contribution of the image forces will be small; at the same time this distance should be short enough so that we can neglect the change in primary field between point a and the surface. Calculations show that these conditions are fulfilled at distances of about 1 μm from the surface.

No potential gradient exists within the conductor, hence the conductor's inner potential $\psi_{in}^{(\alpha)}$ is the potential at any point A inside the conductor (see Section 2.2.1).

The potential difference between points A and a defined as indicated,

$$\chi^{(\alpha)} \equiv \psi_{in}^{(\alpha)} - \psi_{ex}^{(\alpha)}, \tag{2.31}$$

is called the conductor's *surface potential*.[4] The electric field producing

---

[4]For χ, which like φ_G and φ_V is a potential difference, IUPAC and the present book agree in using the same symbol and the same name "surface potential."

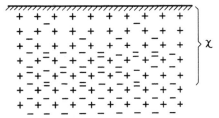

Fig. 2.8. Nonuniform charge distribution in the surface layer of a conductor.

this potential is wholly concentrated in the conductor's surface layer, where an edl is formed on account of nonuniform charge distribution (Fig. 2.8). The value of the surface potential depends on the structure and chemical properties of the conductor surface.

The values of surface and inner potential defined as indicated cannot be measured, in contrast to the outer potential, since they refer to points in different phases.

2.6.2. Work Functions

Conductor/insulator interfaces lack a continuous exchange of free charges, and there is no electrochemical equilibrium. For this reason the work $w_j^{(\alpha,\beta)}$ which is performed in transferring charges from one phase into the other is not zero. The total work $w_j^{(\alpha,0)}$ which must be performed by the external forces in transferring (extracting) a charged particle j from the conductor (α) into vacuum (0) is called this particle's *work function.* Work functions are always positive, since otherwise the free charges would leave the conductor spontaneously. Very often (but not always) electron work functions for metals in vacuum are considered.

Because of the influence of potential gradients, the work function (or work of extraction of a charged particle in vacuum) depends on the position of the point to which the particle is transferred. As in the definition of surface potential, a point (a) situated in the vacuum just outside the conductor is regarded as the terminal point of transfer. It is assumed, moreover, that when the transfer has been completed the velocity of the particle is close to zero, i.e., no kinetic energy is imparted to it.

The work function thus defined is equal to the difference in electrochemical potentials of the particle at point a in vacuum, $\tilde{\mu}_j^{(0)}$, and any point in the conductor, $\tilde{\mu}_j^{(\alpha)}$. In the vacuum where there is no chemical interaction between the particle and the medium we have $\mu_j^{(0)} = 0$, and hence $\tilde{\mu}_j^{(0)} = z_j Q^0 \psi_{ex}^{(\alpha)}$. Using this quantity as well as Eq. (2.31), we find

$$w_j^{(\alpha,0)} = z_j Q^0 \psi_{ex}^{(\alpha)} - \tilde{\mu}_j^{(\alpha)} = -[\mu_j^{(\alpha)} + z_j Q^0 \chi^{(\alpha)}]. \qquad (2.32)$$

Work functions refer traditionally not to one mole of particles (with the charge $z_j F$) but to one particle (with the charge $z_j Q^0$; in the case of electrons, with charge $-Q^0$), and usually are stated in electrical units of electron volt (1 eV = $1.62 \cdot 10^{-19}$ J). In equations of the type of (2.32), therefore, the values of $\tilde{\mu}_j$ and μ_j also refer to one particle.

It follows from Eq. (2.32) that the work function has a chemical and an electrostatic component. Its overall value can be measured while an exact determination of its individual components is not possible. The chemical component depends on the interaction between the charge and the surrounding medium; moreover, it includes the work performed in overcoming the image forces. The sum $\mu_j^{(\alpha)} + z_j Q^0 \chi^{(\alpha)}$ is often called the real potential $\lambda_j^{(\alpha)}$ of the species j in phase α.

The work function of charged particles found for a particular conductor depends not only on its bulk properties (its chemical nature), which govern parameter $\mu_j^{(\alpha)}$, but also on the state of its surface layer, which influences the parameter $\chi^{(\alpha)}$. This has the particular effect that for different single-crystal faces of any given metal the electron work functions have different values. This experimental fact is one of the pieces of evidence for the existence of surface potentials. The work function also depends on the adsorption of foreign species, since this influences the value of $\chi^{(\alpha)}$.

Several methods exist for measuring the electron work functions of metals. In all these methods one determines the level of an external stimulus (light, heat, etc.) required to extract electrons from the metal.

In the method of electron photoemission one determines the lowest frequency of light (ν_0, the "red" limit in the spectrum) at which electrons can be knocked from the metal. The quantum energy $h\nu_0$ of this light gives directly the work of extraction of one electron. When light quanta of higher energies are employed the electrons acquire additional kinetic energy.

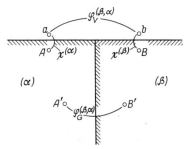

Fig. 2.9. Volta potential, Galvani potential, and surface potentials.

The saturation current I_{sat} observed in thermionic electron emission is related to temperature by the famous Richardson–Fowler law

$$\ln(I_{sat}/T^2) = \text{const} - w_e^{(\alpha,0)}/kT \qquad (2.33)$$

($k \equiv R/N_A$ is the Boltzmann constant). The work function can be determined from the slope of straight lines obtained when plotting $\ln(I_{sat}/T^2)$ against T^{-1}.

The problem of electron work functions for metals in electrolyte solutions will be considered in Section 14.4.1.

2.6.3. Volta Potentials

Consider two conductors, α and β, in mutual contact in a vacuum (Fig. 2.9). Each of them has a certain surface potential; these potentials are $\chi^{(\alpha)}$ and $\chi^{(\beta)}$, respectively. Between the conductors the Galvani potential $\varphi_G^{(\beta,\alpha)}$ is established.

The potential difference between points a and b located in the vacuum just outside of conductors α and β, respectively, is called the *Volta potential* $\varphi_V^{(\beta,\alpha)}$, also the outer or contact potential difference, of this pair of conductors.[5] Taking into account that the potential difference between two points is independent of the path taken between these points, we have

$$\varphi_V^{(\beta,\alpha)} = \chi^{(\alpha)} - \chi^{(\beta)} + \varphi_G^{(\beta,\alpha)}. \qquad (2.34)$$

---

[5]Instead of φ_V, IUPAC recommends $\Delta\psi = \psi^2 - \psi^1$, and uses the term Volta potential difference.

Points a and b are located in the same phase (vacuum); therefore, the Volta potential can be measured, in contrast to what is found for the Galvani potential (between points A' and B') and for the surface potentials (between points a and A, and between points b and B).

For measurements of the Volta potentials one uses a special feature of the electrostatic capacitor. In fact, when the two sides of a capacitor do not (as usual) consist of identical conductors but of different ones, then the charge on the capacitor plates, according to the capacitor relation (2.6), is not related to the difference between the inner potentials of the two conductors, but to their Volta potential (to the difference between the outer potentials). Knowing the value of capacitance of the capacitor and measuring the charge that flows when the plates are made part of a suitable circuit, one can thus determine the Volta potential.

The Volta potential between two metals is related directly to the electron work functions of these metals. Taking into account that for two metals in contact at equilibrium we have $\tilde{\mu}_e^{(\alpha)} = \tilde{\mu}_e^{(\beta)}$ and that $z_e = -1$, we obtain from Eq. (2.32):

$$w_e^{(\beta,0)} - w_e^{(\alpha,0)} = -Q^0[\psi_{ex}^{(\beta)} - \psi_{ex}^{(\alpha)}] = -Q^0\varphi_V^{(\beta,\alpha)}. \qquad (2.35)$$

With the aid of this equation and the results of Volta potential measurements we can determine the electron work function of a second metal when that of the first metal is known.

The fact that there is a potential difference between points close to the surfaces of two conductors in contact implies that the excess charge densities on their exposed surfaces are different. This also implies that when two conductors come in contact there will be a redistribution of free electrons, not only at the actual inner contact surface (which gives rise to the Galvani potential) but also at their exposed surfaces.

Chapter 3
THERMODYNAMICS OF ELECTROCHEMICAL SYSTEMS

3.1. WAYS TO DESCRIBE COMPOSITION

The variable composition of multicomponent phases (solutions, salt melts, metal alloys, etc.) consisting of N components can be unambiguously described by the values of concentration for $N - 1$ components. The concentration of one of the components, for which in the following we shall use the index $k = 0$, can be stated in terms of the values of concentration of the remaining components; in the case of solutions, the solvent is selected as the component for which $k = 0$.

There are different ways to describe composition.

(i) The **molar concentration** or **molarity**, c_k, is the number of moles, n_k, of a given substance k in unit volume of a phase: $c_k = n_k/V$ mol/liter (for the volume V, density ρ, and molar concentration c_k one usually employs the traditional units of liter, kg/liter, and mol/liter admitted as SI units, rather than the more rigorous units of m^3, kg/m^3, and mol/m^3, even though numerical conversion factors will appear in the corresponding equations).

(ii) The molal concentration or **molality**, m_k, is the number of moles of a given substance per kilogram of the solvent: $m_k = n_k/n_0 M_0$ mol/kg (where n_0 and M_0 are the numbers of moles of the solvent and its molar mass in kg/mol).

(iii) The **mole fraction**, Z_k, is the ratio of the number of moles of a given substance to the total number of moles of all substances: $Z_k = n_k/\sum n_k$.

Molar concentrations are of greater practical convenience, since liquid solutions as a rule are prepared according to volume. This way of

expressing the composition though, has the disadvantage that the values
vary with temperature, and they also change when other substances are
added, since these will cause changes in solution volume.

The different ways of describing the composition are interrelated.
It follows from the definitions of the concentrations, and the equality
$\sum n_k M_k = V\rho$, that

$$c_k = \left[n_0 M_0 \rho \Big/ \sum n_k M_k \right] m_k = \left[\rho \sum n_k \Big/ \sum n_k M_k \right] Z_k, \quad (3.1)$$

where M_k are the molar masses (in kg/mol), and the summation is over
all components including the solvent.

For binary solutions, where the solute $k = 1$ and where 1 liter of the
solution contains $\rho - M_1 c_1$ kg of the solvent, said interrelation can be
formulated more simply without a mole count as

$$c_1 = \frac{\rho m_1}{1 + M_1 m_1} = \frac{\rho Z_1}{M_0 + (M_1 - M_0) Z_1}. \quad (3.2)$$

In dilute solutions (with solute concentrations c_k of up to 0.1 mol/liter,
m_k of up to 0.1 mol/kg, or Z_k of up to 0.002, approximately), the following approximate expressions are valid to within 1 %:

$$c_k \approx \rho m_k \approx \rho Z_k / M_0. \quad (3.3)$$

The formal concentrations of the pure solvent are given by $Z_0 = 1$,
$m_0 = 1/M_0$, and $c_0 = \rho/M_0$, respectively, and in the case of water,
for instance, one has $M_0 = 0.018$ kg/mol, $\rho \approx 1$ kg/liter, and hence
$m_0 \approx 55.5$ mol/kg and $c_0 \approx 55.5$ mol/liter.

3.2. CONVENTIONAL AND UNDEFINED PARAMETERS

Often parameters of which the absolute values are not known are contained in physical and thermodynamic equations. Among these parameters
two kinds can be distinguished.

(i) **Conventional parameters** which can be defined only in terms
of differences of the values in two states or points. The value of such a
parameter in any given state depends on a conventional point of reference,
and hence, can only be determined to a constant term. It is physically

meaningless to define absolute values of such parameters. The electrostatic potential is an example.

(ii) **Experimentally undefined parameters** which have a real physical meaning, i.e., they reflect an actual physical phenomenon but cannot be determined from the experimental data (even a thought experiment to measure them cannot be conceived) or by a thermodynamic calculation. In isolated cases such parameters can be calculated on the basis of non-thermodynamic models. The Galvani potential at the interface between two dissimilar conducting phases is an example.

The values of all these parameters are unknown, yet it is useful to include them in physical arguments and equations in order to facilitate the understanding of correlations between different phenomena. The equations used for calculations usually contain sums, differences, or other combinations of such parameters which are measurable.

3.3. THERMODYNAMIC FUNCTIONS IN ELECTROCHEMISTRY

The well-known tools of chemical thermodynamics are used in electrochemical systems to describe equilibria and processes, but some special features arising from the presence of charged particles and potential differences between the phases must be taken into account.

Electrochemical systems are usually discussed under the conditions of constant temperature T and pressure p. Under these conditions the most convenient thermodynamic functions are the Gibbs energy G given by $U + pV - TS$ and the enthalpy H given by $U + pV$ (where U is the internal energy, V is the volume, and S is the entropy).

The values of these functions change when thermodynamic processes take place. Processes in which the Gibbs energy decreases, i.e., for which $\Delta G < 0$ or $dG < 0$, will take place spontaneously, without any specific external action. The Gibbs energy is minimal in the state of equilibrium, and the condition for equilibrium is given by

$$\Delta G = 0 \quad \text{or} \quad dG = 0. \tag{3.4}$$

The heat $\bar{q}$ given off by the system to the environment in the process and the useful work $\bar{w}$ (electrical, mechanical) performed by the system in an external medium are related, according to the first law of thermodynamics, as

$$\bar{q} + \bar{w} = -\Delta H. \tag{3.5}$$

The values of $\bar{q}$ and $\bar{w}$ can be both positive and negative. Parameters $\bar{w}$ and w differ in sign (see Chapter 2).

The partition of an enthalpy change between these terms depends on the way in which the process occurs. When there is a total lack of equilibrium in the process we have $\bar{w} = 0$, and all energy is evolved as heat:

$$\bar{q}_{max} = -\Delta H, \tag{3.6}$$

where $\bar{q}_{max}$ is the heat effect of the reaction (the maximum value of $\bar{q}$).

When there is complete equilibrium we have, according to the second law of thermodynamics,

$$\bar{w}_{max} = -\Delta G. \tag{3.7}$$

Then the heat which is evolved is related, according to Eqs. (3.5) and (3.7), to the entropy change as

$$\bar{q}_{lat} = -\Delta H - (-\Delta G) = -T\Delta S. \tag{3.8}$$

This is known as latent (or entropic) heat.

In real processes the work can have any value between zero and $\bar{w}_{max}$, and the heat can have any value between $\bar{q}_{max}$ and $\bar{q}_{lat}$. (In the following we shall drop the subscript max of these parameters.)

The values of $-\Delta H$ and $-\Delta G$ can be determined by straightforward experiments for a variety of processes. The parameters H and G themselves are defined by convention.

In systems with different components, the values of the thermodynamic functions depend on the nature and number of these components. One distinguishes components forming independent phases of constant composition (the "pure" components) from the components which are part of mixed phases of variable composition (e.g., solutions).

The Gibbs energy is an additive function of all components. For systems consisting of pure components only,

$$G = \sum n_k G_k, \tag{3.9}$$

where G_k is the specific (molar) Gibbs energy of the component k (in J/mol) and n_k is the number of moles of this component in the system.

By convention the values of G_k for the chemical elements in their usual state at 25°C are zero. Then the values of G_k for compounds are given by the values of ΔG of the reaction producing them from the elements. The same is true for the enthalpies. For many compounds, the values of G_k and H_k thus determined are listed in handbooks.

In phases of variable composition, the contribution of each component to the total value of G for a given phase depends not only on the amount of this component present in the phase but also on its concentration. For small changes in composition

$$dG = \sum_k (\partial G/\partial n_k)_{n_l}\, dn_k = \sum \mu_k\, dn_k \qquad (3.10)$$

(where $l \neq k$, i.e., partial derivatives are evaluated where the values of all remaining variables are kept constant). The coefficient

$$\mu_k \equiv (\partial G/\partial n_k)_{n_l} \qquad (3.11)$$

is called the chemical potential of the component k (in J/mol). The values of μ_k are not constant, in contrast to parameter G_k, but depend on the concentration of the given substance in the phase of variable composition, and hence on the presence of other components.

The chemical potential of a gas depends on its partial pressure p_k. When the gas obeys ideal gas laws we have

$$\mu_k = \mu_k^0 + RT \ln p_k. \qquad (3.12)$$

For sufficiently dilute solutions the concentration dependence of chemical potential similarly is given by

$$\mu_k = \mu_k^0 + RT \ln c_k. \qquad (3.13)$$

The same equations are obtained when the concentrations are stated in the units of m_k or Z_k, but the values of constant μ_k^0 then are different. It follows from Eq. (3.3) that

$$\mu_{c,k}^0 = \mu_{m,k}^0 - RT \ln \rho = \mu_{Z,k}^0 + RT \ln(M_0/\rho). \qquad (3.14)$$

Solutions where, as in the case of ideal gases, the concentration dependence of chemical potential obeys Eq. (3.13) have been called ideal solutions. In nonideal solutions (or in other systems of variable composition) the concentration dependence of chemical potential is more complicated.

In phases of variable composition, the values of Gibbs energy are determined by the equation

$$G = \sum n_k \mu_k \qquad (3.15)$$

instead of Eq. (3.9). The chemical potentials of the different components constituting such a phase are interrelated by the *Gibbs–Duhem equation*

$$\sum n_k d\mu_k = 0. \qquad (3.16)$$

This equation is obtained when Eq. (3.10) is subtracted from the total differential of Eq. (3.15). It follows that in a phase containing N components, the values of μ_k of only $N - 1$ of them can be varied arbitrarily (by changing the concentrations).

Since they contain charged components (ions, electrons), electrochemical systems have certain special features.

Influence of the electrostatic potential of a phase. The energy of an ion in a given medium depends not only on chemical forces but also on the electrostatic field, hence the chemical potential of an ion j customarily is called its electrochemical potential and labeled $\tilde{\mu}_j$.

The electrostatic potential energy of an ion when reckoned per mole is given by $z_j F \psi$, where ψ is the electrostatic (inner) potential of the phase containing the ion. Hence the electrochemical potential can be written as the sum of two terms:

$$\tilde{\mu}_j = \mu_j + z_j F \psi, \qquad (3.17)$$

where μ_j is the chemical component of the ion's electrochemical potential (it is commonly called the ion's "chemical potential").

This formulation is somewhat conditional, since it is assumed in it that a concentration change will affect only the value of μ_j, and a potential change will affect only the second term on the right-hand side of (3.17). Actually a potential change is associated with a change in the amount of charge in the edl and implies some change in concentration of the ions in the bulk phase. Hence the potential ψ cannot be varied independently of the concentration, or of μ_j. However, in most cases (other than extremely dilute systems) a change in potential is associated with concentration changes so small that the concentration can practically be regarded as constant.

The Gibbs energy of an electroneutral system is independent of the electrostatic potential. In fact, when substituting into Eq. (3.15) the electrochemical potentials of the ions contained in the system and allowing for the electroneutrality condition, we can readily see that the sum of all terms $n_j z_j F \psi$ is zero. The same is true for any electroneutral subsystem consisting of the two sorts of ion M^{z+} and A^{z-} (particularly when these are produced by dissociation of a molecule of the original compound k into τ_+ cations and τ_- anions), for which

$$\tau_+ \tilde{\mu}_+ + \tau_- \tilde{\mu}_- = \tau_+ \mu_+ + \tau_- \mu_- = \mu_k, \qquad (3.18)$$

where μ_k is the chemical potential of compound k. (In the following, symbol k will only be used for electroneutral compounds or ensembles of ions.)

The electrochemical potential of single ionic species cannot be determined. In systems with charged components, all energy effects and all thermodynamic properties are associated not with ions of a single type but with combinations of different ions. Hence the electrochemical potential of an individual ionic species is an experimentally undefined parameter, in contrast to the chemical potential of uncharged species. From the experimental data only the combined values for electroneutral ensembles of ions can be found. Equally inaccessible to measurements is the electrochemical potential, $\tilde{\mu}_e$, of free electrons in metals, while the chemical potential, μ_e, of the electrons coincides with the Fermi energy and can be calculated (very approximately) by Eq. (1.14).

The energy effects of individual electrode reactions cannot be measured. Any given electrode reaction can occur only in parallel with a second, coupled reaction. Because of the interfering effects of this reaction (which include heat, diffusion, and other fluxes in the electrolyte), it is not possible to experimentally determine the energy effects, and hence the thermodynamic parameters, of an individual electrode reaction. Nor can these parameters be calculated, since this would require a knowledge of the electrochemical potentials of individual ionic species and of the Galvani potential at the electrode/electrolyte interface. All thermodynamic calculations and measurements refer to the current-producing reaction as a whole (this holds true also for reactions involving the reference electrode).

3.4. THERMODYNAMIC ACTIVITY

3.4.1. Definition of the Concept of Activity

By virtue of the function (3.13), concentrations, which are a parameter that can be readily determined, can be used instead of the chemical potentials in the thermodynamic equations for ideal systems. The simple connection between the concentrations and chemical potentials is lost in real systems. In order to facilitate the changeover from ideal to nonideal systems and avoid the use of two different sets of equations, chemical thermodynamics employs another parameter, the *thermodynamic activity* a_k, which has the dimensions of concentration and is defined by the equation

$$\mu_k \equiv \mu_{a,k}^0 + RT \ln a_k, \tag{3.19}$$

where $\mu_{a,k}^0$ is a constant which is independent of composition of a phase and called the standard chemical potential.

The changeover to thermodynamic activities is equivalent to a change of variables in mathematical equations. The relation between parameters μ_k and a_k is unambiguous only when a definite value has been selected for the constant $\mu_{a,k}^0$. For solutes this constant is selected so that in highly dilute solutions where the system approaches an ideal state, the activity will coincide with the concentration: $(\lim a_k)_{c_k \to 0} = c_k$. Hence

$$\mu_{a,k}^0 = \mu_{c,k}^0. \tag{3.20}$$

Depending on the way in which the concentration is expressed, different kinds of activity are obtained: $a_{c,k}$, $a_{m,k}$, or $a_{Z,k}$, which are interrelated just like the corresponding concentrations [Eqs. (3.2) and (3.3)]. In chemical thermodynamics chiefly the molal concentrations (molalities) m_k and molal activities $a_{m,k}$ are used, which are stated in mol/kg. In what follows, the symbols c_k and a_k will be retained to denote concentrations and activities in a general way.

For the solvent, the standard chemical potential is selected so that in an ideal (extremely dilute) solution the activity will coincide with mole fraction, i.e., $(\lim a_0)_{Z_k \to 0} = Z_0$. Hence

$$\mu_{aZ,0}^0 = \mu_{Z,0}^0. \tag{3.21}$$

According to this definition the activity of the pure solvent ($Z_0 = 1$) is unity.

For single-component liquid and solid phases of constant composition the activity is also always taken to be unity.

A system's standard state is defined as the state in which $\mu_k = \mu_k^0$ and hence $a_k = 1$. We must bear in mind that the standard state does not coincide with the limiting state (at low concentrations) when the system becomes ideal. Hence in the standard state the value of activity differs from the value of concentration (except for the solvent).

Using activities instead of chemical potentials has the major advantage that the equations derived for ideal systems can be retained in the same form for real systems, merely with activities in the place of concentrations. For the practical application of these equations we must know the values of activity as a function of concentration.

The degree of departure of a system from the ideal state can be described by another parameter, the *activity coefficient*, which is defined as the ratio of activity to concentration:

$$f_{c,k} \equiv a_{c,k}/c_k. \qquad (3.22)$$

Activity coefficients are dimensionless. Activity coefficients $f_{m,k}$ or $f_{Z,k}$ which differ somewhat in the numerical values are obtained when other ways are selected to describe concentrations and activities.

With standard states selected as indicated above, activity coefficients of any kind will be unity in ideal systems. The degree of departure of a system from the ideal state is described by the departure of the activity coefficients from unity.

3.4.2. The Activities of Ions in Electrolyte Solutions

In electrolyte solutions, nonideality of the system is much more pronounced than in solutions with uncharged species. This can be seen in particular from the fact that electrolyte solutions start to depart from an ideal state at lower concentrations. Hence activities are always used in the thermodynamic equations for these solutions. It is in isolated instances only, when these equations are combined with other equations involving the number of ions per unit volume (e.g., equations for the balance of charges), that concentrations must be used and some error is thus introduced.

In binary electrolyte solutions with an initial concentration c_k and ionic concentrations $c_j = \alpha \tau_j c_k$ [see Eq. (1.17)] the ionic activities can be written as

$$a_j = f_j \alpha \tau_j c_k = \gamma_j \tau_j c_k \tag{3.23}$$

$$(a_+ = \gamma_+ \tau_+ c_k \quad \text{and} \quad a_- = \gamma_- \tau_- c_k),$$

where $\gamma_j \equiv \alpha f_j$ are the so-called stoichiometric activity coefficients in which the degree of dissociation is taken into account.

The stoichiometric activity coefficients are used in combination with the maximum concentration $\tau_j c_k$ of ions calculated under the assumption of complete dissociation. For solutions of strong electrolytes $\gamma_j \approx f_j$.

The activity of an individual ion, like its chemical potential, cannot be determined from experimental data. For this reason the parameters of *electrolyte activity* a_k and *mean ionic activity* $a_\pm$ are used which are defined as follows:

$$a_k \equiv a_\pm^{\tau_k} \equiv a_+^{\tau_+} a_-^{\tau_-}. \tag{3.24}$$

It follows from this definition, with Eq. (3.18), that

$$\mu_k = (\tau_+ \mu_+^0 + \tau_- \mu_-^0) + RT \ln a_k. \tag{3.25}$$

Values of a_k (and of $a_\pm$) can be found experimentally.

Substituting into Eq. (3.24) the (expanded) expression (3.23) for the activities of the individual ions, we obtain

$$a_\pm = (\gamma_+^{\tau_+} \gamma_-^{\tau_-} \tau_+^{\tau_+} \tau_-^{\tau_-})^{1/\tau_k} c_k = \gamma_\pm \lambda c_k. \tag{3.26}$$

Here $\gamma_\pm$ is the mean ionic activity coefficient defined, by analogy with the mean ionic activity, as

$$\gamma_\pm^{\tau_k} \equiv \gamma_+^{\tau_+} \gamma_-^{\tau_-}, \tag{3.27}$$

while λ is the numerical factor

$$\lambda \equiv (\tau_+^{\tau_+} \tau_-^{\tau_-})^{1/\tau_k}. \tag{3.28}$$

For symmetric electrolytes ($\tau_+ = \tau_- = 1$) we find $\lambda = 1$; for 1:2 electrolytes (e.g., Na_2SO_4), 1:3 electrolytes ($AlCl_3$), and 2:3 electrolytes [$Al_2(SO_4)_3$] the corresponding values of λ are 1.587, 2.280, and 2.551.

When activity coefficients f_j rather than $\gamma_\pm$ are used, Eq. (3.26) becomes

$$a_\pm = \alpha f_\pm \lambda c_k. \tag{3.29}$$

Mean ionic activity coefficients of many salts, acids, and bases in binary aqueous solutions are reported, for wide concentration ranges, in special handbooks.

In saturated solutions of electrolytes (particularly those having low solubility), the overall chemical potential, μ_k, of the substance in the solution is equal to the specific Gibbs energy, G_k, of the solid phase in equilibrium with the solution. Hence the values of μ_k and α_k are constant in the solution, and according to Eq. (3.24)

$$a_+^{\tau_+} a_-^{\tau_-} = \text{const} = L(k), \tag{3.30}$$

where the constant $L(k)$ is called the solubility product or ionic product of the substance k.

In aqueous solutions, H^+ and OH^- ions are present owing to the dissociation of water molecules. In dilute solutions, the activity of water is constant. Hence for the activities of these ions an equation of the type of (3.30) is obeyed, too. The ionic product of water is written as K_w. This parameter is given by $K_w = a_{H^+} a_{OH^-}$, and at 25°C has a value of $1.27 \cdot 10^{-14}$ mol^2/liter^2.

3.5. EQUATIONS FOR THE EMF OF GALVANIC CELLS

Let us write ΔG_m for the Gibbs energy change occurring during a chemical reaction m involving the conversion of ν_j moles of reactants and products. We find

$$\Delta G_m = \sum \bar{\nu}_j \mu_j, \tag{3.31}$$

when the concentrations and, hence, the chemical potentials of the components do not markedly change during the reaction (when the degree of conversion is low). For electrode reactions involving electrons, Eq. (3.31) can be written as

$$\pm \Delta \bar{G}_m = \sum \bar{\nu} \tilde{\mu}_j + n \tilde{\mu}_e, \tag{3.32}$$

where the plus sign is used for the anodic, and the minus sign for the cathodic direction of the reaction [cf. Eq. (1.37)].

The value of $\Delta \bar{G}_m$, like that of $\tilde{\mu}_j$, depends on potential. The electrons which are involved in the reaction are at a potential $\psi^{(M)}$ inside the metal, and all ions are at a potential $\psi^{(E)}$ inside the electrolyte. Using the expanded form (3.17) for the electrochemical potential of these components and allowing for Eq. (1.39) as well as for the definition of the Galvani potential, we can transform Eq. (3.32) to

$$\pm \Delta \bar{G}_m = \sum \tilde{\nu}_j \mu_j + n \mu_e - n F \varphi_G. \tag{3.33}$$

It follows from this equation that the Galvani potential in the point of equilibrium of the electrode reaction ($\Delta \bar{G}_m = 0$) is given by

$$\varphi_G = \sum \tilde{\nu} \mu_j / n F + \mu_e / F. \tag{3.34}$$

None of the terms of this equation can be determined experimentally. The equation represents a generalization of Eq. (2.11) for the case where charge transfer between phases is attended by an electrochemical reaction.

Consider the galvanic cell (1.25) where the overall current-producing reaction (1.42) takes place. Allowing for charge balance we can write for the Gibbs energy change of this reaction [when using Eq. (3.32) for electrodes 1 and 2]

$$\Delta G_m = \sum_1 \tilde{\nu}_j \mu_j - \sum_2 \tilde{\nu}_j \mu_j. \tag{3.35}$$

The reaction will occur spontaneously (without an external supply of energy) and the value of ΔG_m will be negative when the cell is operated as a chemical power source, i.e., when the negative electrode to the left (1) is the anode.

The ocv of this cell is the algebraic sum of the Galvani potentials at three interfaces. When each of them is in equilibrium we can find the overall potential difference (which in this case we can call an electromotive force or emf, see Section 2.3) by putting functions (2.11) and (3.34) into Eq. (2.17):

$$\mathcal{E} = \sum_2 \tilde{\nu}_j \mu_j / n F + \mu_e^{(M2)} / F - \sum_1 \tilde{\nu}_j \mu_j / n F - \mu_e^{(M1)} / F$$
$$+ [\mu_e^{(M1)} - \mu_e^{(M2)}] / F \tag{3.36}$$

or, after simplification and allowing for Eq. (3.35),

$$\mathcal{E} = \left[\sum_2 \tilde{\nu}_j \mu_j - \sum_1 \tilde{\nu}_j \mu_j \right] / n F = - \Delta G_m / n F. \tag{3.37}$$

This equation links the emf of a galvanic cell to the Gibbs energy change of the overall current-producing reaction. It is one of the most important equations in the thermodynamics of electrochemical systems. It follows directly from the second law of thermodynamics, since $nF\mathcal{E}$ is the maximum value of useful (electrical) work of the system in which the reaction considered takes place. According to Eq. (3.7) this work is equal to $-\Delta G_m$.

The open cell discussed was considered as an equilibrium cell since equilibrium was established across each individual interface. However, the cell as a whole is not in equilibrium; the overall Gibbs energy of the full reaction is not zero, and when the circuit is closed an electric current flows which is attended by chemical changes, i.e., a spontaneous process sets in.

It is typical that in Eq. (3.36) for the emf, all terms containing the chemical potential of electrons in the electrodes cancel in pairs, since they are contained in the expressions for the Galvani potentials, both at the interface with the electrolyte and at the interface with the other electrode. This is due to the fact that the overall current-producing reaction comprises the transfer of electrons across the interface between two metals in addition to the electrode reactions.

This gives rise to an important conclusion. For nonconsumable electrodes which are not involved in the current-producing reaction, and for which the chemical potential of the electrode material is not contained in the equation for electrode potential, the latter (in contrast to a Galvani potential) depends only on the kind of reaction taking place, but it does not depend on the nature of the electrode itself.

For instance, in the galvanic cell

$$Pt(H_2)|H_2SO_4|(H_2)Pd|Pt \qquad (3.38)$$

which contains two hydrogen electrodes, one made of platinum and the other made of palladium, the Galvani potentials at the two interfaces involving the solution will be different. However, in a thermodynamic sense the cell is symmetric, and current flow does not lead to an overall chemical reaction (hydrogen is evolved at one of the electrodes but ionized at the other). For this reason the Gibbs energy change and the emf are zero and the electrode potentials of the two electrodes are identical.

When reporting electrode potentials and emf one indicates, when necessary, the reactions to which they refer, e.g., $E(Zn^{2+} + 2e^- \rightleftarrows Zn)$ or simply $E(Zn^{2+}, Zn)$ and $\mathcal{E}(Cu^{2+} + Zn \rightleftarrows Cu + Zn^{2+})$.

3.6. CONCENTRATION DEPENDENCE OF ELECTRODE POTENTIALS

3.6.1. The Nernst Equation for Ideal Systems

For ideal systems, the relation between the emf of galvanic cells and the composition of the electrolyte and other phases of variable composition can be found with the aid of Eqs. (3.13) and (3.37):

$$\mathcal{E} = \mathcal{E}_c^0 + (RT/nF)\ln\left(\prod_2 c_j^{\tilde{\nu}_j} \Big/ \prod_1 c_j^{\tilde{\nu}_j}\right) ; \qquad (3.39)$$

the partial pressures p_k will appear here for gaseous substances instead of concentrations c_j [cf. Eq. (3.12)].

In Eq. (3.39), $\mathcal{E}_c^0$ is a constant that depends on the current-producing reaction; it is equal to the emf for values $c_j = 1$. Values of emf mainly depend on the value of this constant, since the second term of this equation, which is the correction term for concentrations, is relatively small, though in certain cases it has an important meaning.

The combination of constants RT/F often appears in electrochemical equations; it has the dimensions of voltage. At 25°C (298.15 K) it has a value of 0.02569 V (or roughly 25 mV). When including the conversion factor for changing natural to common logarithms, we find a value of 0.05916 V (about 59 or 60 mV) for $2.303\,RT/F$ at 25°C. Values for other temperatures can be found by simple conversion, since this parameter is proportional to the absolute temperature.

Electrode potentials are the emf of cells consisting of a reference electrode [electrode 1 in cell (1.25)] and the electrode being examined (electrode 2). The electrolyte composition is constant for the reference electrode, hence the corresponding values of $\nu_j\mu_j$ in Eq. (3.37) for the overall change in Gibbs energy are also constant. As a result, the expression for electrode potential becomes

$$E = -\Delta G_2/nF = E_c^0 + (RT/nF)\ln\prod_j c_j^{\tilde{\nu}_j}, \qquad (3.40a)$$

where the product in the logarithmic term includes the concentrations of only those components which are involved in the electrode reaction at this electrode 2, and which are called the *potential-determining substances*.

This equation can also be written in the forms

$$E = E_c^0 + (RT/nF)\sum \tilde{\nu}_j \ln c_j \qquad (3.40b)$$

$$= E_c^0 + (RT/nF) \ln \left(\prod_{\text{ox}} c_j^{\nu_j} \Big/ \prod_{\text{red}} c_j^{\nu_j} \right). \qquad (3.40b)$$

It follows from these equations that the potential will shift in the positive direction when the concentration of the oxidizing agent or of components entering the reaction together with it ($\tilde{\nu}_j > 0$) is raised, but it will shift in the negative direction when the concentration of the reducing agent or of components entering the reaction together with it ($\tilde{\nu}_j < 0$) is raised.

For metal electrodes of the first kind at which the reaction $M^{z+} + z_+e^- \rightleftarrows M$ takes place, Eq. (3.40) becomes

$$E = E_c^0 + (RT/z_+F) \ln c_{M^{z+}}; \qquad (3.41a)$$

in the particular case of a zinc electrode [reaction (1.27)] we have

$$E = E_c^0 + (RT/2F) \ln c_{Zn^{2+}}. \qquad (3.41b)$$

An equation of this form was suggested by Walther Nernst in 1889.

In the case of amalgam electrodes, e.g., sodium amalgam in NaOH solution, the potential depends on the concentrations of the oxidizing agent (the cations in the electrolyte) as well as of the reducing agent (the metal in the amalgam, $c_{Na}^{(Hg)}$):

$$E = E_c^0 + (RT/F) \ln[c_{Na^+}^{(E)}/c_{Na}^{(Hg)}]. \qquad (3.42)$$

An equation for the very simple redox reaction $Fe^{3+} + e^- \rightleftarrows Fe^{2+}$,

$$E = E_c^0 + (RT/F) \ln(c_{Fe^{3+}}/c_{Fe^{2+}}), \qquad (3.43)$$

was suggested by Franz C. A. Peters in 1898.

At present all equations of this type are known as *Nernst equations*.

3.6.2. Equations for Real Systems

Equations describing the relation between electrode potential and composition of the system in the case of real systems can be written, by analogy with Eq. (3.40), as functions of the component activities:

$$E = E_a^0 + (RT/nF) \sum \tilde{\nu}_j \ln a_j. \qquad (3.44)$$

For the particular case of electrodes of the first kind we have, instead of Eq. (3.41a),

$$E = E_a^0 + (RT/z_+F) \ln a_{M^{z+}}. \tag{3.45}$$

There is a major difficulty which arises when such equations are used in practice, in that the activities of individual ions are unknown, unless the solutions are highly dilute, and in that the ionic components involved in the electrode reaction do not form electroneutral groups. Hence, for practical calculations we must employ values of mean ionic activity $a_\pm$:

$$E = E_a^0 + (RT/nF) \sum \tilde{\nu}_j \ln a_\pm, \tag{3.46}$$

which makes the results somewhat provisional. But this can be tolerated insofar as the experimental values of electrode potentials are also slightly distorted and depend on the activities of other ions, because reference electrodes with other potential-determining ions or liquid-junction potentials are present.

In cases where data for the activity coefficients are altogether unavailable (particularly in multicomponent solutions), concentrations must be used instead of activities [Eq. (3.40)]. This leads to appreciable errors and is admissible only for the purposes of approximate calculations.

Parameter E_a^0 in Eqs. (3.44) and (3.46) is called the *standard electrode potential*; it corresponds to the value of electrode potential which is found when the activities of the components are unity. Values E_a^0 differ somewhat from values E_c^0:

$$E_a^0 - E_c^0 = (RT/nF) \sum \tilde{\nu}_j \ln \gamma_j. \tag{3.47}$$

For a more distinct differentiation between these parameters, E_c^0 is called the *formal electrode potential*.

The values of E_a^0 for electrode reactions (or of E_c^0, when sufficiently reliable activity data are not available) are listed in special tables; some such values are shown in Table 3.1. When using these tables we must bear in mind that the values of E^0 for reactions involving gases have been calculated for partial pressures of 1 atm, which in SI units corresponds to 101,325 Pa (about 0.1 MPa). Hence in the Nernst equation we must use gas pressures in the now obsolete units of atmospheres.

TABLE 3.1. Standard Electrode Potentials (25°C)

Reaction	E^0, V (SHE)	Reaction	E^0, V (SHE)
$Li^+ + e^- \rightleftarrows Li$	-3.045	$HgO + H_2O + 2e^- \rightleftarrows Hg + 2OH^-$	0.098
$K^+ + e^- \rightleftarrows K$	-2.925	$Sn^{4+} + 2e^- \rightleftarrows Sn^{2+}$	0.154
$Ca^{2+} + 2e^- \rightleftarrows Ca$	-2.866	$Cu^{2+} + e^- \rightleftarrows Cu^+$	0.153
$Na^+ + e^- \rightleftarrows Na$	-2.714	$AgCl + e^- \rightleftarrows Ag + Cl^-$	0.2224
$Mg^{2+} + 2e^- \rightleftarrows Mg$	-2.363	$Hg_2Cl_2 + 2e^- \rightleftarrows 2Hg + 2Cl^-$	0.2676
$Al^{3+} + 3e^- \rightleftarrows Al$	-1.662	$Cu^{2+} + 2e^- \rightleftarrows Cu$	0.337
$Ti^{2+} + 2e^- \rightleftarrows Ti$	-1.628	$Fe(CN)_6^{3-} + e^- \rightleftarrows Fe(CN)_6^{4-}$	0.36
$Zn(OH)_2 + 2e^- \rightleftarrows Zn + 2OH^-$	-1.245	$Cu^+ + e^- \rightleftarrows Cu$	0.521
$Mn^{2+} + 2e^- \rightleftarrows Mn$	-1.180	$I_2 + 2e^- \rightleftarrows 2I^-$	0.536
$2H_2O + 2e^- \rightleftarrows H_2 + 2OH^-$	-0.822	$O_2 + 2H^+ + 2e^- \rightleftarrows H_2O_2$	0.682
$Zn^{2+} + 2e^- \rightleftarrows Zn$	-0.763	$Fe^{3+} + e^- \rightleftarrows Fe^{2+}$	0.771
$S + 2e^- \rightleftarrows S^{2-}$	-0.48	$Br_2 + 2e^- \rightleftarrows 2Br^-$	1.065
$Fe^{2+} + 2e^- \rightleftarrows Fe$	-0.441	$O_2 + 4H^+ + 4e^- \rightleftarrows 2H_2O$	1.229
$Cd^{2+} + 2e^- \rightleftarrows Cd$	-0.403	$Cl_2 + 2e^- \rightleftarrows 2Cl^-$	1.358
$Ni^{2+} + 2e^- \rightleftarrows Ni$	-0.250	$PbO_2 + 4H^+ + 2e^- \rightleftarrows Pb^{2+} + 2H_2O$	1.455
$Sn^{2+} + 2e^- \rightleftarrows Sn$	-0.136	$Ce^{4+} + e^- \rightleftarrows Ce^{3+}$	1.61
$2H^+ + 2e^- \rightleftarrows H_2$	0.0000	$F_2 + 2e^- \rightleftarrows 2F^-$	2.87

The overall current-producing reaction in a cell only involves electroneutral sets of species, hence in the equation

$$\mathcal{E} = \mathcal{E}_a^0 + (RT/nF)\ln\left(\prod_2 a_j^{\tilde{\nu}_j} \middle/ \prod_1 a_j^{\tilde{\nu}_j}\right) \qquad (3.48)$$

for the emf of such cells, it is entirely legitimate to replace the true activities of individual ions by mean ionic activities.

Subscripts a and c for quantities E^0 and $\mathcal{E}^0$ will be dropped in the following.

3.6.3. Electrodes of the Second Kind

A silver electrode in aqueous KCl solution is an example of an electrode of the second kind. Here the reaction

$$Ag + Cl^- \rightleftarrows AgCl + e^- \qquad (3.49)$$

takes place. Owing to the low solubility of AgCl, the product of the anodic process precipitates as a solid and the electrolyte is practically always saturated with it. The solubility product $L(AgCl) = a_{Ag^+}a_{Cl^-} = 1.56 \cdot 10^{-10}$ mol^2/liter^2, i.e., the 1 M KCl solution is saturated with AgCl already at a Ag^+ ion concentration of about 10^{-10} mol/liter.

Since the activity of the Ag^+ ions depends on the activity of the anions, the potential of this electrode will depend on the activity of the anions:

$$E = E^0 - (RT/F)\ln a_{Cl^-} \qquad (3.50a)$$

or, in a general form,

$$E = E^0 - (RT/|z_-|F)\ln a_{A^{z-}} = E^0 - (\tau_- RT/z_k F)\ln a_{A^{z-}}. \qquad (3.50b)$$

It is said, therefore, that this electrode is reversible with respect to the anion.

This claim must be examined in more detail. An electrode potential which depends on anion activity still constitutes no evidence that the anions are direct reactants. Two reaction mechanisms are possible at this electrode: a direct transfer of chloride ions across the interface in accordance with Eq. (3.49) or the combination of the electrode reaction, $Ag \rightleftarrows Ag^+ + e^-$ (transfer of the Ag^+ ions), with the ionic reaction in the bulk solution, $Ag^+ + Cl^- \rightleftarrows AgCl$. The overall reaction is the same, hence in both cases Eq. (3.50) is legitimate, yet in the second case the chloride ions are additional, not primary, reactants. Thus, thermodynamic data do not suffice if we want to unravel the true mechanism of an electrode process.

Electrodes of the second kind can formally be regarded as a special case of electrodes of the first kind where the standard state (when $E = E^0$) corresponds not to $a_{Ag^+} = 1$ but to the value $a_{Ag^+} \approx 10^{-10}$ mol/liter which is established in a KCl solution of unit activity. In this case the concentration of the potential-determining cation can be varied by varying the concentration of an anion, which might be called the controlling ion.

The oxides and hydroxides of most metals (other than the alkali metals) are poorly soluble in alkaline solutions, hence, almost all metal electrodes in alkaline solutions are electrodes of the second kind.

3.6.4. Influence of Complex Formation

A picture similar to that just described is seen for metal electrodes when the solution contains a complexing agent, e.g., for the silver electrode when KCN has been added to the $AgNO_3$ solution. Then the complex formation equilibrium

$$Ag^+ + 3CN^- \rightleftarrows [Ag(CN)_3]^{2-} \tag{3.51}$$

will be established in the solution, and as a result the concentration of free Ag^+ ions decreases drastically. The electrode reaction which occurs with anodic current flow can be either the direct formation of complex ions:

$$Ag + 3CN^- \rightleftarrows [Ag(CN)_3]^{2-} + e^-, \tag{3.52}$$

or a primary formation of free Ag^+ ions which are subsequently bound as complex ions according to reaction (3.51). The expression for electrode potential is the same in both cases:

$$E = E^0 + (RT/F) \ln[a_{Ag(CN)_3^{2-}} / a_{CN^-}^3]. \tag{3.53a}$$

Here the value of standard potential $E^0[Ag(CN)_3^{2-}]$ differs from the corresponding value of $E^0[Ag^+]$ for a silver electrode without a complexing agent:

$$E^0[Ag(CN)_3^{2-}] = E^0[Ag^+] + RT \ln K, \tag{3.53b}$$

where K is the equilibrium constant of reaction (3.51).

Thus, by suitable selection of complexing agents one can alter the equilibrium potentials of metal electrodes within wide limits.

3.6.5. The Nernst Equation at Very Low Concentrations

At zero concentration of the potential-determining substances, the values of electrode potential calculated with Eqs. (3.40) or (3.44) tend toward $\pm\infty$, which is physically meaningless. This implies that these equations cannot be used below a certain concentration.

Two kinds of notion exist with respect to the term "low concentrations," viz., a low absolute concentration (highly dilute solutions) and a low equilibrium concentration (as in the formation of complexes or compounds of low solubility). In the latter case, when potential-determining substances start to be withdrawn from the solution they re-form because of the shift in equilibrium, i.e., their potential supply is large.

The Nernst equation is of limited use at low absolute concentrations of the ions. At concentrations of 10^{-7} to 10^{-9} mol/liter and the customary ratios between electrode surface area and electrolyte volume ($S/V \approx 10{-}10^3$ m^{-1}), the number of ions present in the electric double layer is comparable with that in the bulk electrolyte. Hence edl formation is associated with a change in bulk concentration, and the potential will no longer be the equilibrium potential with respect to the original concentration. Moreover, at these concentrations the exchange current densities are greatly reduced, and the potential is readily altered under the influence of extraneous effects. A concentration of the potential-determining substances of 10^{-5} to 10^{-7} mol/liter can be regarded as the limit of application of the Nernst equation.

Such a limitation does not exist for low equilibrium concentrations.

3.7. SPECIAL THERMODYNAMIC FEATURES OF ELECTRODE POTENTIALS

3.7.1. Table of Standard Potentials

The value of electrode potential is a quantitative measure of the redox properties of substances involved in the electrode reaction. An oxidizing agent when undergoing reduction takes up electrons from the cathode. The stronger its oxidizing power, the more positive will be the electrode potential. To the contrary, a reducing agent when undergoing oxidation gives off electrons to the anode, and the potential will be more negative the higher its reducing power.

The series of metals arranged in the order of decreasing reducing, and increasing oxidizing, power (or increasingly positive potential values) used to be called the electromotive series. An example is the series

$$\text{Li, Al, Zn, Cd, Pb, Cu, Hg, Au.} \qquad (3.54)$$

When a metal is immersed into the solution of a salt of another metal further to the right in the electromotive series, the first metal dissolves (is oxidized) while the second metal is deposited (its ions are reduced). Thus, the first metal "displaces" the second from its solution.

Nowadays tables of standard electrode potentials are used instead of the electromotive series. They include electrode reactions not only of metals but also of other substances (Table 3.1).

3.7.2. Electrode Reactions Occurring in Steps

For individual substances, a stepwise oxidation or reduction is possible. For instance, at an iron electrode the reactions $\text{Fe} \rightarrow \text{Fe}^{2+} + 2e^-$ (b) and $\text{Fe}^{2+} \rightarrow \text{Fe}^{3+} + e^-$ (c) are possible aside from the reaction $\text{Fe} \rightarrow \text{Fe}^{3+} + 3e^-$ (a). There is a characteristic value of electrode potential for each of these reactions. But since identical reactants and products are involved in said reactions, the values of electrode potential found for the individual reactions should be interrelated. In fact, in our example

$$E^0(\text{Fe}^{3+}, \text{Fe}) = (\mu^0_{\text{Fe}^{3+}} - \mu^0_{\text{Fe}})/3F + \text{const}, \qquad (3.55a)$$

$$E^0(\text{Fe}^{2+}, \text{Fe}) = (\mu^0_{\text{Fe}^{2+}} - \mu^0_{\text{Fe}})/2F + \text{const}, \qquad (3.55b)$$

$$E^0(\text{Fe}^{3+}, \text{Fe}^{2+}) = (\mu^0_{\text{Fe}^{3+}} - \mu^0_{\text{Fe}^{2+}})/F + \text{const}, \qquad (3.55c)$$

where *const* is related to the reference electrode, and has the same value in all three cases.

It follows that

$$E^0(\text{Fe}^{3+}, \text{Fe}) = 1/3[2E^0(\text{Fe}^{2+}, \text{Fe}) + E^0(\text{Fe}^{3+}, \text{Fe}^{2+})], \qquad (3.56)$$

i.e., the potential of the overall reaction assumes some value intermediate between the potentials of the individual steps. When identical numbers of electrons are involved in the two consecutive steps of a reaction, the potential of the overall reaction will be the mean value of the potentials of the individual steps, e.g.,

$$E^0(Cu^{2+}, Cu) = 1/2[E^0(Cu^+, Cu) + E^0(Cu^{2+}, Cu^+)]. \qquad (3.57)$$

The above correlations are known as *Luther's principle*.

Two cases are possible when intermediate steps exist.

(i) The potential of the first step in the anodic process (or second step in the cathodic process) is more negative than the potential of the other step; e.g., for the iron electrode $E^0(Fe^{2+}, Fe) = -0.441$ V and $E^0(Fe^{3+}, Fe^{2+}) = 0.771$ V. In this case the reaction intermediate may be stable within a certain range of potentials. In our case, in the region more negative than $E^0(Fe^{2+}, Fe)$ only metallic iron is thermodynamically stable, since both ion forms are reduced to the metal. In the region of potentials more positive than $E^0(Fe^{3+}, Fe^{2+})$, the Fe^{3+} ions are stable. In the intermediate region the Fe^{2+} ions are stable, which here can be neither reduced nor oxidized. In this region the spontaneous reaction

$$Fe + 2Fe^{3+} \rightarrow 3Fe^{2+} \qquad (3.58)$$

is possible. For this reason the potential of the Fe^{3+}/Fe system cannot, for all practical purposes, be measured. It can be calculated with the aid of Eq. (3.56), and is -0.037 V.

(ii) There exists a relative position of potentials which is opposite to that in the first case, e.g., for the copper electrode where $E^0(Cu^+, Cu) = 0.521$ V and $E^0(Cu^{2+}, Cu^+) = 0.153$ V. The Cu^+ ions can be produced by anodic oxidation of metallic copper only at potentials more positive than $E^0(Cu^+, Cu)$. Since this region of potentials is *a priori* more positive than $E^0(Cu^{2+}, Cu^+)$ the Cu^+ ions produced will at once be oxidized further to Cu^{2+}. Therefore, Cu^+ ions as the reaction intermediate are thermodynamically unstable at all values of potential. When an aqueous solution is prepared from a salt of univalent copper, the spontaneous disproportionation (dismutation) reaction

$$2Cu^+ \rightarrow Cu + Cu^{2+} \qquad (3.59)$$

is possible.

3.7.3. The pH Dependence of Potentials; Pourbaix Diagrams

Often H^+ or OH^- ions are involved in the electrode reactions, and the electrode potential then depends on the concentration of these ions (or solution pH). Because of the dissociation equilibrium of water, the activities of these ions are interrelated as $a_{H^+} a_{OH^-} = K_w$. For this reason these reactions can be formulated in two ways, e.g., for the hydrogen electrode

$$2H^+ + 2e^- \rightleftarrows H_2 \qquad (3.60a)$$

and

$$2H_2O + 2e^- \rightleftarrows H_2 + 2OH^-. \qquad (3.60b)$$

The former way is used predominantly for acidic solutions, and the latter for alkaline solutions, but thermodynamically the two ways are equivalent (provided of course that the change in pH does not produce a change in direction of the reaction or in the form of individual reactants, as by dissociation). The equations for electrode potential can also be written in two ways, with H^+ ions or with OH^- ions, which again is thermodynamically equivalent. However, the values of E^0 will of course differ, since in one case they refer to $a_{H^+} = 1$ and in the other to $a_{OH^-} = 1$. In the former case the standard potential is written as E_A^0, and in the latter as E_B^0 (from the words "$acid$" and "$base$"). Usually E_A^0 is implied when a subscript is not written (as in Table 3.1).

The expressions for potential of the hydrogen electrode are

$$E = E_A^0 + (RT/2F) \ln(a_{H^+}^2 / p_{H_2}) \qquad (3.61a)$$

$$= E_B^0 + (RT/2F) \ln(a_{H_2O}^2 / p_{H_2} a_{OH^-}^2). \qquad (3.61b)$$

On the scale of the standard hydrogen electrode (SHE) we have $E_A^0 = 0$ for reaction (3.60), by definition. Since $a_{OH^-} = K_w / a_{H^+}$ and $a_{H_2O} \approx 1$, we have for this reaction at 25°C, when allowing for the value of K_w:

$$E_B^0 = (RT/2F) \ln K_w^2 = -0.822 \text{ V}. \qquad (3.62)$$

For many electrodes it is found that one H^+ or OH^- ion is involved in the reaction per electron, hence the electrode potential becomes 0.059 V more negative when the pH is raised by one unit; this is the same potential shift as found for the hydrogen electrode. For such electrodes a special scale of electrode potentials is occasionally employed: These potentials

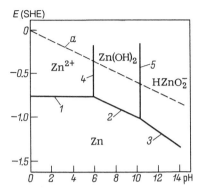

Fig. 3.1. Pourbaix diagram for the zinc electrode in aqueous solutions.

are designated as E_r, and refer to the potential of a reversible hydrogen electrode (*rhe*) in the same solution, i.e., at the given pH. For the electrodes of the type considered, potentials in this scale are independent of solution pH.

When there are changes in the form of reactants and (or) products with solution pH, the values of E^0 and the pH dependence of electrode potential will change accordingly. These changes are clearly illustrated by phase diagrams constructed in the coordinates of E and pH. Diagrams of this kind were suggested in 1945 by Marcel Pourbaix, and became known as *Pourbaix diagrams*.

A simplified Pourbaix diagram for the zinc electrode at pH between 0 and 14 is shown as an example in Fig. 3.1. The vertical axis is that of the values of electrode potential on the SHE scale for activities of the Zn^{2+} and $HZnO_2^-$ ions of 1 mol/kg. The segments of solid lines correspond to the equilibrium potentials of the following electrode reactions:

$$1: Zn \rightleftarrows Zn^{2+} + 2e^- \qquad (pH < 5.8)$$

$$2: Zn + 2OH^- \rightleftarrows Zn(OH)_2 + 2e^- \qquad (5.8 < pH < 10.4)$$

$$3: Zn + 3OH^- \rightleftarrows HZnO_2^- + H_2O + 2e^- \qquad (pH > 10.4)$$

Segments 4 and 5 reflect the chemical equilibria of acid and base dissociation of $Zn(OH)_2$ yielding Zn^{2+} and $HZnO_2^-$ ions, respectively.

The areas bounded by solid lines correspond to regions of thermodynamic stability of certain substances which are named in the diagram. This stability is relative. The dashed line *a* in the diagram corresponds to the equilibrium potential of the hydrogen electrode. Metallic zinc, for which the reaction lines are below the line for the hydrogen electrode, can be oxidized while hydrogen is evolved (see Section 2.5.1).

3.7.4. Electrode Potentials in Nonaqueous Electrolytes

For any type of nonaqueous electrolyte (nonaqueous solutions, melts, solid electrolytes) we can select suitable reference electrodes, measure the potentials of other electrodes, and set up tables of electrode potentials. The order of the reactions (electrodes) as a rule does not strongly differ between the different media. A strong reducing agent such as lithium will have a more negative potential than a weaker reducing agent such as copper, both in water and in the other media.

However, the electrode potentials measured for different types of electrolytes cannot quantitatively be compared with each other, even when the same reference electrode has been used throughout. This is due to the fact that the potential differences at interfaces between dissimilar electrolytes cannot be determined experimentally. For this reason the electrode potentials are measured separately for each type of electrolyte medium.

For a qualitative comparison of the electrode potentials measured in different media, sometimes models and assumptions are employed. Thus, it can be assumed according to a suggestion of Viktor A. Pleskov (1947) that the interaction of the relatively large rubidium (or cesium) ion with water and with different nonaqueous solvents is very slight. Hence the chemical potentials of this ion and also the Galvani potentials at the rubidium/solution interface will be approximately the same for all media. Then a universal potential scale is obtained, to a first approximation, when the potentials of all other electrodes in each medium are referred to rubidium reference electrodes.

3.7.5. Temperature Coefficients of Electrode Potentials

The emf values of galvanic cells and the electrode potentials are usually determined isothermally, when all parts of the cell, and particularly the two electrode/electrolyte interfaces, are at the same temperature. The values of emf will change when this temperature is varied. According to the well-known thermodynamic *Gibbs–Helmholtz equation*, which for electrochemical systems can be written as

$$-d\Delta G_m/dT \equiv nF(d\mathcal{E}/dT) = (\Delta H_m - \Delta G_m)/T = \Delta S_m, \quad (3.63)$$

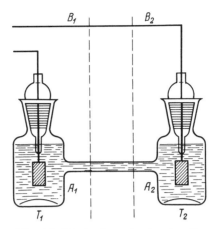

Fig. 3.2. Nonisothermal galvanic cell.

the temperature coefficient of the emf of isothermal cells is related to the entropy change for the current-producing reaction.

As the temperature is varied the Galvani potentials of all interfaces will change, and we cannot relate the measured value of $d\mathcal{E}/dT$ to the temperature coefficient of Galvani potential for an individual electrode. The temperature coefficient of electrode potential likewise depends on the temperature coefficient of Galvani potential for the reference electrode, and hence is not a property of the test electrode alone.

The Gibbs–Helmholtz equation also links the temperature coefficient of Galvani potential for individual electrodes to energy effects or entropy changes of the electrode reactions occurring at these electrodes. However, since these parameters cannot be determined experimentally for an isolated electrode reaction (this is possible only for the full current-producing reaction), this equation cannot be used to calculate this temperature coefficient.

We might try to measure the temperature coefficient of the Galvani potential for an individual electrode under nonisothermal conditions; then only the temperature T_2 of the test electrode would be varied while the reference electrode remains at a constant temperature T_1, and retains a constant value of Galvani potential (Fig. 3.2). However, in this case the emf measured will be distorted by another effect, viz., the variation of electrostatic potential within a given conductor which is caused by a temperature gradient in the conductor (the *Thomson effect*, 1856). Potential gradients will arise even at zero current, both in the electrolyte (between points A_1 and A_2) and in the metallic conductors (between points B_1 and B_2), but cannot be determined.

Thus, the temperature coefficient of Galvani potential of an individual electrode can be neither measured nor calculated. Measured values of the temperature coefficients of electrode potentials depend on the reference electrode employed. For this reason a special scale is used for the temperature coefficients of electrode potential, viz., it is assumed as a convention that the temperature coefficient of potential of the standard hydrogen electrode is zero; in other words, it is assumed that the value of $E^0_A(H^+, H_2)$ is zero at all temperatures. By measuring the emf under isothermal conditions we actually compare the temperature coefficient of potential of other electrodes with that of the standard hydrogen electrode.

Chapter 4
DIFFUSION PROCESSES IN ELECTROCHEMISTRY

4.1. BASIC LAWS OF IONIC DIFFUSION IN SOLUTIONS

In multicomponent systems, such as solutions, diffusion will arise when at least one of the components is nonuniformly distributed, and its direction will be such as to level the concentration gradients.

It follows from experimental data that the diffusion flux density is proportional to the concentration gradient of the diffusing substance:

$$J_{d,j} = -D_j \operatorname{grad} c_j \qquad (4.1)$$

(*Fick's first law*, 1855). The minus sign written in this equation implies that the diffusion flux is in the direction of decreasing concentrations. The proportionality factor D_j is called the diffusion coefficient of the substance concerned (units: m^2/s).

In the diffusion of ions in solutions or other electrolytes, Eq. (4.1) is obeyed only at low concentrations of these ions. At higher concentrations the proportionality between flux and concentration gradient is lost, i.e., the coefficient D_j ceases to be constant.

A possible reason for the departures from Fick's first law is the fact that the diffusion process tends to level chemical potentials (thermodynamic activities) rather than concentrations of the substances involved. Hence the equation is sometimes written as

$$J_{d,j} = -D_{a,j} \operatorname{grad} a_j. \qquad (4.2)$$

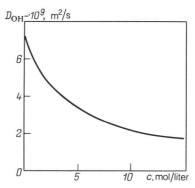

Fig. 4.1. Diffusion coefficients of the OH^- ions in KOH solutions as a function of KOH concentration at 25°C.

However, even in this form the equation does not provide a sufficiently accurate description of the experimental results in solutions unless these are highly dilute, and again coefficient $D_{a,j}$ is not constant when the concentration is varied. This is due to the complexities of diffusion processes, particularly to the fact that the diffusing ions transport solvent molecules which are present in their solvation sheaths, and that these molecules then are transported back. Since Eq. (4.2) has no conspicuous advantages over Eq. (4.1), the latter is used more often when discussing diffusion processes. It is simpler to use in practice, since a knowledge of activity coefficients is not required. All departures from proportionality between the diffusion flux and the concentration gradients are taken into account by assuming that, in real systems, the diffusion coefficients D_j are a parameter that depends on concentration.

In dilute aqueous solutions the diffusion coefficients of most ions and of many neutral substances are similar and have values which at room temperature are within the limits of 0.6×10^{-9} and 2×10^{-9} m²/s. The values as a rule exhibit a marked decrease with increasing solution concentration (Fig. 4.1).

The ionic mobilities u_j depend on the retarding factor θ_j valid for a particular medium [Eq. (1.8)]. It is evident that this factor also influences the diffusion coefficients. We shall assume, in order to find the connection, that the driving force of diffusion is the chemical potential gradient, i.e., in an ideal solution

$$f_d = -\operatorname{grad} \mu_j = -RT \operatorname{grad}(\ln c_j) = -(RT/c_j)\operatorname{grad} c_j \qquad (4.3)$$

(the minus sign indicates that the force acts in the direction of decreasing chemical potentials). With Eq. (1.6) we obtain

$$J_{d,j} = -(RT/\theta_j)\,\text{grad}\,c_j. \qquad (4.4)$$

We see when comparing Eqs. (4.1) and (4.4) that

$$D_j = RT/\theta_j. \qquad (4.5)$$

It follows from relations (1.8) and (4.5), finally, that

$$D_j = u_j(RT/|z_j|F) \qquad (4.6)$$

(the *Einstein relation*, 1905). This equation is valid in dilute solutions. An analogous equation including activity coefficients can be derived, but for the reasons outlined above, it again is not sufficiently accurate in describing the experimental data in concentrated solutions.

By analogy with Eq. (4.6), the diffusion coefficient can also be related to the transport number of a given ion [through Eq. (1.13)] or to its molar conductivity, $\lambda_j \equiv |z_j|Fu_j$:

$$D_j = t_j\sigma RT/z_j^2 F^2 c_j = \lambda_j RT/z_j^2 F^2. \qquad (4.7)$$

The equations reported are of great value insofar as they can be used to evaluate ionic diffusion coefficients from values of u_j, t_j, or λ_j which are more readily measured.

We must remember that the use of Eq. (4.3) is not perfectly justified. This is due to the fact that diffusion is not the result of an external force acting on each individual particle (as, for instance, an electric field in the migration of ions), but of kinetic molecular, thermal motion of the particles. This motion ("random walk") occurs chaotically in all directions of space, but after a time t, any given particle is found a certain distance away from its starting point, in a certain direction. When in a given system we distinguish the two regions (α) and (β) in contact with each other (Fig. 4.2), then thermal motion will have the effect that some number of particles from each region will pass through the interface to the neighboring region. When we further assume that these regions have different particle concentrations, for instance $c_j^{(\alpha)} > c_j^{(\beta)}$, then because of the larger concentration in region α the number of particles reaching the

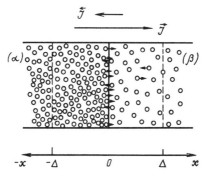

Fig. 4.2. The region of contact between two solutions of different concentration [used in deriving Eq. (4.9)].

interface from this side will be larger than that reaching the interface from the side of region (β).

Suppose the interface considered has the coordinate $x = 0$. We shall write Δ for the mean square displacement of a particle within time t along the x-axis. Consider the region between $x = -\Delta$ and $x = \Delta$. We can assume that within a time t half of the particles initially present in the left-hand zone will pass the interface from left to right, i.e., from α to β (the other half move away from the interface in the other direction). This flux crossing the interface has a density $\vec{J}_j$, which evidently is equal to $c_j^{(\alpha)} \Delta / 2t$. For the flux crossing the interface in the opposite direction we similarly have $\overleftarrow{J}_j = c_j^{(\beta)} \Delta / 2t$. For small Δ and a smooth concentration change along the x-coordinate we have $c_j^{(\alpha)} - c_j^{(\beta)} \approx -(dc_j/dx)\Delta$ (in the present example the value of the derivative is negative). With these values, the resulting flux from left to right is found to be

$$J_j = \vec{J}_j - \overleftarrow{J}_j = -(\Delta^2/2t)dc_j/dx. \tag{4.8}$$

When comparing Eqs. (4.1) and (4.8) we find the important relation between the macrokinetic parameter D_j and the kinetic–molecular parameter Δ, which is given by

$$D_j = \Delta^2/2t \tag{4.9}$$

(the *Einstein–Smoluchowski* equation, 1905).

4.2. DIFFUSION DURING CURRENT FLOW IN ELECTROLYTES

When current flows in an electrolyte solution the concentration, $c_{S,j}$, of a reactant or product close to the electrode surface will change relative to its bulk concentration $c_{V,j}$. For reactants $c_{S,j} < c_{V,j}$, and for products $c_{S,j} > c_{V,j}$. Thus, when allowing for the sign convention adopted (see Section 1.8), the concentration gradient at an electrode will always have a positive value for reactants, and a negative value for the products (both at the anode and cathode, and regardless of the sign of particle charge).

Diffusion fluxes develop as a result of the concentration gradients. The layer of electrolyte where the concentration changes occur and within which the substances are transported by diffusion is called the *diffusion layer*. Its thickness, δ (the diffusion path length), depends on cell design features and on the intensity of convective fluxes (for further details see Section 4.4).

The concept of surface concentration $c_{S,j}$ requires closer definition. At the surface itself the ionic concentrations will change, not only as a result of the reaction but also because of the electric double layer present at the surface. Surface concentration is understood to be the concentration at a distance from the surface small enough as compared to diffusion-layer thickness, yet so large that the effects of the edl are no longer felt. This condition usually is met at points about 1 nm from the surface.

The changes in surface concentrations of the components caused by current flow have two important effects: they produce a change in electrode potential, and they imply that there is an upper limit to the cell currents when the diffusion flux attains its limiting value. The first of these effects will be considered in Section 6.3, and the second will be considered in the present chapter.

The total flux of any substance across the diffusion layer is the sum of terms due to migration, diffusion, and convection. The overall flux density is determined by the balance equation (1.49):

$$\bar{\nu}_j |i|/nF = J_j = J_{d,j} + J_{m,j} + J_{kv,j}. \qquad (4.10)$$

In the present section we consider diffusion processes in electrochemical systems which are not complicated by migration and convection. To exclude migration we shall consider the behavior of uncharged reaction components. We shall assume that the diffusion-layer thickness δ is constant. Such conditions can be realized, e.g., when the electrode is provided

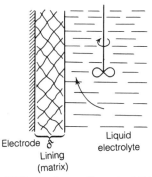

Electrode δ
Lining
(matrix)

Liquid
electrolyte

Fig. 4.3. Diffusion layer of constant thickness.

with a porous lining of thickness δ filled with the electrolyte (Fig. 4.3). In the pores of the lining, convection of the liquid is almost impossible. By vigorous stirring of the solution a concentration of the reactant which is sufficiently close to the starting concentration can be maintained at the outer surface of the lining.

The influence of migration on diffusion will be examined in Section 4.3, and the influence of convection will be examined in Section 4.4.

In electrochemical systems with flat electrodes, all fluxes within the diffusion layers are always linear (one-dimensional). For electrodes of different shape, e.g., cylindrical, linearity will be retained when the thickness δ is markedly smaller than the radius of surface curvature.

Under steady-state conditions, the total flux density must be constant along the entire path (throughout the layer of thickness δ) when the flux is linear. In our case the diffusion-flux density is constant; hence the concentration gradient is also constant within the limits of layer δ and can be described in terms of finite differences as

$$\text{grad}\, c_j \equiv dc_j/dx = \Delta c_j/\delta, \qquad (4.11)$$

where

$$\Delta c_j \equiv c_{V,j} - c_{S,j}.$$

Thus, the equation for the diffusion flux becomes

$$J_{d,j} = -D_j \Delta c_j/\delta = -\varkappa_j \Delta c_j, \qquad (4.12)$$

where $\varkappa_j \equiv D_j/\delta$ is the diffusion-flux constant (units: m/s).

For the case considered here, the balance equation is given by

$$|i| = -(n/\bar{\nu}_j)FD_j(dc_j/dx) = -(n/\bar{\nu}_j)F\varkappa_j\Delta c_j, \qquad (4.13)$$

i.e., the current density is proportional to the concentration difference between bulk and surface.

Limiting current for the reactants. For reactants $\bar{\nu}_j = -\nu_r < 0$, and Eq. (4.13) can be written as

$$|i| = (n/\nu_r)F\varkappa_r\Delta c_r. \qquad (4.14)$$

For a given current density i and known values of the remaining parameters, this equation yields the surface concentration

$$c_{S,r} = c_{V,r} - \nu_r|i|/nF\varkappa_r. \qquad (4.15)$$

The surface concentration decreases with increasing current density. When the current density has attained a certain critical value

$$|i_{l,r}| = (n/\nu_r)F\varkappa_j c_{V,r}, \qquad (4.16)$$

the surface concentration of the reactant has fallen to zero. This current density corresponds to the highest possible concentration gradient, of $c_{V,r}/\delta$; a further increase of the diffusion flux is impossible. The parameter $i_{l,r}$, sometimes written as $i_{d,r}$, is the *limiting diffusion current density* for a given reactant.

Using the expression for the limiting diffusion current density, we can rewrite the surface concentration as

$$c_{S,r} = c_{V,r}(1 - |i/i_{l,r}|). \qquad (4.17)$$

The limiting currents are a typical feature found in galvanic cells but not in circuits consisting purely of electronic conductors.

Limiting current for the products. Here $\bar{\nu}_j = \nu_p > 0$ and

$$|i| = -(n/\nu_p)F\varkappa_p\Delta c_p, \qquad (4.18)$$

$$c_{S,p} = c_{V,p} + \nu_p|i|/nF\varkappa_p. \qquad (4.19)$$

In this case the concentration at the surface will increase, and a limiting condition may arise for different reasons, which are related to attainment of the solubility limits by given substances. The material precipitating will screen the electrode surface and interfere with a further increase in current. The value of the limiting current will depend on the nature of the deposit formed, and is less reproducible than in the previous case; specifically, it may depend on time.

Using Eq. (4.16) we can write the surface concentration as

$$c_{S,p} = c_{V,p}(1 + |i/i_{l,p}|),\qquad(4.20)$$

but here the limiting current does not refer to the given reaction; it refers to the back reaction in which the given species are consumed.

Generally, the surface concentration of any substance (reactant, product) can be described by the expression

$$c_{S,j} = c_{V,j}[1 + (\text{sign } \bar{\nu}_j)|i/i_{l,j}|]\qquad(4.21)$$

or, when allowing for the identity (1.50), by

$$c_{S,j} = c_{V,j}[1 + (\text{sign } \bar{\nu}_j)i/|i_{l,j}|].\qquad(4.22)$$

Key components. Most electrochemical reactions involve several reactants and (or) products. The surface concentrations of all of them change. As the current density is raised, the limiting concentration for one of them will be attained before it is attained for the others. This substance can be called the key component. The actual limiting current attained in the system corresponds to the limiting current of this key component, i.e., is determined by its parameters, and in particular by its concentration.

4.3. IONIC TRANSPORT BY MIGRATION AND DIFFUSION

4.3.1. The System of Balance equations

The equation for the total flux of ions under the simultaneous effects of an electrostatic field **E** [see Eq. (1.9)] and a concentration gradient [see Eq. (4.1)] is

$$J_j = (\text{sign } z_j)c_j u_j \mathbf{E} - D_j \,\text{grad}\, c_j\qquad(4.23)$$

(the *Nernst–Planck equation*, 1890).

Substituting the value of J_j into the balance equation (4.10) and allowing for equality (4.6) which links the parameters u_j and D_j, we obtain

$$|i| = (n/\bar{\nu}_j)FD_j[z_j c_j \mathbf{E}F/RT - \text{grad } c_j]. \qquad (4.24)$$

Equation (4.24) is fulfilled for all the ionic species involved in the reaction. It should also be fulfilled for ions not involved in the reaction ($\bar{\nu}_j = 0$ and $J_j = 0$ but $J_{d,j} = -J_{m,j} \neq 0$). For uncharged reaction components ($z_j = 0$ but $\bar{\nu}_j \neq 0$) the balance equation (4.24) changes into Eq. (4.13).

The total number of balance equations corresponds to the number N of all these components in the electrolyte. The unknowns in these equations are the steady values of field strength $\mathbf{E}$ and gradients $\text{grad } c_j$. The ionic concentrations are interrelated by the electroneutrality condition (1.2); therefore, between the gradients the constraint

$$\sum z_j \text{ grad } c_j = 0 \qquad (4.25)$$

exists, i.e., the gradients of $N - 1$ components are independent. We thus obtain a system of N equations with N unknowns which can be solved. This implies that in the system a steady state actually can be realized, where there is a complete balance with respect to charges and substances and where the parameters (voltage, concentration distribution) have unique values.

4.3.2. Field Strength in the Electrolyte

When concentration gradients are present and we allow for Eq. (4.23), Eq. (1.11) for the total current density becomes

$$i = \sum i_j = F\mathbf{E}\sum |z_j|c_j u_j - F\sum z_j D_j \text{ grad } c_j. \qquad (4.26)$$

Solving this equation for field strength $\mathbf{E}$ and allowing for relation (1.12) we find

$$\mathbf{E} = i/\sigma + (F/\sigma)\sum z_j D_j \text{ grad } c_j = \mathbf{E}_{\text{ohm}} + \mathbf{E}_d. \qquad (4.27)$$

Thus, the field strength in electrolytes during current flow is determined by two components, an ohmic component $\mathbf{E}_{\text{ohm}}$ proportional to current density and a diffusional component $\mathbf{E}_d$ which depends on the concentration gradients. The latter only arises when the D_j values of the individual ions differ appreciably; when they are all identical, $\mathbf{E}_d$ is zero [cf. Eqs. (4.27) and (4.25)]. The existence of the second component is a typical feature of electrochemical systems with ionic concentration gradients. This component can exist even at zero current when concentration gradients are artificially maintained. When a current flows in the electrolyte this component may produce a departure from Ohm's law.

Allowing for Eq. (4.7) we can describe the diffusional field strength in terms of the parameters t_j and λ_j:

$$\mathbf{E}_d = (RT/F) \sum (t_j/z_j) \operatorname{grad} (\ln c_j)$$

$$= (RT/F) \left[\sum (\lambda_j/z_j) \operatorname{grad} c_j \right] / \sum \lambda_j c_j . \qquad (4.28)$$

4.3.3. Transport Numbers in the Diffusion Layer

In Eq. (4.28), the parameter t_j [defined by Eq. (1.13)] is the transport number of ion j in the bulk electrolyte where concentration gradients and diffusional transport of substances are absent.

It must be pointed out that in a diffusion layer where the ions are transported not only by migration but also by diffusion, the effective transport numbers t_j^* of the ions (the ratios between partial currents i_j and total current i) will differ from t_j. In fact, by definition we have $i_j = z_j F J_j$, but according to Eq. (1.24) we have $i = (nF/\bar{\nu}_j)J_j$, where J_j is the total flux of ions j due to diffusion and migration. Hence

$$t_j^* \equiv i_j/i = z_j \bar{\nu}_j/n. \qquad (4.29)$$

Therefore, the effective ionic transport numbers in the diffusion layer depend neither on the mobility nor on the concentration of the ions. Since a solution layer adjacent to the electrode is considered, the effective transport number may even have a negative value, depending on the signs of z_j and $\bar{\nu}_j$ and in accordance with the sign convention outlined in Section 1.8.

4.3.4. First Limiting Case: Excess Base Electrolyte

Often the situation is found where the electrolyte contains only one reacting ion, which has a low concentration, while other ions not involved in the reaction are present in excess concentrations. As the concentration

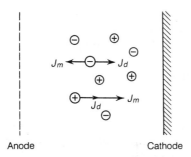

Fig. 4.4. Migration (J_m) and diffusion (J_d) fluxes of anions and cations in cathodic metal deposition from a binary solution.

of such foreign ions is raised, the conductivity σ increases, and in accordance with Eq. (4.27) the field strength E decreases. In the limit, the first term on the right-hand side of Eq. (4.23) becomes small, and the reacting ion will be transported mainly by diffusion. In this case the expression for current density will be the same as Eq. (4.13) for the transport of uncharged particles, and the equation for the limiting current also remains unchanged.

4.3.5. Second Limiting Case: Binary Electrolytes

Consider the binary solution of an electrolyte $M_{\tau+}A_{\tau-}$ of which one ion is involved in the reaction; for the sake of definition, we shall assume that its cation is reduced to metal at the cathode. The cation concentration at the surface will decrease when current flows. Because of the electroneutrality condition, the concentration of anions should also decrease under these conditions, i.e., the total electrolyte concentration c_k should decrease. Considering that the ionic concentrations are $\tau_+ c_k$ and $\tau_- c_k$, respectively, the balance equations for the two ions will be

$$\bar{\nu}_+ |i|/nF = D_+ \tau_+ [z_+ c_k \mathbf{E} F/RT - \operatorname{grad} c_k], \qquad (4.30a)$$

$$\bar{\nu}_- |i|/nF = D_- \tau_- [-|z_-| c_k \mathbf{E} F/RT - \operatorname{grad} c_k]. \qquad (4.30b)$$

The anions, which are not involved in the reaction ($\nu_- = 0$), should not move in the steady state. This implies that the diffusional component of their flux to the surface should be fully compensated by a migration component away from the surface (Fig. 4.4). For the cations, both flux components move toward the surface, i.e., their combined flux is higher in absolute value than the pure diffusion flux. It follows from Eq. (4.30), with $\nu_- = 0$, that the steady field strength is

$$\mathbf{E} = -(RT/|z_-|F)(1/c_k)\,\text{grad}\,c_k = -(RT/|z_-|F)\,\text{grad}\,(\ln c_k). \quad (4.31)$$

This expression is a particular case of Eq. (4.27) and includes an ohmic as well as a diffusional component. Substituting the value of $\mathbf{E}$ into Eq. (4.30a), taking into account that $z_+/|z_-| = \tau_-/\tau_+$ and $c_+ = \tau_+ c_k$, and that in the case being considered $\bar{\nu}_+ = -\nu_+$, we find

$$|i| = (n/\nu_+)F\tau_+[1 + (\tau_-/\tau_+)]D_+\,\text{grad}\,c_k. \quad (4.32)$$

We see that in binary electrolytes, the current increases by a factor of $1 + (\tau_-/\tau_+)$ relative to the pure diffusion current which would be observed (at a given concentration gradient) in the presence of an excess of foreign electrolyte. We shall call

$$\alpha_+ = 1 + (\tau_-/\tau_+) \quad (4.33)$$

the enhancement factor of the total current above the diffusion current. For symmetric electrolytes ($\tau_- = \tau_+$) we have $\alpha_+ = 2$; for electrolytes of the type of $ZnCl_2$ and Ag_2SO_4, the values of α_+ are 3 and 1.5, respectively.

When the anion is the reactant at the anode, then we similarly find

$$\alpha_- = 1 + (\tau_+/\tau_-). \quad (4.34)$$

It follows from Eq. (4.32) that for binary electrolytes the concentration gradient, at a given current density, is constant and independent of the coordinate. Hence, as before, we can change from concentration gradients to concentration differences between bulk and surface:

$$|i| = -(n/\bar{\nu}_j)F\alpha_j\varkappa_j\Delta c_j. \quad (4.35)$$

It follows that for ionic reactants the limiting current is given not by Eq. (4.16) but by the equation

$$|i_{l,\mathrm{r}}| = (n/\nu_\mathrm{r})F\alpha_\mathrm{r}\varkappa_\mathrm{r}c_{V,\mathrm{r}}. \quad (4.36)$$

Equations (4.21) and (4.22) remain valid for the calculation of surface concentrations when this value of the limiting current is used in them.

It should be pointed out that in the case being considered, the effective transport number of the reactant ions in the diffusion layer is unity, and that of the nonreacting ions is zero. This is in harmony with Eq. (4.29); for the reactant ions we should have $z_j \bar{\nu}_j = n$ (otherwise the solution would cease to be binary), and for the nonreacting ions $\bar{\nu}_j = 0$.

4.3.6. The General Case

Generally an electrolyte may contain several ionic reactant species but no obvious excess of a foreign electrolyte. Then a calculation of the migration currents [or coefficients α_j in equations of the type of (4.35)] is very complex and requires computer use.

Often we only need a qualitative estimate, i.e., want to know whether the limiting current is raised or lowered by migration relative to the purely diffusion-limited current, or whether α_j is larger or smaller than unity. It is evident that α_j will be larger than unity when migration and diffusion are in the same direction. This is found in four cases: for cations which are reactants in a cathodic reaction or products in an anodic reaction, and for anions which are reactants in an anodic reaction or products in a cathodic reaction. In the other four cases (for cations which are reactants in an anodic reaction or products in a cathodic reaction, and for anions which are reactants in a cathodic reaction or products in an anodic reaction), we have $\alpha_j < 1$, a typical example being the cathodic deposition of metals from complex anions.

Calculations show that generally when migration is superimposed, the diffusion flux density within the diffusion layer is no longer constant. Hence, depending on the coordinate x, the concentration gradients will also change, and Eq. (4.11) remains valid only for a mean value of the gradient.

4.3.7. Diffusion in Binary Electrolytes at Zero Current

Consider diffusional transport in binary electrolytes at zero current. Suppose that initially a concentration gradient $\mathrm{grad}\, c_k$ had developed through the action of external forces. Diffusional transport of ions in the direction of lower concentrations will start in the system. Generally the diffusion coefficient of the cation differs from that of the anion; therefore, the diffusional fluxes of these ions will also be different. This gives rise to a partial charge separation and development of an electric field which holds back the fast-moving ions but accelerates the slower ones. In the end a steady state is attained in which equivalent amounts of the two ions are transported by a combination of migration and diffusion, as if an electroneutral compound was transported.

From Eq. (4.28) and considering that, according to Eqs. (1.19) and (4.6), in binary solutions the parameter t_j can be written as $z_j D_j / (z_+ D_+ + |z_-|D_-)$, we obtain for the steady-state diffusional field strength

$$\mathbf{E}_d = \frac{D_+ - D_-}{z_+ D_+ + |z_-| D_-} \frac{RT}{F} \operatorname{grad}(\ln c_k). \qquad (4.37)$$

The total flux of the electrolyte as a whole is the stoichiometric sum of fluxes of the cation and anion:

$$J_k = (1/\tau_+)J_+ + (1/\tau_-)J_-. \qquad (4.38)$$

Substituting the ionic fluxes from Eq. (4.23) and using the value obtained for $\mathbf{E}_d$ we can readily show that

$$J_k = -D_k \operatorname{grad} c_k, \qquad (4.39)$$

where D_k is the effective diffusion coefficient of the electrolyte as a whole, which is given by

$$D_k = \frac{(z_+ + |z_-|)D_+ D_-}{z_+ D_+ + |z_-| D_-} = \frac{(\tau_+ + \tau_-)D_+ D_-}{\tau_- D_+ + \tau_+ D_-}. \qquad (4.40)$$

The parameter D_k is some average value of the ionic diffusion coefficients obtained when allowing for the charge on the ions.

If for a given electrolyte the diffusion coefficients of the two ions are identical in value, the effective electrolyte diffusion coefficient will coincide with this value. In this case the diffusional field strength will be zero.

4.4. CONVECTIVE TRANSPORT

Convective transport is the transport of substances with a moving medium, e.g., the transport of a solute in a liquid flow. The convective flux is given by

$$J_{kv,j} = v c_j \qquad (4.41)$$

when v is the linear velocity of the medium and c_j is the concentration of the substance. In electrolyte solutions, the convective flux is always electroneutral because of the medium's electroneutrality.

In electrochemical cells we often find convective transport of reaction components toward (or away from) the electrode surface. In this case the balance equation describing the supply and escape of the components should be written in the general form of (4.10). However, this equation needs further explanation. At any current density during current flow, the migration and diffusion fluxes (or field strength and concentration gradients) will spontaneously settle at values such that condition (4.24) is satisfied. The convective flux, on the other hand, depends on the arbitrary values selected for the flow velocity v and for the component concentrations, i.e., is determined by factors independent of the values selected for the current density. Hence, in the balance equation (4.10) not the total convective flux should appear but only the part that corresponds to the true consumption of reactants from the flux or true product release into the flux. This fraction is defined as the difference between the convective fluxes away from and to the electrode:

$$\Delta J_{kv,j} = v(c'_j - c_j), \qquad (4.42)$$

where c'_j is the concentration of substance j in the flow leaving the electrode (the flow direction away from the electrode is regarded as positive, see Section 1.8).

For the present argument and in what follows, we shall assume that the migrational transport is absent (that we have uncharged reaction components or an excess of foreign electrolyte).

Let us estimate the ratios of diffusion and maximum convective fluxes, $J_{d,j}/J_{kv,j}$ [$= D_j \operatorname{grad} c_j / v c_j$]. The order of magnitude of the concentration gradient is c_j/δ. Therefore

$$J_{d,j}/J_{kv,j} \approx D_j/\delta v. \qquad (4.43)$$

In aqueous solutions $D_j \approx 10^{-9}$ m^2/s; a typical value of δ is 10^{-4} m. It follows that the convective and diffusional transport are comparable already at the negligible linear velocity of 10^{-5} m/s of the liquid flow. At larger velocities convection will be predominant.

In electrochemical systems, two directions of the convective flux are distinguished: normal to the electrode surface in the case of flow-through

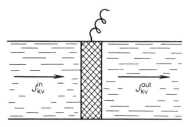

Fig. 4.5. Schematic of a flow-through electrode.

electrodes, and along the electrode surface in the case of flow-by electrodes.

4.4.1. Flow-through Electrodes

A flow-through electrode is shown schematically in Fig. 4.5. The electrode itself can be a screen or porous plate. We shall presuppose for simplicity that all convective fluxes surpass the diffusional and migrational fluxes. Then the balance equation becomes

$$|i| = (n/\bar{\nu}_j)Fv(c'_j - c_j). \tag{4.44}$$

This equation can be used to calculate the concentration changes which occur in the flow as it passes through the electrode. It shows that a limiting convective current

$$|i_{l,r}^{(kv)}| = (n/\nu_r)Fvc_r \tag{4.45}$$

exists, which corresponds to conditions where the concentration of the key component in the flow leaving the electrode has fallen to zero.

4.4.2. Flow-by Electrodes

Flow of the liquid past the electrode is found in electrochemical cells where a liquid electrolyte is agitated with a stirrer or by pumping.

The character of liquid flow near a solid wall depends on the flow velocity v, on the characteristic length L of the solid, and on the kinematic viscosity ν_{kin} (which is the ratio of the usual rheological viscosity η and the liquid's density ρ). A convenient criterion is the dimensionless parameter $\mathrm{Re} \equiv vL/\nu_{kin}$, called the Reynolds number. The flow is laminar when this number is smaller than some critical value (which is about 10^3 for rough surfaces and about 10^5 for smooth surfaces); in this case the liquid

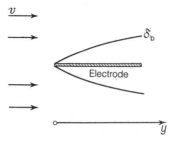

Fig. 4.6. Schematic of a flow-by electrode.

moves in the form of layers parallel to the surface. At high Reynolds numbers (high flow velocities) the motion becomes turbulent, and eddies develop at random in the flow. We shall only be concerned with laminar flow of the liquid.

In the flow, the thin layer of liquid which is directly adjacent to the solid is retained by molecular forces and does not move. The liquid's velocity relative to the solid increases from zero at the very surface to the bulk value v which is attained some distance away from the surface. The zone within which the velocity changes is called the Prandtl or hydrodynamic boundary layer.

Hydrodynamic theory shows that the thickness, δ_b, of the boundary layer is not constant but increases with increasing distance y from the flow's impingement point at the surface (Fig. 4.6); it also depends on the flow velocity:

$$\delta_b \approx v_{kin}^{1/2} y^{1/2} v^{-1/2}. \tag{4.46}$$

It is important to note that even in a strongly stirred solution, a thin layer of stagnant liquid is present directly at the electrode surface, within which convection is absent so that substances involved in the reaction are transported in it only by diffusion and migration. Here the concentration gradient $(\text{grad}\, c_j)_{x=0}$ is steepest and (in the absence of migration) determined by the balance equation

$$\bar{\nu}_j |i| / nF = -D_j (\text{grad}\, c_j)_{x=0}. \tag{4.47}$$

In the bulk, to the contrary, concentration gradients are leveled only on account of convection, and diffusion has practically no effect. In the transition region we find both diffusional and convective transport. The concentration gradient gradually falls to zero with increasing distance from the surface.

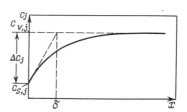

Fig. 4.7. Defining the concept of "diffusion-layer thickness" δ.

Diffusion in a convective flow is called *convective diffusion*. The layer within which diffusional transport is effective (the diffusion layer) does not coincide with the hydrodynamic boundary layer. It is an important theoretical problem to calculate the diffusion-layer thickness δ. Since the transition from convection to diffusion is gradual, the concept of "diffusion-layer thickness" is somewhat vague. In practice this thickness is defined so (Fig. 4.7) that

$$\Delta c_j / \delta = (\text{grad}\, c_j)_{x=0}. \qquad (4.48)$$

This calculated distance δ (or $\varkappa_j$) can then be used in Eq. (4.14) to find the relation between current density and concentration difference.

An analogy exists between mass transfer (which depends on the diffusion coefficient) and momentum transfer between the sliding liquid layers (which depends on the kinematic viscosity). Calculations show that the ratio of thicknesses of the diffusion and boundary layer can be written as

$$\delta / \delta_b \approx (D_j / \nu_{\text{kin}})^{1/3} = \text{Pr}^{-1/3}. \qquad (4.49)$$

The dimensionless ratio ν_{kin}/D_j is called the Prandtl number Pr. In aqueous solutions $D_j \approx 10^{-9}$ m²/s and $\nu_{\text{kin}} \approx 10^{-6}$ m²/s, i.e., Pr $\approx$ 10^3. Thus, the diffusion layer is approximately ten times thinner than the boundary layer. This means that in the major part of the boundary layer, motion of the liquid completely levels the concentration gradients and suppresses diffusion (Fig. 4.8).

Allowing for Eqs. (4.46) and (4.49) we obtain

$$\delta \approx D_j^{1/3} \nu_{\text{kin}}^{1/6} y^{1/2} v^{-1/2}. \qquad (4.50)$$

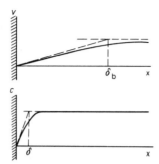

Fig. 4.8. The distributions of flow velocities and concentrations close to the surface of a flow-by electrode.

The gradual increase in thickness δ which occurs with increasing distance y leads to a decreasing diffusion flux. It follows that the current density is nonuniform along the electrode surface.

It is important to note that the diffusion-layer thickness depends not only on hydrodynamic factors but also (through the diffusion coefficient) on the nature of the diffusing species. This dependence is minor, of course, since the values of D_j differ little among the different substances, and in addition are raised to the power one-third in Eq. (4.50).

It follows that convection of the liquid has a twofold influence: it levels the concentrations in the bulk liquid, and it influences the diffusional transport by governing the diffusion-layer thickness. Slight convection is sufficient for the first effect, but the second effect is related in a quantitative way to the convective flow velocity: the higher this velocity is, the thinner will be the diffusion layer and the larger the concentration gradients and diffusional fluxes.

4.4.3. The Rotating Disk Electrode

At the rotating disk electrode (RDE, Fig. 4.9), it is the solid electrode and not the liquid which is driven; but from a hydrodynamic point of view this difference is unimportant. Liquid flows, which in the figure are shown by arrows, are generated in the solution when the electrode is rotated around its vertical axis. The liquid flow impinges on the electrode in the center of the rotating disk, then is diverted by centrifugal forces to the periphery.

Let ω be the angular velocity of rotation; this is equal to $2\pi f$ where f is the disk frequency or number of revolutions per second. The dis-

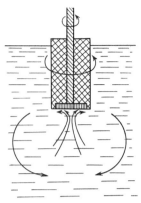

Fig. 4.9. Rotating disk electrode (arrows in
the space below the electrode indicate the
directions of liquid flow).

tance r of any point from the center of the disk is identical with the
distance from the flow impingement point. The linear velocity of any
point on the electrode is ωr. We see when substituting these quantities
into Eq. (4.46) that the effects of the changes in distance and linear velocity mutually cancel, so that the resulting diffusion-layer thickness is
independent of distance.

The constancy of the diffusion layer over the entire surface and thus
the uniform current-density distribution are important features of rotating disk electrodes. Electrodes of this kind are called electrodes with
uniformly accessible surface.

It is seen from the quantitative solution of the hydrodynamic problem
(Veniamin G. Levich, 1948) that for RDE to a first approximation

$$\delta = 1.616 D^{1/3} \nu_{\text{kin}}^{1/6} \omega^{-1/2} \tag{4.51}$$

and, hence,

$$|i| = -0.62(n/\bar{\nu}_j)FD_j^{2/3}\nu_{\text{kin}}^{-1/6}\omega^{1/2}(c_{V,j} - c_{S,j}) \tag{4.52}$$

(the *Levich equation*).

A more exact calculation leads to complex expressions with a number
of correction terms; however, some of these corrections mutually cancel,
so that the accuracy of the equation reported is quite adequate. Equation (4.52) can be used for a quantitative analysis of the experimental

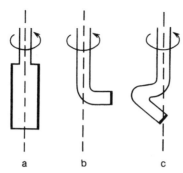

Fig. 4.10. Different forms of rotating electrode. (a) Cylinder; (b) L-shaped; (c) S-shaped.

data as well as for calculating diffusion coefficients from experimental limiting-current values. Equally exact calculations of convective diffusion are not possible for any other electrode type.

In practice, RDE with disk speeds between 1 and 170 rps (or 60 to 10,000 rpm) are used. According to Eq. (4.51), in aqueous solutions this speed range corresponds to thicknesses δ between approximately 60 and 4.5 μm. Whirling liquid flow and wobble of the rapidly spinning electrode cause considerable complications at higher speeds.

In cases where we want to work with drastically higher limiting currents but need no quantitative calculations, we can use other forms of rotating electrodes such as those shown in Fig. 4.10. The turbulent liquid flows which develop at such electrodes give rise to much thinner diffusion layers.

4.4.4. The Rotating Ring–Disk Electrode

In 1959 Alexander N. Frumkin and Lev N. Nekrasov suggested a new electrode type, the rotating ring–disk electrode (RRDE). Here a thin ring used as a second electrode is arranged concentrically around the disk electrode (Fig. 4.11). The gap $r_2 - r_1$ between disk and ring is narrow (less than 1 mm). The primary electrochemical reaction occurs at the disk electrode. The ring electrode is used for the quantitative, and sometimes for a qualitative, determination of reaction products (intermediate and final) which are formed at the disk and dissolve. To this end the potential set at the ring is such that these products will electrochemically react, i.e., will be reduced or oxidized. Using the hydrodynamic theory of convec-

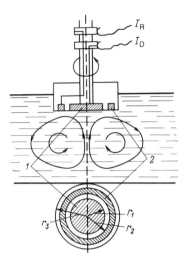

Fig. 4.11. Rotating ring–disk electrode. (1) Disk electrode (current I_D); (2) ring electrode (current I_R).

tive diffusion we can exactly calculate which fraction, N, of the particles released at the disk electrode will reach the surface of the ring electrode and react there. This fraction depends on the ratio of radii of disk and ring, and is usually about 40%. Thus, from the limiting current I_R at the ring electrode we can estimate the rate of formation of the products at the disk electrode.

The RRDE is useful when studying chemical aspects of electrode reactions or the quantitative aspects of electrode reactions with several parallel pathways.

4.4.5. Cells with Natural Convection of the Electrolyte

It follows from Eqs. (4.50) and (4.51) that the diffusion-layer thickness will increase without limits and the diffusion flux will decrease to zero when the electrolyte is not stirred ($v = 0$) or the electrode not rotated ($\omega = 0$). This implies that a steady electric current cannot flow in such cells. But this conclusion is at variance with the experimental data.

This discrepancy arises mainly from the fact that spontaneous liquid flows will always develop in any liquid even without artificial stirring, e.g., under the action of density gradients caused by local temperature or concentration fluctuations. This phenomenon has been termed *natural convection.* Electrochemical reactions reinforce natural convection, since the concentrations of substances involved in the reaction will change near the electrode surfaces, and also since heat is evolved. Gas evolution attending the reactions has a particularly strong effect on natural convection.

Natural convection strongly depends on cell geometry. No convection can arise in capillaries or in the thin liquid layers found in narrow gaps between electrodes. The rates of natural convective flows and the associated diffusion-layer thicknesses depend on numerous factors and cannot be calculated in a general form. Very rough estimates show that the diffusion-layer thickness under a variety of conditions may be between 100 and 500 μm.

Natural convection can be entirely eliminated when electrolytes held in a matrix or porous support are used instead of the free liquids. Natural convection will not develop in the pore space when the individual pores are sufficiently narrow. When such electrolytes are used, the diffusion layer propagates across the entire matrix, i.e., across the full electrode gap.

Chapter 5
PHASE BOUNDARIES (INTERFACES) BETWEEN ELECTROLYTES

5.1. TYPES OF INTERFACES BETWEEN ELECTROLYTES

In practical galvanic cells with more than one electrolyte, pairs of different electrolytes are sometimes in mutual contact forming electrolyte/electrolyte interfaces.

Such an interface will be mechanically stable when at least one of the two paired electrolytes is solid. The interface will also be stable when two immiscible liquid electrolytes are brought together in the form of horizontal layers, and the liquid with the lower density is above the one with the higher density. But when miscible liquids are paired they will start to mix under the effect of hydrodynamic flows, and the interface rapidly disappears. Such interfaces can be stabilized by separating the two electrolytes with a porous diaphragm, which hinders or completely prevents liquid flows, but at the same time does not interfere with conduction (ionic migration) between the electrolytes. In laboratory practice the liquids are often separated by glass stopcocks wetted with the solution. Flowing systems can also be used, such as the system shown in Fig. 5.1, where the liquid with the higher density is fed in from below; after union of the two flows a stable interface is preserved over some distance AB. We shall be concerned only with stable interfaces between electrolytes.

We can distinguish interfaces between similar and dissimilar electrolytes (homogeneous and inhomogeneous systems). In the first case the two electrolytes have the same solvent (medium), but they differ in the nature and (or) concentration of solutes. In the second case the interface separates dissimilar media. An example is the system consisting of salt solutions in water and nitrobenzene.

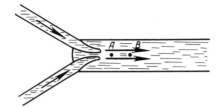

Fig. 5.1. Liquid junction in a flow system.

The interfaces between similar electrolytes are often called liquid junctions, even though this concept includes interfaces between electrolytes which are not liquid, e.g., between gelled aqueous solutions.

All ions which can exist in both phases can diffuse across the electrolyte/electrolyte interface. The diffusion leads to leveling of their electrochemical potentials and eventually brings about an equilibrium distribution between the phases. In a homogeneous system, equilibrium signifies the complete leveling of composition and concentrations in the two phases, i.e., disappearance of the interface as such. Complete equilibration as a rule is a long process. In such systems, therefore, only the phenomena occurring in a predetermined original state prior to equilibration are considered (Section 5.2).

In an inhomogeneous system with immiscible electrolytes, the interface is preserved even after equilibration. In this case the conditions under which equilibria exist are of interest as well (Section 5.3). Such systems as a rule are selective; because of differences in the chemical driving forces, the equilibrium distributions of the components between the two phases are dissimilar. When the selectivity is perfect, some components may exist in only one of the phases and will not transfer to the other.

One of the features found at interfaces between electrolytes [suppose, the two electrolytes (α) and (β)] is the development of a Galvani potential, $\varphi_G^{(\beta,\alpha)}$, between the phases. This potential difference is a component of the total ocv of the galvanic cell [see Eq. (2.18)]. In the case of similar electrolytes it is called the *diffusion potential* φ_d, and can be determined, in contrast to potential differences across interfaces between dissimilar electrolytes.

A special case of interfaces between electrolytes are those involving *membranes*. A membrane is a thin, ion-conducting interlayer (most often solid) separating two similar liquid phases and exhibiting selectivity (Fig. 5.2). Nonselective interlayers, i.e., interlayers uniformly permeable for all components, are called *diaphragms*. Completely selective mem-

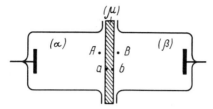

Fig. 5.2. Schematic of an electrochemical cell
with membrane μ.

branes (i.e., membranes which are permeable for some and impermeable
for other substances) are called permselective membranes.

When the original compositions of the outer phases are different, the
permselective membrane will prevent the complete leveling of these compositions. Some equilibrium component distribution between the phases
(α) and (β) will be established, and between points A and B a potential difference called the *membrane potential* (or transmembrane potential)
φ_m will develop. This potential difference is determined by the Galvani
potentials across the two interfaces between the membrane and the outer
phases. Moreover, while the system is not yet in equilibrium a diffusion potential or intramembrane potential φ_d exists inside the membrane
(between the points a and b). Thus, generally

$$\varphi_m \equiv \psi^{(\beta)} - \psi^{(\alpha)} = \varphi_G^{(\mu,\alpha)} - \varphi_G^{(\mu,\beta)} + \varphi_d. \tag{5.1}$$

Since the outer phases are similar, membrane potentials can be measured.

5.2. DIFFUSION POTENTIALS

5.2.1. General Equations

At the interface between the two similar solutions (α) and (β) which
differ merely in composition, a transition layer will develop within which
the concentrations of each component j exhibit a smooth change from their
values $c_j^{(\alpha)}$ in phase (α) to the values $c_j^{(\beta)}$ in phase (β). The thickness δ_{tr}
of this transition layer depends on how this boundary has been realized
and stabilized. When a porous diaphragm is used it corresponds to the
thickness of this diaphragm, since within each of the phases outside the

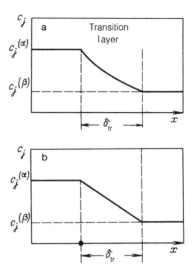

Fig. 5.3. Nonlinear (a) and linear (b) concentration distribution in the transition layer at the interface between two solutions.

diaphragm the concentrations are practically constant owing to the liquid flows.

The concentration distributions within the transition layer also depend on the way in which the boundary was formed. When convection is absent (as, e.g., in the pores of a diaphragm), the ions in the transition layer will be transported by diffusion and migration, and the steady-state concentration distributions in the transition layer, as a rule, will be nonlinear (Fig. 5.3a). In the device shown in Fig. 5.1, mixing of the liquid flows will produce an intermediate layer where the concentration distributions will be almost linear (Fig. 5.3b).

The ionic concentration gradients in the transition layer constitute the reason for development of the diffusion component $\mathbf{E}_d$ of electric field strength [the component arising from the difference in diffusion coefficients between the individual ions; Eqs. (4.27) and (4.28)].

The diffusion potential between the solutions, $\varphi_d^{(\beta,\alpha)} \equiv \psi^{(\beta)} - \psi^{(\alpha)}$, can be calculated by integrating $\mathbf{E}_d$ over the full diffusion-layer thickness from phase (α) to phase (β). Allowing for Eq. (4.28) we have

$$\varphi_d^{(\beta,\alpha)} = -\int_{(\alpha)}^{(\beta)} \mathbf{E}_d \, dx = -\frac{RT}{F} \int_{(\alpha)}^{(\beta)} \sum \frac{t_j}{z_j} \left(\frac{d \ln c_j}{dx} \right) dx. \quad (5.2)$$

This equation, like other equations resting on Fick's law (4.1), is approximate, and becomes less exact with increasing concentration. An analogous expression involving ionic activities instead of concentrations can be derived in a different, quasi-thermodynamic way. But in the following we shall mainly use concentrations, since the accuracy of such equations is not improved with the use of activities.

Equation (5.2) can be integrated in a general form, only when $\mathbf{E}_d$ (or t_j) is known as a function of coordinate x, or (since these parameters depend on the component concentrations) when the concentration distribution can be specified.

Different solutions corresponding to different forms of concentration distribution are known for Eq. (5.2). It is assumed in all cases, when integrating this equation, that the diffusion coefficients (or mobilities) of all ions are independent of the concentrations.

Nonlinear concentration distributions. The mathematical analysis of this model is very complicated. Max Planck showed in 1890 that for 1:1 electrolytes, the diffusion potential at such a boundary is determined by the equation

$$\varphi_d = (RT/F)\ln\xi, \qquad (5.3)$$

where ξ is the solution of the transcendental equation

$$\frac{\xi U_+^{(\beta)} - U_-^{(\alpha)}}{U_+^{(\beta)} - \xi U_-^{(\alpha)}} = \frac{\ln[C^{(\beta)}/C^{(\alpha)}] - \ln\xi}{\ln[C^{(\beta)}/C^{(\alpha)}] + \ln\xi} \frac{\xi C^{(\beta)} - C^{(\alpha)}}{C^{(\beta)} - \xi C^{(\alpha)}}; \qquad (5.4)$$

here $U_+^{(\beta)} \equiv \sum_+ u_j c_j^{(\beta)}$, $C^{(\alpha)} \equiv \sum c_j^{(\alpha)}$, etc. (the *Planck equation*).

Linear concentration distributions. A much simpler mathematical solution is obtained when a linear concentration distribution is assumed for each component:

$$c_j(x) = c_j^{(\alpha)} + \gamma_j x, \qquad (5.5)$$

where $\gamma_j \equiv [c_j^{(\beta)} - c_j^{(\alpha)}]/\delta_{tr}$ is the constant value (within the transition layer) of the given component's concentration gradient.

After substituting this function into Eq. (4.28) (in the version containing the parameter λ_j) we can write the equation for φ_d in a form that is convenient for integration:

$$\varphi_d = \frac{RT}{F} \int_0^{\delta_{tr}} \frac{[\sum(\lambda_j/z_j)\gamma_j]\,dx}{[\sum\lambda_j c_j^{(\alpha)}] + (\sum\lambda_j\gamma_j)\,x}. \qquad (5.6)$$

Substituting values for γ_j and integrating we obtain

$$\varphi_d = \frac{RT}{F} \frac{\sum(\lambda_j/z_j)[c_j^{(\beta)} - c_j^{(\alpha)}]}{\sum \lambda_j[c_j^{(\beta)} - c_j^{(\alpha)}]} \ln \frac{\sum \lambda_j c_j^{(\beta)}}{\sum \lambda_j c_j^{(\alpha)}} \qquad (5.7)$$

(the *Henderson equation*, 1907).

Because of its simplicity, the Henderson equation is used more often when calculating diffusion potentials than the Planck equation, despite the fact that the basic assumptions do not correspond to the usual experimental conditions.

5.2.2. Equations for Particular Cases

Binary solutions containing the same electrolyte in different concentrations. For a binary solution $t_j = u_j/(u_+ + u_-)$, i.e., the transport numbers do not depend on concentration and have constant values throughout the transition layer. Taking into account that $z_+\tau_+ = |z_-|\tau_- = z_k$ we can write the ratio t_j/z_j as $(\text{sign } z_j)\tau_j t_j/z_k$. Substituting this expression into Eq. (5.2) and taking into account that $c_j = \tau_j c_k$ we obtain, after integration,

$$\varphi_d = (\tau_-^2 t_- - \tau_+^2 t_+)(RT/z_k F) \ln[c_k^{(\beta)}/c_k^{(\alpha)}] \qquad (5.8)$$

or, for a symmetric electrolyte ($\tau_+ = \tau_- = 1$ and $z_+ = |z_-| = z$),

$$\varphi_d = (t_- - t_+)(RT/zF) \ln[c_k^{(\beta)}/c_k^{(\alpha)}]. \qquad (5.9)$$

An equation of this kind was derived by Walther Nernst in 1889. The same equation can be obtained as a particular case of the Planck equation (5.3) and Henderson equation (5.7). This means that in the present system the nature of the electrolyte/electrolyte interface has no effect on the values of φ_d; the concentration distribution will be linear even when the transition layer is formed by diffusion.

In this particular case activities are sometimes used instead of concentrations. Integration of the corresponding version of Eq. (5.2) yields the solution

$$\varphi_d = (RT/z_k F)\{\tau_- t_- \ln[a_-^{(\beta)}/a_-^{(\alpha)}] - \tau_+ t_+ \ln[a_+^{(\beta)}/a_+^{(\alpha)}]\}. \qquad (5.10)$$

747-4450

renew

1 wk

#	Column 1	Column 2	Column 3
8	QA KF - KZ	QA Q - QA	QA KF - QA
8			
7	KF	HG - KF	HG - KF
7	HG	HG	HG
6	HG	HF - HG	HF - HG
6	HF	HF	HF
5	HF	HF	HF
5	HE	HE - HF	HE - HF
4	HD - HE	HD	HE - HF
4	HD	HD	HD - HE
3	HD	HD	HD
3	HD	HD	HD
2	HC - HD	HC	HD
2	HB	HB - HC	HD
1	ENCYCLOPEDIAS	A - HB SMALL BUSINESS	HC - HD
1	ENCYCLOPEDIAS	TEST BOOKS CAREER BOOKS	HB - HC

A - H B SCIENCE FAIR

B/S/T RANGE FINDER

SECTION AISLE	A REFERENCE	B REFERENCE	C CIRCULATING	SECTION AISLE
17	Z	UA - Z	UB - Z	17
17	TX	TX - UA	TX - UB	17
16	TX	TS - TX	TX	16
16	TL - TP	TP - TS	TP - TX	16
15	TL	TK - TL	TL - TP	15
15	TH - TJ	TJ - TK	TJ - TL	15
14	TD - TH	TA - TD	TC - TJ	14
14	SD - SF	SF - TA	SF - TA	14
13	SB - SD	RT - SB	SB - SF	13
13	RG - RJ	RJ - RT	RM - SB	13
12	RD - RG	RC - RD	RC - RM	12
12	RA - RB	RB - RC	RC	12
11	R - RA	QP - R	RA - RC	11
11	QL	QL - QP	QL - RA	11
10	QH - QL	QH	QH - QL	10
10	QD	QD - QH	QE - QH	10

Thus, even in this very simple case the diffusion potential cannot be calculated exactly, since this would require a knowledge of activity values for the individual ions.

Binary solutions of z:z electrolytes having a common ion and the same concentration. As an example, we consider the $(\alpha)KA{\mid}MA(\beta)$ system, where $c_{KA} = c_{MA}$. We can readily show, when allowing for Eq. (1.21), that in this case the general Henderson equation changes to

$$\varphi_d = (RT/zF)\ln[\Lambda^{(\beta)}/\Lambda^{(\alpha)}] \qquad (5.11)$$

(the *Lewis–Sargent equation*, 1909). The Planck equation leads to the same expression, hence even in this case the final result is independent of the boundary type.

5.2.3. Values of Diffusion Potentials for Different Interfaces

Table 5.1 lists values of φ_d for interfaces between aqueous solutions of different composition calculated from Eqs. (5.7), (5.8), or (5.11) as well as values determined experimentally (in cases where this could be done with sufficient accuracy). A minus sign indicates that the potential of phase (β) is more negative than that of phase (α). We can see from the table that the calculated values agree quite well with those measured. The values of φ_d are small; their absolute values are not over 10 mV when the ions have similar diffusion coefficients. But the diffusion potentials attain several tens of millivolt when the solutions contain H^+ or OH^- ions, which have diffusion coefficients several times higher than those of other ions. In this case the phase containing the higher concentration of H^+ ions is negatively charged relative to the other phase. For systems involving OH^- ions the diffusion potentials have the opposite sign.

5.2.4. Ways of Reducing Diffusion Potentials

Aqueous solutions of the salts KCl and NH_4NO_3 are of interest inasmuch as here the mobilities (and also the diffusion coefficients) of the anion and cation are very similar. The higher the concentration of these salts the larger is the contribution of their ions to transition layer composition and, as can be seen from Table 5.1, the lower the diffusion potentials will be at interfaces with other solutions. This situation is often used for a drastic reduction of diffusion potentials in cells with transference. To

TABLE 5.1. Diffusion Potentials Calculated for Interfaces between Aqueous Solutions of Different Composition from Eqs. (5.7), (5.8), or (5.11), and Similar Potentials Determined Experimentally ($\varphi_d = \psi^{(\beta)} - \psi^{(\alpha)}$)

Composition of phases		φ_d, mV	
(α)	(β)	calculation	experiment
0.005 M HCl	0.04 M HCl	−33.3	−
0.005 M KCl	0.04 M KCl	1.0	−
0.01 M NaCl	0.01 M LiCl	1.3	1.1
0.01 M HCl	0.01 M KCl	27.5	25.7
0.1 M KCl	0.1 M NaCl	4.9	6.4
0.1 M HCl	0.1 M KCl	28.5	26.8
0.1 M HCl	4.2 M KCl	3.2	−
4.2 M KCl	0.1 M KCl	−1.9	−
0.1 M HCl ¦ 4.2 M KCl ¦ 0.1 M KCl		1.3	1.1

this end one interposes between the two solutions a third solution, usually saturated KCl solution (which is about 4.2 mol/liter):

$$M|(\alpha)|KCl, aq(\varepsilon)|(\beta)|M. \qquad (5.12)$$

In laboratory practice "salt bridges" (Fig. 5.4) are often used to connect the vessels holding the two solutions. The reduction of the overall diffusion potential is particularly marked when solutions (α) and (β) are either both acids or both bases. In this case the residual diffusion potentials at interfaces $(\beta)/(\varepsilon)$ and $(\alpha)/(\varepsilon)$ compensate each other to an appreciable degree.

The diffusion-potential reduction thus attained is entirely satisfactory for many measurements not demanding high accuracy. However, this approach is not feasible for the determination of the accurate corrected ocv values of cells with transference which are required for thermodynamic calculations.

5.3. DISTRIBUTION OF THE IONS BETWEEN DISSIMILAR ELECTROLYTES

Two dissimilar electrolytes (α) and (γ) that are in contact may contain ions existing in both phases and freely transferring across the interface (the permeating ions, in the following provisionally labeled K^{z+}, M^{z+}, A^{z-},

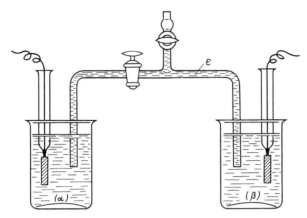

Fig. 5.4. Salt bridge.

etc.) as well as ions existing only in one of the phases (e.g., anions Q^{z-} and R^{z-}). For the permeating ions the equilibrium condition can be written, in accordance with Eq. (2.11), as

$$\varphi_G^{(\gamma,\alpha)} = -\Delta\mu_j^0/z_j F + (RT/z_j F)\ln[a_j^{(\alpha)}/a_j^{(\gamma)}]. \qquad (5.13)$$

When several kinds of permeating ion are present in the system, condition (5.13) must be fulfilled for each of them. But since the potential difference $\varphi_G^{(\gamma,\alpha)}$ between the phases can have just one unique value, the sum of terms on the right-hand side of this equation must also be the same for all permeating ions. It follows that

$$\varkappa_j^{1/z_j}[a_j^{(\alpha)}/a_j^{(\gamma)}]^{1/z_j} = \varkappa_k^{1/z_k}[a_k^{(\alpha)}/a_k^{(\gamma)}]^{1/z_k} = \ldots = \lambda, \qquad (5.14)$$

where $\varkappa_k \equiv \exp(-\Delta\mu_j^0/RT)$ is a constant that is typical for each kind of ion, and $\lambda \equiv \exp(F\varphi_G^{(\gamma,\alpha)}/RT)$ is a constant independent of the ion which is often called the distribution coefficient.

This thermodynamic equation defines the equilibrium distribution of all permeating ions between the two phases. For quantitative calculations, the conditions of electroneutrality of the phases must be taken into account in addition to this equation.

Consider a few examples where the equilibrium concentrations of components of one of the phases, e.g., (γ), are calculated when their concentrations in the other phase (α) are given. Assumptions will be made which simplify the notation; we assume that the activities of the ions appearing in the thermodynamic equations are equal to their concentrations, and we consider systems where the ions have like valences z.

Under these assumptions Eq. (5.14) becomes

$$\varkappa_j[c_j^{(\alpha)}/c_j^{(\gamma)}] = \ldots = \varkappa_l^{-1}[c_l^{(\gamma)}/c_l^{(\alpha)}] = \ldots = \lambda^z \tag{5.15}$$

(here j are the cations and l are the anions for which $z_l = -z$).

For the particular case where two ions of like sign are involved in the transfer (e.g., the two ions K^{z+} and M^{z+}), this equation becomes

$$c_M^{(\gamma)}c_K^{(\alpha)}/c_K^{(\gamma)}c_M^{(\alpha)} = \sigma_{M/K}, \tag{5.16}$$

while for transfer of a cation M^{z+} and an anion A^{z+} it becomes

$$c_M^{(\gamma)}c_A^{(\gamma)}/c_M^{(\alpha)}c_A^{(\alpha)} = \sigma_{MA}, \tag{5.17}$$

where $\sigma_{M/K} \equiv \varkappa_M/\varkappa_K$ and $\sigma_{MA} \equiv \varkappa_M\varkappa_A$ are the *selectivity coefficients* (which are a quantitative measure for the degree of selectivity in a given system).

Unlike constants $\varkappa_j$, the values of $\sigma_{M/K}$ and σ_{MA} can be determined from experimental data [e.g., with the aid of Eqs. (5.16) and (5.17) using the analytical values for the ionic concentrations in the two phases].

The equation for the potential difference between the phases becomes

$$\varphi_G^{(\gamma,\alpha)} = (RT/zF)\ln\varkappa_j + (RT/zF)\ln[c_j^{(\alpha)}/c_j^{(\gamma)}], \tag{5.18}$$

where j is any of the permeating ions (for the anions $z < 0$).

Using this equation we can calculate the concentration-dependent changes in φ_G (absolute values of φ_G and $\varkappa_j$ cannot be determined experimentally).

Example: Consider the system

$$(\alpha)MA \mid MA + MR(\gamma). \tag{5.19}$$

Ions M^{z+} and A^{z-} for which the concentrations, $c^{(\alpha)}$, in phase (α) are given are freely exchanged between the phases. Solving Eq. (5.15) jointly with the electroneutrality equation $c_M^{(\gamma)} = c_A^{(\gamma)} + c_R$ we find

$$c_j^{(\gamma)} = \sqrt{c_R^2/4 + \sigma_{MA}[c^{(\alpha)}]^2} \pm c_R/2 \tag{5.20}$$

(the plus sign is for the cation, the minus sign for the anion). It also follows that

$$c_M^{(\gamma)} c_A^{(\gamma)} = \sigma_{MA} [c^{(\alpha)}]^2. \tag{5.21}$$

Example: Consider the system

$$(\alpha)KQ + MQ | KR + MR(\gamma). \tag{5.22}$$

In this system two kinds of cation are exchanged between the phases. The electroneutrality conditions are: $c_K^{(\alpha)} + c_M^{(\alpha)} = c_Q$ and $c_K^{(\gamma)} + c_M^{(\gamma)} = c_R$. Solving Eq. (5.15) jointly with these electroneutrality equations we find for the concentration of cations M^{z+}:

$$c_M^{(\gamma)} = c_R c_M^{(\alpha)} / [c_M^{(\alpha)} + \sigma_{K/M} c_K^{(\alpha)}], \tag{5.23}$$

and an analogous equation for the cations K^{z+}. The expression for the potential difference between the phases becomes

$$\varphi_G^{(\gamma,\alpha)} = (RT/zF)\ln(\varkappa_M/c_R) + (RT/zF)\ln[c_M^{(\alpha)} + \sigma_{K/M} c_K^{(\alpha)}]. \tag{5.24}$$

5.4. DISTRIBUTION OF IONS IN CELLS WITH MEMBRANE

Consider the system which consists of two similar solutions (α) and (β) which are separated by a membrane impermeable for at least one of the solution components, Y. We write $\mu[Y]$ for such a membrane.

5.4.1. Equilibrium Systems

For any initial composition of the two solutions, an equilibrium distribution of the species between the membrane, each of the solutions, and hence also between the two solutions, is attained after some time as a result of transfer of the solution components (other than Y) through the membrane. It will suffice, then, to consider the equilibrium between the outer solutions, and disregard the membrane except for its blocking of the transfer of species Y between the solutions.

The condition of complete equilibrium for any permeating ion in phases (α) and (β) can be written in a form similar to Eq. (5.15). Since the phases were assumed to be similar, the standard chemical potentials of the ions will be the same in the two phases: $\Delta\mu_j^0 = 0$, and hence $\varkappa_j = 1$. On this basis we find

$$[c_j^{(\alpha)}/c_j^{(\beta)}] = \ldots = [c_l^{(\beta)}/c_l^{(\alpha)}] = \ldots = \lambda^z, \tag{5.25}$$

$$\varphi_m = (RT/zF)\ln[c_j^{(\alpha)}/c_j^{(\beta)}] \tag{5.26}$$

where φ_m is the membrane potential (by definition).

The equilibrium conditions for homogeneous systems with membranes were first formulated in this form by Frederick G. Donnan in 1911. Hence, such equilibria are often called *Donnan equilibria*, and the membrane potentials associated with them are called *Donnan potentials*. Sometimes these terms are used as well for the equilibria arising at junctions between dissimilar solutions (Section 5.3).

Example: Consider the system

$$(\alpha)MA\!\mid\!\mu[R^-]\!\mid\!MA + MR(\beta). \tag{5.27}$$

This system is the analog of system (5.19) for an equilibrium between dissimilar phases, when one of the phases contains the nonpermeating anion R^{z-} while ions M^{z+} and A^{z-} are freely exchanged between the phases. For the distributions of these ions we find, accordingly,

$$c_j^{(\beta)} = \sqrt{c_R^2/4 + [c^{(\alpha)}]^2} \pm c_R/2 \tag{5.28}$$

(the plus sign is for the cation, the minus sign for the anion). Equation (5.21) for the product of concentrations of the permeating ions in the two phases takes on the simple form

$$c_M^{(\beta)}c_A^{(\beta)} = [c^{(\alpha)}]^2. \tag{5.29}$$

5.4.2. Quasi-equilibrium Systems

Complete equilibration of two solutions separated by a membrane is a very slow process. Often quasi-equilibrium systems are used, where there is no equilibrium between the outer solutions (their composition is that arbitrarily given at the outset), though each of these solutions is in equilibrium with an adjacent thin membrane surface layer; there is no equilibrium within the membrane between these surface layers.

We shall write (μ) and (η) for the membrane surface layers adjacent to solutions (α) and (β), respectively. Using the equations reported in Section 5.3 we can calculate the ionic concentrations in these layers as well as the potential differences $\varphi_G^{(\mu,\alpha)}$ and $\varphi_G^{(\eta,\beta)}$ between the phases. According to Eq. (5.1), the expression for the total membrane potential additionally contains the diffusion potential φ_d within the membrane itself, where equilibrium is lacking. Its value can be found with the equations of Section 5.2 if the values of $c_j^{(\mu)}$ and $c_j^{(\eta)}$ are first calculated.

Example: Consider the system

$$(\alpha)\text{MA}|\mu[\text{A}^{z-}]|\text{MA} + \text{KA}(\beta). \tag{5.30}$$

Since the membrane is permeable for cations but not for the anions A^{z-}, it should intrinsically contain anions R^{z-}. When these are fixed their concentration c_Q will remain the same everywhere. Hence in layers (μ) and (η) the overall cation concentration should also be the same, and the diffusion potential (which is caused by a possible difference in cation mobilities) is extremely small. In the left-hand part of the membrane system the concentration of cations M^{z+} in each of the phases is equal to the given (invariant) concentration of anions A^{z-} or R^{z-}, respectively; the potential difference between the phases is determined, according to Eq. (5.18), by the cation concentration ratio. The right-hand part of the membrane system corresponds to the system (5.22) where phase (β) now takes the place of phase (α), and phase (η) takes that of phase (γ). As a result we obtain for the membrane potential

$$\varphi_m = (RT/zF)\ln\{[c_M^{(\beta)} + \sigma_{K/M}c_K^{(\beta)}]/c_M^{(\alpha)}\} + \varphi_d. \tag{5.31}$$

5.5. GALVANIC CELLS WITH TRANSFERENCE

Galvanic cells which include at least one electrolyte/electrolyte interface (which may be an interface with a membrane) across which ions can be transported by diffusion are called *cells with transference*. For the electrolyte/electrolyte interfaces considered in earlier sections, cells with transference can be formulated, e.g., as

$$M|(\alpha)|(\beta)|M, \tag{5.32}$$

$$M|(\alpha)|(\gamma)|M, \tag{5.33}$$

$$M|(\alpha)|(\mu)|(\beta)|M, \tag{5.34}$$

where (α) and (β) are similar, (α) and (γ) are dissimilar phases, (μ) is the membrane, and M is an electrode reversible with respect to the ions M^{z+} present in all electrolyte phases considered.

The ocv values of these cells can be measured experimentally. We shall write them as

$$\mathcal{E} = \mathcal{E}^* + \varphi^{(E)}, \tag{5.35}$$

where $\varphi^{(E)}$ is the potential difference between the electrolytes or membrane potential, and $\mathcal{E}^*$ is the potential difference between the two electrodes, i.e., the corrected ocv value [see Eq. (2.19)]. It is only when one of the two terms on the right-hand side can be calculated independently that the other term can be determined from experimental data of $\mathcal{E}$.

After attainment of the ionic equilibrium distribution, cell (5.32) becomes a symmetric cell (the electrolyte/electrolyte interface disappears) for which the ocv is zero. It is important to note that even for cell (5.33) the ocv becomes zero after equilibration, but the electrolyte/electrolyte interface does not disappear here. This implies that at equilibrium the value of $\varphi^{(E)}$ is exactly compensated by the potential difference $\mathcal{E}^*$. In fact, the current flowing in such a cell is not associated with any chemical or concentration changes, or changes in Gibbs energy of the system. The values of $\varphi^{(E)}$ and $\mathcal{E}^*$ cannot be determined individually for this inhomogeneous cell.

For all kinds of nonequilibrium cells and also for cell (5.34) in the quasi-equilibrium state, nonzero ocv values are measured.

For the homogeneous cells (5.32) and (5.34), the values of $\varphi^{(E)}$ can be calculated with the aid of the equations reported in Sections 5.2 and 5.4, but the results are only approximate owing to the assumptions made in deriving these equations. Under the assumptions made, the values of $\mathcal{E}^*$ [$= \Delta E(M^+, M)$] obey the equation

$$\mathcal{E}^* = (\tau_+ RT/z_k F)\ln[a_+^{(\beta)}/a_+^{(\alpha)}]. \tag{5.36}$$

For an exact calculation of $\mathcal{E}^*$ we should know the ratio of activities of the potential-determining ions in the two electrolytes. Some error arises when the activity of an individual ion is replaced by the mean-ionic activity. Therefore, the individual components of the right-hand part of Eq. (5.35) can be determined for homogeneous cells, but the accuracy is limited.

For thermodynamic calculations one often uses the so-called concentration cells with transference, which contain two binary solutions of like nature but different concentration, e.g.,

$$M|(\alpha)MA, aq \vdots MA, aq(\beta)|M. \tag{5.37}$$

Using Eq. (5.10) for the diffusion potential and Eq. (5.36) for the value of $\mathcal{E}^*$ we obtain

$$\mathcal{E} = t_-(\tau_k RT/z_k F)[\ln a_\pm^{(\beta)}/a_\pm^{(\alpha)}] = t_-(RT/z_k F)\ln[a_k^{(\beta)}/a_k^{(\alpha)}] \tag{5.38}$$

for the overall ocv of such a cell. This expression contains accessible values of mean-ionic activity or activity of the electrolyte as a whole. For ideal systems where activities can be replaced by concentrations,

$$\mathcal{E} = t_-(RT/z_k F)\ln[c_k^{(\beta)}/c_k^{(\alpha)}]. \tag{5.39}$$

When electrodes reversible with respect to the anions are employed in concentration cells with transference, analogous expressions are obtained for the ocv, but they contain t_+ instead of t_-.

These values of ocv are usually compared with similar values for concentration cells without transference where an electrolyte/electrolyte interface is not present. An example of such a cell is the twin cell

$$M|(\alpha)MA, aq|A|Cu|A|MA, aq(\beta)|M, \tag{5.40}$$

where A is an electrode reversible with respect to the anion.

When considering how the potentials of all four electrodes depend on the activities of the potential-determining ions M^{z+} and A^{z-}, we obtain for the overall ocv

$$\mathcal{E} = (RT/z_k F)\ln[a_k^{(\beta)}/a_k^{(\alpha)}]. \tag{5.41}$$

This expression shows that by comparing the ocv values of (5.38) and (5.41) which have been measured in concentration cells of the same type with and without transference, we can find the ionic transport numbers in binary solutions.

Chapter 6
POLARIZATION OF ELECTRODES

6.1. BASIC CONCEPTS

6.1.1. Electrochemical Reaction Rates

For thermodynamic reasons, an electrochemical reaction can occur only within a definite region of potentials, viz., a cathodic reaction at electrode potentials more negative, and an anodic reaction at potentials more positive, than the equilibrium potential of that reaction. This condition only implies a possibility that the electrode reaction will occur in the corresponding region of potentials, but provides no indication whether the reaction actually will occur, and if so, what its rate will be. The answers are provided not by thermodynamics but by electrochemical kinetics.

The concept of *"electrochemical reaction rate"* needs explanation. The reaction rate, i.e., the amount of reactant converted in unit time, is proportional to the current. But the current does not depend on intrinsic properties of the galvanic cell; it is impressed and can be varied arbitrarily between zero and the limiting value that is typical for a given system. Therefore, the effective rate is not an indicative figure for the electrode reaction. However, current flow gives rise to electrode polarization, which means a shift of potential away from the equilibrium value. The magnitude of polarization depends both on current density (cd) and on the nature of the reaction. For a given value of current density, some reactions exhibit high polarization and others exhibit low polarization. The term "slow reaction" is used for reactions associated with high polarization; for them, low ("normal") values of polarization can only be attained at very low current densities. Low values of polarization are typical for "fast reactions." Thus, the value of current density at a particular value of polarization, or the value of polarization at a particular value of current density, quantitatively characterize the relative rate of an electrochemical reaction.

In the case of redox reactions, polarization also depends on the nature of the nonconsumable electrode at which a given reaction occurs (for the equilibrium potential, to the contrary, no such dependence exists). Hence the term "reaction" will be understood as "reaction occurring at a specified electrode."

6.1.2. Electrode Polarization

In the electrochemical literature, the concept of "electrode polarization" has three meanings:

(i) the phenomenon of change in electrode potential under current flow,

(ii) an operation performed by the experimenter aiming at obtaining a potential change by passing current of a suitable strength and direction, and

(iii) the quantitative measure,[1] ΔE, of the shift of electrode potential E_i relative to the equilibrium value E_0 which occurs under current flow [cf. Eq. (2.28)].

In the case of anodic currents ΔE has positive values; in the case of cathodic currents it has negative values. In expressions such as "high polarization" and "the polarization increases," the absolute value of polarization is implied in the case of cathodes.

The value of polarization defined by Eq. (2.28) is referred to a calculated value of equilibrium potential of the reaction, rather than to the electrode's effective open-circuit potential, when the latter is not the equilibrium potential. Sometimes a thermodynamic calculation of the equilibrium potential is not possible, for instance when several electrode reactions occur simultaneously. In this case one either uses, conditionally, the concept of a polarization which via Eq. (2.28) refers to the effective open-circuit potential, or (since the latter is often irreproducible) one simply discusses electrode potentials at specified current densities rather than the potential shift away from some original value.

When currents flow in galvanic cells, the polarization phenomena which arise at any one of the two electrodes are independent of the prop-

---

[1]The value of polarization depends on the reaction occurring at the electrode as well as on the nature of the electrode (see Section 6.1.4). For the quantity ΔE, the term "overpotential" (η) is widely used abroad.

erties of the second electrode, and of the processes occurring there. Therefore, when studying these phenomena one considers the behavior of each
electrode individually.

6.1.3. Overall and Partial Reaction Currents

It has been shown in Section 2.2.4 that at the equilibrium potential,
the net (external) current i is zero, but partial current densities $\overrightarrow{i}$ and $\overleftarrow{i}$ of
the anodic and cathodic reaction exist for which the relation $\overrightarrow{i} = |\overleftarrow{i}| = i^0$
holds, where i^0 is the exchange current density (we recall that the net
current for the cathode and the current of the cathodic partial reaction have
negative values). The value of $\overrightarrow{i}$ increases, and that of $|\overleftarrow{i}|$ decreases, when
the potential is made more positive; but $|\overleftarrow{i}|$ increases and $\overrightarrow{i}$ decreases
when the potential is made more negative. The net current density i is
the algebraic sum of the partial current densities:

$$i = \overrightarrow{i} + \overleftarrow{i}. \tag{6.1}$$

When anodic polarization is appreciable the reverse (cathodic) partial
cd becomes exceedingly low and we practically can assume that $i \approx \overrightarrow{i}$;
when cathodic polarization is appreciable we can assume that $i \approx \overleftarrow{i}$.

Thus, the total range of potential can be divided into three regions:
one region at low values of polarization (to both sides of the equilibrium
potential), where the two partial reactions occur at comparable rates, and
two regions at high values of anodic and cathodic polarization, where the
reverse partial reactions can be neglected.

6.1.4. The Different Types of Polarization

Different reasons exist for the development of polarization attending
current flow at electrodes, and hence different types of polarization.

Electrode reactions are heterogeneous, since they occur at interfaces
between dissimilar phases. During current flow the surface concentrations
$c_{S,j}$ of the substances involved in the reaction change relative to the initial (bulk) concentrations $c_{V,j}$. Hence the value of equilibrium potential
defined by the Nernst equation changes, and a special type of polarization
arises where the shift of electrode potential is due to a change in equilibrium potential of the electrode. The surface concentrations which are
established are determined by the balance between electrode reaction rates
and the supply or elimination of each substance by diffusion [Eq. (4.21)].

Hence this kind of polarization, ΔE_d, is called *diffusional concentration polarization* or simply concentration polarization, also concentration overpotential (here we must take into account that another type of concentration polarization or concentration overpotential exists which is not tied to diffusion processes; see Section 13.5).

Other kinds of polarization are caused by specific features in the various steps of the electrochemical reaction which produce a potential shift relative to the effective equilibrium potential (i.e., that which already accounts for the prevailing values of surface concentrations). These kinds of polarization, which may differ in character, are jointly termed *activation polarization* or activation overpotential. The absolute value of activation polarization, $|\Delta E_k|$, is sometimes incorrectly called the overvoltage (this term should be reserved for the complete cell, see Section 2.5.3).

When concentration changes affect the operation of an electrode while activation polarization is not present (Section 6.3), the electrode is said to operate in the diffusion mode (under diffusion control), and the current is called a diffusion current i_d. When activation polarization is operative while marked concentration changes are absent (Section 6.2), the electrode is said to operate in the kinetic mode (under kinetic control), and the current is called a reaction or kinetic current i_k. When both types of polarization are operative (Section 6.4) the electrode is said to operate in the mixed mode (under mixed control).

The polarization equation describes polarization as a function of current density. In the case of concentration polarization the form of the polarization equation is unrelated to the nature of the reaction or the electrode. In the case of activation polarization the parameters of the polarization equations decisively depend on the nature of the reaction. At identical values of current density and otherwise identical conditions, the values of polarization for different reactions will vary within wide limits, from less than 1 mV to more than 2 or 3 V. However, these equations still have common features. A relatively simple set of equations is obtained for simple redox reactions of the type

$$\text{Red} \rightleftarrows \text{Ox} + ne^- \qquad\qquad (6.2)$$

[e.g., reaction (1.31) between Fe^{2+} and Fe^{3+} ions] for which it is typical that (i) the reaction only involves the transfer of electrons (one or several), and substances other than the principal components are not involved, (ii) the stoichiometric numbers of the components are unity, (iii) the reaction rate is proportional to the reactant concentration, (iv) the reaction occurs in a single step without the formation of intermediates, and (v) solid or gaseous phases are not produced or consumed in the reaction. Simple polarization functions are observed for some of the more complicated reactions, but in most of them the equations are highly complex.

Here we shall consider the polarization equations for simple redox reactions and for reactions which are similar to them. The special features of more complex reactions will be discussed in Chapters 13 to 15.

6.2. THE LAWS OF ACTIVATION POLARIZATION

6.2.1. Polarization Equations

At high values of polarization (the exact limits of the corresponding region will be indicated below) the relation between activation polarization and current density can often be written in the form

$$\pm\Delta E = a + b\ln|i| = a + b'\log|i|, \tag{6.3}$$

where a and b are constants (in volts), and $b' = 2.303b$ (here and in many later equations, the plus sign refers to the anode and the minus sign to the cathode; since the argument of a logarithm cannot have a negative value, the equations are written with $|i|$ instead of i).

Such a "semilogarithmic" polarization function was first established by Julius Tafel in 1905 for cathodic hydrogen evolution at a number of metal electrodes, and is known in the electrochemical literature as the *Tafel equation.*

Because of the logarithmic relation, polarization depends more strongly on parameter a than on parameter b. The parameter a, which is the value of polarization at a current density of 1 A/m^2, assumes values which for different reactions range from 0.03 to 2–3 V. Parameter b, which is the Tafel slope, changes within much narrower limits; in many cases at room temperature $b \approx 0.05$ V and $b' \approx 0.115$ V (or roughly 0.12 V).

For reasons which will become clear later on, the slopes are often written in the form

$$b \equiv RT/\beta F \quad \text{and} \quad b' = 2.303 RT/\beta F \tag{6.4}$$

where β is a dimensionless coefficient termed the *transfer coefficient.* For the particular value of $b = 0.05$ V mentioned above, this coefficient has a value of 0.5.

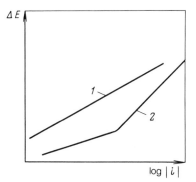

Fig. 6.1. Polarization curves in the region of
high polarization: (1) Tafel; (2) complex.

Equation (6.3) can also be written in a form where the current density
is a function of polarization:

$$i = \pm B \exp(\pm \beta F \Delta E / RT), \qquad \text{where } B \equiv \exp(-aF/RT) \quad (6.5)$$

In a number of cases the potential of the electrode under current flow
is used instead of electrode polarization in the polarization equations for
high values of polarization:

$$\pm E_i = a^* + b \ln |i| \quad \text{or} \quad i = \pm B^* \exp(\pm \beta F E_i / RT). \quad (6.6)$$

Now, in contrast to the earlier relations, the values of constants a^* and B^*
depend not only on the reaction being considered but also on the reference electrode against which the potential is measured; only the value of
constant b remains unchanged.

The polarization relations found in the region of high polarization are
usually plotted semilogarithmically as ΔE against $\log |i|$ (Fig. 6.1). These
plots are straight lines, the so-called Tafel lines (curve 1 in Fig. 6.1), when
relation (6.3) holds.

More complicated polarization functions are found at many real electrodes in the region of high polarization. Sometimes several Tafel sections
can be distinguished in an actual polarization curve (curve 2 of Fig. 6.1);
each of these sections has its own characteristic values of parameters a
and b (or B and β).

In the region of low polarization the values of activation polarization
are usually proportional to current density:

$$\Delta E = \rho i. \quad (6.7)$$

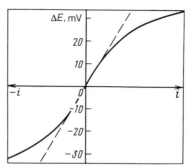

Fig. 6.2. Polarization curves in the region of low polarization.

The proportionality factor ρ (units $\Omega \cdot m^2$) formally has the same function as the electric resistance (per unit cross-sectional area) in Ohm's law, hence sometimes it is called the reaction resistance. However, this "resistance" is not ohmic.

Equation (6.7) applies to anodic and cathodic currents. At low values of polarization the parameter ρ usually has the same value for anodic and cathodic currents, and the slope of the ΔE vs. i straight-line plots does not change at the coordinate origin (Fig. 6.2).

6.2.2. Influence of Reactant Concentrations

The specific rate of an electrode reaction depends not only on electrode polarization but also on the reactant concentrations. In the region of high polarization, this function is usually described in terms of the kinetic equation

$$i = \pm nFk\tilde{c}\exp(\pm\beta FE_i/RT), \qquad (6.8)$$

where k is the reaction rate constant, which depends on the reference electrode selected, and $\tilde{c} \equiv \prod c_j^{p_j} \equiv c_1^{p_1} c_2^{p_2} \ldots$ is the product of concentrations of the substances influencing the rate. The units in which rate constant k is stated depend on the numerical values of the p_j.

This equation is also used in the form

$$\pm E_i = a^* - b \sum p_j \ln c_j + b \ln |i| \qquad (6.9)$$

(in the following we shall write the quantities a^* and E_i without indices).

Parameter p_j defines the kinetic reaction order with respect to substance j. When $p_j = 1$ the reaction is a first-order reaction with respect to this substance, i.e., the reaction rate is proportional to its concentration. The exponent p_j can be larger or smaller than unity (fractional reaction orders). Sometimes it is even negative, i.e., the reaction rate decreases with increasing concentration of a given substance. The reaction rate is usually a function only of the reactant concentrations, though in individual cases the products also have an effect (usually inhibiting the reaction). Exponents p_j for the reactants may or may not coincide with the values of the stoichiometric coefficients ν_j. Sometimes the reaction rate is influenced by substances which are neither reactants nor products.

In the case of simple redox reactions of the type of (6.2), the kinetic equation becomes

$$i = \pm nFkc_j \exp(\pm \beta FE/RT); \tag{6.10}$$

here the constant k has the units of m/s.

Changes in reactant concentrations are important not only for reaction rates but also for the values of equilibrium potential. It is important to remember, therefore, whether electrode polarization or electrode potential is used in the kinetic equations. For instance, if for a certain anodic reaction the equation

$$i = nFkc_1^{p_1} \exp(\beta FE/RT) \tag{6.11}$$

is valid, while for the concentration dependence of equilibrium potential the expression

$$E_0 = \text{const} - (\nu_1 RT/nF) \ln c_1 \tag{6.12}$$

holds (the reactant in an anodic reaction is a reducing agent, hence the minus sign), the equation to be used with polarization is

$$i = nFk'c_1^{p_1^*} \exp(\beta F\Delta E/RT), \tag{6.13}$$

where $p_1^* \equiv p_1 - \nu_1/n$.

In formula (6.11) the influence of potential on reaction rate is clearly distinguished from that of concentration, while in formula (6.13) the influence of concentration appears twice, first explicitly and then implicitly through the equilibrium potential contained in the value of polarization. For this reason Eq. (6.11) is preferred when analyzing the influence of concentrations.

Thus, all electrochemical reactions can be characterized by the form of kinetic relation and by the set of coefficients k, β, p_j, etc. Particular values of the coefficients always hold for specific reactions, hence the corresponding indices should be appended to the coefficients. In the following we shall use a subscript m for an anodic reaction m (e.g., k_m, β_m, $p_{m,j}$), but we shall use a subscript $-m$ (k_{-m}, β_{-m}, $p_{-m,j}$) for the same reaction when it occurs in the opposite (cathodic) direction.

6.2.3. General Kinetic and Polarization Equations

In the region of high polarization the kinetic equations for partial current densities $\vec{\imath}$ and $\overleftarrow{\imath}$ coincide with the equation for the net anodic i_a or cathodic i_c current density, respectively:

$$\vec{\imath} = i_a = nFk_m\tilde{c}_m \exp(\beta_m FE/RT), \tag{6.14}$$

$$\overleftarrow{\imath} = i_c = -nFk_{-m}\tilde{c}_{-m} \exp(-\beta_{-m} FE/RT). \tag{6.15}$$

When the laws of the partial reactions are preserved throughout the whole range of potentials, a general kinetic equation can be written (which is valid both for the anodic and the cathodic currents):

$$i = \vec{\imath} + \overleftarrow{\imath} = nF[k_m\tilde{c}_m \exp(\beta_m FE/RT) - k_{-m}\tilde{c}_{-m} \exp(-\beta_{-m} FE/RT)]. \tag{6.16}$$

The expression for the exchange current density ($i^0 = \vec{\imath} = |\overleftarrow{\imath}|$) at the equilibrium potential E_0 becomes

$$i^0 = nFk_m\tilde{c}_m \exp(\beta_m FE_0/RT) = nFk_{-m}\tilde{c}_{-m} \exp(-\beta_{-m} FE_0/RT). \tag{6.17}$$

The kinetic equations (6.14) and (6.15), which are valid for the partial cd at all values of polarization and for the net cd at high anodic and cathodic polarization, can be written, with expression (6.17) for the exchange cd, as

$$i_a = \vec{\imath} = i^0 \exp(\beta_m F\Delta E/RT), \tag{6.18}$$

$$i_c = \overleftarrow{\imath} = -i^0 \exp(-\beta_{-m} F\Delta E/RT), \tag{6.19}$$

while

$$i = i^0[\exp(\beta_m F\Delta E/RT) - \exp(-\beta_{-m} F\Delta E/RT)] \tag{6.20}$$

is an equation which holds for the net cd over the entire potential range. It follows when Eqs. (6.5) and (6.18) are compared that the value of the empirical constant B coincides with that of exchange cd i^0, while the constant a in the Tafel equation is given by

$$a = -(RT/\beta F) \ln i^0. \qquad (6.21)$$

Equations (6.14) to (6.16), which contain the rate constants, the electrode potential, and the concentrations, are equivalent to Eqs. (6.18) to (6.20), which contain the exchange cd and the electrode's polarization. But in the second set of equations the concentrations do not appear explicitly; they enter the equations through the values of exchange cd and equilibrium potential. By convention, the equations of the former type will be called kinetic, and those of the latter type will be called polarization equations.

Polarization equations are convenient when (i) the measurements are made in solutions of a particular constant composition and (ii) the equilibrium potential is established at the electrode, and the polarization curve can be measured both at high and low values of polarization. The kinetic equations are more appropriate in other cases, when the equilibrium potential is not established (e.g., for noninvertible reactions, or when the concentration of one of the components is zero), and also when the influence of component concentrations on reaction kinetics is of interest.

6.2.4. Relations between the Parameters of the Forward and Reverse Process

Different electrode reactions will occur independently, and their kinetic coefficients are unrelated. But for the forward and reverse process of a given reaction such a correlation should exist, since at the equilibrium potential the corresponding partial current densities assume equal values.

Solving Eq. (6.17) for E_0 we find

$$E_0 = \frac{RT}{(\beta_m + \beta_{-m})F} \ln(k_{-m}/k_m) + \frac{RT}{(\beta_m + \beta_{-m})F} \ln(\tilde{c}_{-m}/\tilde{c}_m). \quad (6.22)$$

This equation can be compared with the thermodynamic Nernst equation for the equilibrium potential of the system concerned, which when written in terms of the component concentrations is

$$E_0 = E^0 + (RT/nF) \ln \prod_j c_j^{\tilde{\nu}_j}. \tag{6.23}$$

This comparison yields the following important relations between the kinetic coefficients of the forward and reverse process:

$$\beta_m + \beta_{-m} = n, \tag{6.24}$$

$$k_{-m}/k_m = \exp(nFE^0/RT), \tag{6.25}$$

$$p_{-m,j} - p_{m,j} = \tilde{\nu}_j. \tag{6.26}$$

These equations show that, while the kinetic coefficients of an individual reaction can assume any value, the coefficients of its forward and reverse process are always interrelated. The relation between the standard equilibrium potential E^0 and the rate constants k_m and k_{-m} is analogous to the well-known physicochemical relation between equilibrium constant K and the rate constants of the forward and reverse process.

In one-electron reactions ($n = 1$), coefficients β often assume values $\beta_m = \beta_{-m} = 0.5$, even though Eq. (6.24) is consistent with any value of β which is in the interval of $0 \geq \beta_m \geq 1$.

6.2.5. Relation between the Kinetic Parameters in the Regions of Low and High Polarization

In the region of low polarization where $|\Delta E| < RT/F$, the exponential terms of Eq. (6.20) can be expanded into series, and it will suffice to retain the first two terms of each series: $\exp(\pm y) \approx 1 \pm y$. As a result, when allowing for Eq. (6.24) we obtain

$$i = i^0(nF/RT)\Delta E. \tag{6.27}$$

Thus, while the relation between the partial current densities and potential is exponential, in the region of low polarization a linear relation is obtained between polarization and the net cd owing to a superposition of the currents of the forward and reverse process. At $|\Delta E| = 10$ mV the error introduced by the above approximation will be between 1 and 20% depending on the relative values of β_m and β_{-m}; it becomes even smaller with decreasing polarization. Hence we can by convention consider the interval of polarization values between -10 and 10 mV as that of low polarization where the linear relation (6.7) is valid.

It follows from Eq. (6.27) that the kinetic parameter ρ of Eq. (6.7) is related to the exchange cd as

$$\rho \equiv (d\Delta E/di)_{\Delta E=0} = (RT/nF)(1/i^0). \qquad (6.28)$$

In the regions of high anodic and cathodic polarization we can use Eqs. (6.18) and (6.19), respectively. The error introduced when the reverse process is neglected is 5% for a polarization of 80 mV (for $n = 1$), and it decreases as the polarization increases further. The value of 80 mV can be regarded as the lower limit of the region of high polarization. For reactions with $n = 2$ this limit drops to 40 mV.

In the intermediate regions of moderate polarization (between 10 and 80 mV) we must use the polarization equation (6.20) in its general form.

An analysis of Eq. (6.20) shows that for $n = 1$ and $\beta = 0.5$ the following relation exists between polarization and the ratio i/i^0:

i/i^0	0.04	0.1	0.4	1	4	10
$|\Delta E|$, mV	1	2.5	10	24	72	110

Thus, at current densities lower than 4% of i^0, the activation polarization is very low (less than 1%) and can practically be neglected. The linear section of the polarization curve extends up to current densities which are 40% of i^0. At current densities higher than $4i^0$, the semilogarithmic polarization relation is observed.

Figure 6.3a shows curves where the partial current densities $\vec{i}$ and $\overleftarrow{i}$ and the net current densities i_a or i_c are plotted against polarization. Figure 6.3b shows the same curves plotted semilogarithmically. We clearly see the region of low polarization values where ΔE is a linear function of i, the regions of high polarization values where the relation is semilogarithmic, and also the corresponding intermediate regions of moderate polarization values.

The straight lines for the partial cd $\vec{i}$ and $\overleftarrow{i}$ in Fig. 6.3b intersect at the equilibrium potential ($\Delta E = 0$). The value of cd corresponding to the point of intersection is that of the exchange cd i^0, according to Eq. (6.17). It follows that the exchange cd can be determined when the linear sections of the anodic or cathodic polarization curve, which have been measured experimentally and plotted as log i against ΔE, are extrapolated to the equilibrium potential. Moreover, according to Eq. (6.27) the exchange cd can be determined from the slope of the polarization curve near the equilibrium potential when the curve is plotted as i against ΔE.

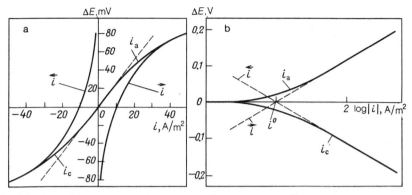

Fig. 6.3. Potential dependence of the anodic ($\vec{\imath}$) and cathodic ($\overleftarrow{\imath}$) partial current densities as well as of the anodic (i_a) and cathodic (i_c) net current densities.

6.2.6. Concentration Dependence of the Exchange Current Density

Substituting into Eq. (6.17) the relation (6.23) between equilibrium potential and component concentrations, we obtain

$$i^0 = nFk_m \exp(\beta_m FE^0/RT) \prod c_j^{(p_{m,j}+\beta_m \nu_m/n)} \tag{6.29}$$

or the analogous equation with the parameters k_{-m}, β_{-m}, and $p_{-m,j}$. Allowing for Eqs. (6.24) to (6.26) we can transform these expressions to

$$i^0 = nFk_m^0 \prod c_j^{(1/n)(\beta_m p_{-m,j}+\beta_{-m} p_{m,j})}, \tag{6.30}$$

where

$$k_m^0 = k_m^{(\beta_{-m}/n)} k_{-m}^{(\beta_m/n)}. \tag{6.31}$$

The standard rate constant k_m^0 characterizes both the rate of the forward process and that of the reverse process. Its value is independent of the reference electrode selected, in contrast to what holds true for the values of k_m and k_{-m}, and it is also independent of the component concentrations, in contrast to what holds true for the exchange cd. Therefore, this constant is an unambiguous characteristic of the kinetic properties exhibited by a given electrode reaction. Sometimes the parameter i^{00} is used for the same purposes; it is the standard exchange cd defined as the value of exchange cd at component concentrations of unity, and numerically it is equal to nFk_m^0.

We can see from Eq. (6.30) that the exchange cd of the reaction increases with increasing concentrations of all reaction components. In the case of reactions of the type of (6.2) the expression for the exchange cd assumes the simpler form

$$i^0 = nF k_m^0 c_{\text{Red}}^{(\beta_{-m}/n)} c_{\text{Ox}}^{(\beta_m/n)}. \tag{6.32}$$

6.2.7. Brief Notation

A more compact notation, which is defined below, will be used in order to simplify some of the kinetic equations to be discussed.

(a) A general kinetic reaction parameter which includes the rate constant and the potential dependence of reaction rate:

$$h_m \equiv k_m \exp(\beta_m F E / RT); \qquad h_{-m} \equiv k_{-m} \exp(-\beta_{-m} F E / RT). \tag{6.33}$$

The particular values of h_m and h_{-m} at the equilibrium potential E_0 will be designated as h_m^0 and h_{-m}^0, respectively. Parameters h_m and h_{-m} do not depend on the reference electrode selected, in contrast to constants k_m and k_{-m}.

(b) The system's nonequilibrium factor

$$\gamma_m \equiv h_m / h_m^0 \equiv \exp(\beta_m F \Delta E / RT),$$
$$\gamma_{-m} \equiv h_{-m} / h_{-m}^0 \equiv \exp(\beta_{-m} F \Delta E / RT). \tag{6.34}$$

At the equilibrium potential $\gamma_m = \gamma_{-m} = 1$.

Using this notation we can write the general kinetic equation as

$$i = nF(h_m \tilde{c}_m - h_{-m} \tilde{c}_{-m}), \tag{6.35}$$

the polarization equation as

$$i = i^0(\gamma_m - \gamma_{-m}), \tag{6.36}$$

and the expression for the exchange cd as

$$i^0 = nF h_m^0 \tilde{c}_m = nF h_{-m}^0 \tilde{c}_{-m}. \tag{6.37}$$

6.3. DIFFUSIONAL CONCENTRATION POLARIZATION

6.3.1. Solutions with Excess Foreign Electrolyte

Under the effect of pure concentration polarization, when activation polarization is absent, the electrode potential retains an equilibrium value, but this is a value tied to the variable nonequilibrium values of surface concentrations $c_{S,j}$:

$$E = E^0 + (RT/nF) \sum \bar{\nu}_j \ln c_{S,j}. \qquad (6.38)$$

It follows that concentration polarization is defined by the expression

$$\Delta E = (RT/nF) \sum \bar{\nu}_j \ln(c_{S,j}/c_{V,j}) \qquad (6.39)$$

(equations for concentration polarization are often used in conjunction with the diffusion and kinetic equations, hence it will be more convenient here to use concentrations rather than activities).

The surface concentrations which are attained as a result of balance between the electrode reaction rates and the rates of supply or escape of components by diffusion and migration are given by Eq. (4.22). Hence the overall expression for concentration polarization becomes

$$\Delta E = (RT/nF) \sum \bar{\nu}_j \ln[1 + (\text{sign } \bar{\nu}_j)i/|i_{l,j}|]. \qquad (6.40)$$

This expression, which contains i rather than $|i|$ as a parameter, is valid both for anodic and cathodic ($i < 0$) currents.

For simple reactions of the type of (6.2), when $\nu_{Ox} = \nu_{Red} = 1$, the equation becomes much simpler:

$$\Delta E = (RT/nF) \ln \frac{1 + i/|i_{l,Ox}|}{1 - i/i_{l,Red}}. \qquad (6.41)$$

This last equation is readily solved for i:

$$\frac{1}{i} = \frac{1}{|i_{l,Ox}|[\exp(nF\Delta E/RT) - 1]} + \frac{1}{i_{l,Red}[1 - \exp(-nF\Delta E/RT)]}. \qquad (6.42)$$

In the region of low polarization where two terms in the series expansion of the exponentials are sufficient, it follows from this equation that

$$\Delta E = (RT/nF)(1/|i_{l,\text{Ox}}| + 1/i_{l,\text{Red}})i, \qquad (6.43)$$

i.e., we obtain a linear relation between current density and polarization. Comparison with Eq. (6.7) reveals that in the case of pure concentration polarization, the parameter ρ is related to the limiting current density:

$$\rho = (RT/nF)(1/|i_{l,\text{Ox}}| + 1/i_{l,\text{Red}}). \qquad (6.44)$$

Curve 1 of Fig. 6.4a shows a plot of cd against ΔE which corresponds to Eq. (6.42) with $n = 2$. At zero polarization the current is zero. Under anodic polarization the current tends toward a limiting value given by $i_{l,\text{Red}} = nF \varkappa_{\text{Red}} c_{V,\text{Red}}$ [cf. Eq. (4.16)]. We can see from Fig. 6.4b that the surface concentration $c_{S,\text{Red}}$ then falls to zero, while the value of $c_{S,\text{Ox}}$ increases to $c_{V,\text{Ox}} + (\varkappa_{\text{Red}}/\varkappa_{\text{Ox}})c_{V,\text{Red}}$ [cf. Eq. (4.20)]. Under cathodic polarization, similarly, the current tends toward a limiting value of $i_{l,\text{Ox}}$, the surface concentration of Ox falls to zero, and the surface concentration of Red increases to $c_{V,\text{Red}} + (\varkappa_{\text{Ox}}/\varkappa_{\text{Red}})c_{V,\text{Ox}}$.

Curve 1 in Fig. 6.4a is symmetric relative to the inflection point A. In this point $i = 1/2(i_{l,\text{Red}} - |i_{l,\text{Ox}}|)$, hence this point has been termed the half-wave point. According to Eqs. (4.16) and (4.20), the values of surface concentrations $c_{S,\text{Red}}$ and $c_{S,\text{Ox}}$ in this point are half the limiting values listed above. Substituting these values into the Nernst equation we obtain for the potential, $E_{1/2}$, of this point (the *half-wave potential*):

$$E_{1/2} = E^0 + (RT/nF)\ln(\varkappa_{\text{Red}}/\varkappa_{\text{Ox}}). \qquad (6.45)$$

The half-wave potential is independent of component concentrations. But because of the change in equilibrium potential, the value of polarization in this point depends on the component concentrations. The half-wave potential is usually close to E^0, since $\varkappa_{\text{Red}}$ and $\varkappa_{\text{Ox}}$ differ little in most cases.

Using the parameter $E_{1/2}$ (and allowing for the Nernst equation) we can rewrite Eq. (6.41) in the rather convenient form

$$E_i = E_{1/2} + (RT/nF)\ln \frac{|i_{l,\text{Ox}}| + i}{i_{l,\text{Red}} - i}. \qquad (6.46)$$

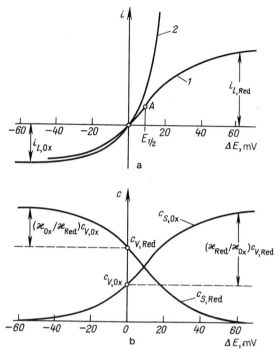

Fig. 6.4. (a) Curves of concentration polarization; (b) plots of the surface concentrations against polarization for the polarization curve 1; (1) $c_{V,\text{Red}} \approx c_{V,\text{Ox}}$; (2) $c_{V,\text{Red}} \gg c_{V,\text{Ox}}$.

Curves showing the concentration polarization for different concentration ratios are presented in Fig. 6.5.

Let us consider a few particular cases.

(i) The concentration of one of the components, and hence its limiting current density, is zero. In this case the Nernst equation is not applicable for the equilibrium potential; therefore, we must use a kinetic equation written in terms of potential rather than polarization. When an oxidizing agent is not present in the solution ($c_{V,\text{Ox}} = 0$), only anodic currents are possible in the system, and these produce an oxidizing agent. It then follows from Eq. (6.46) that

$$E = E_{1/2} + (RT/nF)\ln[i/(i_{l,\text{Red}} - i)] \qquad (6.47)$$

(Fig. 6.5, curve 1). Similarly, when a reducing agent is not present in the

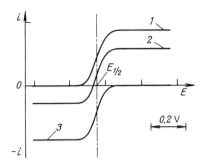

Fig. 6.5. Curves of concentration polarization (i vs. E): (1) $c_{V,Ox} \approx 0$; (2) $c_{V,Ox} < c_{V,Red}$; (3) $c_{V,Red} = 0$.

system ($c_{V,Red} = 0$) only cathodic currents are possible (curve 3), and

$$E = E_{1/2} - (RT/nF)\ln|i/(i_{l,Ox} - i)|. \tag{6.48}$$

(ii) The concentration of one of the components, e.g., the reducing agent, and its limiting current density are large, so that practically $c_{S,Red} =$ const, or a solid component with constant concentration (such as metallic zinc in the reaction: $Zn^{2+} + 2e^- \rightleftarrows Zn$) is involved in the reaction. In this case Eq. (6.41) becomes

$$\Delta E = (RT/nF)\ln(1 + i/|i_{l,Ox}|) \tag{6.49}$$

(Fig. 6.4, curve 2), and the polarization curve is of unusual shape in the region of high anodic cd where $i \gg |i_{l,Ox}|$ (the oxidizing agent is the anodic reaction product, hence this relation is possible). In this region

$$\Delta E = A + (RT/nF)\ln i, \qquad A \equiv -(RT/nF)\ln|i_{l,Ox}|. \tag{6.50}$$

i.e., in contrast to other cases of concentration polarization we obtain a linear relation between polarization and the logarithm of current density. This function is the analog of Eq. (6.3) with a coefficient b which has the value RT/nF.

A similar expression is obtained when a component which has constant concentration is the oxidizing agent.

6.3.2. Binary Electrolyte Solutions

The trends of behavior described above are found in solutions containing an excess of foreign electrolyte which by definition is not involved in the electrode reaction. Without this excess of foreign electrolyte, additional effects arise which are most distinct in binary solutions. An appreciable diffusion potential φ_d arises in the diffusion layer because of the gradient of overall electrolyte concentration which is present there. Moreover, the conductivity of the solution will decrease and an additional ohmic potential drop φ_{ohm} will arise when an electrolyte ion is the reactant and the overall concentration decreases. Both of these potential differences are associated with the diffusion layer in the solution, and strictly speaking are not a part of electrode polarization. But in polarization measurements, the potential of the electrode usually is defined relative to a point in the solution which, though not far from the electrode, is outside the diffusion layer. Hence, in addition to the true polarization ΔE, the overall potential drop across the diffusion layer, $\varphi = \varphi_d + \varphi_{\mathrm{ohm}}$, is included in the measured value of polarization, ΔE_{meas}.

Consider, as an example, the cathodic deposition of metal from a binary solution of the electrolyte $M_{\tau_+} A_{\tau_-}$ of concentration c_k. The concentration changes from $c_{V,k}$ to $c_{S,k}$ within the diffusion layer. The anion is not discharged, and its distribution in the diffusion layer is determined by the equilibrium condition of $\tilde{\mu}_{S,-} = \tilde{\mu}_{V,-}$. Substituting into this expression the electrochemical potential as a function of electrode potential and of anion concentration $c_- = \tau_- c_k$ [Eqs. (3.13) and (3.17)] and taking into account that $-z_- \tau_- = z_k$, we find

$$\varphi = -(\mu_{S,-} - \mu_{V,-})/z_- F = \tau_-(RT/z_k F) \ln(c_{S,k}/c_{V,k}). \qquad (6.51)$$

Comparing this expression with the value of concentration polarization according to Eq. (6.39) and taking into account that in this case $\nu_+ = 1$ and $n = z_+ = z_k/\tau_+$, we readily notice that in binary solutions the measured value of polarization is $1 + (\tau_-/\tau_+)$ times higher than that found, for the same concentration gradient of the reactant ion, when there is an excess of foreign electrolyte (we recall that according to the results of Section 4.3, the limiting cd in binary solutions are also higher by the same factor).

We can also calculate the individual values of φ_d (with the equations reported in Section 5.2) and of φ_{ohm}, by determining the concentration distribution in the diffusion layer, and from this, the distribution of solution

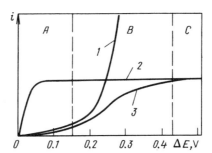

Fig. 6.6. Plots of i against ΔE for activation (1), concentration (2), and combined (3) polarization.

conductivity. The resulting combined value of φ coincides with the value determined from Eq. (6.51).

6.4. SUPERPOSITION OF CONCENTRATION AND ACTIVATION POLARIZATION

The kinetic and polarization equations described in Sections 6.1 and 6.2 have been derived under the assumption that the component concentrations do not change during the reaction. Therefore, the current density appearing in these equations is the kinetic current density i_k. Likewise, the current density appearing in the equations of Section 6.3 is the diffusion current density i_d. When the two types of polarization are effective simultaneously the real current density i (Fig. 6.6, curve 3) will be smaller than current densities i_k and i_d (Fig. 6.6, curves 1 and 2), for a given value of polarization.

Consider the joint effects of concentration and activation polarization in the instance of a simple reaction of the type of (6.2) for which the partial cd are proportional to the concentrations of the corresponding reactants. We shall assume in addition that an excess of foreign electrolyte is present in the solution. Allowing for the concentration changes we can write the polarization equation (6.36) as

$$i = i^0[\gamma_m(c_{S,\text{Red}}/c_{V,\text{Red}}) - \gamma_{-m}(c_{S,\text{Ox}}/c_{V,\text{Ox}})]. \qquad (6.52)$$

Replacing the surface concentrations in accordance with Eq. (4.22), we obtain

$$i = i^0[\gamma_m(1 - i/i_{l,\text{Red}}) - \gamma_{-m}(i/|i_{l,\text{Ox}}|)] \qquad (6.53)$$

or, solving for i, we have

$$i = \frac{i^0(\gamma_m - \gamma_{-m})}{1 + \gamma_m(i^0/i_{l,\text{Red}}) + \gamma_{-m}(i^0/|i_{l,\text{Ox}}|)}. \qquad (6.54)$$

In the region of low polarization where $\gamma_m \approx 1 + \beta_m F \Delta E/RT$ and $\gamma_{-m} \approx 1 - \beta_{-m} F \Delta E/RT$, this equation becomes

$$\Delta E = (RT/nF)(1/i^0 + 1/i_{l,\text{Red}} + 1/|i_{l,\text{Ox}}|)i, \qquad (6.55)$$

which for the kinetic parameter ρ yields the expression

$$\rho = (RT/nF)(1/i^0 + 1/i_{l,\text{Red}} + 1/|i_{l,\text{Ox}}|). \qquad (6.56)$$

Equation (6.56) is a generalization of Eqs. (6.28) and (6.44); it shows that the formal resistance is the sum of reaction resistance (the first term in brackets) and diffusion resistance (the second and third term).

Equation (6.54) directly yields the important relation

$$1/i = 1/i_k + 1/i_d, \qquad (6.57)$$

where i_k is defined by Eq. (6.36) and i_d by Eq. (6.42).

In those cases where $i_d \gg i_k$ (region A in Fig. 6.6), the real current density i practically coincides with the kinetic current density: $i \approx i_k$, and the electrode reaction is kinetically controlled. When $i_d \ll i_k$ (region C), we practically have $i \approx i_d$, and the reaction is diffusion-controlled. When i_d and i_k have comparable values, the electrode operates under mixed control (region B). The relative values of these current densities depend on the kinetic parameters (i^0, h_m, β_m, and others) and on potential (the values of γ_m and γ_{-m}).

Often it is claimed that at a given current density, the total polarization ΔE is the sum of pure concentration polarization ΔE_d and pure activation polarization ΔE_k. This is true only in the region of low polarization, where the values of polarization are proportional to current density. In other regions it is not true. In fact, the total polarization defined by Eq. (6.53) (Fig. 6.6, curve 3) is larger than the sum of the individual types of polarization which, for the same current density, are

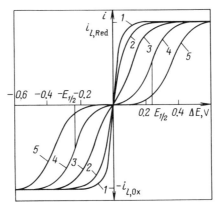

Fig. 6.7. Plots of i against ΔE for combined polarization with $i_{l,\text{Red}} = |i_{l,\text{Ox}}|$ and different ratios i^0/i_l: (1) ∞; (2) 1; (3) 0.1; (4) 0.01; (5) 0.001 (the values of $E_{1/2}$ are indicated in the case of curve 4).

defined by Eqs. (6.20) and (6.41) (curves 1 and 2). This is due to the fact that concentration changes affect activation and concentration polarization in different ways.

Figures 6.7 and 6.8 show polarization curves which correspond to the equations obtained, in the particular case where $\beta_m = \beta_{-m} = 0.5$ ($n = 1$) and $i_{l,\text{Red}} = |i_{l,\text{Ox}}|$ (since $\varkappa_{\text{Red}}$ and $\varkappa_{\text{Ox}}$ have similar values, this implies that $c_{V,\text{Red}} \approx c_{V,\text{Ox}}$). Curve 1 of Fig. 6.7 shows the case of pure concentration polarization ($i^0 \to \infty$), the other curves show the influence of decreasing exchange cd (decreasing reaction rate constants) which is revealed at constant values of the limiting current densities. Curve 1 of Fig. 6.8 corresponds to pure kinetic control ($i_{l,j} \to \infty$), while the other curves for which the exchange cd has the same value show the influence of a gradual decrease in limiting diffusion cd caused by decreasing diffusional transport constants $\varkappa_j$ (e.g., when the electrode is rotated more and more slowly, but not when the concentration is reduced, since this would alter the exchange cd).

It follows from the figures and also from an analysis of Eq. (6.54) that in the particular case being discussed, electrode operation is practically purely diffusion-controlled at all potentials when $i^0/i_j > 5$. By convention, reactions of this type are called reversible (or reactions thermodynamically in equilibrium). When this ratio is decreased a region of mixed control arises at low current densities. When the ratio falls be-

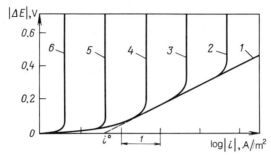

Fig. 6.8. Plots of ΔE against log i for combined polarization at i^0 = const and different ratios i^0/i_l: (1) 0; (2) 0.001; (3) 0.01; (4) 0.1; (5) 1; (6) 10.

low 0.05 we are then in a region of practically purely kinetic control. In the case of reactions for which the ratio has values of less than 0.02, the kinetic region is not restricted to low values of polarization but partly extends to high values of polarization. By convention, such reactions are called irreversible. We must remember that even highly irreversible reactions (with very low values of i^0/i_l) remain kinetically controlled, only up to current densities which are about 10% of the limiting current density. At higher cd the influence of diffusion processes always becomes evident as we approach the limiting diffusion cd, and electrode operation then is under mixed control.

It will be more convenient sometimes to describe the boundaries of the different regions in terms of the overall reaction rate constant k_m^0 [Eq. (6.31)] and the diffusional transport constant $\varkappa_j$. In our example, we can replace the ratio i^0/i_l by the ratio $k_m^0/\varkappa_j$ since $c_{V,\text{Red}} \approx c_{V,\text{Ox}} = c_V$ and, hence, $i^0 = nFk_m^0c_V$.

The values of $\varkappa_j$ which can be realized experimentally vary between $5 \cdot 10^{-6}$ m/s (natural convection) and $2 \cdot 10^{-4}$ m/s (rotating disk electrode at f = 10,000 rpm). Therefore, reactions for which $k_m^0 > 10^{-3}$ m/s will remain reversible whatever the stirring intensity. Such reactions are called completely reversible ("very fast"). Reactions with $k_m^0 < 10^{-7}$ m/s will always be irreversible, and are called completely irreversible ("very slow"). In the region of intermediate values of the constant, the character of the reaction will depend on stirring conditions. With other values of β_m and β_{-m} and of ratios $i_{l,\text{Red}}/|i_{l,\text{Ox}}|$ or $c_{V,\text{Red}}/c_{V,\text{Ox}}$, the boundaries between the different regions of electrode operation will shift slightly, but the overall picture of the phenomena remains the same.

Figure 6.7 shows a typical special feature of the polarization curves. In the case of reversible reactions (curve 1), the anodic and cathodic branch of the curve form a single step or "wave." In the case of irreversible reactions, two independent waves develop, an anodic and a cathodic one, each having its own inflection or half-wave point. The differences between the half-wave potentials of the anodic and cathodic waves will be larger the lower the ratio i^0/i_l.

For an anodic reaction occurring under mixed control in the region of high polarization ($\gamma_m \gg \gamma_{-m}$), Eq. (6.54) becomes

$$i = i_{l,\text{Red}} i^0 \gamma_m / (i_{l,\text{Red}} + i^0 \gamma_m) \tag{6.58}$$

or, when we solve for γ_m and change to polarization,

$$\Delta E = A' + (RT/\beta_m F) \ln[i/(i_{l,\text{Red}} - i)], \tag{6.59}$$
$$A' \equiv (RT/\beta_m F) \ln(i_{l,\text{Red}}/i^0).$$

Similarly for a cathodic reaction under the same conditions,

$$\Delta E = A'' - (RT/\beta_{-m} F) \ln|i/(i_{l,\text{Ox}} - i)|, \tag{6.60}$$
$$A'' \equiv (RT/\beta_{-m} F) \ln(i^0/|i_{l,\text{Ox}}|).$$

These equations resemble Eqs. (6.47) and (6.48) for pure diffusion control, but differ from them in that the prelogarithmic factors are n/β_m (n/β_{-m}) times larger.

For an investigation of the influence exerted by the concentrations of the reaction components, a kinetic equation of the form of (6.35) must be used. Substituting into this equation the values of the surface concentrations we obtain

$$i/nF = h_m c_{V,\text{Red}}(1 - i/i_{l,\text{Red}}) - h_{-m} c_{V,\text{Ox}}(1 + i/|i_{l,\text{Ox}}|) \tag{6.61}$$

or, when solving for i and taking into account that $i_{l,j} = nF\varkappa_j c_{V,j}$,

$$i/nF = \frac{h_m c_{V,\text{Red}} - h_{-m} c_{V,\text{Ox}}}{1 + h_m/\varkappa_{\text{Red}} + h_{-m}/\varkappa_{\text{Ox}}}. \tag{6.62}$$

When one of the components has zero concentration the corresponding term will disappear from the numerator of Eq. (6.62).

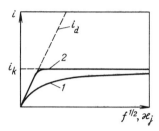

Fig. 6.9. Plots of the total current density against $f^{1/2}$ (or $\varkappa_j$) for first-order (1) and zeroth-order (2) reactions.

For a cathodic reaction occurring under mixed control in the region of high polarization ($h_m \ll h_{-m}$), we obtain from Eq. (6.62):

$$E = A - (RT/\beta_{-m}F)\ln|i/(i_{l,\text{Ox}} - i)|, \qquad (6.63)$$

$$A \equiv (RT/\beta_{-m}F)\ln(k_{-m}/\varkappa_{\text{Ox}}).$$

This equation resembles Eq. (6.60), but here k_{-m} is the reaction rate constant at the potential of the reference electrode used to measure the potential E^0, while in Eq. (6.60) the exchange cd i^0 is that at the equilibrium potential. An analogous equation is found for an anodic reaction occurring under the same conditions.

Curve 1 in Fig. 6.9 shows the influence of constant $\varkappa_j$ (or of parameters $\omega^{1/2}$ or $f^{1/2}$ which are proportional to it) on the current density at constant potential for a reaction with an intermediate value of k_m^0. Under diffusion control (low values of f) the current density increases in proportion to $f^{1/2}$. Later its growth slows down, and at a certain disk speed kinetic control is attained where the current density no longer depends on disk speed. The figure also shows curves for the kinetic current density i_k and the diffusion current density i_d.

All the equations reported above were derived for first-order reactions with respect to the reactant. The laws change when different reaction orders are involved. In particular, the plots of i against $f^{1/2}$ will be different in shape. At zero reaction order (Fig. 6.9, curve 2), since concentration changes can have no effect on reaction rate, the real current will either be a pure diffusion current (at low values of f) or a pure kinetic current, and the curve has a sharp break at $i_d = i_k$.

Chapter 7
TRANSIENT PROCESSES

7.1. EVIDENCE FOR TRANSIENT CONDITIONS

In electrochemical systems, a steady state during current flow implies that a time-invariant distribution of the concentrations of ions and neutral species, of potential, and of other parameters is maintained in any section of the cell. The distribution may be nonequilibrium, and it may be a function of current, but at a given current it is time-invariant.

The steady state is disturbed and the system exhibits transient behavior, when at least one of its parameters is altered under an external stimulus ("perturbation"). Transitory processes which adjust the other parameters set in ("response"), and at the end produce a new steady state. The time of adjustment (transition time, relaxation time) is an important characteristic of the system.

In electrochemical systems, transient processes are of major practical significance, because they are an efficient route for studying electrode reactions and phenomena (see Chapter 9). In addition, transient measurements are useful for analytical purposes (see Chapter 20).

The polarization functions described in Chapter 6 are valid in the steady state. These laws are not obeyed initially, when the current has just been turned on. Let us look more closely at the notion that "the current is turned on." In the steady state the current (or current density) and electrode polarization are inseparably linked; polarization can be the result of current flow, or vice versa. Things are different in the transient state. Here we can initially set either the current or the electrode potential. When "the current is turned on," i.e., when the current is set to a certain value (*galvanostatic conditions*) from the very beginning, then during a transition time the electrode potential will change from its initial value

147

at zero current to its final steady-state value. If to the contrary a certain value of electrode potential is set (*potentiostatic conditions*) the current will change during the transition time by certain laws. Electrochemical systems can be perturbed in other ways, too, e.g., by applying a potential which varies with time according to a particular law (*potentiodynamic conditions*) or by applying an alternating current. The different possibilities for perturbation will be considered in more detail in Chapter 9.

During the transition time a variety of processes of adjustment take place: development or change of an ohmic potential gradient, a change in edl charge density, the development of concentration gradients in the electrolyte, etc. Each of these processes has its own rate and its own characteristic time of adjustment.

Ohmic potential gradients are established practically instantaneously across conductors, certainly within times shorter than the response time of the fastest measuring devices, which is about 1 ns. They are caused by formation of a double layer, the charge of which is located on the opposite faces of the conductor in question.

Electrode polarization is associated with a change in edl charge density at the electrode surface. Other changes in surface state of the electrode are possible, too, e.g., the adsorption or desorption of different components, which also involve a consumption of electric charge. By convention, we shall describe this set of nonfaradaic processes as charging of the electrode surface.

The net current crossing the electrode at any time is the algebraic sum of the faradaic and various nonfaradaic currents. During the transition time part of the net current is consumed for surface-layer charging, and is not available for the primary electrode reaction. This part of the current is called the *charging current* (I_{ch}). It is highest at the start of the transition period, but toward the end of this period it falls to zero. The transition time of charging, t_{ch}, depends on the value of current and on the system, and may vary within wide limits (between 0.1 ms and 1 s).

Concentration gradients in the electrolyte layer next to the electrode surface will develop or change as a result of the primary electrode reaction. Therefore, the current associated with these changes is faradaic, though it is also transient and falls to zero when adjustment of the concentration profile is complete. These processes, unlike other transient processes, can be described in a quantitative way (Sections 7.2 and 7.3). The transition times of such processes as a rule are longer than 1 s.

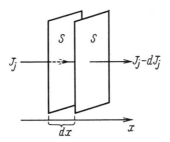

Fig. 7.1. Concerning the derivation of Eq. (7.1).

The different processes of adjustment have different values of the transition time; therefore, which of them become evident will depend on the method used in the measurements, and in particular on the time interval between perturbation and measurement of the response. It thus becomes possible to study the different transitory processes individually.

7.2. TRANSIENT DIFFUSION TO ELECTRODES OF LARGE SIZE

Under transient conditions the concentration distribution depends not only on the coordinate but also on time. The relevant functions can be found by considering the linear diffusion occurring along the x-axis in a volume element dV bounded by the two planes S which are a distance dx apart (Fig. 7.1); it is obvious that $dV = S\,dx$. The rate of concentration change $\partial c_j/\partial t$ in this volume is given by the ratio of $-S\,dJ_j$ (the decrease in total flux seen when advancing through the volume) to dV. Using Eq. (4.1) we thus have

$$(\partial c_j/\partial t)[= (\partial J_j/\partial x)] = D_j(\partial^2 c_j/\partial x^2) \qquad (7.1)$$

(Fick's second law). To solve this differential equation we additionally must know the boundary conditions, which depend on the conditions used in the measurements.

Solutions for a number of typical cases will be reported below. To simplify our task we shall use the assumption that reactant migration is not observed (a large excess of foreign electrolyte) and that the diffusion coefficients D_j do not depend on concentration.

7.2.1. Galvanostatic Conditions

At time $t = 0$ an electric current of constant strength begins to flow in the system. At this time the uniform initial concentration distribution is still not disturbed, and everywhere in the solution, even close to the electrode surface, the concentration is the same as the bulk concentration $c_{V,j}$. Hence the first boundary condition (for any value of x) is given by

$$c_j = c_{V,j} \quad \text{for } t = 0. \tag{7.2}$$

According to Eq. (4.47), the current density is related to the concentration gradient at the surface. Since the current density is constant, the gradient will be also constant and at all times t

$$\partial c_j / \partial x = -|i|\bar{\nu}_j / nFD_j \quad \text{for } x = 0. \tag{7.3}$$

The solution of differential equation (7.1) with boundary conditions (7.2) and (7.3) is

$$c_j(x,t) = c_{V,j} + \frac{|i|\bar{\nu}_j}{nFD_j}\left[2\sqrt{\frac{D_j t}{\pi}}\exp\left(-\frac{x^2}{4D_j t}\right) - x\,\text{erfc}\left(\frac{x}{2\sqrt{D_j t}}\right)\right]. \tag{7.4}$$

Here erfc(u) is the error function complement, a mathematical function of the argument u given by $1 - \text{erf}(u)$ where erf(u) is the error function or Euler–Laplace integral. This integral in turn is defined by the expression

$$\text{erf}(u) \equiv (2/\sqrt{\pi})\int_0^u \exp(-y^2)dy. \tag{7.5}$$

Variable y in the expression under the integral sign is an auxiliary variable; the value of the integral only depends on the limits of integration, i.e., on the value of u. The numerical values of the error function vary from zero for $u = 0$ to an upper limit of unity for $u \to \infty$ (this value is practically attained even for $u \approx 2$). Plots of functions erf(u) and erfc(u) are shown in Fig. 7.2.

It follows from Eq. (7.4) that at the electrode surface ($x = 0$) the concentration varies with time according to the function

$$c_{S,j} = c_{V,j} + 2|i|(\bar{\nu}_j/nF)\sqrt{t/\pi D_j}. \tag{7.6}$$

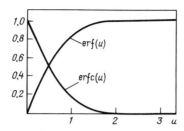

Fig. 7.2. Plots of functions erf(u) and erfc(u).

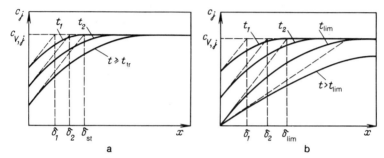

Fig. 7.3. The distributions of reactant concentration near the electrode after different times of galvanostatic operation: (a) $t_{tr} < t_{lim}$; (b) $t_{tr} > t_{lim}$.

The concentration change near the electrode surface gradually reaches solution layers further away from the electrode. In these layers, the rate of concentration change is the same as at the electrode, but there is a time lag. The concentration distributions found at different times are shown in Fig. 7.3. The diffusion-layer thickness δ_{tr} defined by Eq. (4.48) gradually increases with time; it follows from Eqs. (7.3) and (7.6) that

$$\delta_{tr} = 2\sqrt{D_j t/\pi}. \tag{7.7}$$

This equation holds only at short times when the thickness δ_{tr} is small as compared to the steady-state diffusion-layer thickness δ_{st} which will be attained under given experimental conditions, particularly when the solution is stirred. As soon as δ_{tr} attains the value of δ_{st}, the transitory processes end and a steady state is attained; there is no further change in concentration distribution with time (Fig. 7.3a). It follows from Eq. (7.7) that the transition time of the transient process

$$t_{tr} = \pi \delta_{st}^2 / 4 D_j. \tag{7.8}$$

In cases where the values of δ_{st}, and thus of t_{tr}, are large enough (without significant convection in the electrolyte solution), another limiting state is attained which is typical for galvanostatic conditions, and where the reactant concentration at the surface falls to zero (Fig. 7.3b). For the time t_{lim} required to attain this state (with $\bar{\nu}_j = -\nu_j$ for the reactant), Eq. (7.6) yields

$$t_{lim} = (nF/\nu_j)^2 \pi D_j c_{V,j}^2 / 4i^2. \tag{7.9}$$

When the surface concentration has fallen to zero, further current flow and the associated increase in δ_{tr} lead to a decrease in concentration gradient and in current (Fig. 7.3b, the curve for $t > t_{lim}$). Therefore, at $t > t_{lim}$ the original, constant current density can no longer be sustained. It follows that a steady state can only exist under the condition $t_{tr} < t_{lim}$.

When, after the attainment of zero surface concentration, a constant current density is artificially maintained from outside, the electrode potential will shift to a value such that a new electrochemical reaction involving other solution components can start (for instance, in aqueous solution the evolution of hydrogen or oxygen).

It follows from Eq. (7.9) that at a given concentration $c_{V,j}$ the product $i^2 t_{lim}$ is constant and independent of the set current density.

Using the value found for time t_{lim}, we can write expressions for the surface concentrations of reactants and products replacing Eq. (7.6) as

$$c_{S,r} = c_{V,r}[1 - (t/t_{lim})^{1/2}], \tag{7.10}$$

$$c_{S,p} = c_{V,p} + c_{V,r}(\nu_p D_r^{1/2}/\nu_r D_p^{1/2})(t/t_{lim})^{1/2}. \tag{7.11}$$

7.2.2. Potentiostatic Conditions

Equation (7.2) remains valid as the first boundary condition in this case. The surface concentrations, $c_{S,j}$, of the reactants will remain constant, in accordance with the Nernst equation, when the electrode potential is held constant during current flow (and activation polarization is absent). Hence the second boundary condition can be formulated as

$$c_{S,j} = \text{const} \qquad \text{for } x = 0. \tag{7.12}$$

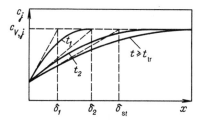

Fig. 7.4. The distributions of reactant concentration near the electrode after different times of potentiostatic operation; $t_1 < t_2 < t_{\text{lim}}$.

With these boundary conditions, and with the notation of $\Delta c_j \equiv c_{V,j} - c_{S,j}$, the differential equation (7.1) has the solution

$$c_j(x, t) = c_{S,j} + (2\Delta c_j/\sqrt{\pi})\,\text{erf}(x/2D_j t). \qquad (7.13)$$

Differentiating this equation with respect to x and setting $x = 0$ we find

$$(\partial c_j/\partial x)_{x=0} = \Delta c_j/\sqrt{\pi D_j t}, \qquad (7.14)$$

and for the current density

$$|i| = -(n/\bar{\nu}_j)FD_j(dc_j/dx)_{x=0} = -(n/\bar{\nu}_j)F\sqrt{D_j/\pi t}\,\Delta c_j. \qquad (7.15)$$

We see from these equations that the current density decreases with the inverse square root of time.

From Eq. (7.14) we obtain an expression for the effective transient diffusion-layer thickness:

$$\delta_{\text{tr}} \approx \sqrt{\pi D_j t}. \qquad (7.16)$$

The concentration distributions found at different times after the start of current flow are shown in Fig. 7.4. It is a typical feature of the solution obtained that the variable parameters x and t do not appear independently but always as the ratio $x/\sqrt{t}$. Like Eq. (7.16), this indicates that the diffusion front advances in proportion to the square root of time. This behavior arises because, as the diffusion front advances toward the bulk solution, the concentration gradients become flatter and thus diffusion slows down.

The transient process continues until the thickness δ_{tr} has attained the value of the diffusion-layer thickness δ_{st} corresponding to the applicable experimental conditions. Hence we obtain for the duration of this transitory process:

$$t_{tr} = \delta_{st}^2 / \pi D_j. \tag{7.17}$$

7.2.3. Use of Alternating Currents

In electrochemical measurements, a method often employed is that of applying an alternating current of low amplitude to the electrode. For sinusoidal ac $i = i_m \sin \omega t$ (i_m is the amplitude and ω the angular frequency), the concentration gradient at the surface varies according to the law

$$(\partial c_j / \partial t)_{x=0} = -(\tilde{\nu}_j / nF)(i_m / D_j) \sin \omega t \qquad \text{for } x = 0 \tag{7.18}$$

(during the anodic half period when $0 < \omega t < \pi$, the oxidized component is the reaction product for which $\partial c_j / \partial x < 0$).

With the boundary conditions (7.2) and (7.18), the differential equation (7.1) has the solution

$$c_j(x, t) = c_{V,j} + (\tilde{\nu}_j / nF)(i_m / \sqrt{\omega D_j}) \exp(-x / \sqrt{2D_j / \omega})$$
$$\times \sin[\omega t - x / \sqrt{2D_j / \omega} - \pi / 4]. \tag{7.19}$$

It follows from this equation that an ac passing through the solution gives rise to periodic concentration variations having the same frequency and reaching out from the electrode into the bulk solution with an amplitude decreasing in proportion to $\exp(-x / \sqrt{2D_j / \omega})$, i.e., the faster the higher the ac frequency.

For the surface concentration (at $x = 0$) we find

$$c_{S,j} = c_{V,j} + (\tilde{\nu}_j / nF)(i_m / \sqrt{\omega D_j}) \sin(\omega t - \pi / 4). \tag{7.20}$$

It follows from this equation that the effective diffusion-layer thickness can be described as

$$\delta \approx \sqrt{D_j / \omega}. \tag{7.21}$$

Diffusion is transient in each of the half periods, but the effects produced in one half period (e.g., an increase in concentration) are compensated by the effects produced in the following half period (in this example, a decrease in concentration). Therefore, the time-average thickness of the diffusion layer remains unchanged and the state of the system as a whole is quasi-steady.

7.3. TRANSIENT DIFFUSION TO ELECTRODES OF FINITE SIZE

In Chapter 4 and in the preceding section, cases of linear diffusion to a flat electrode were considered where the particles advance toward the entire surface along parallel lines normal to the surface. This implies that the electrode considered was large enough for edge effects ("lateral" diffusion to peripheral sections of the electrode) to be practically negligible. "Large enough" means, in this case, that the linear dimensions of the electrode (its width, height, and radius of curvature) were large as compared to the diffusion-layer thickness.

When this condition is not met and substances diffuse toward the surface along pathways converging on the surface, the total diffusion flux evidently will be larger.

Consider the specific example of a spherical electrode having the radius a. We shall assume that diffusion to the spherical surface occurs uniformly from all sides (spherical symmetry). Under these conditions it will be convenient to use a spherical coordinate system having its origin in the center of the sphere. Because of this symmetry, then, all parameters have distributions which are independent of the angle in space and can be described in terms of the single coordinate r, i.e., the distance from the center of the sphere.

In this coordinate system, Fick's second diffusion law becomes

$$\partial c_j / \partial t = D_j [(2/r) \partial c_j / \partial r + \partial^2 c_j / \partial r^2] \tag{7.22}$$

[this equation can be derived in exactly the same way as Eq. (7.1)].

Consider the case of transient diffusion at constant potential (constant surface concentration). The first boundary condition, (7.2), is preserved and the second boundary condition can be written (for any time t) as

$$c_{S,j} = \text{const} \quad \text{for } r = a. \tag{7.23}$$

With these boundary conditions, the differential equation (7.1) has the solution (for $r \geq a$)

$$c_j(r,t) = \Delta c_j(2a/r\sqrt{\pi})\,\mathrm{erf}[(r-a)/2\sqrt{D_jt}] + \Delta c_j[1-(a/r)]. \quad (7.24)$$

Differentiating this expression with respect to r and setting $r = a$ we find for the concentration gradient at the surface:

$$(\partial c_j/\partial r)_{r=a} = \Delta c_j[1/\sqrt{\pi D_j t} + 1/a], \qquad (7.25)$$

and for the current density:

$$|i| = -(n/\bar{\nu}_j)FD_j\Delta c_j[1/\sqrt{\pi D_j t} + 1/a]. \qquad (7.26)$$

We see that the expression for the current consists of two terms. The first term depends on time, and completely coincides with Eq. (7.15) for transient diffusion to a flat electrode. The second term is time-invariant. The first term is predominant initially, at short times t, where diffusion follows the same laws as for a flat electrode. During this period the diffusion-layer thickness is still small compared to radius a. At longer times t the first term decreases and the relative importance of the current given by the second term increases. At very long times t the current tends, not to zero as in the case of linear diffusion without stirring (when δ_{st} is large), but to a constant value. For the characteristic time required to attain this steady state (i.e., the time when the second term becomes equal to the first), we can write

$$t_{tr} \approx a^2/\pi D_j. \qquad (7.27)$$

At this time δ_{tr} has the approximate value of $a/2$.

It is a typical feature of the diffusion processes at electrodes of small size, which are reached by converging diffusion fluxes, that a steady state can be attained even without convection (e.g., in gelled solutions). Such electrodes, which have dimensions comparable to typical values of δ, are called microelectrodes.

Chapter 8
TYPES OF ELECTRODES USED

In the present chapter we shall look at certain types of electrodes used in laboratory practice and in industry.

According to their use, indicator and current-carrying electrodes are distinguished. The latter are intended for productive use of an electrode reaction, viz., for producing certain substances (in electrolyzers) or electrical energy (in chemical power sources). The current-carrying electrodes in electrolyzers include the working electrodes (WE) at which the desired products are formed, and auxiliary electrodes (AE) which merely serve to pass current through the working electrode.

Indicator electrodes are used, both for analytical purposes (in determining the concentrations of different substances from values of the open-circuit potential or from characteristic features of the polarization curves) and for the detection and quantitative characterization of various phenomena and processes (as electrochemical sensors or signal transducers). One variety of indicator electrode are the reference electrodes (RE), which have stable and reproducible values of potential and thus can be used to measure the potentials of other electrodes.

In practice, often several electrode reactions will occur simultaneously at electrodes. In this case secondary (or side) reactions are distinguished from the principal reaction ("principal" for a given purpose). In the electrolytic production of substances and when generating electrical energy on account of the Gibbs energy of a current-producing reaction, one usually tries to suppress all side reactions, so that the principal (desired) reaction will occur with the highest possible efficiency or current yields approaching 100%.

8.1. NONCONSUMABLE ELECTRODES

A nonconsumable electrode remains chemically inactive in redox reactions occurring at it, hence the equilibrium potentials of such reactions are independent of the electrode material. Yet for efficient reactions it is very important to select an appropriate nonconsumable electrode. The electrode material should meet a number of requirements. It should be stable chemically and electrochemically under the prevailing conditions, and should not corrode. It should be catalytically active, and secure a sufficiently high rate of the desired electrode reaction. Its catalytic activity should be selective, and all potential side reactions should be inhibited to the largest possible extent. Which electrode material is the best will depend on the nature of the reaction and on the reaction conditions.

The reactants in redox reactions are substances (both ionic and neutral) specifically introduced into the bulk electrolyte or the electrolyte region next to the electrode. When they are not present, the electrode should be ideally polarizable (see Section 2.5) and should not undergo dissolution within a certain potential range. Reactions in which the solvent is electrolytically decomposed should occur at it only at very negative or positive potentials; in aqueous solutions, for instance, cathodic hydrogen or anodic oxygen is evolved:

$$2H_2O + 2e^- \rightarrow 2OH^- + H_2, \qquad (8.1)$$

$$2H_2O \rightarrow 4H^+ + 4e^- + O_2. \qquad (8.2)$$

Reaction (8.1) is possible at potentials more negative than the potential of an equilibrium hydrogen electrode, and reaction (8.2) is possible at potentials more positive than the equilibrium potential of the oxygen electrode, which is 1.23 V relative to the potential of a reversible hydrogen electrode in the same solution (same pH). Therefore, the region of thermodynamic stability of water is restricted to a band of 1.23 V. Hydrogen and oxygen evolution actually occur at different rates and with different polarization at the various materials used as electrodes. At some materials polarization is quite significant, so that the region of kinetic stability of water is much wider.

Electrochemical redox reactions can be realized in a pure form, only in the region of potentials where water is stable at the electrode selected. Beyond this region the current yields of these reactions will decrease because of the simultaneous evolution of hydrogen or oxygen.

The analogous effects are found in solutions prepared with other solvents, which will also be reduced or oxidized electrochemically outside a certain region of potentials.

Platinum and other metals of the platinum group are rather universal materials for use as nonconsumable electrodes in cathodic and anodic reactions. These metals are stable over a wide range of potentials, both in acidic and in alkaline aqueous solutions, even in the presence of oxidizing agents and other aggressive chemicals. Their catalytic activity is high in many electrochemical reactions. Their defect is a relatively low polarization in hydrogen evolution, i.e., a narrow region of kinetic stability of the solvent. These metals are also active in other undesired reactions, i.e., they are insufficiently selective in a number of cases.

Platinum is widely used in the laboratory for making indicator and current-carrying electrodes. In the past platinum electrodes had been used in industry as anodes for synthesizing a number of inorganic and organic substances. In view of the high price of platinum, today one tries to replace these electrodes by other materials, but in many cases an adequate substitute for platinum is not available.

Cheaper and more readily available are the carbon and graphitic carbon materials, which can be used both in acidic and alkaline solutions. Their catalytic activities as a rule are inferior to those of the platinum metals. Their chemical stability decreases as positive potentials approaching the region of anodic oxygen evolution are attained, and processes start which lead to an electrochemical oxidation and to attack of their surfaces. At more negative potentials these materials have a satisfactory chemical stability. Graphite electrodes were widely used as anodes in electrolytic chlorine production, and it is only recently that they are replaced by more stable metal oxide electrodes.

For anodic reactions occurring at more positive potentials, the selection of suitable materials is difficult because of the ease of oxidation of most metal surfaces in this region of potentials. In a number of cases nonconsumable electrodes made of electronically conducting oxides are satisfactory, e.g., of lead(IV) oxide, nickel(III,IV) oxide, etc. Anodes made of oxides of titanium or other metals and activated with small amounts of ruthenium oxide or platinum metals became popular recently.

Mercury, tin, lead, and similar metals can be used as electrode materials for cathodic reactions. They exhibit high polarization in cathodic hydrogen evolution. Therefore, other substances can be reduced (without marked hydrogen evolution) at potentials much more negative than that of an equilibrium hydrogen electrode in the same solution. Nickel is often used as a cathode material in alkaline solutions; it is catalytically rather active, but its polarization in the hydrogen evolution reaction is low.

8.2. REACTING ELECTRODES

The most common example of reacting electrodes is that of metal electrodes in electrolytes containing ions of the same metal (electrodes of the first kind). Reactions

$$M \rightleftarrows M^{z+} + z_+ e^- \tag{8.3}$$

occurring at such electrodes are simple in their chemistry but complex in their mechanisms. In particular, anodic currents destroy the lattice of metals, and cathodic currents rebuild it.

Metal electrodes of the first kind are widely used in electrometallurgy for the cathodic production of different metals (zinc, sodium, and others), and for the electrochemical refining (purification) of metals by anodic dissolution and subsequent cathodic deposition. Cathodic metal deposition is the basis of all electroplating and electroforming. Anodic metal dissolution is used for the electrochemical machining and surface treatment of metals. For many of these processes (and particularly in electrometallurgy) molten salts rather than aqueous solutions are used as the electrolytes.

In aqueous solutions, simple ions of most metals are stable only when the solution is acidic (alkali metal ions are an exception). In neutral or alkaline solutions these ions hydrolyze, either partly to yield complex metal hydroxy ions, or completely to yield hydroxides:

$$M^{z+} \xrightarrow{+OH^-} [MOH]^{(z-1)+} \xrightarrow{+OH^-} [M(OH)_2]^{(z-2)+} \ldots \xrightarrow{+OH^-} M(OH)_z. \tag{8.4}$$

The electrode becomes an electrode of the second kind when the hydrolysis products are poorly soluble. At sufficiently high pH values the electrode reactions occur as

$$M + z_+ OH^- \rightleftarrows M(OH)_{z_+} + z_+ e^-. \tag{8.5}$$

In acidic solutions, the equilibrium electrode potential of the metal is independent of pH (the H^+ and OH^- ions are not involved in the electrode reaction), but in alkaline solutions the equilibrium potential depends on pH. This dependence can be described most clearly in the form of Pourbaix diagrams (see Section 3.7.3).

At reacting electrodes of the second kind, processes occur which are even more complicated than those occurring at electrodes of the first kind; the crystal structure of the solid reactants changes completely in the reaction, since the lattice of one of the components decomposes and that of another component is formed.

Electrodes of the second kind are widely used in chemical power sources (primary and secondary batteries) as well as in reference electrodes.

Electrodes where the materials involved in the reaction are oxides or other solid chemical substances, both in the reduced and oxidized state, are also considered as reacting electrodes. An example is the lead dioxide electrode which is used as the positive electrode in lead–acid batteries. The electrode reaction occurring at it can be written as

$$PbO_2 + 3H^+ + HSO_4^- + 2e^- \rightleftarrows PbSO_4 + 2H_2O. \qquad (8.6)$$

Under certain conditions reacting electrodes (metal or nonmetal) behave like nonconsumable electrodes. Thus, when a metal electrode immersed into a solution not containing ions of this metal is cathodically polarized, the only reaction will be hydrogen evolution at a metal surface remaining unchanged as a function of time. During anodic polarization of an oxide electrode (in its higher oxidation state) oxygen evolution is observed. In other regions of potential these electrodes become reacting electrodes.

8.3. REFERENCE ELECTRODES

One distinguishes practical and standard reference electrodes (RE). A standard RE is an electrode system of particular configuration, the potential of which, under specified conditions, is conventionally taken as zero in the corresponding scale of potentials, i.e., as the point of reference used in finding the potentials of other electrodes. Practical RE are electrode systems having a sufficiently stable and reproducible value of potential which are used in the laboratory to measure the potentials of other electrodes. The potential of a practical reference electrode may differ from the conventional zero potential of the standard electrode, in which case the potential of the test electrode is converted to this scale by calculation.

The standard RE universally adopted is the standard hydrogen electrode (SHE). At this electrode the equilibrium of the hydrogen evolution and ionization reaction (1.33) is established, and the electrode is under standard-state conditions for this reaction, i.e., it is in contact with a gas atmosphere containing hydrogen at a partial pressure of 1 atm and with a binary electrolyte solution which contains hydrogen ions and has a mean ionic activity $a_\pm$ of 1 mol/kg. The potential E of this electrode according to Eq. (3.61a) is identical with the value of the standard potential E^0 which, on the hydrogen scale, is adopted as zero.

It is important to point out that this definition applies at any temperature: It is assumed as a convention that for the standard hydrogen electrode, not only the potential but even the temperature coefficient of potential is zero (see Section 3.7.5).

In the past the potential of a hydrogen electrode in "one-normal" acid solution, i.e., with a hydrogen ion concentration of 1 mol/liter, had been regarded as the point of reference, and this electrode was called the normal hydrogen electrode. Nowadays this designation is sometimes used, though incorrectly, for the standard hydrogen electrode.

Hydrogen electrodes and different versions of electrodes of the second kind are in use as practical reference electrodes. They are designed as half cells. A version with hydrogen RE is shown in Fig. 8.1. Here a piece of platinum foil on which a layer of highly disperse (spongy) platinum has been deposited electrolytically serves as the electrode. Part of it is in contact with the gas phase, the other part dips into the electrolyte solution. Hydrogen is passed through the cell and solution for 20 to 30 min prior to measurements in order to remove all air including that dissolved in the solution. Gas and solution should both be carefully purified. The value of potential which is established under these conditions is highly reproducible (to $\pm 1 \cdot 10^{-5}$ V) and time-invariant. The slightest contamination with mercury or arsenic compounds, hydrogen sulfide, and certain other compounds strongly reduces the stability (they "poison" the electrode).

The potential of the hydrogen RE depends on the hydrogen partial pressure and on the hydrogen ion activity [Eq. (3.61a)]. The gas phase always contains water vapor in addition to hydrogen; its equilibrium pressure p_{H_2O} depends on temperature. Therefore, the hydrogen partial pressure p_{H_2} is given by the difference $p_0 - p_{H_2O}$, where p_0 is the total pressure of the gas mixture (usually equal to the ambient pressure), and the potential of the hydrogen RE is given by

$$E = (RT/2F) \ln[a_\pm^2/(p_0 - p_{H_2O})] \qquad (8.7)$$

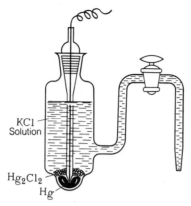

Fig. 8.1. Half cell with hydrogen reference electrode.

(by definition $E^0 = 0$ in the SHE scale; all pressures are stated in atmospheres, see the remark made in Section 3.6.2).

The electrolytes most often used are 0.1 to 1 M sulfuric acid solutions. The 3.4 m solution where $a_\pm \approx 1$ and the hydrogen electrode is the standard electrode has no advantages over solutions with other concentrations. When a reduction of diffusion potentials is required one can use sulfuric acid solutions with concentrations between 0.01 and 15 M and even alkaline solutions, e.g., 0.01 to 5 M KOH solutions. But since in alkaline solutions the ionic activity arises from OH^- rather than H^+ ions, it will be more convenient here to use the value $E_B^0(H_2O, OH^-)$ as the standard electrode potential [cf. Eq. (3.61b)], which at 25°C is −0.822 V, and which in contrast to the value $E_A^0(H^+, H_2)$ depends on temperature. Hydrogen RE are not used in unbuffered neutral solutions, since here the hydrogen ion concentration (and activity) readily changes, so that the potential is much less stable.

The potential of an electrode of the second kind is determined by the activity (concentration) of anions, or more correctly by the mean ionic activity of the corresponding electrolyte:

$$E = E^0 - \tau_-(RT/z_k F)\ln a_\pm \qquad (8.8)$$

[cf. Eq. (3.50)]. The most common among electrodes of this type are the calomel RE. In them, a volume of mercury is in contact with KCl solution which has a well-defined concentration and is saturated with calomel, Hg_2Cl_2, a poorly soluble mercury salt. The E^0 value of such an electrode is 0.2676 V (all numerical values refer to 25°C, and potentials are reported

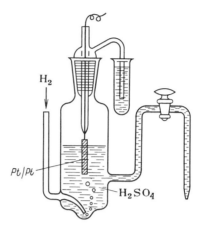

Fig. 8.2. Half cell with calomel reference electrode.

in the SHE scale). Three kinds of calomel electrode are in practical use; they differ in the KCl concentration and, accordingly, in the values of ionic activity and potential:

decimolar [decinormal], $c_{KCl} = 0.1\ M$, $E = 0.3337$ V,

molar [normal], $c_{KCl} = 1.0\ M$, $E = 0.2801$ V,

saturated (SCE), $c_{KCl} \approx 4.2\ M$, $E = 0.2412$ V.

Because of solubility changes, the saturated calomel RE has a large temperature coefficient (0.65 mV/K). Its main advantages are the ease of preparation (an excess of KCl is added to the solution) and the low values of diffusion potential at interfaces with other solutions (see Section 5.2). The potentials of calomel RE can be reproduced to $\pm 1 \cdot 10^{-4}$ V. These electrodes are very convenient for measurements in neutral solutions (particularly chloride solutions).

A half cell with calomel RE is shown in Fig. 8.2. Mercury purified by distillation and mixed with solid calomel is placed on the bottom of the cell. A length of platinum wire fused into glass tubing is fully immersed into the mercury in order to make electrical contact. The cell and connecting tubes are filled with KCl solution of the chosen concentration.

Similar designs are used for other RE on the basis of poorly soluble mercury compounds: (a) the mercury–mercurous sulfate RE with H_2SO_4 or K_2SO_4 solutions saturated with Hg_2SO_4, for which $E^0 = 0.6151$ V; (b) the mercury–mercuric oxide RE, for measuring electrode potentials

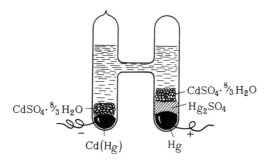

$CdSO_4 \cdot {}^{8}\!/_3 H_2O$

$CdSO_4 \cdot {}^{8}\!/_3 H_2O$

Hg_2SO_4

$Cd(Hg)$

Hg

Fig. 8.3. A Weston standard cell.

in alkaline solutions, with KOH solution saturated with HgO, for which $E_A^0 = 0.098$ V and $E_B^0 = 0.920$ V.

The silver–silver chloride RE consists of a small length of silver wire or piece of silver sheet coated with a thin layer of silver chloride (this layer can be deposited by anodic polarization of the silver in chloride-containing solution) and dipping into HCl or KCl solutions of defined concentration; its $E^0 = 0.2224$ V.

In alkaline solutions, sometimes the cadmium–cadmium oxide RE is used; its design is the same as that of the silver–silver chloride RE (a thin layer of cadmium oxide is formed on the surface of metallic cadmium). This electrode is quite simple to make and manipulate, but its potential is not very stable; $E_B^0 = +0.013$ V.

In selecting reference electrodes for practical use one should consider two criteria: reducing the diffusion potentials and eliminating interference of RE components with the system being studied. Thus, mercury-containing RE (calomel or mercury–mercuric oxide) are inappropriate for measurements in conjunction with platinum electrodes, since the mercury ions readily poison platinum surfaces. Calomel RE are also inappropriate for systems sensitive to chloride ions.

As a voltage (emf) standard in electrochemical measurements the so-called standard (or normal) cells are used, which are galvanic cells with highly reproducible, stable ocv values. The best-known is Edward Weston's standard cell proposed in 1892 and officially adopted for metrological purposes in 1908. These cells are sealed into H-shaped glass vessels (Fig. 8.3). The positive electrode is mercury in contact with a paste of crystalline Hg_2SO_4 and also $CdSO_4 \cdot \frac{8}{3}H_2O$, and the negative electrode is 6–12% cadmium amalgam in contact with crystals of $CdSO_4 \cdot \frac{8}{3}H_2O$. Saturated $CdSO_4$ solution is used as the electrolyte. These cells have an ocv of 1.01864 ± 0.00002 V at 20°C.

Chapter 9
ELECTROCHEMICAL RESEARCH TECHNIQUES

Experimental studies in electrochemistry deal with the bulk properties of electrolytes (conductivity etc.), equilibrium and nonequilibrium electrode potentials, the structure, properties, and condition of interfaces between different phases (electrolytes and electronic conductors, other electrolytes, or insulators), and the nature, kinetics, and mechanism of electrochemical reactions.

Electrochemical as well as nonelectrochemical techniques are used when studying these aspects. The electrochemical techniques are commonly used, too, in chemical analysis, in determining the thermodynamic properties of various substances, and for other purposes.

Methods with and without current flow can be distinguished among the common electrochemical techniques. In the methods without current flow one mainly measures the ocv of galvanic cells, while in the methods with current flow one studies the phenomena which occur in systems with steady or transient current flow.

The nonelectrochemical techniques include chemical (determining the identity and quantity of reaction products), radiotracer, optical, and other methods. Sometimes these methods are combined with the electrochemical methods, for instance, when studying the optical properties of an electrode surface while this is polarized.

In the present chapter we consider the methods used to measure ocv and electrode potentials and to study the kinetics of electrode reactions. Other methods will be described in Chapters 10 and 12.

Electrochemical measurements usually concern not a galvanic cell as a whole but one of the electrodes, the working electrode (WE). However, a complete cell including at least one other electrode is needed to measure

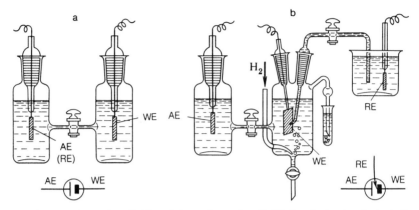

Fig. 9.1. Two-electrode cell (a) and three-electrode cell (b) for electrochemical studies.

the WE potential or allow current to flow. In the simplest case a two-electrode cell (Fig. 9.1a) is used for electrochemical studies.[1] The second electrode is used, either as the reference electrode (RE) or as an auxiliary electrode (AE) to allow current to flow. In some cases these two functions can be combined; viz., when the surface area of the auxiliary electrode is much larger than that of the working electrode, so that the current densities at the AE are low, it is practically not polarized, and thus can be used as RE.

Three-electrode cells (Fig. 9.1b) are more common for measurements involving current flow; they contain both an AE and a RE. No current flows in the circuit of the reference electrode, which therefore is not polarized. However, the ocv value which is measured includes the ohmic potential drop in the electrolyte section between the working and reference electrode. In order to reduce this undesired contribution from ohmic losses, one brings the connecting tube with the electrolyte of the reference electrode right up to the surface of the working electrode. To reduce screening of this surface the end of the tube is drawn out to a fine capillary, the *Luggin capillary*. The design and placement of this capillary are very important for measuring accuracy. It is necessary to bring the tip of the capillary as close as possible to the electrode, yet the capillary should

---

[1] Ways to symbolically represent two- and three-electrode cells in electric circuit diagrams are also shown in Fig. 9.1.

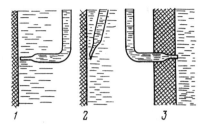

Fig. 9.2. Different designs of Luggin capillaries.

not screen the segment of electrode surface being measured or disturb the uniform current distribution. Different capillary designs are shown in Fig. 9.2. The degree of surface screening is more important in version 1 than in version 2; in version 3 there is practically no screening effect.

During the experiments the electrochemical cell is usually placed into a thermostat. When required the liquid electrolyte is agitated with a stirrer or a rotating WE is used.

Impurities can have an important influence on the properties of electrode/electrolyte electrochemical systems; even minor quantities of foreign material (both organic and inorganic) readily adsorb at the interface and strongly affect its properties. Therefore, the purity requirements for the chemicals used in electrochemical studies are very high. The chances for the electrode surface to become contaminated by impurities before and during the experiments must be reduced to the maximum possible extent. The measurements are conducted in cells made of glass, PTFE, or other materials not releasing impurities into the system being studied. The electrode and electrolyte must be kept from contacting rubber (stopcocks, tubing) and similar materials. The preliminaries should include a careful purification of the chemicals used (multiple distillation of the solvents, recrystallization and calcination of the salts), cleaning of solid electrode surfaces (degreasing, etching, trimming), and washing of the cell.

In many cases the system being studied must not come in contact with air oxygen, hence the measurements are conducted in an atmosphere of hydrogen or inert gas (argon, helium, etc.), and one must monitor the complete exclusion of air (or other undesirable gases) from the system. In order to create the gas atmosphere needed one adds a gas-inlet pipe to the lower part of the cell, and a pipe with a water seal as the gas outlet to the upper part (see Fig. 9.1b).

9.1. VOLTAGE AND ELECTRODE POTENTIAL MEASUREMENTS (POTENTIOMETRY)

An important step in measurements of electrode potentials is that of selecting a suitable reference electrode. Reference electrodes with electrolytes of the same nature and same (or similar) composition as that at the working electrode are used in order to reduce the liquid-junction potentials. During the measurements both electrodes must be at the same temperature.

An accuracy of 10^{-3} to 10^{-4} V is needed ordinarily in ocv measurements; an accuracy of 10^{-5} V is needed when determining thermodynamic parameters.

The device used to measure the ocv should not cause any current flow in the galvanic cell. Currents of 10^{-4} to 10^{-3} A would arise when connecting an ordinary permanent-magnet moving-coil voltmeter to the cell; with a circuit resistance of 100 Ω, the ohmic voltage drop would cause the results to be off by several tens of millivolts even without counting the error due to polarization. Therefore, electronic voltmeters with digital readout and very low current drain (10^{-8} to 10^{-14} A) are used for such measurements at present. In the past, compensating potentiometers based on the Wheatstone-bridge principle were used for these purposes.

9.2. STEADY-STATE POLARIZATION MEASUREMENTS

9.2.1. Special Technical Features

Steady-state measurements of polarization characteristics can be made when all transitory processes associated with changes in current or potential have ended. Here one does not count the relatively slow changes in the system's condition such as a gradual decrease in reactant concentration. The currents measured in the steady state are purely faradaic.

Steady-state measurements can be made over a wide range of current densities, e.g., between 10^{-4} and 10^4 A/m^2, provided the system being studied will sustain such current densities. The measurements become difficult at lower current densities because of longer transitory processes. For instance, when the electrode's edl capacitance is 0.2 F/m^2, approximately 10 min would be required at a nonfaradaic current density of 10^{-5} A/m^2 to shift the potential by 30 mV (where the charge to be supplied is 6 mC/m^2). The distorting effects caused by low-level impurities in

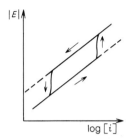

Fig. 9.3. Schematic shape of polarization curves
during an anodic and cathodic potential scan.

the system strongly increase during longer measurements. Ohmic factors (in the gap between the electrode and the tip of the Luggin capillary) and local temperature rise affect the measurements at high current densities.

In steady-state measurements at current densities such as to cause surface-concentration changes, the measuring time should be longer than the time needed to set up steady concentration gradients. Microelectrodes or cells with strong convection of the electrolyte are used to accelerate these processes.

Steady-state measurements can be made "pointwise" or continuously. In the first case the level of perturbation (current or potential) is varied discontinuously, and at some time after the end of transitory processes the response is measured. In the second case the perturbation level is continuously varied, but so slowly as not to disturb the system's steady state.

It is basically irrelevant in steady-state measurements in which direction the polarization curves are recorded, i.e., whether the potential is moved in the direction of more positive (anodic scan) or more negative (cathodic scan) values. But sometimes the shape of the curves is seen to depend on scan direction, i.e., the curve recorded in the anodic direction does not coincide with that recorded in the cathodic direction (Fig. 9.3). This is due to changes occurring during the measurements in the properties of the electrode surface (e.g., surface oxidation at anodic potentials) and producing changes in the kinetic parameters.

9.2.2. Galvanostatic and Potentiostatic Circuits

Steady-state measurements can be made both under galvanostatic and potentiostatic conditions. It is irrelevant for the results of the measurements whether the current or the potential was set first. But in certain

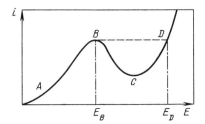

Fig. 9.4. Polarization (i vs. E) curve with
falling section.

cases in which the polarization (i vs. E) curve is nonmonotonic and includes a "falling" section (BC in Fig. 9.4), the potentiostatic method has
important advantages, since it allows the potential to be set to any point
along the curve and the corresponding current measured. But when the
galvanostatic method is used an increase in current beyond point B causes
a jump in potential to point D, i.e., the potential changes discontinuously
from the value E_B to the value E_D, and the entire intermediate part of
the curve is inaccessible.

Galvanostatic conditions can be realized very simply by connecting in
series with the cell an external power source having a much higher voltage
($\mathcal{E}_{\text{ext}}$) than the cell's ocv ($\mathcal{E}_{\text{cell}}$). The current I in the circuit is adjusted
with the aid of a high-resistance rheostat R, whereupon its value will be
($\mathcal{E}_{\text{ext}} - \mathcal{E}_{\text{cell}})/R$. Small fluctuations in cell voltage $\mathcal{E}_{\text{cell}}$ under transitory
conditions have little effect in view of the large value of $\mathcal{E}_{\text{ext}}$, so that there
is little change in current during the transition period. In practice various
special electronic instruments are used, viz., galvanostats maintaining the
current highly accurately constant whatever the changes in cell voltage.

When an external power source of low voltage but sufficiently high
output rating (almost no change in voltage under current drain) is used
and a resistor R is not present, the voltage at the cell terminals will be the
same as that at the terminals of the power source, i.e., practically constant
regardless of the level of current. This is the case of constant-cell-voltage
operation.

Potentiostatic conditions are realized with electronic potentiostats. The
potential of the working electrode is continuously monitored with the aid
of a reference electrode. When the potential departs from a set value,
the potentiostat will automatically adjust the current flow in the cell so
as to restore the original value of potential. Important characteristics of
potentiostats are their risetime and the maximum currents which they can

deliver to the cell. Modern high-quality potentiostats have risetimes of 10^{-5} to 10^{-6} s.

9.2.3. Calculation of the Kinetic Parameters

Measurements must be made under kinetic control or at least under mixed control of electrode operation, if we want to determine the kinetic parameters of electrochemical reactions. When the measurements are made under purely kinetic control, i.e., when the kinetic currents i_k are measured directly, the accuracy with which the kinetic parameters can be determined will depend only on the accuracy with which the polarization relation is measured. Measurements under mixed control require quantitative estimates of the diffusion currents i_d, e.g., through use of a rotating disk electrode. The overall accuracy here depends on the ratio of i_k to i_d and on the accuracy with which i_d was determined. For instance, with a ratio of $i_k/i_d = 3$ an error of $\pm 5\%$ in the determination of i_d according to Eq. (6.57) will produce an additional error of $\pm 15\%$ in the calculation of the kinetic current. Hence, this ratio should be regarded as the limit.

For a reliable calculation of coefficients β_m and β_{-m} from the potential dependence of kinetic currents, experimental data are needed in which the kinetic currents are varied by at least an order of magnitude. It follows that in at least one point the ratio i_k/i_d should not be higher than 0.3. In the case considered in Section 6.4 where $i_{l,\text{Red}} = |i_{l,\text{Ox}}|$, this corresponds to values of i^0/i_l or $k_m^0/\varkappa_j$ which are not higher than 0.15. The highest value of $\varkappa_j$ typically found in aqueous solutions is about $2 \cdot 10^{-4}$ m/s. It follows that steady-state methods can yield reliable kinetic parameters only for reactions in which $k_m^0 \leq 3 \cdot 10^{-5}$ m/s. At a component concentration of 10^{-2} M, this corresponds to exchange current densities $i^0 \leq 30$ A/m^2.

The possibilities for a determination of kinetic parameters for very slow reactions (e.g., with $k_m^0 < 10^{-12}$ m/s) are also limited, this time on account of the difficulties encountered in measurements at very low current densities.

In the region of mixed kinetics, graphical methods are often used in determining the kinetic current and corresponding parameters. In measurements with the rotating disk electrode, the experimental data are plotted as i^{-1} against $\omega^{-1/2}$ (Fig. 9.5). According to Eq. (4.52), the values of $\omega^{-1/2}$ are proportional to $1/\varkappa_j$ or $1/i_{l,j}$, hence experimental points

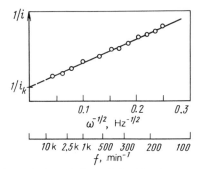

Fig. 9.5. Calculation of the kinetic current from experimental data obtained with a rotating disk electrode.

when thus plotted will fall on a straight line [see Eq. (6.57)]. The point of intersection of the straight line with the vertical axis ($\omega \to \infty$) immediately yields $1/i_k$, i.e., the value of the kinetic current.

In measurements under the conditions mentioned when deriving Eqs. (6.59), (6.60), or (6.63), it will be convenient to plot the experimental data as E against $\log|i/(i_{l,j} - i)|$. From the slope of the resulting straight line we can find the coefficient β_m, from the ordinate of the half-wave point (where $i = i_{l,j}/2$ and the logarithmic term becomes zero) we can find the value of constant A, and from this constant we can find the value of i^0 or h_m.

9.3. TRANSIENT (PULSE) MEASUREMENTS

Transient measurements are made before transitory processes have ended, hence the current in the system consists of the faradaic and of nonfaradaic components. Such measurements are made in order to determine the kinetic parameters of fast electrochemical reactions (by measuring the kinetic currents under conditions when the contribution of concentration polarization still is small) and also in order to determine the properties of electrode surfaces, in particular the edl capacitance (by measuring the nonfaradaic current).

In transient measurements, essentially a certain perturbation is applied to the electrode and then the response is recorded as a function of time. Usually the transition times are short (fractions of a second), and the

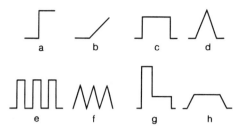

Fig. 9.6. Different types of perturbation for transient measurements.

transient measurements are performed very rapidly with automated data acquisition.

The transient techniques can be grouped according to different criteria (Fig. 9.6), as follows:

(i) According to the shape of the perturbation: a step (static perturbation: curve a), when the perturbing parameter is raised discontinuously to a new level, then remains constant; or a continuous (dynamic) variation of the same parameter occurring with a rate sufficiently high to upset the system's steady state; when the rate of variation is constant (curve b) the variation of the perturbation level is called linear (linear scan).

(ii) According to the number of perturbations: single perturbations (curves a and b); pulses (perturbation in the form of an excursion): square or rectangular (curve c) or triangular (curve d); cyclic perturbations (repetitive excursions: curves e and f); and complex pulses resulting from the combination of several of the kinds of pulse already mentioned: double-step (curve g) and trapezoidal (curve h).

(iii) According to the parameter controlled: when this is the potential we speak of potentiostatic and potentiodynamic methods; when it is the current we speak of galvanostatic and galvanodynamic methods (galvanodynamic methods are rarely used); in the coulostatic method a certain amount of charge is delivered to the electrode (e.g., from a capacitor) and the potential–time variation is followed.

For a practical realization of the above transient modes, special programming units or signal synthesizers are used which are integrated in the potentiostatic or galvanostatic equipment.

For the individual kinds of transient measuring techniques, special names exist but their terminology lacks uniformity. The potentiostatic techniques where the time-dependent current variation is determined are often called chronoamperometric, and the galvanostatic techniques where the potential variation is determined are called chronopotentiometric. For the potentiodynamic method involving linear potential scans, the term "voltammetry" is used, but often this term is used for other transient methods as well.

9.3.1. The Potentiostatic Method (Chronoamperometry)

A certain potential is applied to the electrode with the potentiostatic equipment, and the variation of current is recorded as a function of time. At the very beginning a large current flows, which is largely due to charging of the electrode's edl as required by the potential change. The maximum current and the time of edl charging depend, not only on the electrode system and size but also on the parameters of the potentiostat used. When this process has ended, mainly the faradaic component of current remains, which in particular will cause the changes in surface concentrations described in Section 7.2.

In a reversible process which occurs under diffusion control, the time-dependent drop of the faradaic current is due to the gradual increase in diffusion-layer thickness. According to Eq. (7.15) we have, for reactants ($\bar{\nu}_j = -\nu_j$):

$$|i_d| = (n/\nu_j)F\sqrt{D_j/\pi t}\,\Delta c_j. \qquad (9.1)$$

In an irreversible reaction which occurs under kinetic or mixed control, the boundary condition can be found from the requirement that the reactant diffusion flux to the electrode be equal to the rate at which the reactants are consumed in the electrochemical reaction [cf. Eqs. (6.35) and (7.15)]:

$$(1/\nu_j)D_j(\partial c_j/\partial x)_{x=0} = h_m c_{S,j}. \qquad (9.2)$$

With these boundary conditions, the differential transient-diffusion equation (7.1) has the solution

$$|i| = (n/\nu_j)F h_m \Delta c_j \exp(\lambda^2)\,\mathrm{erfc}(\lambda), \qquad (9.3)$$

where $\lambda \equiv h_m t^{1/2}/D_j^{1/2}$ is a dimensionless parameter.
It follows from Eqs. (9.1) and (9.3) that

$$|i/i_d| = \sqrt{\pi}\,\lambda \exp(\lambda^2)\,\mathrm{erfc}(\lambda). \qquad (9.4)$$

We can see when analyzing this equation that the right-hand side is smaller than unity and increases with increasing λ. For $\lambda \geq 5$ it tends toward unity, i.e., the reaction is practically reversible under the given conditions. Therefore, the kinetic reaction parameters (λ, and hence h_m) can be determined from the current decay curve, only when $\lambda < 5$, i.e., when $h_m t^{1/2} < 5 D^{1/2}$.

The parameters of reactions for which $h_m < 3 \cdot 10^{-4}$ m/s can be determined when the minimum time at which reliable measurements of the faradaic current are possible is assumed to be a few tenths of a second.

9.3.2. Potentiodynamic Method (Voltammetry)

A linear potential scan (lps) is applied to the electrode with the aid of the potentiodynamic equipment, i.e., a potential which has a constant rate of variation $v \equiv |dE/dt|$:

$$E = E_{\text{in}} \pm vt, \qquad (9.5)$$

where E_{in} is the initial potential (at $t = 0$); the plus sign is for the anodic direction, the minus sign for the cathodic direction.

The current is recorded as a function of time. Since the potential also varies with time, the results are usually reported as the potential dependence of current, or plots of i against E (hence the name voltammetry).

Curve 1 in Fig. 9.7 schematically shows the polarization curve recorded for an electrochemical reaction under steady-state conditions, and curve 2 shows the corresponding kinetic current i_k (the current in the absence of concentration changes). Unless the potential scan rate v is very low, there is no time for attainment of the steady state, and the reactant surface concentration will be higher than it would be in the steady state. For this reason the current will also be higher than the steady value (section AB of curve 3), though it is still lower than the purely kinetic current. As the time increases the concentration drop will become ever more important, and at some point this factor will begin to predominate over the accelerating effect of potential, i.e., the current will begin to decrease (section BC). Therefore, in the i vs. E curve, a maximum appears which is typical for this technique and has the coordinates i_{max} and E_{max}.

A theoretical analysis of the functions obtained is mathematically difficult, hence a simplified analysis is given here.

In reversible reactions, reactant surface concentrations are related to potential in an unambiguous way. For reactions of type $\text{Ox} + ne^- \rightleftarrows \text{Red}$,

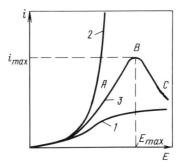

Fig. 9.7. Polarization curves: (1) steady-state; (2) without concentration polarization; (3) potentiodynamic.

it follows from the Nernst equation and Eq. (9.5) that

$$c_{S,Ox}/c_{S,Red} = \exp[nF(E_{in} - E^0)/RT]\exp(\pm nFvt/RT). \quad (9.6)$$

A differential equation with these boundary conditions was solved independently by Augustin Ševčík and John E. B. Randles in 1948. The expression obtained for the current is

$$i = \pm nF(nFD_j v/RT)^{1/2}c_{V,j}P(u), \quad (9.7)$$

where $u \equiv \pm nF(E - E_{1/2})/RT$, $E_{1/2}$ is the half-wave potential in the curve of concentration polarization [Eq. (6.45)], and j is the corresponding reactant, e.g., Red for anodic currents.

The function $P(u)$ is complicated; it is dimensionless and can be found numerically. Its shape as shown in Fig. 9.8 at once determines the shape of the i vs. E curve. The value of P in the maximum is 0.446. Hence we obtain for the current in the maximum at 25°C (after substituting numerical values for F, R, and T):

$$i_{max} = \pm 2.69 \cdot 10^5\, n^{3/2}(D_j v)^{1/2}c_{V,j}, \quad (9.8)$$

where the numerical factor has the units of C/mol · V$^{1/2}$.

The potential (in volts) corresponding to the maximum of function P is shifted relative to $E_{1/2}$:

$$E_{max} = E_{1/2} \pm 0.028/n. \quad (9.9)$$

Like $E_{1/2}$, the value of E_{max} is independent of the initial concentrations.

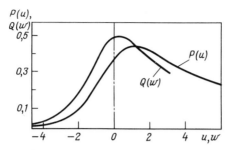

Fig. 9.8. Functions $P(u)$ and $Q(w)$.

In an irreversible reaction the boundary condition can be written, by analogy with Eq. (9.2) (and allowing for the fact that in the case being considered we have $\nu_j = 1$),

$$D_j(\partial c_j/\partial x)_{x=0} = k'_m c_{S,j} \exp(\beta_m F v t / RT), \qquad (9.10)$$

where $k'_m \equiv k_m \exp(\pm \beta_m F E_{in}/RT)$.

The differential diffusion equation has the solution

$$i = \pm n F (\beta_m F D_j v / RT)^{1/2} c_{V,j} Q(w), \qquad (9.11)$$

where $E_a \equiv E^0 - (RT/2\beta_m F)\ln[\pi \beta_m F D_j v / RT(k^0_m)^2]$ and $w \equiv \pm \beta_m F(E - E_a)/RT$.

The dimensionless function $Q(w)$ (see Fig. 9.8) also goes through a maximum, where its value is 0.496. Hence we obtain for the maximum current:

$$i_{max} = \pm 3.0 \cdot 10^5 \, n \beta_m^{1/2} (D_j v)^{1/2} c_{V,j}. \qquad (9.12)$$

The numerical factor in this equation is larger than that in Eq. (9.9), yet the current for an irreversible reaction is lower than that for a reversible reaction, since $\beta_m < n$ (for $\beta_m = 0.5n$, it is 78% of the current for a reversible reaction).

Solutions (9.7) and (9.11) are valid when the initial scan potential, E_{in}, in an anodic scan is at least 0.1 to 0.2 V more negative (in a cathodic scan: more positive) than $E_{1/2}$ or E_a, i.e., when the current in the system is still very low. Under these conditions the current during the scan is still independent of the value selected for E_{in}.

In a roundtrip potential scan the values of E_{max} corresponding to the anodic and cathodic direction are different. For reversible reactions the difference is minor, according to Eq. (9.9), viz., only $0.056/n$ V regardless of the component concentrations and of the potential scan rate v.

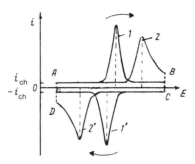

Fig. 9.9. Anodic (1, 2) and cathodic (1', 2') potentiodynamic i vs. E curves for a reversible (1, 1') and an irreversible (2, 2') reaction (horizontal lines above and below the axis of E refer to the charging current in the absence of reactants).

It is typical for irreversible reactions that the difference between these potentials is much larger (Fig. 9.9); the gap between the maxima increases with decreasing value of the reaction rate constant and increasing scan rate v.

During the measurements a nonfaradaic current i_{ch},

$$i_{ch} = dQ_S/dt = C(dE/dt) = \pm Cv, \qquad (9.13)$$

which is consumed for charging the electrode, is superimposed upon the faradaic current. If in the region of potentials considered the edl capacitance C is approximately constant, the charging current will also be constant. This implies that the experimental i vs. E curve is shifted vertically relative to the curve which corresponds to the above equations. The influence of charging currents is particularly clearly seen in i vs. E curves recorded with a triangular scan (Fig. 9.9). At the instant when the scan direction is changed (point B) the charging current changes sign (point C). At this potential, therefore, a discontinuous change in current is observed which corresponds to double the charging current. In repetitive triangular scans an analogous change in current is seen between the terminal (D) and initial (A) point of the curve.

9.3.3. The Galvanostatic Method (Chronopotentiometry)

A particular constant current density is applied to the electrode, and the potential variation is followed as a function of time. When there is no

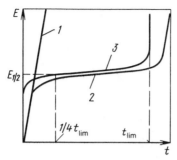

Fig. 9.10. Galvanostatic curves: (1) without reaction; (2) with a reaction; (3) corrected for the charging current.

electrode reaction the entire current is a nonfaradaic charging current i_{ch}. According to Eq. (9.13), the slope of the E vs. t curve (Fig. 9.10, curve 1) is determined by edl capacitance.

When an electrode reaction takes place the applied current is divided between the nonfaradaic components and a faradaic component. Because of the latter, there is a gradual decrease in surface concentration of the reactant [according to Eq. (7.6)]. When the time t_{tr} required for diffusion to change from transient to steady is large as compared to the transition time t_{lim} [Eq. (7.9)], the reactant's surface concentration will fall to zero within the time t_{lim} (see Fig. 7.3).

Chronopotentiometry is conducted so that *a priori* $t_{tr} > t_{lim}$. The values of t_{tr} are limited by two factors: (a) convection in the liquid (even natural), which lowers the steady-state diffusion-layer thickness δ_{st}, and (b) the finite size of the electrode [Eq. (7.27)]. To reduce the influence of the first factor we can reduce natural convection (by avoiding vibration of the cell, by thermostating the solution, etc.); in this way we can extend t_{tr} to about 200–300 s. In order for this value not to be reduced by the second factor the size of the electrode, according to Eq. (7.27), should be at least 10^{-3} m. But, in order to limit t_{lim} to a value of, say, 60 s, it is necessary, according to Eq. (7.9), to have $|i|/c_{V,j} \geq 0.5$ A·m/mol; in this case a minimum (faradaic) current density of 50 A/m^2 must be used when the concentration is 0.1 M. In practice the measurements are performed with values of t_{lim} between 1 and 60 s.

Consider the shape of the E vs. t relation for the cathodic reaction Ox + $ne^- \rightarrow$ Red, and assume that the initial product concentration

$c_{V,\text{Red}} = 0$. Assume, further, that the share of the nonfaradaic current is small and all of the applied current can be regarded as faradaic.

In reversible reactions the electrode potential is determined by the values of $c_{S,\text{Ox}}$ and $c_{S,\text{Red}}$. Prior to current flow the potential is highly positive, since $c_{S,\text{Red}} = c_{V,\text{Red}} = 0$. When the current is turned on, the changes in surface concentrations are determined by Eqs. (7.10) and (7.11). Substituting these values into the Nernst equation and taking into account that in our case $c_{V,\text{Red}} = 0$ and $\nu_{\text{Red}} = \nu_{\text{Ox}} = 1$ we obtain

$$E = E_{1/4} + (RT/nF)\ln[(t_{\text{lim}}^{1/2} - t^{1/2})/t^{1/2}], \qquad (9.14)$$

where $E_{1/4} \equiv E^0 + (RT/nF)\ln(D_{\text{Ox}}/D_{\text{Red}})^{1/2}$ is the potential at $t = t_{\text{lim}}/4$ (when the logarithmic term becomes zero), which is analogous to the half-wave potential in Eq. (6.45).

The relation between E and t is S-shaped (curve 2 in Fig. 9.10). In the initial part we see the nonfaradaic charging current. The faradaic process starts when certain values of potential are attained, and a typical potential "arrest" arises in the curve. When zero reactant concentration is approached the potential again moves strongly in the negative direction (toward potentials where a new electrode reaction will start, e.g., cathodic hydrogen evolution). It thus becomes possible to determine the transition time t_{lim} precisely. Knowing this time we can use Eq. (7.9) to find the reactant's bulk concentration or, when the concentration is known, its diffusion coefficient.

When the nonfaradaic current is not small enough the appropriate correction must be included when constructing the curves. At constant current, the charge consumed is proportional to time; therefore, we can graphically correct by subtracting at each potential the time t_{ch} spent for charging of the electrode (or actually: the charge) from the current value of time t (curve 3).

Equations (7.6) or (7.10), which do not depend on the mode of electrode operation, remain valid for irreversible reactions. Substituting the value of $c_{S,\text{Ox}}$ into the kinetic equation (6.10) for a cathodic process at significant values of polarization we obtain, after transformations,

$$E = E_{(t=0)} + (RT/\beta_{-m}F)\ln[1 - (t/t_{\text{lim}})^{1/2}], \qquad (9.15)$$

where $E_{(t=0)} \equiv (RT/\beta_{-m}F)\ln(nFc_{V,\text{Ox}}k_{-m}/|i|)$.

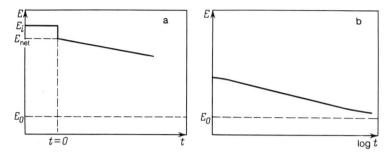

Fig. 9.11. Potential decay curves recorded after switching off the current; (a) at short times; (b) at long times.

When plotting the experimental data as E against $\ln[1 - (t/t_{\text{lim}})^{1/2}]$ we obtain straight lines. The values of β_{-m} can be determined from their slopes; the values of $E_{(t=0)}$ and hence k_{-m} can be obtained by extrapolating these plots to $t = 0$.

One of the advantages of the galvanostatic method is that an ohmic error, even if present when measuring the potential, will remain constant during the measurements.

A version of the galvanostatic method is that where the current is turned off (or "a current $i = 0$ is applied") and the polarization decay curve is measured. Consider an electrode which up to the time $t = 0$, when the current was turned off, had the potential E_i at the net current density i_{net}. When the current is turned off the ohmic voltage drop in the electrolyte gap between the electrode and the tip of the Luggin capillary vanishes, so that the potential instantaneously shifts to the value E_{net} (Fig. 9.11). After that the electrode potential returns (falls) relatively slowly to its open-circuit value, for which a certain nonfaradaic charging current is required. Since $i_f + i_{\text{ch}} = i = 0$ after the current has been turned off, the charging current must be compensated by a faradaic current, and the electrochemical reaction will continue to the end of the transitory process, rather than cease immediately.

Assume that the current i_{net} is sufficiently small and will not cause any marked concentration changes. We shall also assume that the faradaic current in a major part of the potential decay curve obeys the kinetic equation (6.6), i.e., at time t and the potential E_t

$$i_t = i_{\text{net}} \exp[\pm(E_t - E_{\text{net}})/b] \qquad (9.16)$$

(the plus sign is for an anodic reaction). Substituting the expressions for currents i_{ch} and i_t into the relation $i_{ch} = -i_t$ we find, in the case of an anodic reaction,

$$C(dE/dt) = -i_{net}\exp[(E_t - E_{net})/b].\qquad(9.17)$$

Integrating this equation between the limits of $t = 0$ and t, taking into account that at $t = 0$ the potential $E = E_{net}$, and performing simple transformations we obtain an equation for the potential decay curve:

$$E_t = E_{net} - b\ln(1 + i_{net}t/bC),\qquad(9.18)$$

which at short times ($t < bC/i_{net}$) changes to the linear relation

$$E_t = E_{net} - i_{net}t/C\qquad(9.19a)$$

(Fig. 9.11a), and at long times to the logarithmic relation

$$E_t = E_{net} - b\ln(i_{net}/bC) - b\ln t\qquad(9.19b)$$

(Fig. 9.11b). From the slope of the latter relation we can determine the slope b, while from the second term on the right-hand side of Eq. (9.19b) or from the slope of the linear relation we can find the electrode's capacitance for the time during which the electrochemical reaction took place at it.

9.4. ALTERNATING-CURRENT MEASUREMENTS

9.4.1. Electrode Impedance Measurements

When alternating current is used for the measurements, a transient state arises at the electrode during each half period, and the state attained in any half period changes to the opposite state during the next half period. These changes are repeated according to the ac frequency, and the system will be quasi-steady on the whole, i.e., its average state is time-invariant.

For measurements, an alternating current component $I_\sim = I_m\sin\omega t$ with the amplitude I_m and angular frequency ω ($\omega = 2\pi f$, where f is the ac frequency) is passed through the electrode (alone or in addition to a direct current). Alternating potential (polarization) changes $\Delta E_\sim$ having the same frequency and an amplitude ΔE_m are the response. Sometimes alternating potential components are applied, and the resulting alternating current component is measured. In all cases the potential changes are small in amplitude ($\leq$ 10 mV).

For an electrode behaving like a pure (ohmic) resistance R, the relation between the instantaneous values of current and the changes in potential at all times would be $\Delta E_\sim / I_\sim = \Delta E_m / I_m = R$. This is actually not found, but instead a phase shift α analogous to that observed in electric circuits containing reactive elements appears between alternating current and alternating polarization. In electrochemical systems the potential changes always lag the current changes: $\Delta E_\sim = \Delta E_m \sin(\omega t - \alpha)$, which corresponds to an electric circuit with capacitive elements. Thus, the ac behavior of an electrode cannot be described in terms of a simple polarization resistance R (even if variable) but only in terms of an impedance $\bar{Z}$ characterized by two parameters, the modulus of impedance $Z \equiv \Delta E_m / I_m$ and the phase shift α. The reciprocal of impedance, $\bar{Y} \equiv 1/\bar{Z}$, is known as admittance or ac conductance.

A model for the ac response of real electrodes is the simple electric equivalent circuit consisting of a resistance R_s and capacitance C_s connected in series (Fig. 9.12a). It follows from the rules for ac circuits that for this combination

$$Z = [R_s^2 + (1/\omega C_s)^2]^{1/2}, \qquad \tan\alpha = 1/\omega C_s R_s. \qquad (9.20)$$

We can also use a circuit with a resistance R_p and capacitance C_p connected in parallel (Fig. 9.12b). For this circuit

$$Z = [(1/R_p)^2 + (\omega C_p)^2]^{1/2}, \qquad \tan\alpha = \omega C_p R_p. \qquad (9.21)$$

The elements of these two circuit versions are interrelated as

$$\begin{aligned}
R_s &= R_p[1 + (\omega C_p R_p)^2]^{-1}, & C_s &= C_p[1 + (\omega C_p R_p)^{-2}]; \\
R_p &= R_s[1 + (\omega C_s R_s)^{-2}], & C_p &= C_s[1 + (\omega C_s R_s)^2]^{-1}.
\end{aligned} \qquad (9.22)$$

For the calculations the capacitive (reactive) impedances $X_s \equiv 1/\omega C_s$ and $X_p \equiv 1/\omega C_p$ are often used instead of capacitances C_s and C_p. The impedance (admittance) of an ac circuit can be stated in terms of a complex number where the real and imaginary part are the resistive and reactive part of impedance (admittance), respectively. All calculations concerning these circuits can be performed following the rules for manipulation of complex numbers. The impedance of the series network (Fig. 9.12a) can

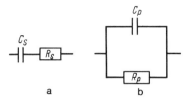

Fig. 9.12. Equivalent circuits with a resistance and capacitance in series (a) and in parallel (b).

be written as

$$\bar{Z} = R_s - jX_s \qquad (9.23a)$$

(here j is the imaginary unit). An analogous expression can be written for the admittance of the parallel network (Fig. 9.12b):

$$\bar{Y} = (1/R_p) + j(1/X_p) = K_p + j\omega C_p \qquad (9.23b)$$

(K_p is the conductance, which is equal to $1/R_p$).

Thus, the behavior of an electrode at any particular frequency ω can be described by any of the following pairs of parameters: Z and α, R_s and C_s (or X_s), R_p and C_p (or X_p).

But when considered over a wide range of frequencies, the properties of a real electrode do not match those of the equivalent circuits shown in Fig. 9.12; the actual frequency dependence of Z and α does not obey Eqs. (9.20) or (9.21). In other words, the actual values of R_s and C_s or R_p and C_p are not constant but depend on frequency. In this sense the equivalent circuits described are simplified. In practice they are only used for recording of the original experimental data. The values of R_s and C_s or R_p and C_p found experimentally for each frequency are displayed as functions of frequency. In a subsequent analysis of these data more complex equivalent circuits are explored which might describe the experimental frequency dependence, and where the parameters of the individual elements remain constant. It is the task of theory to interpret the circuits obtained and find the physical significance of the individual elements.

In the measurements, one commonly determines the impedance of the entire cell and not that of an individual (working) electrode. The cell impedance $\bar{Z}_{cell}$ (Fig. 9.13) is the series combination of impedances of the working electrode ($\bar{Z}_{WE}$), auxiliary electrode ($\bar{Z}_{AE}$), and electrolyte ($\bar{Z}_e$, practically equal to the electrolyte's resistance R_e). Moreover, between

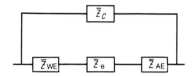

Fig. 9.13. Equivalent circuit for the impedance of a galvanic cell.

parallel electrodes a capacitive coupling develops which represents an impedance $\bar{Z}_C$ parallel to the other impedance elements. The experimental conditions are selected so that $\bar{Z}_C \gg \bar{Z}_{WE} \gg \bar{Z}_{AE}$. To this end the surface area of the auxiliary electrode should be much larger than that of the working electrode, and these electrodes should be sufficiently far apart. Then the measured cell impedance $\bar{Z}_{cell}$ is practically given by $\bar{Z}_{WE} + \bar{Z}_e$.

Bridge schemes are used to measure impedance (Wheatstone bridge, Fig. 9.14). The test cell with its impedance R_{cell} is connected as one arm of the bridge, while the other arm contains sets of precision capacitors and resistors connected in series ($R_{s,2}$ and $C_{s,2}$) [or parallel ($R_{p,2}$ and $C_{p,2}$)], which allow any value of impedance $\bar{Z}_2$ to be selected for this arm. The third and fourth arm have constant, precisely known values of impedance $\bar{Z}_3$ and $\bar{Z}_4$ (usually standard resistors are used). Potentiometer P is used to compensate the cell's (dc) ocv. An ac of preselected frequency is made to flow across the bridge with the aid of signal generator G. During measurements the impedance of the second arm is adjusted until the ac detected in null detector ND has dropped to zero. The condition of complete bridge balance (equal instantaneous values of potential in points A and B at any phase of the current) is

$$\bar{Z}_{cell} : \bar{Z}_2 = \bar{Z}_3 : \bar{Z}_4. \tag{9.24}$$

Knowing the values of all other parameters we can thus find the cell's impedance. When a symmetric bridge is used (where the values of impedance in the third and fourth arm are identical), the unknown cell impedance components will be equal to the values of $R_{s,2}$ and $C_{s,2}$ or $R_{p,2}$ and $C_{p,2}$ in the second bridge arm.

Ordinarily the measurements are made over a frequency range from 20 Hz to 50 kHz. Certain difficulties arise in measurements extended over wider ranges of frequency. However, methods suitable for measurements at very low frequencies, down to 1 mHz, have recently been developed

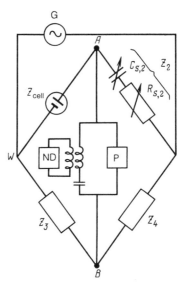

Fig. 9.14. Bridge arrangement for impedance
measurements.

which can be used to obtain additional information concerning the properties of the electrochemical systems.

9.4.2. Impedance in the Case of Irreversible Reactions

Consider the ac behavior of an electrode at which an electrochemical
reaction occurs under kinetic control at low polarization, i.e., under conditions where the polarization equation (6.7) is obeyed. The impedance
of a cell containing such an electrode corresponds to the equivalent circuit shown in Fig. 9.15a. The impedance of the working electrode is
in series with the electrolyte resistance R_e. The current passing through
the electrode has a faradaic and a nonfaradaic component. The former
gives rise to periodic potential changes as described by Eq. (6.7). The
corresponding branch of the equivalent circuit can be represented as a
resistance R_f given by the ratio of the (specific) polarization resistance ρ
to the electrode's surface area: $R_f = \rho/S$. The nonfaradaic charging current is required for the periodic changes in the amounts of charge in the
electric double layer necessitated by the potential variation. It depends on
the total capacitance C_σ of the edl, which is SC (C is the specific capacitance). The charging current in the circuit is independent of the faradaic

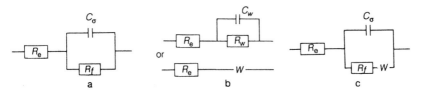

Fig. 9.15. Electrode impedance with kinetic (a), diffusional (b), and combined (c) reaction control (W is the Warburg impedance).

current, since the circuit element with capacitance C_σ is in parallel with the resistance R_f.

In order to find the relation between the values of R_s and C_s measured experimentally in terms of the circuit of Fig. 9.12a and the parameter values in the circuit of Fig. 9.15a we must first of all convert [with the aid of Eq. (9.23)] the parameters of the circuit with parallel elements R_f and C_σ into the parameters of a circuit with a resistance and capacitance in series, and to the value of resistance obtained we must add R_e. As a result we have

$$R_s = R_e + R_f[1 + (\omega C\rho)^2]^{-1}; \qquad C_s = C_\sigma[1 + (\omega C\rho)^{-2}] \qquad (9.25)$$

(here the obvious equality of $R_f C_\sigma = \rho C$ was used).

We can see here that at very low frequencies R_s tends toward the sum $R_e + R_f$, and C_s tends toward infinity. At very high frequencies R_s becomes equal to R_e, and C_s becomes equal to C_σ. Therefore, by extrapolating the experimental data to zero and to infinite frequency we basically can find the kinetic reaction parameter R_f (or ρ) and the edl capacitance as well as the electrolyte's ohmic resistance.

In many cases the extrapolation of experimental data is difficult. To make this extrapolation more accurate, we can use different ways of plotting the experimental results. For instance, when R_e has first been determined at high frequencies, we can plot the experimental data as $(R_s - R_e)^{-1}$ against f^2. It follows from Eq. (9.25) that then the experimental points will fall onto a straight line which is readily extrapolated to zero frequency where an intercept $1/R_f$ is produced on the vertical axis. Figure 9.16a shows a realistic example for this extrapolation of the experimental data. With the parameters chosen for this example ($S = 10^{-4}$ m^2, $R_f = 1$ Ω, $R_e = 0.5$ Ω, and $C_\sigma = 5 \cdot 10^{-5}$ F), an ac frequency of at least 50 kHz is required for the prior determination of R_e.

Another coordinate system has been found very convenient for extrapolation of the experimental data, viz., plots of the capacitive component

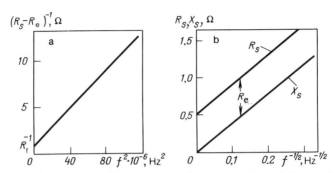

Fig. 9.16. Examples for an extrapolation of experimental data to determine equivalent circuit parameters under kinetic (a) and diffusional (b) reaction control.

of impedance, X_s, against the resistive component, R_s (Kenneth Stewart Cole and Robert Hugh Cole, 1941; Margaretha Sluyters-Rehbach and Johannes Hendricus Sluyters, 1960). In the case discussed, the resulting *impedance diagram* has the typical form of a semicircle with the center on the horizontal axis (Fig. 9.17a). This is readily understood when the term $\omega C\rho$ is eliminated from the expressions for R_s and C_s in Eq. (9.25). Then we obtain, after simple transformations,

$$X_s^2 + (R_s - R_e - R_f/2)^2 = (R_f/2)^2, \qquad (9.26)$$

which is the analytical equation for a semicircle of radius $R_f/2$ having its center in the point of $X_s = 0$ and $R_s = R_e + R_f/2$ when the coordinates of X_s and R_s are used (and these parameters have positive values). Experimental data falling on a semicircle are readily extrapolated to the semicircle's intersections with the horizontal axis. Point A corresponds to zero frequency, and point B corresponds to infinite frequency. The abscissas of these points are $R_e + R_f$ and R_e, respectively. The frequency is not explicitly apparent in this coordinate system. The corresponding values of the frequencies are stated at the individual points when required.

9.4.3. Impedance in the Case of Reversible Reactions

As an example, consider a simple reaction of the type of (6.2) taking place under pure diffusion control. At all times the electrode potential, according to the Nernst equation, is determined by the reactant concen-

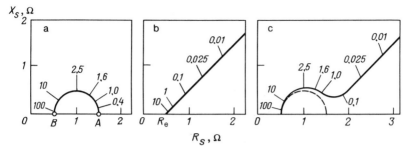

Fig. 9.17. Impedance diagrams for electrodes under kinetic (a), diffusional (b), and combined (c; very simple case) control of electrode operation (numbers indicate the frequencies f in kHz).

trations at the electrode surface. It has been shown in Section 7.2.3 that periodic changes in the surface concentrations which can be described by Eq. (7.20) are produced by ac flow. We shall assume that the amplitude of these changes is small, i.e., that $\Delta c_j \ll c_{V,j}$. In this case we can replace $\ln(c_{S,j}/c_{V,j})$ by $\Delta c_j/c_{V,j}$ in Eq. (6.39) for electrode polarization. With this substitution and using Eq. (7.20) we obtain

$$\Delta E = (RT/nF)[\Delta c_{Ox}/c_{Ox} - \Delta c_{V,Red}/c_{V,Red}]$$
$$= (RT/n^2 F^2)(i_m/\sqrt{\omega}) \left[1/c_{V,Ox}\sqrt{D_{Ox}} + 1/c_{V,Red}\sqrt{D_{Red}} \right]$$
$$\times \sin(\omega t - \pi/4). \tag{9.27}$$

We see that in this case the phase shift is $\pi/4$ (45°). This phase shift corresponds to the circuit shown in Fig. 9.15b, which includes the resistance R_w and a capacitance C_w for which $X_w = R_w$, hence $\tan \alpha = 1$ and $Z_w = \sqrt{2}\,R_w$ (it does not matter in this case whether the capacitance and resistance are connected in parallel or in series). It follows from Eq. (9.27) that

$$R_w = X_w = A_w/\sqrt{\omega} \quad \text{and} \quad C_w = 1/A_w\sqrt{\omega}, \tag{9.28}$$

where $A_w \equiv (RT/n^2 F^2)[1/c_{V,Ox}\sqrt{D_{Ox}} + 1/c_{V,Red}\sqrt{D_{Red}}]$ (in units $\Omega \cdot m^2/s^{1/2}$).

The impedance Z_W with its components R_W and C_W is known as the *Warburg diffusion impedance*, and constant A_W as the Warburg constant. In the equivalent circuits for electrochemical reactions, a Warburg impedance is represented by the symbol $-W-$.

In the Warburg impedance, parameters R_W and C_W are not constant but depend on frequency according to Eq. (9.28). Figure 9.16b shows plots of the values of R_s and X_s against $f^{-1/2}$ as an example ($A_W/S = 10 \; \Omega/s^{1/2}$, $R_f = 0$, the other parameters have the same values as in the example above). The plots are parallel straight lines in these coordinates, according to Eq. (9.28). The line for X_s goes through the coordinate origin, while that for R_s is shifted upward, since the measured values of R_s in addition to R_W contain the frequency-independent electrolyte resistance R_e.

When the data are plotted in the coordinates of X_s and R_s, the diagram for the present example is a straight line rising at an angle of 45°and producing an intercept corresponding to resistance R_e on the horizontal axis (Fig. 9.17b).

9.4.4. Impedance in the Case of More Complex Reactions

In the case of reactions which are not completely irreversible (or not completely reversible), we must account for both the kinetic factors (e.g., the polarization resistance R_f) and the concentration changes (the Warburg impedance). The simplest equivalent circuit for this case is shown in Fig. 9.15c, while Fig. 9.17c shows the impedance diagram for this circuit ($A_W/S = 10 \; \Omega/s^{1/2}$, $R_f = 1 \; \Omega$, the other parameters have the same values as in the earlier examples). We see that the character of the plot changes with frequency. At high frequencies (short half periods) the relative concentration changes are insignificant, and the behavior of the electrodes is determined mainly by the reaction kinetics; plots of X_s against R_s contain the semicircular segment that is typical for this case. The contribution of the concentration changes increases with decreasing ac frequency, and below a certain frequency a linear segment arises which corresponds to the Warburg impedance and constitutes evidence for slow diffusion processes.

In many cases the plots are even more complex, and a theoretical interpretation is difficult. Often the plots of X_s against R_s are evaluated only in a qualitative way, and segments with kinetic semicircles or diffusional straight lines are considered separately.

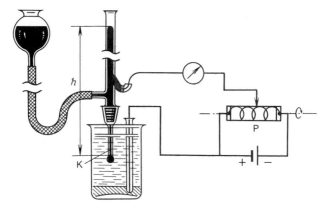

Fig. 9.18. Dropping mercury electrode.

9.5. POLAROGRAPHY

9.5.1. The Dropping Mercury Electrode

The dropping mercury electrode (DME) set up as shown in Fig. 9.18 is often employed for electrochemical measurements in laboratory practice. Under the pressure of a mercury column of height h, mercury flows with the constant volume flow rate w from a glass capillary K. The drop forming at the capillary tip grows, and it tears away when at a time t_{dr} after the start of formation it has attained a certain mass. After its detachment a new drop starts to form and grow, and the cycle is repeated. The growing drop, while suspended, is used as the cathode for reduction of various substances (since mercury readily undergoes anodic dissolution, it cannot be used as an "inert" anode). Usually a potentiostatic circuit is used, i.e., during the drop's life its potential is kept constant.

Regarding, provisionally, the drop as spherical and neglecting the influence of its neck we can write for drop radius r_{dr} and surface area S at time t:

$$r_{dr} = (3/4\pi)^{1/3} w^{1/3} t^{1/3} = 0.620 w^{1/3} t^{1/3},$$

$$S = (36\pi)^{1/3} w^{2/3} t^{2/3} = 4.836 w^{2/3} t^{2/3}. \tag{9.29}$$

The flow rate w can be varied by changing the mercury column height. The drop time t_{dr} is sometimes controlled with an electrically actuated hammer which taps the capillary when needed (before natural drop detachment under the pull of gravity). A typical value of flow rate w in the equipment commonly used is $1.5 \cdot 10^{-10}$ m^3/s (2 mg of mercury per second), and a typical drop time is 3 s. Therefore, the drop's radius at the time of detachment is about 0.5 mm, and its surface area about 3 mm^2.

The electrolyte solution is stirred when a drop falls off, and diffusion processes begin anew at each new drop. It follows from Eq. (7.16) that the diffusion-layer thickness δ during drop life remains smaller than the drop radius, hence the diffusion processes at DME are transient.

The surface of the growing drop advances into the electrolyte with a linear velocity of

$$v \equiv dr_{\mathrm{dr}}/dt = (w/36\pi)^{1/3}t^{-2/3} = w/4\pi r_{\mathrm{dr}}^2. \qquad (9.30)$$

While expanding, the drop pushes against the closest layer of solution, and the motion is transmitted to layers further away. The volume of the spherical shells taken up by these layers increases with increasing distance r from the center of the shells, hence their linear velocity decreases. For a point A at distance r_A the velocity relative to the stationary coordinate origin is given by $w/4\pi r_A^2$. Thus the liquid element in this point actually moves toward the advancing drop surface with a relative linear velocity

$$v = (w/4\pi)[1/r_{\mathrm{dr}}^2 - 1/r_A^2]. \qquad (9.31)$$

It follows that the liquid is in convective motion toward the electrode surface. Its velocity decreases in the approach to the surface and finally becomes zero, yet this motion influences the diffusional transport. The problem of convective diffusion toward the growing drop was solved in 1934 by Dioníz Ilkovič under certain simplifying assumptions. It follows from the calculation that under potentiostatic conditions

$$J_j = -\sqrt{7/3}(D_j/\pi t)^{1/2}\Delta c_j \qquad (9.32)$$

i.e., the diffusion flux density resulting in the presence of convection is higher than the transient diffusion flux density in the absence of convection [cf. Eq. (7.15)] by a factor of $\sqrt{7/3} = 1.53$.

Charge for formation of an edl corresponding to the set potential E must constantly be supplied to the DME because of the continuous renewal and growth of the surface. The charge, Q_{dl}, present on each side of the edl depends on the potential difference ψ_0 between the two layers of charge (this parameter will be defined more closely in Section 12.5).

According to Eq. (2.6) we have $Q_{dl} = SC\psi_0$ (where C is the edl's specific capacitance). Therefore, apart from any electrochemical reaction that may occur, a charging current i_{ch} is continuously recorded at the electrode:

$$I_{ch} = (dS/dt)C\psi_0 = 3.224w^{2/3}t^{-1/3}C\psi_0. \qquad (9.33)$$

Depending on the sign of ψ_0 the charging current can be positive or negative. It is readily measured in solutions where no electrochemical reactions take place and hence no faradaic current is recorded.

Thus, the conditions of DME operation are transient since the drop surface grows and moves, diffusion is transient, and a nonfaradaic current is present. The current passing through the electrode will increase from a minimum value at drop birth to a maximum value at the time of drop detachment. However, after detachment of each drop the processes are exactly repeated, i.e., all parameters when averaged over drop life will be time-invariant. For this reason the DME is often employed as a quasi-steady electrode and its averaged parameters are used.

The averaged surface area of the electrode can be determined from the expression

$$\bar{S} = (1/t_{dr}) \int_0^{t_{dr}} S(t)\, dt = 2.902w^{2/3}t_{dr}^{2/3}, \qquad (9.34)$$

the averaged charging current from the equation

$$\bar{I}_{ch} = (1/t_{dr}) \int_0^{t_{dr}} I_{ch}(t)\, dt = 4.836w^{2/3}t_{dr}^{-1/3}C\psi_0. \qquad (9.35)$$

When an electrochemical reaction takes place the faradaic current I_f (both the instantaneous and the averaged one) can be found from the difference between the measured current I_{meas} and the charging current: $I_f = I_{meas} - I_{ch}$ (subscript f for the faradaic current will be dropped from now on).

The instantaneous faradaic current $I(t)$ recorded at the electrode at time t is given by $i(t)S(t)$, and the averaged current by

$$\bar{I} = (1/t_{dr}) \int_0^{t_{dr}} i(t)S(t)\, dt. \qquad (9.36)$$

In purely diffusion-controlled operation, the current density i, which is related to the diffusion flux density of the reactant ($\bar{\nu}_j = -\nu_j$) as

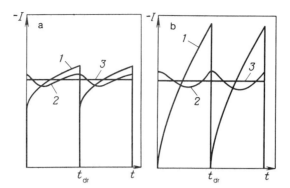

Fig. 9.19. Current–time relation during the growth and detachment of a mercury drop under diffusional (a) and kinetic (b) control of the reaction: (1) true current; (2) current measured with a sluggish instrument; (3) averaged current.

$|i| = -(n/\nu_j)FJ_j$, is determined by Eq. (9.32). Substituting the expressions for $i(t)$ and $S(t)$ into Eq. (9.36) and integrating we obtain

$$|\bar{I}| = 3.572(n/\nu_j)FD_j^{1/2}w^{2/3}t_{dr}^{1/6}\Delta c_j \qquad (9.37)$$

(the *Ilkovič equation*).

In the other limiting case of a reaction under pure kinetic control, the current density i at a given potential E is constant, and the expression for the averaged current becomes

$$\bar{I} = i\bar{S} = 2.902iw^{2/3}t_{dr}^{2/3}. \qquad (9.38)$$

Figure 9.19 shows plots of the instantaneous current against drop age for the different operating conditions, together with the corresponding values of the averaged currents. Usually sluggish instruments are used to record the current; these only sense the averaged current or display minor current fluctuations.

The DME has the advantage of continuous renewal, i.e., impurities will not accumulate on its surface or in its volume, so that the measurements have a high reproducibility. Another advantage are the high values of the averaged transient diffusion constants $\bar{\varkappa}_j$ defined by the expression

$$\bar{\varkappa}_j \equiv (|\bar{I}|/\bar{S})(\nu_j/nF\Delta c_j).$$

These values are about ten times higher than those of steady diffusion, $\varkappa_j$, to a spherical electrode having the same radius as the DME at the time of drop detachment [cf. Eq. (7.26)].

9.5.2. Classical Polarography

The first version of a polarographic technique was put forward in 1922 by the Czech scientist Jaroslav Heyrovský, who made quasi-steady polarization measurements at a DME under potentiodynamic conditions, using a slow linear potential scan (with v between 1 and 20 mV/s). With special instruments (polarographs), one can automatically record the resulting I vs. E curves (polarograms). These measurements became very popular particularly for analytical purposes. For the development of this method, Heyrovský was awarded the Nobel prize in 1959.

In the classical version one uses a two-electrode cell with DME and a mercury AE (the pool) at the bottom of the cell (see Fig. 9.18). The latter, which has a large surface area, is practically not polarized. The current at the DME is low and causes no marked ohmic potential drop in the solution. Hence in order to change the DME potential by $|\Delta E|$ it will suffice to vary the external voltage $\mathcal{E}_{ext}$ applied to the cell by the same value $|\Delta \mathcal{E}_{ext}|$. During the measurements, I vs. $\mathcal{E}$ rather than I vs. E curves are recorded.

For polarization one uses a very simple galvanostatic arrangement with low resistance in the external circuit (Fig. 9.18). The voltage of the power source is controlled and varied with the aid of a low-resistance potentiometric device P. Usually this device is a wire wound in 10-20 turns around a drum, which is rotated. The brush in the center does not rotate but can move along a guide in a direction parallel to the drum axis. When the drum is rotated the brush slides along the wire, thus a smooth variation of voltage $\mathcal{E}_{ext}$ is obtained. A linear voltage scan rate is obtained through slow, uniform drum rotation.

During the measurements the DME potential is moved in a negative direction relative to the AE potential (otherwise the mercury would start to dissolve anodically), a cathodic current flows, and cathodic reactions (metal ion discharge or the reduction of other compounds) take place.

The discharge of metal ions at mercury and formation of the corresponding amalgams are reversible in most cases. The equilibrium potential of an amalgam electrode is determined by an equation of the type of (3.42). Concentration polarization arises as soon as a cathodic current

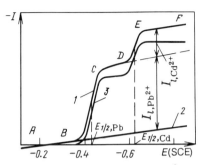

Fig. 9.20. Polarogram measured in $1 \cdot 10^{-3}$ M PbCl$_2$ + $5 \cdot 10^{-4}$ M CdCl$_2$ + 1 M KCl solution.

flows; metal accumulates in the surface layer of the mercury, forming amalgams, while the concentration of the corresponding metal ions in the solution layer at the surface decreases. The shape of the polarization curve is determined by Eq. (6.48) (the amalgam bulk concentration is zero); it is that of a typical wave, the *polarographic wave* (see Fig. 6.5, curve 3).

Figure 9.20 shows a polarogram recorded in $1 \cdot 10^{-3}$ M PbCl$_2$ + $5 \cdot 10^{-4}$ M CdCl$_2$ + 1 M KCl solution (curve 1) as well as a curve (2) for the charging current recorded in the base electrolyte, 1 M KCl. The polarographic curve obtained after subtracting the charging current (curve 3) corresponds to the faradaic reaction current. In practice the curve for the faradaic current is constructed, not by measuring in each case a charging curve in the base electrolyte but by assuming that this curve is the linear continuation of the initial section AB in the polarogram.

The first wave (segment BCD) corresponds to lead ion discharge with lead amalgam formation. The current of cadmium ion discharge is superimposed on the limiting current of this reaction in the region of the second wave (segment DEF). Discharge of the base-electrolyte ions (alkali metal or hydrogen ions) starts at even more negative potentials (to the right of point F).

Characteristic parameters of each wave are the value of the half-wave potential $E_{1/2}$, which is defined by Eq. (6.45), and the wave "height" [the value of the limiting diffusion current, $I_{l,Ox}$, determined by the Ilkovič equation (9.37) for $c_{S,j} = 0$]. The value of $E_{1/2}$ depends on the nature of the ions undergoing discharge, and can be used to identify them. From the value of the limiting current we can determine the concentration of the reactant ions in the solution.

Thus, polarography offers possibilities for a qualitative and quantitative analysis of different solution constituents, even when they are present simultaneously.

When the electrode reaction at the DME is irreversible (which is so in particular for many redox reactions), the shape of the curve will be determined by the kinetic relation for mixed control, Eq. (6.63). This equation is similar to Eq. (6.48), and the resulting curve also has the shape of a polarographic wave. However, here the half-wave potential is equal to the value of the constant A, i.e., is no longer determined by the nature of the components involved in the reaction (their standard equilibrium potential) but by the kinetic parameters of the reaction. Moreover, the slope of the wave in its central part in irreversible reactions [which is determined by the factor $(RT/\beta_{-m}F)^{-1}$] is lower than that in reversible reactions [which is determined by the factor $(RT/nF)^{-1}$]. It is seen here that polarography can also be used to determine the kinetic parameters of electrochemical reactions.

Polarograms are sometimes distorted by so-called polarographic maxima, where the current in individual segments of the I vs. E curves is much higher (several times) than the limiting diffusion current. A number of reasons exist for the development of these maxima.

Polarographic maxima of the first kind develop in dilute solutions, and are caused by a nonuniform potential distribution over the electrode. This leads to different values of surface tension (excess surface energy, see Section 12.3) in different parts of the surface and hence to tangential motions of the liquid mercury surface, viz., a contraction of parts with high excess surface energy at the expense of expansion of parts with low excess surface energy. Convective diffusion of the reactants to the surface drastically increases as a result of these motions. These maxima only develop at potentials some distance away from the potential of zero charge (see Section 12.4.3), and their usual shape is that of current spikes.

Maxima of the second kind, to the contrary, are observed more often in concentrated solutions. They extend over a wider range of potentials but are not as high. They attain their maximum height near the point of zero charge. They are caused by tangential motions of the mercury drop surface which develop as the mercury flows from the capillary, descends within the drop, but then ascends along the drop surface.

The polarographic maxima of the first and second kind can be eliminated by adding surface-active organic substances to the solution. The changes in adsorption of these substances which develop during the tangential surface motions have a strong retarding effect on these motions.

Along with their advantages, dropping mercury electrodes have disadvantages. In particular, they cannot be used at potentials more positive than the equilibrium potential of mercury in a given solution, hence they are unsuitable for most anodic reactions. Also, work with DME is inconvenient, e.g., in the field, and it is for this reason that solid electrodes are often used for polarographic measurements. Conditions must then be set up which will provide a quantitative control of diffusion of the substance in solution, so that the concentration can be calculated from the limiting current. This is possible in two cases: with microelectrodes [where a steady state is established quite rapidly, see Eq. (7.27)] and with rotating disk electrodes. These electrodes are suitable for measurements, both in the classical polarographic mode and in different new modes. It must be pointed out that nowadays the term "polarography" is reserved for DME measurements, while the measurements involving other kinds of electrode are called voltammetric.

Improvements of the polarographic method and the development of various other voltammetric methods for analytical purposes will be considered in Chapter 20.

Part 2
PROPERTIES OF ELECTROLYTES AND INTERFACES

Chapter 10
AQUEOUS ELECTROLYTE SOLUTIONS

Aqueous solutions of acids, bases, and salts are the ionic conductors most widely used and most thoroughly studied. The importance of other types of ionic conductors has increased in recent times, but aqueous solutions are still preeminent. Their significance goes far beyond electrochemistry as such; they can be found in practically all spheres of human activity. They are of exceptional importance in the form of intracellular fluids in the biological and physiological processes of all living beings. They are of equally great importance in the form of natural waters in the oceans, rivers, and underground for geomorphological processes.

Aqueous electrolyte solutions have been a subject of exhaustive studies for over a century. Numerous attempts were made to construct theories which could link the general properties of solutions to their internal structure, and predict properties as yet unknown. Modern theories of electrolyte solutions are most intimately related to many branches of physics and chemistry. The electrochemistry of electrolyte solutions is a large branch of electrochemistry sometimes regarded as an independent science.

10.1. THE PROPERTIES AND STRUCTURE OF WATER

The conductivity of water depends on its degree of purification. The theoretical value of this parameter is $3.8 \cdot 10^{-6}$ S/m at 20°C. The purest water with $\sigma = 4.3 \cdot 10^{-6}$ S/m was produced in 1894 by Friedrich Kohlrausch after 42 operations of vacuum distillation in special equipment. The conductivity of ordinary distilled water is $(80–100) \cdot 10^{-6}$ S/m, owing to absorption of CO_2 from the atmosphere and the formation of H^+ and HCO_3^- ions (the so-called equilibrium water).

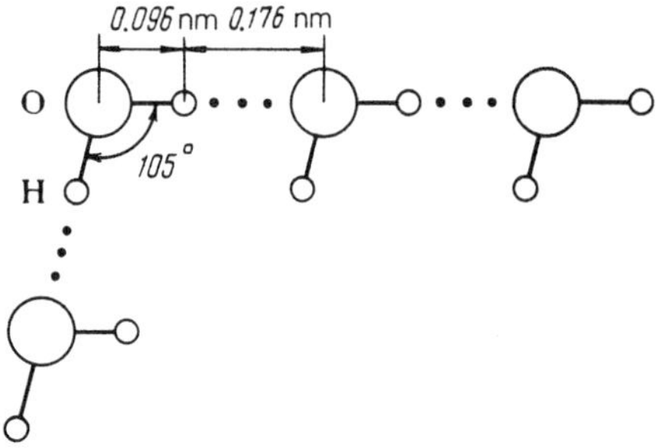

Fig. 10.1. Structure of water molecules and hydrogen bonding between neighboring molecules.

The O–H bond length in water molecules is 0.096 nm, and the angle between the two O–H bonds is about 105° (Fig. 10.1). Water molecules have an effective radius of 0.138 nm. They are dipolar; the negative charge is closer to the oxygen atom, and the positive charge is closer to the hydrogen atoms. The dipole moment, μ_{H_2O}, has a value of $6.2 \cdot 10^{-30}$ C·m (1.87 D). The relative permittivity (dielectric constant) of water, ε_{H_2O}, at 25°C is 78.5.

A typical feature of water molecules is their tendency to form hydrogen bonds: a hydrogen atom can interact with an oxygen atom of a neighboring molecule, and thus link two water molecules. Hydrogen bonds are directed; the two oxygen atoms and the hydrogen atom linking them are situated along a straight line (see Fig. 10.1). The distance between an oxygen atom and the hydrogen atom of the neighboring molecule is 0.176 nm. Hydrogen bonding can hold together many molecules at the same time. The bond energy involved is not large; it is about 20 kJ/mol (of bonds), so that the bonds are not stable in time, and the molecules continuously part and reassociate.

Considering the radius of water molecules stated above, the density of ice could be almost twice as large as it actually is, were the water molecules to form a close-packed lattice. This implies that actually the water molecules are rather loosely packed and have gaps (voids) between each other. It can be seen by X-ray analysis that in the ice lattice, each oxygen atom is surrounded by a first sphere of four like atoms occupying the corners of a tetrahedron, the reason being that hydrogen bonds are spatially directed. A bonding hydrogen atom is located between each pair of oxygen atoms. The resulting structure is a loose hexagonal lattice (of

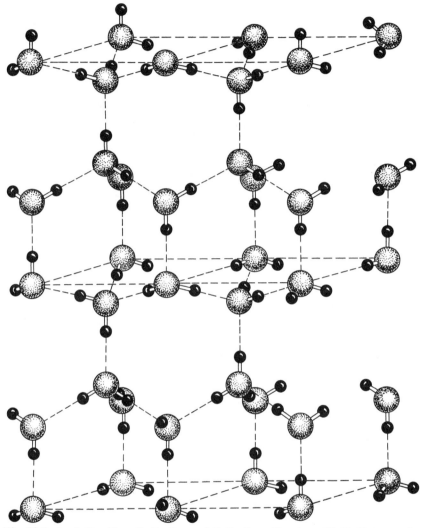

Fig. 10.2. Ice lattice [from L. Pauling and P. Pauling, *Chemistry*, W. H. Freeman, San Francisco (1975)].

the tridymite type) containing relatively large voids with diameters of 0.52 to 0.69 nm (Fig. 10.2).

The structure of liquid water is still not completely known. It is seen from X-ray studies that a large fraction of the hydrogen bonds of ice are preserved and individual lattice fragments with tridymite structure

("icebergs") are formed. Thermal motion has the effect that these fragments continuously decay, rearrange, and reform ("flickering clusters").
Part of the molecules which at any given time are not involved in hydrogen bonding can reside in the lattice voids. In this way it can be
explained that liquid water has a higher density than ice. The fraction of
the bonded molecules decreases with increasing temperature. Dissolution
of the framework structure or changes in this structure require a certain
expenditure of energy, which is the energy of water reorganization.

10.2. THERMODYNAMIC PROPERTIES OF SOLUTIONS

When solutions are formed, the system's Gibbs energy always decreases, i.e., this process is spontaneous. The enthalpy can change in both
directions upon dissolution, and the heat of solution, $\bar{q}_s$, can be positive
or negative. The enthalpy change depends on solution concentration. We
distinguish the first heat of solution or heat of solution to infinitely dilute
solution (which is the heat evolved per mole of solute when dissolving the
first quantities of solute in a very large amount of solvent) from the heat
of formation of a solution of given concentration; the difference between
these quantities is the heat of dilution.

Of great importance for the development of solution theory were the
studies of the so-called colligative solution properties detected in the 1870s
and 1880s by the Frenchman François Marie Raoult, the Dutchman Jacobus Hendricus van't Hoff, and others. These are properties which depend, not on the chemical nature of the solutes but only on their concentration. Three such colligative properties exist:

1. *Osmotic pressure of the solvent.* The general thermodynamic equation for a solvent's osmotic pressure Π in solutions is

$$\Pi = -(RT/\bar{V}_0)\ln a_0, \tag{10.1}$$

where $\bar{V}_0$ is the solvent's partial molar volume.

In ideal solutions the solvent's activity a_0 is equal to its mole fraction Z_0. By definition, the parameter Z_0 can be replaced by $(1 - Z_k)$,
where Z_k is the mole fraction of solute k. The logarithmic part can be
expanded into a series of which only the first term is retained: $\ln(1 - Z_k) \approx -Z_k$, when the solution is sufficiently dilute; in dilute solutions
$Z_k = n_k/(n_0 + n_k) \approx n_k/n_0$. The solvent's partial volume can prac-
tically be regarded as equal to the solution's specific volume, so that

$n_k/n_0\bar{V}_0 \approx c_k$. Hence for an ideal dilute solution

$$\Pi = RTc_k \ (van't\ Hoff's\ law). \tag{10.2}$$

2. *Relative lowering of the solvent's vapor pressure.* The equilibrium vapor pressure of the solvent over a solution is proportional to the solvent's activity: $p_0 = a_0 p_0^0$ (where p_0^0 is the vapor pressure of the pure solvent). For an ideal solution where a_0 is replaced by $(1 - Z_k)$, we have

$$(p_0^0 - p_0)/p_0^0 = Z_k \approx M_0 m_k \ (Raoult's\ law). \tag{10.3}$$

3. *Elevation of boiling point and depression of freezing point of the solution.* Within certain limits, the change in temperature of these phase transitions obeys the equation

$$\pm\Delta T = -(RT^2/\bar{q}_{ph})\ln a_0, \tag{10.4}$$

where $\bar{q}_{ph}$ is the (molar) heat of the corresponding phase transition. For an ideal dilute solution we obtain

$$\pm\Delta T = (RT^2/\bar{q}_{ph})Z_k \approx (RT^2/\bar{q}_{ph})M_0 m_k. \tag{10.5}$$

It follows from these equations that in ideal solutions, said effects depend only on the concentration but not on the nature of the solute. These relations hold highly accurately in dilute solutions of nonelectrolytes (up to about 10^{-2} M). It is remarkable that Eq. (10.2) coincides, both in its form and in the numerical value of constant R, with the equation of state for an ideal gas. It was because of this coincidence that the concept of ideality of a system was transferred from gases to solutions. As in an ideal gas, there are no chemical and other interactions between solute particles in an ideal solution.

In contrast to nonelectrolyte solutions, in the case of electrolyte solutions the colligative properties appreciably depart from the values following from the above equations, even in highly dilute electrolyte solutions which otherwise can be regarded as ideal (anomalous colligative properties).

The thermodynamic properties of real solutions can be described in terms of solvent activity coefficients, which in nonideal solutions depart from unity. However, this parameter is not very convenient, since often the departure from unity is minor, even though the solution's nonideality is pronounced. In 1909 Niels Bjerrum introduced another thermodynamic parameter, the solvent's osmotic coefficient, to describe the properties of real solutions.

The solvent's real osmotic coefficient, g (a dimensionless parameter), is defined by the expression

$$\mu_0 \equiv \mu_0^0 + gRT \ln Z_0. \tag{10.6}$$

After some transformations (analogous to those reported above) we obtain for dilute solutions:

$$\mu_0 \approx \mu_0^0 - gRTM_0 m_k. \tag{10.7}$$

The solvent's practical (molal) osmotic coefficient, Φ (dimensionless), is defined for all concentrations by the expression

$$\mu_0 \equiv \mu_0^0 - \Phi RT M_0 m_k. \tag{10.8}$$

In dilute solutions the numerical values of parameters Φ and g coincide (and in the particular case of ideal solutions $g = \Phi = 1$).

We can see when comparing Eqs. (3.19), (10.6), and (10.8) that the parameters are interrelated as

$$\ln a_0 = \ln f_0 Z_0 = g \ln Z_0 = -\Phi M_0 m_k = (\mu_0^0 - \mu_0)/RT. \tag{10.9}$$

10.3. ELECTROLYTIC DISSOCIATION

10.3.1. Early Ideas

At the beginning of the 19th century, the first theories were advanced to explain the two major features of electrolyte solutions known at the time: conduction, and the fact that chemical reactions could occur at electrodes during current flow.

The first theory was that of Theodor Freiherr von Grotthuss, a Lithuanian physicist who in 1805 introduced the concept that water molecules are dipolar. According to his hypothesis, in an electric field the water molecules will align in chainlike fashion (Fig. 10.3). The molecules forming the terminal chain links in contact with the electrodes will decompose, evolving hydrogen and oxygen, respectively. The remnants of the broken molecules will combine with fragments of neighboring molecules. The

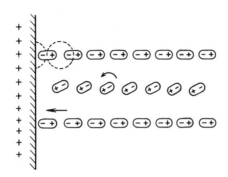

Fig. 10.3. The Grotthuss model for conduction in aqueous solutions.

new molecules after rotating through an angle of 180° are again aligned in the field, and the cycle is repeated. Hence a relay-type transfer of particles is accomplished; the H or O atoms, which eventually are discharged, are transferred from the bulk solution to the electrode surfaces through chains of water molecules.

A clear idea about independent charged particles (atoms or atom groups) existing in solutions was formulated in 1833 by Michael Faraday. He introduced the new, now current terms of "ion" (from the Greek word for "wanderer"), "anion," "cation," etc. Faraday first pointed out that the moving ions at once secure the transport of electricity (charges) and the transport of the substance reacting at the electrode. He assumed, however, that the ions are formed from uncharged molecules only upon application of the electric field, which gave rise to the term "electrolyte," i.e., "one untied or dissolved by electricity."

The first ideas that ions might form spontaneously (without an electric field effect) were formulated in the 1850s. In 1857 Rudolf Clausius thought that ions could form spontaneously during collisions of the solute molecules, but gave them a very short lifetime, and assumed their fraction among the total number of molecules to be insignificant.

10.3.2. Arrhenius' Theory of Electrolytic Dissociation

A theory close to modern concepts was developed by the Swede Svante Arrhenius. The first version of the theory was outlined in his doctoral dissertation of 1883, the final version in a classical paper published at the end of 1887. This theory took up J. H. van't Hoff's suggestions, published one year earlier, that ideal gas laws could be used for the osmotic pressure in solutions. It had been found that anomalously high values of osmotic pressure, which cannot be ascribed to nonideality, sometimes occur even in highly dilute solutions. Van't Hoff had introduced an empirical correction factor i larger than unity, the so-called isotonic coefficient or van't Hoff factor,

$$\Pi = iRTc_k, \tag{10.10}$$

into the equation for the osmotic pressure of such solutions in order to explain the anomaly. Values of the factor i determined experimentally depend on the nature and concentration of the solute. In dilute solutions they sometimes approach integers between 2 and 4.

Arrhenius was the first to point out that conductivity and a departure of colligative properties from the normal values always occur together. He concluded from this observation that both effects have the same origin.

The chief points of Arrhenius' theory are as follows.

(i) In electrolyte solutions the molecules spontaneously dissociate into ions, so that the solution becomes conductive. Different electrolytes exhibit different degrees of dissociation, α, which will influence the actual values of molar conductivity Λ; the two parameters are interrelated as

$$\alpha = \Lambda/\Lambda^0, \tag{10.11}$$

where Λ^0 is the limiting value of Λ at complete dissociation.

(ii) Because of dissociation and the resulting increase in the total number of particles in solution, the parameters of the colligative properties assume higher values. These values are proportional to the total concentration, c_σ, of particles (ions and undissociated molecules) in the solution, which for a binary electrolyte is given by $[1 + \alpha(\tau_k - 1)]c_k$. The isotonic coefficient i is the ratio of c_σ and the concentration c_j that would be observed in the absence of dissociation:

$$i = 1 + \alpha(\tau_k - 1) \quad \text{or} \quad \alpha = (i - 1)/(\tau_k - 1). \tag{10.12}$$

In the case of complete dissociation, $\alpha = 1$ and $i = \tau_k$, i.e., the isotonic coefficient assumes integer values.

(iii) For any given electrolyte that is dissolved, the degree of dissociation increases as the solution is made more dilute.

Thus, quantitative criteria which could be tested experimentally had now been formulated for the first time in the theory of electrolytic dissociation, in contrast to the earlier theories. The good agreement between

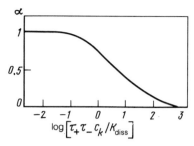

Fig. 10.4. Degrees of dissociation of a 1:1 electrolyte as a function of solution concentration.

degrees of dissociation calculated from independent measurements of two different properties with Eqs. (10.11) and (10.12) was a fundamental and rather convincing argument for this theory and contributed to its success (in 1903 Arrhenius was awarded the Nobel prize in chemistry for its development).

The third point of Arrhenius' theory was amplified in 1888 by Wilhelm Ostwald. He introduced the idea of an equilibrium between the ions and the undissociateu molecules:

$$M_{\tau_+}A_{\tau_-} \rightleftarrows \tau_+ M^{z+} + \tau_- A^{z-}, \tag{10.13}$$

which obeys the laws of chemical equilibria:

$$[M^{z+}]^{\tau_+}[A^{z-}]^{\tau_-}/[M_{\tau_+}A_{\tau_-}] = K_{\text{diss}}, \tag{10.14}$$

where K_{diss} is the dissociation constant, which for a given system (of solute and solvent) is solely a function of temperature.

Using the expressions for the concentrations of ions and undissociated molecules we can write this equation as well in the form

$$\alpha^2 \tau_+ \tau_- c_k/(1-\alpha) = K_{\text{diss}}. \tag{10.15}$$

Thus, the degree of dissociation is a function of solute concentration,

$$\alpha = (K_{\text{diss}}/2\tau_+\tau_- c_k)\left[\sqrt{1 + (4\tau_+\tau_- c_k/K_{\text{diss}})} - 1\right], \tag{10.16}$$

i.e., the higher the concentration the lower the value of α (Fig. 10.4).

Substituting the value of α into Eq. (10:11) we obtain the relation between Λ and concentration:

$$\Lambda = (K_{\text{diss}}\Lambda^0/2\tau_+\tau_-c_k)\left[\sqrt{1 + (4\tau_+\tau_-c_k/K_{\text{diss}})} - 1\right]. \qquad (10.17)$$

At low concentrations when $c_k \ll K_{\text{diss}}/4\tau_+\tau_-$ and the value of α approaches unity, relation (10.17) becomes (after series expansion of the square root)

$$\Lambda = \Lambda^0 - (4\tau_+\tau_-\Lambda^0/K_{\text{diss}})c_k. \qquad (10.18)$$

At higher concentrations when $c_k \gg K_{\text{diss}}/4\tau_+\tau_-$ and the value of α is very low we obtain accordingly

$$\Lambda = \Lambda^0\sqrt{K_{\text{diss}}/\tau_+\tau_-c_k}. \qquad (10.19)$$

The relations described by Eqs. (10.15) to (10.19) became known as *Ostwald's dilution law.*

Soon after inception of the theory of electrolytic dissociation it was shown that two types of compounds exist which can dissociate upon dissolution in water (or other solvents):

(i) Compounds forming ionic crystals (e.g., NaCl). In them the ions exist even prior to dissolution, but are held in lattice sites owing to electrostatic interaction. In ionic lattices covalent bonds between the ions are practically inexistent. These lattices disintegrate during dissolution and the ions become mobile (free). Such substances are called ionophors.

(ii) Compounds consisting of molecules with covalent bonding (e.g., HCl). They form ions only upon dissolution as a result of interaction with the solvent. They are called ionogens.

The concept of degree of dissociation made it possible to distinguish strong electrolytes (where the values of α in the solution are close to unity) and weak electrolytes (where the values of α are low). This distinction is somewhat arbitrary, since according to Eq. (10.16) the degree of dissociation depends on solution concentration.

Experimental data show that, at the usual concentrations (10^{-3} to 10 M), most salts and also the hydroxides of alkali metals are strong electrolytes. This is true also for some inorganic acids: HCl, $HClO_4$, and others. Weak electrolytes are the organic acids and the hydroxides of metals other than alkali. Few electrolytes of the intermediate type (with moderate values of α) exist; in particular, certain transition-metal halides such as $ZnCl_2$, ZnI_2, $CdCl_2$ are in this category.

The theory of electrolytic dissociation also provided the possibility for a transparent definition of the concept of acids and bases. According to the concepts of Arrhenius, an acid is a substance which upon dissociation forms hydrogen ions, and a base is a substance which forms hydroxyl ions. Later these concepts were extended (see Section 10.9).

10.3.3. Further Development of the Theory of Electrolytic Dissociation

The theory of electrolytic dissociation was not at once universally recognized, despite the fact that it could qualitatively and quantitatively explain certain fundamental properties of electrolyte solutions. For many scientists the reasons for spontaneous dissociation of stable compounds were obscure. Thus, an energy of about 770 kJ/mol is required to break up the bonds in the lattice of NaCl, and about 430 kJ/mol are required to split the H–Cl bonds during formation of hydrochloric acid solution. Yet the energy of thermal motions in these compounds is not above 10 kJ/mol. It was the weak point of Arrhenius' theory that this mismatch could not be explained.

Between 1865 and 1887 Dmitri I. Mendeleev developed another, so-called chemical theory of solutions. According to this theory the dissolution process is the chemical interaction between the solutes and the solvent. Upon dissolution of salts, dissolved hydrates are formed in the aqueous solution which are analogous to the solid crystal hydrates. In 1889 Mendeleev criticized Arrhenius' theory of electrolytic dissociation. Arrhenius in turn did not accept the idea that hydrates exist in solutions.

It was found in later work that it is precisely the idea of ionic hydration that is able to explain the physical nature of electrolytic dissociation. The energy of interaction between the solvent molecules and the ions that are formed is high enough to break up the lattices of ionophors or the chemical bonds in ionogens (for more details see Section 10.6). The significance of ionic hydration for the dissociation of electrolytes had first been pointed out by Ivan A. Kablukov in 1891.

According to modern views, the basic points of the theory of electrolytic dissociation are correct and were of exceptional importance for the development of solution theory. However, there are a number of defects. The quantitative relations of the theory are applicable to dilute solutions of weak electrolytes (up to 10^{-3}–10^{-2} M). Deviations are observed at higher concentrations; the values of α calculated with Eqs. (10.11) and (10.12) do not coincide; the dissociation constant calculated with Eq. (10.15) varies with concentration, etc. For strong electrolytes the quantitative relations of the theory are altogether inapplicable, even in extremely dilute solutions.

Arrhenius' theory contains the idea that the effects of the ions on the colligative properties are additive, i.e., that interactions between the ions are absent. However, a simple calculation shows that at distances of less than 10 to 20 nm between the ions, marked electrostatic interaction will arise. These are the distances between ions in solutions having concentrations of 10^{-3}–10^{-4} M. The fact that electrostatic forces must be taken into account was pointed out in 1894 by Johannes Jacobus van Laar.

At the beginning of the 20th century the idea was put forward that in solutions of strong electrolytes the degree of dissociation is not simply high but dissociation of the solute is complete, i.e., equilibrium between ions and undissociated molecules does not exist. This point is particularly evident for ionophors, which in the solid state do not possess individual molecules and for which it is unlikely that undissociated molecules should appear in a solution.

Hence, the theory of electrolyte solutions subsequently developed in two directions: (a) studies of weak electrolyte solutions, in which a dissociation equilibrium exists, and where because of the low degree of dissociation the concentration of ions and the electrostatic interaction between the ions are minor; and (b) studies of strong electrolyte solutions, in which electrostatic interaction between the ions is observed.

Of great importance for the development of solution theory was the work of Gilbert N. Lewis, who introduced the concept of activity in thermodynamics (1907) and in this way greatly eased the analysis of phenomena in nonideal solutions. Substantial information on solution structure was gathered too, when the conductivity (Section 10.4) and the activity coefficients (Section 10.7) were analyzed as functions of solution concentration.

10.4. CONDUCTIVITY OF ELECTROLYTE SOLUTIONS

10.4.1. Methods of Measurement

All methods of determining the conductivity of electrolyte solutions rest on measurements of the ohmic resistance of a symmetric galvanic cell

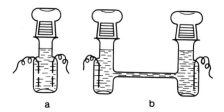

Fig. 10.5. Cells for conductivity measurements in solutions of low (a) and high (b) values of conductivity.

with platinum electrodes and the test solution through which a current is made to flow.

Current flow in the cell is attended by polarization of the electrodes, and this polarization is superimposed on the ohmic voltage drop. The most perfect method was proposed by Kohlrausch in 1868, and it is used even today. He employed electrodes of platinized platinum (through the increase in true surface area of the electrodes, the current density and polarization are drastically reduced) and alternating current (500 to 2000 Hz, which reduces the influence of concentration polarization and other kinds of polarization; see Section 9.4).

In this method a bridge circuit is used (see Fig. 9.14). Since cells have a certain capacitance, bridge balance requires compensation not only of their resistance but also of their reactive impedance. For higher sensitivity and accuracy in the measurements the setup should be as symmetric as possible (identical impedance in the bridge arms, symmetric disposition of the individual circuit elements and connecting leads).

The cell resistance measured is inversely proportional to the desired solution conductivity: $R_{cell} = K_{cell}/\sigma$, and directly proportional to the specific resistance. The factor K_{cell}, the cell constant, depends on the size and mutual disposition of the electrodes (the cross section and length of the liquid column between them). It is determined not by calculation but by calibration of the cell with a standard electrolyte solution having a precisely known conductivity, e.g., KCl solution of a particular concentration. The resistance of the cell with solution should be between 0.1 and 50 kΩ, which can be attained by suitable cell design: electrodes close together when the conductivity is low (Fig. 10.5a), and electrodes far apart with a reduced cross section of the solution column between them when the conductivity is high (Fig. 10.5b).

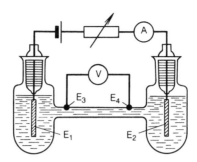

Fig. 10.6. Arrangement with four-electrode cell for measuring the conductivity of solutions.

Often the conductivity of solutions is measured by a different method involving four-electrode cells (Fig. 10.6). Outer electrodes E_1 and E_2 are used to produce current flow in the cell, while the ohmic voltage drop in the solution is measured with the aid of the rigidly mounted indicator electrodes E_3 and E_4. Since no current flows in the circuit of the indicator electrodes they are not polarized, and dc can be used for the measurements.

The temperature coefficient of conductivity is about 2 to 3% per degree for electrolyte solutions. For precision measurements the cells are thermostated to 0.002–0.005 K.

When all precautions are taken the accuracy and reproducibility of the measurements can be pushed to 0.01%. In highly dilute solutions (below 10^{-4} M) the accuracy decreases because of the increasing contribution of the water's own conductivity.

10.4.2. Dependence of Conductivity on Solution Concentration

In the case of binary solutions it is convenient to use the molar conductivity $\Lambda \equiv \sigma/c_k = \alpha z_k F(u_+ + u_-)$ considered in Section 1.3.2, which reflects the variation of the values of α and u_j with concentration. In highly dilute solutions when $\alpha \to 1$ and $u_j \to u_j^0$, parameter Λ tends toward its limiting value of Λ^0, which only depends on the nature of the electrolyte. The values of Λ decrease with increasing concentration, since both the degree of dissociation and the ionic mobilities decrease. Writing f_u (the mobility factor) for the degree of reduction of the combined ionic mobilities $(u_+ + u_-)/(u_+^0 + u_-^0)$, we can formulate Λ as

$$\Lambda = \alpha f_u \Lambda^0. \tag{10.20}$$

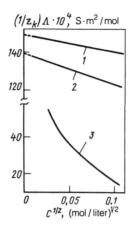

Fig. 10.7. Molar conductivities as functions of concentration for solutions of KCl (1), $BaCl_2$ (2), and CH_3COOH (3).

The limiting value of f_u in highly dilute solutions is unity. The decrease of this value which occurs with increasing concentration is due to the development and increase in ion–ion interactions.

In binary solutions of strong electrolytes for which $\alpha = 1$ over the entire concentration range, the decrease in Λ is due solely to the decrease in factor f_u. In 1900 Kohlrausch found that in such solutions the value of Λ decreases linearly with increasing $\sqrt{c_k}$ (Kohlrausch's *square-root law*):

$$\Lambda = \Lambda^0 - k\sqrt{c_k}, \qquad (10.21)$$

where k is an empirical constant which depends on the charge but not on the nature of the ions (Fig. 10.7, curves 1 and 2).

This law holds in dilute solutions up to a concentration of $\approx 10^{-2}\ M$. Using it we can find the limiting value, Λ^0, by extrapolation of experimental data plotted as Λ against $\sqrt{c_k}$ (values of Λ for the most dilute solutions in which measurements are still possible sometimes differ markedly from the limiting values Λ^0).

The square-root law could not be explained in terms of the Arrhenius concepts. It was only in 1926 that an interpretation could be offered in terms of the theory of ion–ion interaction (Section 10.8), and a method for calculating constant k was proposed.

In solutions of weak electrolytes the ionic concentrations are not high, hence, to a first approximation we can neglect the influence of ion–ion interaction, and assume that parameter Λ changes with concentration only as a result of changes in the degree of dissociation. This approach was employed when deriving Eq. (10.11) and the Eq. (10.17) following from it.

Thus, the concentration dependence of Λ in dilute solutions of weak electrolytes (10.18) differs from that in dilute solutions of strong electrolytes (10.21).

In concentrated electrolyte solutions more complex functions are found. Because of the drastic decrease of f_u and (or) α which occurs with increasing concentration, in a number of cases not only the molar conductivity Λ will decrease but even the conductivity σ itself (see Fig. 1.3).

10.4.3. Calculation of the Molar Conductivities of Ions

According to Eq. (1.21), the limiting value Λ^0 can be written as a combination of limiting values of the molar conductivities of the two ions: $\Lambda^0 = \tau_+\lambda_+^0 + \tau_-\lambda_-^0$. Below we report the values of Λ^0 (in units of 10^{-4} S·m^2/mol) at 25°C for solutions of a number of strong electrolytes (potassium and sodium salts):

	MF	MCl	MClO$_4$	CH$_3$COOM	M$_2$SO$_4$	(COOM)$_2$
M = K	128.9	149.85	140.86	114.4	307.0	295.3
M = Na	105.5	126.45	117.45	91.0	260.2	248.5

We can see from these data that the differences between the values of $(1/\tau_+)\Lambda^0$ for potassium and sodium salts having the same anion are independent of the anion; they always amount to $23.4 \cdot 10^{-4}$ S·m^2/mol.

It was on the basis of such observations that Kohlrausch in 1900 formulated his law of independent migration of ions, according to which the limiting value of molar conductivity λ_j^0 of any ion is independent of the other ion present in the solution. This law is valid too in multicomponent systems with any number of ion types. It reflects the condition that there are no interactions between the ions and that each ion moves independently of other ions in an electric field.

Experience shows that this law is strictly valid only at very low concentrations; at higher concentrations the actual values of λ_j (and the values of u_j which are proportional to them) depend not only on the bulk concentration but also on the nature of the other ions present in the solution.

Thus, in 1 M solutions of KCl and KNO$_3$, the values of λ_{K^+} are $47.7 \cdot 10^{-4}$ and $41.3 \cdot 10^{-4}$ S $\cdot$ m^2/mol, respectively.

A direct determination of the Λ^0 values by extrapolation of experimental data is impossible in solutions of weak electrolytes, since this would require measurements to be made in overly dilute solutions. Therefore, these values of Λ^0 are found with the aid of Eq. (1.21) from λ_j^0 values determined for solutions of strong electrolytes. Thus, according to measurements in HCl and CH$_3$COONa solutions (strong electrolytes), the λ_j^0 values for the H$^+$ and CH$_3$COO$^-$ ion are $349.8 \cdot 10^{-4}$ and $40.9 \cdot 10^{-4}$ S $\cdot$ m^2/mol, respectively. We thus find for a solution of the weak electrolyte CH$_3$COOH that $\Lambda^0 = 390.7 \cdot 10^{-4}$ S $\cdot$ m^2/mol. This value is much higher than the Λ values measured experimentally in CH$_3$COOH solutions (Fig. 10.7, curve 3).

For a determination of the individual values of the limiting molar conductivities of the cation and anion from the combined value of Λ^0, Eq. (1.21) must be supplemented by an equation linking λ_+^0 and λ_-^0 with parameters that can be determined experimentally. A suitable parameter are the ionic transport numbers t_j, and particularly their limiting values t_j^0 for $c_j \to 0$. (Ways to measure transport numbers are described in Section 10.5.) By definition, in binary solutions λ_j^0 and t_j^0 are interrelated as

$$\tau_+ \lambda_+^0 : \tau_- \lambda_-^0 = t_+^0 : t_-^0. \tag{10.22}$$

Values of λ_j^0 for various ions at 25°C which were determined with the aid of Eqs. (1.21) and (10.22) are reported in Table 10.1.

We can see that for all ions other than H$^+$ and OH$^-$ the values of $\lambda_j^0 / |z_j|$ (the "ionic equivalent conductance," see Section 1.3.2) vary within the relatively narrow limits given by $(60 \pm 20) \cdot 10^{-4}$ S $\cdot$ m^2/mol. The same holds true for the ionic mobilities $u_j^0 = \lambda_j^0 / |z_j| F$, which have values within limits given by $(6 \pm 2) \cdot 10^{-8}$ m^2/V $\cdot$ s.

According to Stokes' law, when a sphere of radius r moves in a liquid with viscosity η, the drag coefficient is given by

$$\theta = 6 \pi \eta r. \tag{10.23}$$

We cannot immediately apply this equation to ionic motion in solutions since the ions and the water molecules are of comparable size and the water, therefore, cannot be regarded as a continuum. The concept of Stokes radii, $r_{St,j}$, of the ions is often used, however; provisionally, these

TABLE 10.1 Limiting Values of Molar Conductivity λ_j^0 of the Ions at 25°C and Stokes Radii $r_{St,j}$ of the Same Ions

Cation	$\lambda_j^0 \cdot 10^4$, S · m²/mol	$r_{St,j}$, nm	Anion	$\lambda_j^0 \cdot 10^4$, S · m²/mol	$r_{St,j}$, nm
H⁺	349.8	0.026	OH⁻	198.8	0.046
Li⁺	38.72	0.237	F⁻	55.4	0.166
Na⁺	50.14	0.183	Cl⁻	76.35	0.120
K⁺	73.55	0.125	Br⁻	78.14	0.117
NH_4^+	73.55	0.125	I⁻	76.84	0.117
$N(CH_3)_4^+$	44.92	0.204	NO_3^-	71.46	0.128
$N(C_4H_9)_4^+$	19.5	0.471	ClO_4^-	67.36	0.136
Ag⁺	61.90	0.184	HCOO⁻	71.46	0.128
Mg^{2+}	106.1	0.346	CH_3COO^-	40.9	0.224
Ca^{2+}	119.0	0.308	SO_4^{2-}	160.0	0.229
Cu^{2+}	113.2	0.324	CO_3^{2-}	138.6	0.265
Zn^{2+}	107.0	0.343			
Ce^{3+}	209.4	0.395			
Al^{3+}	189.0	0.437			

radii are calculated from Eqs. (10.23) and (1.8) and the mobility values under the assumption that Stokes' law is applicable (see Table 10.1).

The reasons for the elevated values of molar conductivity and mobility of the H⁺ and OH⁻ ions will be discussed in Section 10.6.

10.4.4. Anomalous Effects

Between 1927 and 1929 it was discovered that unusual conductivity behavior will occur under certain extreme conditions.

Thus, in electrostatic fields having very high field strengths **E** ($> 10^6$–10^7 V/m), the conductivities σ of electrolyte solutions (and also the values of Λ) were found to become higher (the *Wien effect*), whereas according to Ohm's law [Eq. (1.3)] these parameters should not depend on **E**. This anomalous increase in conductivity becomes more pronounced with increasing concentration and charge of the ions. The effect is observed in weak and in strong electrolytes. In the latter, the values of Λ asymptotically approach the limiting values Λ^0 with increasing **E**.

Moreover, the values of σ and Λ increase when very high ac frequencies (above 1 MHz) are used for the measurements (the *Debye–Falkenhagen effect* or frequency dispersion of conductivity). With increasing frequency, the values of Λ again tend toward limiting values, which in this case are somewhat lower than Λ^0.

The two effects described have been interpreted qualitatively and quantitatively with the theory of ion–ion interaction (see Section 10.8.4).

10.5. IONIC TRANSPORT NUMBERS

Transport numbers t_j are a measure not of an individual property of the ion j but of the importance of this ion for migrational charge transport in a given electrolyte solution (here we shall only be concerned with the ionic transport numbers found in the absence of concentration gradients; cf. Section 4.3.3). According to Eq. (1.19), in binary solutions the transport number of one ion depends on the mobility of the second ion; and according to Eq. (1.13), in multicomponent solutions it depends in addition on the relative concentrations of the solution components.

It has been shown in Section 10.4.3 that the transport numbers for binary solutions can be used in conjunction with experimental data for the molar conductivities Λ^0 to calculate very important data for individual ions, viz., their molar conductivities λ_j^0 and, from them, the mobilities u_j^0. Values of the transport numbers are also useful in other calculations.

Several methods exist which allow transport numbers in solutions to be relatively accurately determined.

10.5.1. The Hittorf Method

In the method which was developed by Wilhelm Hittorf between 1853 and 1859, the transport numbers are determined by measuring concentration changes in the solution near the electrodes.

During current flow in a galvanic cell, the amount of component j present near each electrode will change for two reasons: the component's involvement in an electrode reaction, and its transport by migration (it is assumed that diffusion and convection effects are absent). For an amount of charge Q passing through the cell, the change in the number of moles, Δn_j, associated with the first factor can be determined by Eq. (1.44); the change associated with the second factor can be found with the aid of Eq. (1.47). Hence, we have for the total change in the number of moles of component j,

$$\Delta n_j = (Q/F)[(\bar{\nu}_j/n) - (\operatorname{sign} i)(t_j/z_j)].\qquad(10.24)$$

This equation can be used to calculate t_j when the values of Δn_j and Q have been determined experimentally.

The transport numbers are measured in binary solutions of strong electrolytes. Electrodes in which the electrode reactions involve either the cation or the anion are selected. These reactions should have 100% current yields, i.e., side reactions and, in particular, reactions involving gas evolution must be excluded.

Consider the concentration changes that occur near a cathode of metal M where the electrode reaction $M^{z+} + z_+e^- \rightarrow M$ takes place (an electrode of the first kind). Here $i < 0$, $n = z_+$, $\bar{\nu}_+ = -1$, and $\bar{\nu}_- = 0$. From Eq. (10.24) we find (taking into account that $1 - t_+ = t_-$)

$$\Delta n_+ = -(Q/z_+F)t_-, \qquad \Delta n_- = -(Q/|z_-|F)t_-.\qquad(10.25)$$

A change Δn_k in the total number of moles of electrolyte is associated with the change $\Delta n_k = (1/\tau_+)\Delta n_+ = (1/\tau_-)\Delta n_-$ in the number of moles of ions. Then

$$\Delta n_k = -(Q/z_kF)t_-.\qquad(10.26)$$

Sometimes it is difficult to select a suitable electrode. Electrodes of the first kind are used for cations such as Cu^{2+}, Ag^+, and the like, since their electrode reactions are not attended by side reactions such as hydrogen evolution. Silver electrodes of the second kind, Ag/AgX, are most often used when halide ions X^- are the anions; at these electrodes, cathodic reactions $AgX + e^- \rightarrow Ag + X^-$ take place for which

$$\Delta n_k = (Q/z_kF)t_+.\qquad(10.27)$$

When electrodes of the first or second kind are used as anodes the changes in the amounts of substance will evidently be equal in magnitude but opposite in sign.

We can see from these equations that the solution concentration changes occurring near each electrode are proportional to the transport number of the ion that does not react at that electrode.

For the measurements, a cell consisting of three interconnected compartments is used (Fig. 10.8). During current flow the solution concentration will change only in the outer compartments housing the electrodes.

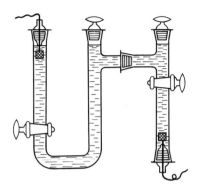

Fig. 10.8. Cell for measuring transport numbers according to Hittorf.

The central compartment is free of concentration gradients, hence no diffusion occurs in it, and the ions only move because of migration. The cell is bent and carefully thermostated in order to reduce convective flows. Minute currents are used in order to avoid heating in the cell.

When current flow has been arrested the solutions are withdrawn from each compartment separately, without any mixing, and the mass and concentration of each solution sample are determined. Because of changes in solution density it is more convenient to use molalities m_k. Knowing the concentrations before (m_k^0) and after (m_k') the experiment, and calculating the mass of water, M_{aq}, in each solution sample from the total mass of the solution, we can compute the changes in the numbers of moles of reactant:

$$\Delta n_k = (m_k' - m_k^0)M_{aq}. \tag{10.28}$$

Here the assumption is made that the amount of water present in any compartment does not change during the experiment.

Using the Hittorf method we can determine ionic transport numbers to $\pm 0.1\%$. It is an advantage of the method that concentrated solutions can be used. In dilute solutions ($< 10^{-2}$ M) the accuracy falls because of the difficulties met when measuring minor concentration changes.

The values of t_j determined by this method are called effective or Hittorf transport numbers.

10.5.2. Moving Boundary Method

This method was suggested by Oliver Lodge in 1886 and improved by Lewis G. Longsworth et al. in 1923. In a vertical, cylindrical cell

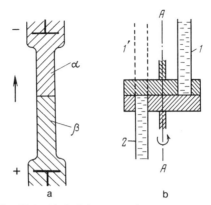

Fig. 10.9. (a) Cell for measuring transport numbers with the moving-boundary method, (b) the turning device used to set up a sharp boundary between the two solutions.

(Fig. 10.9a), a sharp boundary is set up between two electrolyte solutions, the test solution (α) of an electrolyte $M_{\tau_+}A_{\tau_-}$ for which one wants to find the transport numbers, and the indicator solution (β) of electrolyte $N_{\tau'_+}A_{\tau'_-}$ having the anion in common with the former electrolyte.

To set up such a boundary one uses a turning device with two ground-glass disks (Fig. 10.9b). Tubes 1 and 2 are filled in advance with the corresponding solutions. By turning the upper disk around axis AA, tube 1 is brought in position $1'$ and the sharp boundary is set up.

When current flows in the cell in the direction indicated in the drawing, anions will migrate downwards, and cations M^{z+} and $N^{z'+}$ will migrate upwards. The first of these cations is called the leading ion; the other is called the trailing ion. Under experimental conditions such that the rates of migration, v_M and v_N, of the two kinds of cation are the same, the boundary between the solutions will remain sharp and move upwards with the same velocity, $v_b = v_M = v_N$. This velocity can be measured optically (e.g., by observing the differences in refractive indices of the two solutions).

The rate of migration of cations in solutions of strong electrolytes is given by [taking into account Eqs. (1.7), (1.18), and (1.19)]

$$v_j^{(\text{sol})} = u_j \mathbf{E}^{(\text{sol})} = u_j i / \sigma^{(\text{sol})} = i t_j^{(\text{sol})} / F z_k c_k^{(\text{sol})} \qquad (10.29)$$

[the superscript sol stands for solution (α) or (β)]. Hence, the condition for equal velocities of ions M^{z+} and $N^{z'+}$ can be written as

$$z_{MA} c_{MA}^{(\alpha)} : z_{NA} c_{NA}^{(\beta)} = t_M^{(\alpha)} : t_N^{(\beta)}. \tag{10.30}$$

This expression is called *Kohlrausch's regulating function*. When the values are such that this function is obeyed, the transport number of the leading ion can be calculated from the speed of the boundary:

$$t_M^{(\alpha)} = (z_{MA} F c_{MA}^{(\alpha)} / i) v_b. \tag{10.31}$$

It is difficult in experiments to preselect the concentration of the indicator solution so that the regulating function will be obeyed exactly. Hence, a somewhat lower concentration is adopted, which leads to higher velocity v_N. During current flow the ions $N^{z'+}$ (and, because of electroneutrality, the anions as well) will accumulate in the solution layer next to the boundary until the concentration in it has attained the value required by Eq. (10.30).

Interdiffusion of the ions [of ion M^{z+} from (α) to (β) and of ion $N^{z'+}$ in the opposite direction] causes gradual blurring of the boundary between the solutions. However, it can be arranged that during current flow the boundary remains sharp (of course, when there is no convection). To this end an indicator solution is selected such that $u_M > u_N$ and hence $t_+^{(\alpha)} > t_+^{(\beta)}$. It follows from Eq. (10.29), in this case, that when the regulating function is obeyed one has $\sigma^{(\alpha)} > \sigma^{(\beta)}$ and $E^{(\alpha)} < E^{(\beta)}$. When an ion M^{z+} has diffused into solution (β) it will be "hurried" toward the boundary by the field, since it has a higher mobility and since the field strength in solution (β) is higher than that in solution (α). An ion $N^{z'+}$ which has diffused into solution (α) will be delayed, and in the end "taken over" by the moving boundary. Thus, even here a peculiar mechanism of automatic adjustment is operative.

The moving boundary method is used when investigating dilute solutions. In the concentration range of 10^{-3}–10^{-1} M the measuring accuracy is higher than in the Hittorf method.

10.5.3. Concentration Dependence of Ionic Transport Numbers

For binary solutions of most salts the ionic transport numbers have values between 0.4 and 0.6, i.e., do not appreciably depart from a mean value of 0.5. This must actually be expected, since the cations and anions differ slightly in their mobilities. It is only in acids and bases where the mobilities of the H^+ and OH^- ions are appreciably higher than those of the other ions that the H^+ and OH^- ions have transport numbers between 0.75 and 0.85.

One can observe a dependence of the transport numbers on solution concentration. The values t_j found for a concentration of 0.2 M may differ by 5 to 10% from the limiting value t_j at very low concentrations. As a rule, the transport number decreases with increasing concentration for an ion with $t_j < 0.5$, and for the other ion, with $t_j > 0.5$, it increases.

Sometimes an unusual dependence is found when the solution concentration is raised, and transport numbers are obtained which at first glance seem physically meaningless, viz., negative values or values larger than unity. This is typical for solutions where the ions tend to associate or form complexes with other ions or molecules. Hittorf already observed that in CdI_2 solution the transport number of the cadmium ion decreased from 0.445 to -0.15, and that of the iodide ion accordingly increased from 0.555 to 1.15, when the concentration was raised from 0.0025 to 1 M. He offered the correct explanation for this observation, by noting that complex, cadmium-containing anions are formed at higher concentrations following the reaction $Cd^{2+} + 4I^- \rightleftarrows CdI_4^{2-}$. These anions carry cadmium not to the cathode but to the anode, which distorts the results of the measurements. Analogous effects are observed in solutions of weak electrolytes, e.g., weak acids HA where aggregates of the type of $A^- \cdot HA$ may form which cause an additional transport of HA molecules to the anode.

10.5.4. Effective and True Transport Numbers

The transport numbers measured by the methods described above are falsified to some extent because of hydration of the ions, which has the effect that each ion when moving transfers a certain number of water molecules. Since cations are usually more strongly hydrated than anions, a certain amount ΔM_{aq} of water will be transferred during the experiment from the anode to the cathode compartment. The assumption used in the calculations following Eq. (10.28) that the mass of water in each compartment remains unchanged was therefore unjustified. The effective transport numbers t_j calculated under this assumption differ from the true transport numbers T_j calculated when allowing for water transfer.

If, after the experiment, the amount of water in the cathode compartment is M'_{aq}, then before the experiment it was $M'_{aq} - \Delta M_{aq}$. Therefore,

$$\Delta n_k = m'_k M'_{aq} - m^0_k(M'_{aq} - \Delta M_{aq}) = (\Delta n_k)_{exp} + m^0_k \Delta M_{aq}, \quad (10.32)$$

where $(\Delta n_k)_{exp}$ is the value calculated with Eq. (10.28).

Allowing for Eq. (10.26) we have, when an electrode of the first kind was used as the cathode,

$$T_- = t_- - (z_k F/Q)m^0_k \Delta M_{aq}, \quad (10.33)$$

and when an electrode of the second kind was used,

$$T_+ = t_+ + (z_k F/Q)m^0_k \Delta M_{aq}. \quad (10.34)$$

The same equations are obtained for the anode.

It follows from these equations that in highly dilute solutions the effective and true transport numbers coincide; they differ increasingly with increasing concentration.

One must know the amount of water transferred, ΔM_{aq}, in order to calculate true transport numbers. Basically, this can be determined experimentally from the concentration change of a neutral substance additionally introduced into the solution, e.g., sugar or urea, which are not involved in electrode reactions and do not migrate in the electric field. However, the correction thus obtained is not very accurate; experience shows that the neutral substance sometimes is "caught" by the moving ions. The water transfer caused by hydration could be calculated if one knew the number, h_j, of water molecules associated with each ion j. Unfortunately, in most cases the hydration numbers are not known with sufficient accuracy.

Thus, difficulties arise in the exact determination of true transport numbers. Fortunately, though, this is not particularly urgent inasmuch as the transfer of water of hydration is an integral quality of ionic motion, and all the phenomena mentioned above depend on effective rather than true transport numbers.

10.6. IONIC SOLVATION (HYDRATION) IN SOLUTIONS

Ionic solvation is the interaction between ions and solvent molecules which leads to the formation of relatively strong aggregates, the solvated

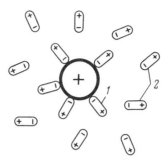

Fig. 10.10. Solvation of an ion: (1) nearest; (2) farther.

ions. In aqueous solutions the terms "ionic hydration" and "hydrated ions" are used as well.

Ions not solvated are unstable in solutions; between them and the polar solvent molecules, electrostatic ion–dipole forces, sometimes also chemical forces of interaction, arise which produce solvation. That it occurs can be felt from a number of effects: the evolution of heat upon dilution of concentrated solutions of certain electrolytes (e.g., sulfuric acid), the precipitation of crystal hydrates upon evaporation of solutions of many salts, the transfer of water during the electrolysis of aqueous solutions (see Section 10.5), and others. Solvation gives rise to larger effective radii of the ions and thus influences their mobilities.

Solvated ions have a complicated structure. The solvent molecules nearest to the ion form the so-called primary, or nearest, solvation sheath (Fig. 10.10). Owing to the small distances, the ion–dipole interaction in this sheath is strong and the sheath is stable. It is unaffected by thermal motion of the ion or solvent molecules, and when an ion moves it carries along its entire primary shell. In the secondary or farther shells, interactions are weaker; one notices an orientation of the solvent molecules under the effect of the ion. The disturbance caused by the ions among the solvent molecules becomes weaker with increasing distance and also with increasing temperature.

Thus, ionic solvation is associated with a substantial rearrangement of solvent structure; its primary structure is broken where the ion is located, and its molecules undergo reorganization (reorientation) within a certain volume around the ion.

The ion's solvated state can be described in two ways: (a) in terms of the energy effects: the heat, $\bar{q}^{(s)} = -\Delta H^{(s)}$, work, $\bar{w}^{(s)} = -\Delta G^{(s)}$, and entropy, $\Delta S^{(s)}$, of solvation, and (b) in terms of physical parameters: the solvation number h_j (the number of solvent molecules associated with ion j) and the radius $r^{(s)}$ of the solvated ion. These parameters refer to the primary sheath.

10.6.1. Solvation Energies of Electrolytes

Energy effects associated with the dissolution of a given substance (which in the following will be distinguished with the superscript "d") can be determined experimentally. They depend on the system's initial and final state, but not on the path taken by the process. Hence, for calculations, the device of thermodynamic cycles is often used, where the true path of the process is replaced by another path (which may even be a path that actually cannot be realized) for which the energy effects of the individual intermediate steps can be determined.

The dissolution process can be performed mentally in two steps: (a) from the original substance, ions are formed in the gas phase (or in vacuum) where they are sufficiently far apart so that their electrostatic interaction can be excluded, and (b) transfer of these ions from the gas phase into the solvent. The first step is independent of the solvent. For ionophors, it implies breakdown of the lattice, and for ionogens, it implies breaking the chemical bonds in the original molecule and ionizing the resulting atoms or atom groups. Provisionally, this step will be called "break-up" (superscript "b"). The energy effects of the second step are related to ionic solvation (superscript "s"). The total heat of solution of a compound is the sum of heats of the two steps:

$$\bar{q}_k^{(d)} = \bar{q}_k^{(b)} + \bar{q}_k^{(s)}. \tag{10.35}$$

This equation is used in calculating heats of solvation of electrolytes. The heat of solution can be determined highly accurately by calorimetry (with an error of $< \pm 0.1\%$). This heat is relatively small, and the values are between -100 and $+40$ kJ/mol. Different methods exist to approximately calculate the break-up energies on the basis of indirect experimental data or models. Unfortunately, the accuracy of these calculations is much lower, i.e., not better than $\pm 5\%$.

Determining lattice break-up energies from experimental data. The process of lattice break-up can be split into individual steps for which the energies can be measured. Thus, breaking up the NaCl lattice to form free ions in the gas phase can be described (with a *Born–Haber cycle*) as

$$\text{NaCl (s)} \underset{\text{(1)}}{\Bigg[} \quad \begin{array}{l} \xrightarrow{\text{(1)}} \text{Na (s)} \xrightarrow{\text{(2)}} \text{Na (g)} \xrightarrow{\text{(4)}} \text{Na}^+ \text{ (g)} \\[2mm] \xrightarrow{} \tfrac{1}{2}\text{Cl}_2 \text{ (g)} \underset{\text{(3)}}{\longrightarrow} \text{Cl (g)} \underset{\text{(5)}}{\longrightarrow} \text{Cl}^- \text{ (g)} \end{array} \qquad (10.36)$$

Calorimetrically or otherwise, heats can be determined for the steps: (1) decomposition of NaCl into the elements, $\bar{q}^{(1)} = -411$ kJ/mol; (2) evaporation of sodium, $\bar{q}^{(2)} = -109$ kJ/mol; (3) dissociation of the chlorine molecules, $\bar{q}^{(3)} = -121$ kJ/mol. From spectroscopic data one can determine the work required in the steps: (4) ionization of the sodium atoms, $\bar{w}^{(4)} = -496$ kJ/mol; (5) ionization of the chlorine atoms, $\bar{w}^{(5)} = +365$ kJ/mol. Neglecting the difference between heats and work in individual steps (the error thus introduced is within $\pm 5\%$ in the present case) we obtain for the total heat of crystal break-up: $\bar{q}^{(b)}_{\text{NaCl}} = \sum \bar{q}^{(i)} = -772$ kJ/mol.

Calculation of the electrostatic energy of lattice break-up as proposed by Max Born in 1919. Consider a pair of ions with charges $z_+ Q^0$ and $z_- Q^0$ (Q^0 is the elementary charge) in vacuum, a distance r apart. Electrostatic attraction forces $-z_+ |z_-| (Q^0)^2 / 4\pi\varepsilon_0 r^2$ are operative between the ions (the direction of these forces is conditionally taken as negative). An extreme approach of the ions is prevented by repulsion forces operating over very short distances. They can be described as B/r^{n+1}, where B and n are constants. The value of constant n can be found from the lattice compressibility, and is between 5 and 12. Therefore, the total force consists of two terms:

$$f = -z_+ |z_-| (Q^0)^2 / 4\pi\varepsilon_0 r^2 + B/r^{n+1}. \qquad (10.37)$$

Constant B can be found from the evident condition that at the equilibrium distance $r = r^0$ between the ions, the total force of interaction should be zero. We thus find the final expression for the total interaction force as

$$f = -[z_+ |z_-| (Q^0)^2 / 4\pi\varepsilon_0 r^2][1 - (r^0/r)^{n+1}]. \qquad (10.38)$$

The distance r^0 can be determined by X-ray diffraction.

The work $\bar{w}_1$ to be performed by the system against the electrostatic forces when the pair of charges is drawn apart to infinite distance can be found by integrating the force over path r from $r = r^0$ to $r \to \infty$.

In order to convert from an ion pair to the true lattice we must multiply the value of $\bar{w}_1$ by N_A (to obtain one mole of ion pairs) and take into account that each ion interacts not just with one ion of opposite charge but with all surrounding ions of either sign. This consideration gives rise to an additional numerical factor, the Madelung constant K_M, which depends on the lattice type. For the face-centered cubic NaCl lattice $K_M = 1.748$, for the body-centered CsCl lattice $K_M = 1.763$, and for the fluorite-type lattice (CaF$_2$) $K_M = 5.039$.

Thus, the general expression for the lattice break-up energy found as described above is

$$\bar{w}^{(b)} = K_M N_A \bar{w}_1 = K_M N_A \int_{r^0}^{\infty} f \, dr$$

$$= -K_M N_A [z_+ |z_-|(Q^0)^2 / 4\pi\varepsilon_0 r^0][1 - (1/n)]. \quad (10.39)$$

For the NaCl lattice, parameter r^0 has a value of 0.286 nm, and parameter n is close to 7.5. Substituting these values into Eq. (10.39) we find a value of -762 kJ/mol for the lattice energy, i.e., a value close to the experimental value of the heat of break-up that we had mentioned.

Determination of the break-up energies of ionogens. Through the analysis of molecular spectra one can determine the bond energies, i.e., the energies required to break a given molecule into atoms (or atom groups). Through the analysis of atomic spectra on the other hand, one can calculate the energy of ion formation from atoms. The energy of breaking an HCl molecule into free ions H$^+$ and Cl$^-$ thus determined is -432 kJ/mol.

Table 10.2 lists values for the break-up energies and first heats of solution in water for various compounds, as well as values for the hydration energies of these compounds calculated from Eq. (10.35). In view of what we have said, we can estimate that values of $\bar{q}_k^{(s)}$ are accurate to $\pm 5\%$. We see that the ionic solvation energies are high (several hundred kJ/mol), hence they can compensate the rather high break-up energies of ionophors and ionogens.

10.6.2. Solvation Energies of Individual Ions

If in an electrolyte solution the solvation energy of each ion is independent of the second ion's identity, the solvation energy of the electrolyte

TABLE 10.2 Heats of Break-up $\bar{q}^{(b)}$, First Heats of Solution $\bar{q}^{(d)}$, and Heats of Solvation $\bar{q}^{(s)}$ for Different Ionophors in Water (in kJ/mol)

Ionophor	$\bar{q}^{(d)}$	$\bar{q}^{(b)}$	$\bar{q}^{(s)}$	Ionophor	$\bar{q}^{(d)}$	$\bar{q}^{(b)}$	$\bar{q}^{(s)}$
LiF	−4.6	1031.0	1026.4	KI	−20.5	637.9	617.4
NaF	−0.4	912.1	911.7	RbF	+26.4	780.2	806.6
NaCl	−3.8	773.1	769.3	RbBr	−21.8	658.8	637.1
NaBr	+0.8	741.3	742.1	RbI	−26	622.0	596.1
KF	+17.6	810.4	828.0	CsF	+37.7	744.7	782.3
KCl	−17.2	702.8	685.6	CsI	−33.1	604.8	571.8
KBr	−20.1	678.5	658.4				

can be written as

$$\bar{q}_k^{(s)} = \tau_+ \bar{q}_+^{(s)} + \tau_- \bar{q}_-^{(s)}. \tag{10.40}$$

Above, the solvation energy has been defined as the energy set free upon transfer of ions of a given type from the gas phase into the solution. During this transfer the ions cross the phase boundary between gas and solution where the solution's surface potential $\chi \equiv \psi^{(sol)} - \psi^{(g)}$ is effective. During the crossing an additional energy $\bar{q}_j^{(\chi)} \approx \bar{w}_j^{(\chi)} = -z_j F \chi$ is evolved (per mole of the ions). Hence, two kinds of ionic solvation energy are distinguished: the chemical, $\bar{q}_j^{(s,chem)}$, which only characterizes the ion–solvent interaction, and the real, $\bar{q}_j^{(s,real)}$, which in addition includes crossing of the surface layer by the ions. These two parameters are interrelated as follows:

$$\bar{q}_j^{(s,real)} = \bar{q}_j^{(s,chem)} - z_j F \chi. \tag{10.41}$$

For the electrolyte as a whole the effects associated with cations and anions crossing the surface layer mutually cancel, and Eq. (10.40) is valid, both with the chemical and with the real ionic solvation energies.

The question arises, though, whether the solvation energies of individual ions can be determined experimentally. It is evident that if this value was known at least for one kind of ion, then from the solvation energies of the corresponding compounds we could determine the values of $\bar{q}_j^{(s)}$ for any other ion.

A few ways exist for theoretically calculating the solvation energies of individual ions, or calculating them from indirect experimental data.

Theoretical calculation of ionic solvation energies. A first method for such calculations was suggested by Max Born in 1920. In this method the solution is regarded as a homogeneous continuum with the relative permittivity ε. The ion's transfer from vacuum into the solution is mentally split into three steps: (1) removal of the electric charge from the ion in vacuum, (2) transfer of the uncharged particle from vacuum into the solution, and (3) restitution of charge to the particle in solution. Since only electrostatic forces, and not chemical forces, are considered, the work performed in the second step is zero. In calculating the work of discharging and charging, the assumption is made that the particle is a sphere of radius r_j. When this sphere carries a charge Q its surface ($r = r_j$) will have a potential $\psi(r_j)$ given by $Q/4\pi\varepsilon_0\varepsilon r_j$, according to Eq. (2.4). The work $dw = -d\bar{w} = \psi(r_j)dQ$ must be performed in order to increase the charge of the sphere by an amount dQ. We therefore find, for the work performed in the third step, which involves the charging of the ion in solution from $Q = 0$ to $Q = z_j Q^0$,

$$\bar{w}^{(3)} = - \int_0^{z_j Q^0} \psi(r_j)dQ = -z_j^2 (Q^0)^2/8\pi\varepsilon_0\varepsilon r_j. \qquad (10.42)$$

The work of discharge of the ion in vacuum (the first step) is determined by an analogous expression involving the value $\varepsilon = 1$.

Thus, for the total work required to transfer the ion from vacuum into the solution we obtain (per mole of ions)

$$\bar{w}_j^{(s)} = N_A \sum \bar{w}_j^{(i)} = [N_A z_j^2 (Q^0)^2/8\pi\varepsilon_0 r_j][1 - (1/\varepsilon)]. \qquad (10.43)$$

To convert from the work to heat of transfer, we use the Gibbs–Helmholtz equation (3.63). The only temperature-dependent parameter in Eq. (10.43) is ε. The expression for the heat of transfer becomes

$$\bar{q}_j^{(s)} = [N_A z_j^2 (Q^0)^2/8\pi\varepsilon_0 r_j][1 - (1/\varepsilon) - (T/\varepsilon^2)d\varepsilon/dT]. \qquad (10.44)$$

Comparison with experimental data shows that Born's equation (10.44) yields high values of $\bar{q}_j^{(s)}$. A defect of Born's model is that the solvent is regarded as a continuum with the unchanged bulk value of the parameter ε, even at short distances from the ion. However, the solvent molecules in the first solvation sheath are strongly oriented, and one cannot disregard the real structure of this sheath. In the model developed by John D.

Bernal and Ralph H. Fowler in 1933 and in models of other workers, the solvent around the ion is conditionally split into two regions, a first shell containing h_j solvent molecules, and the remainder which, as in Born's model, is regarded as a continuum.

Let μ_0 be the dipole moment of a solvent molecule and r_0 its radius. The electrostatic energy of interaction between the ion and h_j solvent molecules in the primary shell when computed per mole of ions can be written as follows [cf. Eq. (2.7)]:

$$\bar{w}^{(\text{prim})} = N_A h_j \mu_0 z_j Q^0 / 4\pi\varepsilon_0 (r_j + r_0)^2. \tag{10.45}$$

The interaction with the remainder of the solvent is determined by Born's equation, but taking into account that the molecules of this part of the solvent are found at distances larger than $r_j + 2r_0$.

In addition to these interactions one must also take into account that reorganization of the solvent molecules requires the expenditure of some energy $\bar{w}^{(\text{reorg})}$. Calculations show that this energy for water has values of -60 to -120 kJ/mol.

Calculation of the solvation energy from experimental data. The solvation energies of individual ions can be calculated from experimental data for the solvation energies of electrolytes when certain assumptions are made. If it is assumed that an ion's solvation energy only depends on its crystal radius (as assumed in Born's model), then these energies should be the same for ions K^+ and F^-, which have similar values of these radii (0.133 ± 0.002 nm). It follows that in aqueous solutions $\bar{q}_{K^+}^{(s)} = \bar{q}_{F^-}^{(s)} = (1/2)\bar{q}_{KF}^{(s)} = -414.0$ kJ/mol. With the aid of these values we can now determine the values for other ions. According to another hypothesis, it is the ions Cs^+ and I^- which have identical solvation energies: $\bar{q}_{Cs^+}^{(s)} = \bar{q}_{I^-}^{(s)} = (1/2)\bar{q}_{CsI}^{(s)} = -285.9$ kJ/mol. Here the smaller radius of the cesium ion (0.169 nm, as compared to 0.215 nm for the iodide ion) compensates the asymmetry of the water molecules.

Values for the heats of hydration of a number of ions which were calculated by the above methods on the basis of theoretical models and experimental data, are reported in Table 10.3. We see that there is a certain general agreement but in individual cases the discrepancies are large, due to inadequacies of the theoretical concepts used in the calculations.

TABLE 10.3 Heats of Hydration of Individual Ions (in kJ/mol) and Hydration Numbers h_j

Ion	r_{cryst}, nm	Calculation		Experimental data		h_j
		Eq. (10.44)	improved	from KF	from CsI	
Li^+	0.060	1162.4	670.1	612.4	529.9	5–6
Na^+	0.095	734.6	498.9	497.7	415.2	6–7
K^+	0.133	524.5	378.8	414.0	331.5	4
Rb^+	0.148	473.4	342.8	392.6	310.1	3
Cs^+	0.169	412.7	300.5	368.4	285.9	1–2
F^-	0.136	513.2	329.0	414.0	496.5	2–5
Cl^-	0.184	385.5	237.8	271.6	354.1	0–3
Br^-	0.195	357.9	216.0	244.4	326.9	0–2
I^-	0.216	323.1	187.1	203.4	285.9	0–1

10.6.3. Solvation Numbers

Different methods are available for determining the solvation number h_j and (or) the radius of the primary solvation sheath, e.g., (1) by comparing the values of the true and apparent ionic transport numbers, (2) by determining the Stokes radii of the ions [Eq. (10.23)], or (3) by measuring the compressibility of the solution [the compressibility decreases in the presence of ions owing to the decrease in specific volume of the water (electrostriction of water)]. These methods are not highly accurate.

The values of h_j for different ions have values between 0 and 15 (see Table 10.3). As a rule it is found that the solvation number will be larger the smaller the true (crystal) radius of the ion. Hence, the overall (effective) sizes of different hydrated ions tend to become similar. This is the explanation why different ions in solution have similar values of the mobilities or diffusion coefficients. The solvation numbers of cations (which are relatively small) are usually higher than those of anions. Yet for large cations, of the type of $N(C_4H_9)_4^+$, the hydration number is zero.

10.6.4. Hydration of Protons

The behavior of protons in aqueous solutions strongly differs from that of other ions, and also from that of protons themselves in organic solvents. The proton's hydration energy (about 1100 kJ/mol) and its mobility in aqueous solutions are two to four times higher than the corresponding parameters of other ions.

These special features are explained by an interaction between the proton and one of the water molecules, which is not merely electrostatic but also covalent. This yields a new chemical species, the hydroxonium ion H_3O^+. The existence of such ions was demonstrated in the gas phase by mass spectrometry, and in the solid phase by X-ray diffraction and NMR. The $H^+–H_2O$ bond has an energy of 712 kJ/mol, which is almost two-thirds of the proton's hydration energy.

The species H_3O^+ are subject to further hydration in the usual manner. Their primary sheath contains three water molecules linked through electrostatic forces, and in part through hydrogen bonds, i.e., the ion with its primary solvation sheath can be formulated as $H_9O_4^+$.

The $H^+–H_2O$ bond has the special feature that, although the bond energy is high, the proton will readily hop from one water molecule in the hydration complex to a neighboring water molecule. This hop is a quantum motion, and will occur only when the water molecules have a favorable mutual orientation. It will occur predominantly in the direction of an electric field that may be present in the solution. Therefore, in solutions with hydroxonium ions two transport mechanisms exist: (a) transport of hydrated hydroxonium ions (which is analogous to the transport of other kinds of ion) and (b) transport of nonhydrated protons along the relatively immobile framework of oriented water molecules. The second mechanism resembles that of charge transport suggested by Grotthuss in the early 1800s. As a result of the joint operation of these two mechanisms, protons have a higher mobility than other ions in aqueous solutions.

The elevated mobility of the hydroxyl ions is explained in analogous fashion.

10.7. ACTIVITY OF REAL ELECTROLYTE SOLUTIONS

10.7.1. Ways to Describe the Thermodynamics of Real Solutions

The thermodynamic properties of real electrolyte solutions can be described with different parameters: the solvent's activity a_0, the solvent's osmotic coefficients g or Φ, the solute's activity a_k, the mean ionic activities $a_\pm$, as well as the corresponding activity coefficients.

In electrolyte solutions one must allow for the increase in the total number of particles upon dissociation, and clearly distinguish between the analytical concentrations c_k of the original compounds, which disregard their dissociation, and the real concentrations c_j of ions and undissociated molecules. The total concentrations c_σ, m_σ, and Z_σ appearing in the equations for the colligative properties are those of all the particles in the solution, i.e., are linked to the real concentrations. For binary solutions of strong electrolytes $m_\sigma = \tau_1 m_1$. Therefore, if we take an example, Eq. (10.9) linking the solvent's activity to its practical osmotic coefficient becomes, in the case of such solutions,

$$- \ln a_0 = \Phi M_0 \tau_1 m_1. \tag{10.46}$$

The activities of the solvent and solutes are interrelated by the Gibbs–Duhem equation (3.16), which for binary solutions can be written as

$$n_0 d \ln a_0 + n_1 d \ln a_1 = 0. \tag{10.47}$$

Using this equation we basically can calculate the value of a_1 when the values of a_0 for different concentrations are known. Dividing both sides of the equation by n_1 and taking into account that $m_1 = n_1/n_0 M_0$ we find

$$-d \ln a_1 = (1/M_0 m_1)d \ln a_0. \tag{10.48}$$

When calculating the values of $\ln a_1$ we must integrate this equation from $m_1 = 0$ to m_1. Straightforward integration is not possible, since at $m_1 = 0$ the value of the right-hand side of this equation tends toward infinity. Hence, other variables must be used for the integration. The practical osmotic coefficient Φ can be used here. Differentiating Eq. (10.46) we obtain

$$-d \ln a_0 = M_0 \tau_1 (\Phi dm_1 + m_1 d\Phi). \tag{10.49}$$

Substituting this value of $d \ln a_0$ into Eq. (10.48), integrating from 0 to m_1, and taking into account that for $m_1 = 0$ the coefficient $\Phi = 1$, we find

$$(1/\tau_1) \ln a_1 (= \ln a_\pm) = \Phi - 1 + \int_0^{m_1} \Phi d \ln m_1. \tag{10.50}$$

10.7.2. Ways to Determine the Activity of Electrolyte Solutions

Two approaches exist for determining the activity of an electrolyte in solution: (a) by measuring the solvent's activity, and subsequently converting it to electrolyte activity via the Gibbs–Duhem equation, and (b) by directly measuring the solute's activity.

The solvent's activity can be determined by measuring the saturation vapor pressure above the solution. Such measurements are rather tedious, and their accuracy at concentrations below 0.1 to 0.5 M is not high enough to produce reliable data; therefore, this method is only used for concentrated solutions. The activity can also be determined from the freezing-point depression or boiling-point elevation of the solution. These temperature changes must be ascertained with an accuracy of about 0.0001 K, which is quite feasible. This method is mainly used for solutions with concentrations not higher than 1 M.

Direct measurements of solute activity are based on studies of the equilibria in which a given substance is involved. The parameters of these equilibria (the distribution coefficients, equilibrium constants, and emf of galvanic cells) are determined at different concentrations. Then these data are extrapolated to very low concentrations where the activity coincides with concentration and the activity coefficient becomes unity.

Consider two examples of equilibria of this kind.

Electrochemical cell without transference. Assume that we want to determine the activities of HCl solutions of different concentration. We assemble a galvanic cell with hydrogen and calomel electrode:

$$Pt, \ (H_2)|HCl(m_1)|Hg_2Cl_2, \ Hg|Pt. \tag{10.51}$$

The emf of this cell is defined by the equation

$$\mathcal{E} = \mathcal{E}^0 - (RT/F)\ln a_1. \tag{10.52}$$

Taking into account that in this case $a_1 = \gamma_{\pm}^2 m_1^2$, we can write this equation also as

$$\mathcal{E} + (2RT/F)\ln m_1 = \mathcal{E}^0 - (2RT/F)\ln \gamma_{\pm}. \tag{10.53}$$

Experimentally we can determine the values of $\mathcal{E}$ at different values of m_1. Then we construct a plot of the left-hand side of Eq. (10.53) against concentration, and extrapolate it to zero concentration, i.e., into the region of ideal solutions where $\ln \gamma_{\pm} = 0$ (Fig. 10.11). We thus determine the value of $\mathcal{E}^0$. Knowing this, we can readily find the values of $\gamma_{\pm}$ for any solution concentration investigated when using Eq. (10.53).

The accuracy of this method depends on the correct extrapolation of the experimental data. The error associated with the extrapolation

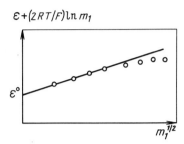

$\varepsilon + (2RT/F)\ln m_1$

ε°

$m_1^{1/2}$

Fig. 10.11. Extrapolation of experimental data for the determination of activity coefficients.

can be reduced by plotting the experimental data, not as a function of concentration but as a function of square root of concentration. It will be shown below that in this case the experimental data for dilute solutions fall onto a straight line which can be extrapolated more accurately to zero concentration than a curve.

Electrochemical cells with transference. This version of the method has the advantage that only one kind of reversible electrode is needed; on the other hand, the transport numbers of the electrolyte ions involved must be known in the concentration range studied.

Thus, the concentration cell with transference

$$\text{Pt, } (H_2)|HCl(m_1^{(\alpha)})|HCl(m_1^{(\beta)}|(H_2)Pt \qquad (10.54)$$

(or the analogous cell with two calomel electrodes) is used to determine the activities of HCl solutions. In one series of measurements the emf's of this cell are determined as a function of concentrations $m_1^{(\beta)}$, but at constant concentration $m_1^{(\alpha)}$. It follows from Eq. (5.38) that in this case

$$d\ln a_{\pm}^{(\beta)} = (F/2RT)(d\varepsilon/t_-). \qquad (10.55)$$

Integrating this equation from the value of $m_1^{(\alpha)}$ to different given values of $m_1^{(\beta)}$ and taking into account that $a_{\pm}^{(\beta)} = \gamma_{\pm}^{(\beta)} m_1^{(\beta)}$, we find

$$(F/RT)\int_\alpha^\beta (1/t_-)d\varepsilon - 2\ln m_1^{(\beta)} = 2\ln \gamma_{\pm}^{(\beta)} - \ln a_1^{(\alpha)}. \qquad (10.56)$$

The integral on the left-hand side can be evaluated (e.g., graphically) when t_- and ε are known as functions of the concentration $m_1^{(\beta)}$. Plotting (as in the previous case) the function on the left-hand side of Eq. (10.56) against $[m_1^{(\beta)}]^{1/2}$ and extrapolating this plot to $m_1^{(\beta)} = 0$ we find the value of $\ln a_1^{(\alpha)}$. Knowing this parameter, we can calculate $\gamma_\pm^{(\beta)}$ for any of the solutions studied.

Proceeding as described we find the stoichiometric activity coefficients $\gamma_\pm$ (= $\alpha f_\pm$) of the ions. A separate determination of α and $f_\pm$ (for weak electrolytes) is difficult.

10.7.3. Concentration Dependence of the Activity Coefficient

The activities have by now been determined for binary solutions of most electrolytes. As a rule, the values determined by different methods are in good mutual agreement (the scatter is not over 0.5%). These data are reported in special tables listing coefficients $\gamma_\pm$ as functions of molalities m_k.

Figure 10.12 shows such functions for binary solutions of a number of strong electrolytes and also, for the purposes of comparison, for solutions of certain nonelectrolytes (coefficients f_k). We can see that in electrolyte solutions the values of the activity coefficients vary within much wider limits than in solutions of nonelectrolytes. In dilute electrolyte solutions the values of $\gamma_\pm$ always decrease with increasing concentration. For many (but not all) electrolytes they go through a minimum and then increase with increasing concentration. In a number of cases very high values of $\gamma_\pm$ are attained in concentrated solutions. In other cases these values vary relatively little or decrease monotonically. The highest value, $\gamma_\pm = 1457$, was obtained for 5.5 m $UO_2(ClO_4)_2$ solution, and the lowest (among strong electrolytes), $\gamma_\pm = 0.0168$, for 2.5 m $CdCl_2$ solution.

The individual differences between electrolytes mainly appear in concentrated solutions; in dilute solutions distinct common traits are exhibited. When the experimental data are plotted as $\log \gamma_\pm$ against $c_k^{1/2}$, a linear relation is observed in very dilute solutions, as can be seen from Fig. 10.12:

$$- \log \gamma_\pm = k\sqrt{c_k}. \tag{10.57}$$

It is typical that in this region the curves of electrolytes of the same valence type practically coincide, i.e., at a given concentration the activity coefficients only depend on the electrolyte's valence type and not on its identity.

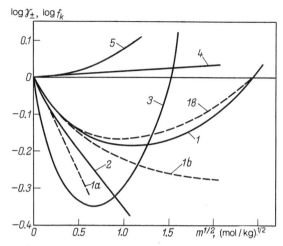

Fig. 10.12. Mean ionic activity coefficients as functions of molality for solutions of: (1) NaCl; (2) KNO₃; (3) CaCl₂. Activity coefficients as functions of molality for solutions of: (4) glycerol; (5) glycine. Dashed lines: curves for solutions of 1:1 electrolytes calculated (1a) via Eq. (10.79), (1b) via Eq. (10.83), and (1c) via Eq. (10.84).

In activity studies in multicomponent systems, Gilbert N. Lewis and Merle Randall found in 1921 that in the case of dilute solutions, when a foreign electrolyte is added, the activity change of the substance studied depends only on the concentration and valence type of the substance added, but not on its identity. For a quantitative characterization of solutions they introduced the concept of ionic strength of a solution,

$$I_c \equiv (1/2) \sum c_j z_j^2, \qquad (10.58)$$

where c_j are the real concentrations of the ions (not counting the undissociated molecules).

They formulated the *ionic-strength principle* according to which "in dilute solutions, the activity coefficient of a given strong electrolyte is the same in all solutions of the same ionic strength."

In 1922 Johannes Nicolaus Brønsted established an empirical relation for the activity coefficients in dilute electrolyte solutions:

$$-\log \gamma_\pm = z_+ |z_-| h \sqrt{I_c}, \qquad (10.59)$$

where h is a constant which does not depend on the identity or valence type of the electrolyte and has the approximate value of 0.50 (liter/mol)$^{1/2}$.

Equation (10.57) can be regarded as a particular case of the more general equation (10.59), since for binary solutions of strong electrolytes $I_c = (1/2)(\tau_+ z_+^2 + \tau_- z_-^2)c_k$. For binary 1:1, 1:2, and 2:2 electrolytes the values of I_c are c_k, $3c_k$, and $4c_k$, respectively. It follows that the coefficients k in Eq. (10.57) have values of h, $3.46h$, and $8h$ for the same electrolytes.

10.7.4. Physical Meaning of Activity Coefficients

The departure of a system from the ideal state is due to interaction forces between the individual particles contained in the system.

The dependence of chemical potential of a species on its concentration can be written as

$$\mu_j = \mu_{a,j}^0 + RT \ln c_j + RT \ln f_j. \tag{10.60}$$

The last term on the right-hand side,

$$w_{\text{int},j} \equiv RT \ln f_j, \tag{10.61}$$

(in kJ/mol) represents the energy contributed by this interaction; it is zero for ideal solutions.

When repulsion forces exist between the particles, the chemical potential of the corresponding species will increase (an additional energy $w_{\text{int},j} > 0$ must be expended in order to place a particle into a given volume), and hence, the activity coefficient will be larger than unity. When attraction forces are present the activity coefficient will be smaller than unity.

The ions in solution are subject to two kinds of forces: those of interaction with the solvent (solvation) and those of electrostatic interaction with other ions. The interionic forces decrease as the solution is made more dilute and the mean distance between the ions increases; in highly dilute solutions their contribution is small. However, solvation occurs even in highly dilute solutions, since each ion is always surrounded by solvent molecules. When determining the ionic activities one picks the standard state on the basis of condition (3.20), and regards a solution diluted to the limit as ideal, despite the fact that solvation-type interaction is present in it. This implies that the solvation energy which, to a first approximation, is independent of concentration, is included in the standard chemical potential μ_j^0, and has no influence on the activity.

Therefore, the activity coefficients in solutions are mainly determined by the energy of electrostatic interaction $w_{e,j}$ between the ions. It is only in concentrated solutions, when solvation conditions may change, that changes in (but not the existence of) solvation energy must be included, and nonelectrostatic interactions between the ions must also be accounted for.

10.8. PHYSICAL THEORIES OF ION–ION INTERACTIONS

It is the aim of physical solution theories to quantitatively calculate ion–ion interactions, i.e., to theoretically calculate the activity coefficients.

In an ionic lattice the energy of electrostatic interaction between the ions is high as compared to the thermal energy RT, hence, the ions are rigidly fixed in space and orderly arranged. In dilute solutions of nonelectrolytes, there are practically no interaction forces between the dissolved particles, and at any given time the relative positions of the particles are random and disordered owing to thermal motion. In electrolyte solutions the situation is intermediate: relatively weak electrostatic forces having an energy comparable with that of thermal motion exist between the ions. A certain degree of short-range order is observed in them; at short distances from a given ion, ions of opposite sign are more likely to be found. As a result attraction forces are predominant between the ions at short distances, and hence $w_{e,j} < 0$.

The first attempt at statistically calculating the distribution of ions in a solution while allowing for electrostatic interaction and thermal motion was made by S. Roslington Milner in 1912. The mathematical procedures used by him were very complicated.

In 1918, Jnanendra Chandra Ghosh proposed to calculate the energy of electrostatic interaction of the ions while assuming the ions in the solution to have a rigid arrangement resembling that in crystals, though allowing for the actual interionic distances. The function obtained by him is close to the experimental function (10.57), though with a different exponent of concentration. However, the model used is physically unsound, since the distorting influence of thermal motion of the ions on their distribution in the solution is not taken into account in it.

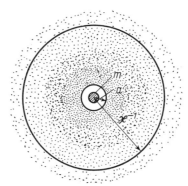

Fig. 10.13. Ionic atmosphere.

An appreciable advance in the theory of electrostatic interaction between ions in solution was made in 1923 by Peter Debye and Erich Hückel, who introduced the concept of an *ionic atmosphere* to characterize the averaged distribution of the ions. In its initial form the theory was applied to fully dissociated electrolytes, hence, it was named the theory of strong electrolytes.

10.8.1. The Ionic Atmosphere

In the Debye–Hückel (DH) theory, the interaction of a particular central ion m with all other ions in the solution, including other ions of the same sort, is analyzed. Because of electrostatic forces in the immediate vicinity of the central ion, an excess of ions of the opposite sign is found which gives rise to a certain space charge compensating the charge of the central ion. This space charge surrounding the central ion (Fig. 10.13) is called an ionic atmosphere or ionic cloud.

The ionic atmosphere has a blurred (diffuse) structure. Because of thermal motion one cannot attribute precise locations to its ions relative to the central ion; one can only define a probability to find them at a certain point, or define a time-average ionic concentration at that point (the charge of the ionic atmosphere is "smeared out" around the central ion). In the DH theory, the interaction of the central ion with specific (discrete) neighboring ions is replaced by its interaction with the ionic atmosphere, i.e., with a continuum.

The most important parameters of the ionic atmosphere are the charge density Q_V and the electrostatic potential ψ at the different points. Each of these parameters is understood as the time-average value. These values only depend on distance r from the central ion but not on a direction in space. For such a system it is convenient to use a polar (spherical) coordinate system having its origin at the point where the central ion is located; then each point can be described by a single and unique coordinate, r. The properties of the surrounding space can be described in terms of the distribution of space charge density, $Q_V(r)$, of the ionic atmosphere, the distribution of electrostatic potential, $\psi(r)$, and other parameters.

Because of thermal motion of the central ion, the coordinate system considered is in continuous three-dimensional motion. Thus, when a cation moves, a certain average density of negative charge, $Q_V(r)$, will always exist at points having a distance r from it. However, at any stationary point of the solution which has been defined, e.g., relative to the vessel walls, positive and negative charges are equally likely to materialize. Therefore, at stationary points the time-average charge density is zero and the average potential is constant and independent of the coordinate, while at points in the moving coordinate system the charge densities and potential are not zero and the potential depends on the coordinate. It follows that an ionic atmosphere only exists in the moving coordinate system of the central ion.

The total charge, Q_{atm}, of the ionic atmosphere can be calculated by integrating the charge density over its total volume. Since the system is electroneutral, the total charge of the ionic atmosphere must be equal in absolute value and opposite in sign to the central ion's charge $z_m Q^0$. The charge density is constant in an elementary volume $dV = 4\pi r^2\,dr$ enclosed between the two concentric spherical surfaces with radii r and $r + dr$. Therefore,

$$Q_{\text{atm}} = -z_m Q^0 = \int_{(V)} Q_V\,dV = 4\pi \int_0^\infty Q_V(r) r^2\,dr. \qquad (10.62)$$

The electrostatic potential ψ at each point is reckoned relative to the solution's constant average potential; the latter is assumed to be zero. The total value of potential $\psi_0(r)$ can be written as the sum of two components, one due to the central ion, $\psi_m(r)$, and one due to the ionic atmosphere, ψ_{atm}, i.e.,

$$\psi_0 = \psi_m(r) + \psi_{\text{atm}}(r). \qquad (10.63)$$

The potential set up by the central ion is determined by Eq. (2.4). It is one of the tasks of the physical theory of ion–ion interaction to calculate the potential of the ionic atmosphere.

10.8.2. The Debye–Hückel Limiting Law

In the space surrounding the central ion, the total value of electrostatic potential ψ_0 and the charge density Q_V are related by Poisson's equation (2.3). In the coordinate system used here, this equation takes the form

$$(1/r^2)[d(r^2 d\psi_0/dr)/dr] = -Q_V/\varepsilon_0\varepsilon. \qquad (10.64)$$

A second equation linking parameters ψ_0 and Q_V is needed so that they may be calculated individually. Debye and Hückel assumed in order to derive such an equation that the concentration of ions c_j is determined by the Boltzmann distribution law

$$c_j = c_j^0 \exp(-w_{\text{pot}}/RT) = c_j^0 \exp(-z_j F\psi_0/RT), \qquad (10.65)$$

where c_j^0 is the concentration averaged over the entire volume, and $w_{\text{pot}} \equiv z_j F\psi_0$ is the potential energy of the ions in a given point.

Space charge arises because the character of cation distribution differs from that of anion distribution (since the signs of z_j are different). According to Eq. (1.1),

$$Q_V = F\sum z_j c_j = F\sum z_j c_j^0 \exp(-z_j F\psi_0/RT), \qquad (10.66)$$

where the summation extends over all ions including $j = m$.

When combining Eqs. (10.64) and (10.66) we obtain a second-order nonlinear differential equation for $\psi_0(r)$ which is mathematically very difficult to solve. Therefore, in DH theory a simplified equation is used: the exponential terms of Eq. (10.66) are expanded into series and only the first two terms of each series are retained. When we include the condition of electroneutrality (1.2) and use the ionic strength I_c we can write this equation as

$$Q_V = F\sum z_j c_j^0(1 - z_j F\psi_0/RT) = -(2F^2 I_c/RT)\psi_0. \qquad (10.67)$$

This simplification is admissible only for sufficiently small values of $z_j F\psi_0/RT$; for instance, it is correct to within 10% for values of $|\psi_0| < RT/2z_j F = 12.5/z_j$ mV. Though Eq. (10.67) is limited in its range of application because of this restriction, it is conditionally used as well at higher values of $|\psi_0|$.

Combining Eqs. (10.64) and (10.67) we find the basic differential equation of the Debye–Hückel theory:

$$(1/r^2)[d(r^2 d\psi_0/dr)/dr] = \varkappa^2 \psi_0, \qquad (10.68)$$

where the parameters which are independent of coordinates are gathered in the constant $\varkappa$ (in units of m^{-1}) defined by

$$\varkappa^2 \equiv 2F^2 I_c / RT \varepsilon_0 \varepsilon. \qquad (10.69)$$

This equation has the general solution

$$\psi_0(r) = (C_1/r)\exp(\varkappa r) + (C_2/r)\exp(-\varkappa r). \qquad (10.70)$$

To find the values of integration constants C_1 and C_2 we must formulate the boundary conditions. At large distances from the central ion $(r \to \infty)$ the value of ψ_0 is zero, hence $C_1 = 0$. Equation (10.62) can be used as the second boundary condition. Substituting the value of ψ_0 [the second term on the right-hand side of Eq. (10.70)] into the expression for Q_V [Eq. (10.67)] and integrating by parts from $r = 0$ to $r = \infty$ we find

$$C_2 = z_m Q^0 / 4\pi \varepsilon_0 \varepsilon. \qquad (10.71)$$

Thus,

$$\psi_0(r) = (z_m Q^0 / 4\pi \varepsilon_0 \varepsilon r)\exp(-\varkappa r). \qquad (10.72)$$

We can find the potential of the ionic atmosphere by subtracting, in accordance with Eq. (10.63), the value of potential of the central ion from the overall value of potential $\psi_0(r)$:

$$\psi_{\text{atm}}(r) = (z_m Q^0 / 4\pi \varepsilon_0 \varepsilon r)[\exp(-\varkappa r) - 1]. \qquad (10.73)$$

It follows for the potential of the ionic atmosphere at the point where the central ion is located $(r = 0)$ that

$$\psi_{\text{atm}}(0) = -z_m Q^0 / [4\pi \varepsilon_0 \varepsilon (1/\varkappa)]. \qquad (10.74)$$

We can see from this equation that the potential ψ_{atm} at the point $r = 0$ has the value that would exist if there were at distance $1/\varkappa$ a point charge $-z_m Q^0$ or, if we take into account the spherical symmetry of the system, if the entire ionic atmosphere having this charge were concentrated on a spherical surface with radius $1/\varkappa$ around the central ion. Therefore, the parameter $r_D \equiv 1/\varkappa$ with the dimensions of length is called the effective thickness of the ionic atmosphere or *Debye radius* (*Debye length*). This is one of the most important parameters describing the ionic atmosphere under given conditions.

According to Eq. (10.69), the Debye radius depends on I_c [when writing $B \equiv F\sqrt{2/RT\varepsilon_0\varepsilon}$ and $B' \equiv 1/B$] as

$$\varkappa = BI_c^{1/2} \quad \text{or} \quad r_D = B'I_c^{-1/2}. \tag{10.75}$$

When stating I_c in mol/liter (not mol/m^3) and including a conversion factor of $10^{3/2}$, we find the parameter values of $B = 3.29 \cdot 10^9$ liter$^{1/2}$/m $\cdot$ mol$^{1/2}$ and $B' \equiv 1/B = 3.04 \cdot 10^{-10}$ m $\cdot$ mol$^{1/2}$/liter$^{1/2}$ for aqueous solutions at 25°C. The values of r_D increase with decreasing ionic strength (concentration) of the solution; they are 0.3, 3, and 30 nm for values of I_c of 1, 10^{-2}, and 10^{-4} M.

An equation for quantitatively determining the interaction energy $w_{e,m}$ between the central ion and the ionic atmosphere can be derived with the aid of the following thought experiment. We assume that in the initial state the central ion is deprived of its charge ($Q_m = 0$), and hence neither an ionic atmosphere nor the potential of this atmosphere exist. We now charge the ion gradually. At all times the value of $\psi_{\text{atm}}(0)$, according to Eq. (10.74), will be proportional to $-Q_m$, i.e., $\psi_{\text{atm}}(0) = -kQ_m$. An energy $dw_{e,m} = \psi_{\text{atm}}(0)dQ_m = -kQ_m dQ_m$ must be expended for each consecutive charge increment dQ_m. The energy corresponding to complete charging of the ion from $Q_m = 0$ to $Q_m = z_m Q^0$ is determined (per mole of ions) by the expression

$$w_{e,m} = -N_A k \int_0^{Q_m} Q_m dQ_m = -N_A k Q_m^2/2 = (N_A/2)z_m Q^0 \psi_{\text{atm}}(0); \tag{10.76}$$

it is evidently equal to the interaction energy sought.

With Eqs. (10.61), (10.74), and (10.75) the expression for the activity coefficient of ion m becomes

$$- \log f_m = A z_m^2 I_c^{1/2}, \quad \text{where } A \equiv N_A(Q^0)^2 B/(2.303 \cdot 8\pi\varepsilon_0\varepsilon). \tag{10.77}$$

For aqueous solutions at 25°C the constant $A = 0.51$ (liter/mol)$^{1/2}$.

This is an equation for calculating the activity coefficient of an individual ion m, i.e., a parameter inaccessible to experimental determination. Let us, therefore, change to the values of mean ionic activity. By definition [cf. Eq. (3.27)]

$$- \log f_\pm = -(1/\tau_k)(\tau_+ \log f_+ + \tau_- \log f_-). \tag{10.78}$$

Substituting into (10.78) the values of f_+ and f_- according to Eq. (10.77) and taking into account that the electroneutrality condition of $\tau_+ z_+ = \tau_- |z_-|$ yields the equality $\tau_+ z_+^2 + \tau_- z_-^2 = \tau_k z_+ |z_-|$, we finally obtain

$$- \log f_\pm = A z_+ |z_-| I_c^{1/2}. \tag{10.79}$$

We can see from Fig. 10.12, curve 1a, that this equation describes the experimental data in very dilute solutions of strong electrolytes, viz., for 1:1 electrolytes approximately up to 10^{-2} M; for other electrolytes the concentration limit is even lower. It correctly conveys the functional dependence on charge of the ions and ionic strength of the solution (as well as the lack of dependence on individual properties of the ions); it can moreover be used to calculate the value of the empirical constant h in Eq. (10.59).

The behavior expressed by Eq. (10.79) became known as the *Debye–Hückel limiting law*. Its derivation was one of the first instances in physical chemistry where with the aid of models built at the molecular level, it was possible to formulate an equation of state of a real system with which the properties of this system could be calculated without the use of empirical ("fitting") constants. This was the basis for the triumph of this theory, which had great significance for the general development of electrolyte solution theory, even though its range of application was limited to very low concentrations. Subsequent theoretical work focussed on interpreting the departures of concentrated solutions from the properties predicted by the limiting law. The DH theory is always one of the criteria for other, more general theories which, in the limit of low concentrations, should lead to the equations of the DH limiting law.

10.8.3. Second and Third Approximation of the Theory

The great success of DH theory provoked numerous attempts of further improvement and extension to more concentrated solutions.

In the derivation reported in Section 10.8.2, known as the first approximation, ion size was disregarded: all ions were treated as point charges. This is reflected in Eq. (10.62), where the integration was started from $r = 0$, i.e., it was assumed that other ions can closely approach the central ion and that all these ions have zero radius.

In the second approximation Debye and Hückel introduced the idea that the centers of the ions cannot come closer than a certain minimum distance a which depends on ion size: the ions were now treated as entities with a finite radius. The mathematical result of this assumption are charge densities Q_V which are zero for $r < a$, and Eq. (10.62) is integrated from $r = a$ to $r = \infty$. This produces a change in the value of the integration constant C_2, viz., instead of Eq. (10.71) we obtain

$$C_2 = \frac{z_m Q^0 \exp(a\varkappa)}{4\pi\varepsilon_0\varepsilon(1 + a\varkappa)}. \tag{10.80}$$

The equation for the activity coefficient is derived by repeating the earlier derivation of Eqs. (10.72) to (10.79). The expression for the distribution of potential of the ionic atmosphere becomes

$$\psi_{\text{atm}}(r) = (z_m Q^0/4\pi\varepsilon_0\varepsilon r)\{[\exp \varkappa(a - r)]/(1 + a\varkappa) - 1\}. \tag{10.81}$$

Within a spherical space of radius a by definition $Q_V = 0$, so that the value of potential of the ionic atmosphere here is constant and equal to that at point $r = a$:

$$\psi_{\text{atm}}(0) = \psi_{\text{atm}}(a) = -z_m Q^0 \varkappa/4\pi\varepsilon_0\varepsilon(1 + a\varkappa). \tag{10.82}$$

As a result we obtain for the mean ionic activity coefficient

$$-\log f_\pm = \frac{A z_+|z_-|\sqrt{I_c}}{1 + a\varkappa} = \frac{A z_+|z_-|\sqrt{I_c}}{1 + aB\sqrt{I_c}}. \tag{10.83}$$

For highly dilute solutions the value of $a\varkappa$ ($= aB\sqrt{I_c}$) is small as compared to unity, and the solution of Eq. (10.83) coincides with the limiting law. In more concentrated solutions, in agreement with experiment, the values of $f_\pm$ calculated by Eq. (10.83) are larger than the values obtained from the limiting law.

In practical applications of this equation one must pick values for constant a. To a first approximation it can be regarded as equal to the sum of the radii of two solvated ions. It is not clear, however, whether the solvation sheaths of approaching ions would not be deformed. Moreover, in deriving Eq. (10.83) it was assumed without sufficient reasoning that the constant a for a given central ion will be the same for different ions present in the ionic atmosphere.

Therefore, constant a is regarded not as a physical parameter which can be determined in independent ways but as a fitting factor which in each case is picked so that Eq. (10.83) will give the best agreement with the experimental data. The equation is thus converted from a theoretical to a semiempirical one. In a number of cases good agreement with experiment can be attained up to concentrations of 0.1 M with values of $a = 0.3$–0.4 nm (Fig. 10.12, curve 1b).

At higher concentrations this equation is no longer suitable for calculating $f_\pm$. In particular, it cannot explain why $f_\pm$ goes through a minimum and increases so strongly at high concentrations.

In solutions of nonelectrolytes where the particles do not interact electrostatically, $\log f_k$ often increases linearly with increasing concentration: $\log f_k = b'c$ (Fig. 10.12, curve 4). By analogy with this behavior, it has been proposed, as a way of accounting for nonelectrostatic interaction forces between the ions, to supplement Eq. (10.83) with an additional term $b'c$ or bI_c:

$$-\log f_\pm = \frac{Az_+|z_-|\sqrt{I_c}}{1 + aB\sqrt{I_c}} - bI_c, \qquad (10.84)$$

where b is an empirical constant.

Values of b are usually small, e.g., $b \approx 0.1z_+|z_-|$ liter/mol. This expression provides a good description for the increase in $f_\pm$ values at higher concentrations. Thus, for aqueous NaCl solutions Eq. (10.84) describes the experimental data up to $c_k = 4$ M (Fig. 10.12, curve 1c) with values of $a = 0.4$ nm and $b = 0.055$ liter/mol.

Equation (10.84) is known as the third approximation of the Debye–Hückel theory. Numerous attempts have been made to theoretically interpret it. In these attempts, either individual simplifying assumptions which had been made in deriving the equations are dropped or additional factors are included. The inclusion of ionic solvation proved to be the most important point. In concentrated solutions, solvation leads to binding of a significant fraction of the solvent molecules, hence certain parameters may change when solvation is taken into account. For instance, in a binary solution containing 1 mole of electrolyte and n_0 moles of the solvent with a molecular mass M_0, the molality m_1 by definition is $1/n_0 M_0$ mol/kg. The solution contains $n_\sigma = \tau_+ + \tau_-$ moles of ions; for their solvation,

$h_\sigma = \tau_+ h_+ + \tau_- h_-$ moles of solvent are consumed. When the molality is referred not to the mass of all solvent but only to the mass of free solvent, the value is $m_1' = 1/(n_0 - h_\sigma)M_0$ mol/kg (parameters which allow for solvation carry a prime). There will be a corresponding difference between mean ionic activity coefficients $f_\pm$ and $f_\pm'$.

The influence of these and some other factors was analyzed in 1948 by Robert A. Robinson and Robert H. Stokes. The equation derived by them is

$$-\log f_\pm = \frac{A z_+ |z_-| \sqrt{I_c}}{1 + aB\sqrt{I_c}} - \frac{h_\sigma \Phi M_0 m_1}{2.303} + \log[1+(h_\sigma - \tau_1)m_1 M_0]. \quad (10.85)$$

The sum of the last two terms corresponds to the term bI_c in Eq. (10.84) [when m_1 is small the logarithmic term can be converted to $(h_\sigma - \tau_1)m_1 M_0$, i.e., will be proportional to m_1].

10.8.4. Ion–Ion Interaction and Conductivity

In the classical theory of conductivity of electrolyte solutions, independent ionic migration is assumed (see Section 10.4). However, in real solutions the mobilities u_j and molar conductivities λ_j of the individual ions depend on the total solution concentration, a situation which, for instance, is reflected in Kohlrausch's square-root law. The values of said quantities also depend on the identities of the other ions. All these observations point to the influence of ion–ion interaction on the migration of the ions in the solution.

The ideas concerning the ionic atmosphere can be used for a theoretical interpretation of these phenomena. There are at least two effects associated with the ionic atmosphere, the electrophoretic effect and the relaxation effect, both lowering the ionic mobilities. Formally this can be written as

$$\lambda_j = \lambda_j^0 - \Delta\lambda_{j,\text{ep}} - \Delta\lambda_{i,\text{rel}}, \quad (10.86)$$

where λ_j^0 is the limiting value of molar conductivity of an ion in the absence of ion–ion interaction (highly dilute solutions), $\Delta\lambda_{j,\text{ep}}$ and $\Delta\lambda_{j,\text{rel}}$ are the changes produced in this parameter by the electrophoretic and relaxation effect, respectively.

The *electrophoretic (cataphoretic) effect* arises because the central ion and its ionic atmosphere, which differ in the sign of charge, will move in

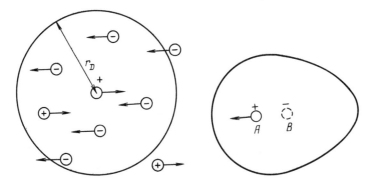

Fig. 10.14. Electrophoretic effect.   **Fig. 10.15.** Relaxation effect.

opposite directions in an electric field (Fig. 10.14). The countermovement of the ionic atmosphere (as the surrounding medium) slows down the motion of the central ion. Usually the value of the ionic atmosphere's own "conductivity" λ_{atm} is adopted as the value of $\Delta\lambda_{j,ep}$. Conditionally the ionic atmosphere is regarded as a sphere with the radius r_D. The values of r_D approach the size of colloidal particles, for which Stokes' law applies when they move in an electric field (electrophoresis or cataphoresis). We then have on the basis of Eqs. (1.7), (10.23), and (10.75):

$$\Delta\lambda_{j,ep} = \lambda_{atm} = z_j^2 F^2 / N_A 6\pi\eta r_D = z_j^2 F^2 B\sqrt{I_c}/N_A 6\pi\eta. \quad (10.87)$$

The *relaxation effect* arises because a certain time, t_{rel}, is required for the formation or collapse of an ionic atmosphere around the central ion. When an ion moves in the electric field its ionic atmosphere lags somewhat behind as it were; its center (Fig. 10.15, point B) is at a point where the central ion had been a little earlier. The configuration of the ionic atmosphere around the central ion (point A) will no longer be spherical but elongated (ovoid). Because of this displacement of the charges the ionic atmosphere has an electrostatic effect upon the central ion which acts in a direction opposite to this ion's motion. A rigorous calculation of this effect was made in 1927 by Lars Onsager. His solution was

$$\Delta\lambda_{j,rel} = \omega F^2 \lambda_j^0 B\sqrt{I_c}/N_A RT 8\pi\varepsilon_0\varepsilon, \quad (10.88)$$

where ω is a numerical factor; its value is 0.1953 for symmetric electrolytes.

Thus, Eq. (10.86) can be written as

$$\lambda_j = \lambda_j^0 - (k_{ep} z_j^2 + k_{rel} \lambda_j^0)\sqrt{I_c}, \qquad (10.89)$$

where

$$k_{ep} = F^2 B / N_A 6\pi\eta \quad \text{and} \quad k_{rel} = \omega F^2 B / N_A RT 8\pi\varepsilon_0\varepsilon. \qquad (10.90)$$

It follows (when taking into account that $\tau_+ z_+ = \tau_- |z_-| = z_k$) that the molar conductivity of binary electrolyte solutions is given by

$$\Lambda = \Lambda^0 - [k_{ep} z_k (z_+ + |z_-|) + k_{rel}\Lambda^0]\sqrt{I_c}. \qquad (10.91)$$

These equations are known as the *Debye–Hückel–Onsager equations.*

In the case of binary solutions, Eq. (10.91) coincides with the empirical equation (10.21), both in form and in value of the numerical constant k. Therefore, the empirical square-root law can be quantitatively explained on the basis of the theory of ion–ion interaction.

An analysis of Eq. (10.91) shows that the electrophoretic effect accounts for about 60–70% of the decrease in solution conductivity, and the relaxation effect for the remaining 40–30%.

The ideas outlined also provide an explanation for the anomalous conductivity effects described in Section 10.4. At high values of the electric field strength the distance between the central ion and the center of the ionic atmosphere becomes larger than r_D, i.e., the central ion "pulls out" as it were from its ionic atmosphere. Then the electrophoretic and relaxation effects will disappear and the value of Λ will approach Λ^0 (the Wien effect). The retarding effect of relaxation phenomena will decrease when high-frequency electric fields of alternating direction are applied; when the period between sign changes is shorter than the relaxation time t_{rel}, the central ion is practically not displaced relative to the center of the ionic atmosphere. But, electrophoretic phenomena persist. Therefore, at high field frequencies the value of Λ increases somewhat but does not reach the value of Λ^0 (the Debye–Falkenhagen effect).

10.8.5. Further Development of Electrolyte Solution Theory

Despite its attractions, the Debye–Hückel theory proved unable to quantitatively explain the behavior of electrolyte solutions that are not very dilute. The semiempirical or empirical constants subsequently introduced into the equations only formally improved the range of application of these equations. This indicates that the physical premises of the theory are overly simplified, and particularly: (i) the individual ions are considered from a molecular point of view but the solvent is treated macroscopically, as a continuum; (ii) interactions between the ions and the water molecules, and the distortion of water structure close to an individual ion, are disregarded; and (iii) the possibility is ignored that in addition to the electrostatic forces, chemical forces will develop between ions short distances apart.

The first ideas concerning the role of pairwise electrostatic interaction between ions were advanced in 1924 by Vladimir K. Semenchenko. A quantitative theory of the formation of so-called *ion pairs* was formulated in 1926 by Niels Bjerrum.

According to the basic ideas concerning ionic atmospheres, the ions contained in them are in random thermal motion uncoordinated with the displacements of the central ion. But at short distances between the central ion m and an oppositely charged ion j of the ionic atmosphere, electrostatic attraction forces will develop which are so strong that these two ions are no longer independent but start to move together in space like one particle, viz., the ion pair. The total charge of the ion pair is $(|z_m| - |z_j|)Q^0$; when the two ions carry identical numbers of charges the ion pair will be electrically neutral but constitute a dipole.

Ion pair formation lowers the concentrations of free ions in the solution, and hence the conductivity of the solution. It must be pointed out that ion pair formation is not equivalent to the formation of undissociated molecules or complexes from the ions. In contrast to such species, ions in an ion pair are linked only by electrostatic and not by chemical forces. During ion pair formation a common solvation sheath is set up, but between the ions thin solvation interlayers are preserved (Fig. 10.16). The ion pair will break up during strong collisions with other particles (i.e., not in all collisions). Therefore, ion pairs have a finite lifetime, which is longer than the mean time between individual collisions.

For ion pair formation the electrostatic attraction energy, described by $w_e = N_A |z_m z_j| (Q^0)^2 / 4\pi\varepsilon_0\varepsilon r$ (per mole of ion pairs), should be larger than the ion pair's mean thermal energy, i.e., at least $2RT$. This condition yields, for the critical distance of ion pair formation in aqueous solutions at 25°C,

$$r_{cr} = N_A |z_m z_j| (Q^0)^2 / 8\pi\varepsilon_0\varepsilon RT = |z_m z_j| \, 0.357 \text{ nm}. \qquad (10.92)$$

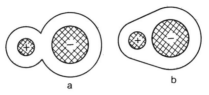

Fig. 10.16. Deformation of solvation sheaths when ions approach: (a) under the effect of electrostatic forces; (b) under the effect of chemical forces.

Ion pairs can only form when the distance of closest approach, a, of the two ions is less than r_{cr}. For 1:1 electrolytes for which $r_{cr} = 0.357$ nm this condition is not always fulfilled, but for others it is.

For a calculation of the fraction of ions forming ion pairs, one must determine the probability that ion j will be at distances between a and r_{cr} from the central ion m (any of the two ions in the pair can be regarded as the central ion). According to the Boltzmann equation, the probability for finding ion j at a distance r (i.e., within the volume element $dV = 4\pi r^2 dr$) is

$$P_j \equiv (1/N_j)dN_j/dr = 4\pi r^2 \exp(-z_j F\psi_m/RT)$$

$$= 4\pi r^2 \exp[-|z_m z_j|Q^0 F/4\pi\varepsilon_0\varepsilon r RT]. \tag{10.93}$$

A plot of this probability against distance r is shown in Fig. 10.17. We can see that the probability increases drastically at short distances, when electrostatic attraction between the ions is strong. The probability falls with increasing distance, goes through a minimum, then rises again because of the progressively increasing volume dV of the corresponding element of space. It is easy to show that the minimum is at the distance r_{cr}. Therefore, ions corresponding to the left-hand branch of the curve form ion pairs, while the ions which correspond to the right-hand branch do not. Integrating the area under the P_j vs. r curve from $r = a$ to $r = r_{cr}$ (the shaded area in the figure) we can determine the fraction of ions associated in ion pairs.

Table 10.4 lists the fractions of paired ions calculated for different electrolytes by Bjerrum in the way shown. It is seen that these fractions increase with increasing concentration of the solutions. They are small in aqueous solutions of 1:1 electrolytes. In nonaqueous solutions which have lower values of permittivity ε than water, the values of r_{cr} and the fractions of paired ions are higher. In some cases the values of r_{cr} coincide

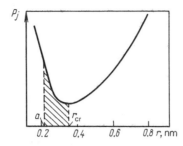

Fig. 10.17. Probability of finding an ion of opposite charge at a particular distance from the central ion (according to Bjerrum).

TABLE 10.4 Fraction of Ions (%) Forming Ion Pairs

c, mol/liter	\multicolumn{4}{c}{a, nm}			
	0.282	0.232	0.176	0.101	0.282	0.232
	\multicolumn{4}{c}{$|z_+z_-| = 1$}			$|z_+z_-| = 2$
0.005	0.2	0.4	0.7	1.6	6.8	10
0.01	0.5	0.8	1.2	3.0	17.1	20
0.02	0.8	1.3	2.2	5.3	27.4	33
0.05	1.7	2.8	4.6	10.5	58.2	71
0.1	2.9	4.8	7.2	16.3	99.2	100
0.2	4.8	7.9	12.1	24	100	
0.5	9.0	14	20.4	36		
1.0	13.8	20.6	28.6	45.7		
2.0	20.4	28.9	38.3	55.4		

with the statistical mean distance between the ions, i.e., association of the ions is complete.

Bjerrum's theory of ion pairs qualitatively correctly explains a number of experimental data, but cannot be fully used in quantitative calculations, particularly because of the provisional character of the quantities a and r_{cr} (the integration limits).

Ion-pair formation (or the formation of triplets etc.) is a very simple kind of interaction between ions of opposite charge. As the electrolyte concentration increases and the mean distance between ions decreases, electrostatic forces are no longer the only interaction forces. Aggregates within which the ions are held together by chemical forces have certain special features, viz., shorter interatomic distances and a higher degree of desolvation (cf. Fig. 10.16) than found in ion pairs, and can form a common solvation sheath instead of the individual sheaths. These aggregates are distinctly seen in spectra, and in a number of cases their concentrations can be measured spectroscopically.

A more general theory of solutions would require detailed notions of solution structure and of all kinds of interaction between the particles (ions and solvent molecules) in the solution. Numerous experimental and theoretical studies have been carried out, and some progress has been made, but a sufficiently universal theory which could describe all properties in not very dilute electrolyte solutions has not been developed so far.

10.9. IONIC REACTIONS AND EQUILIBRIA

The ions in electrolyte solutions can chemically react in diverse ways, both with each other and with neutral solvent or solute molecules. These homogeneous ionic reactions (ionic reactions occurring in bulk solution) are highly important in chemistry, physics, and biology.

There are two types of ionic reaction: (i) recombination, without changes in oxidation state of the particles, and (ii) redox reactions, with a change in oxidation state.

Recombination reactions in turn can be subdivided into two groups: (a) association–dissociation reactions, in particular the dissociation of a number of salts which are weak electrolytes or, conversely, the formation of molecules from the ions, e.g.,

$$ZnBr_2 \rightleftarrows Zn^{2+} + 2Br^-, \qquad (10.94)$$

and also the formation or decay of coordinatively bonded chemical complexes, e.g.,

$$Ag^+ \underset{}{\overset{+NH_3}{\rightleftarrows}} Ag(NH_3)^+ \underset{}{\overset{+NH_3}{\rightleftarrows}} Ag(NH_3)_2^+ \underset{}{\overset{+NH_3}{\rightleftarrows}} \ldots; \qquad (10.95)$$

(b) ionolytic reactions which involve the transfer of fragments (an individual ion or ionic aggregate) between reactant particles. Very common in aqueous solutions are the protolytic reactions involving proton transfer, e.g.,

$$H_2O + CO_3^{2-} \rightleftarrows HCO_3^- + OH^-. \qquad (10.96)$$

Protolytic reactions include the dissociation of acids and certain bases.

In systems where ionic reactions are possible, ionic equilibrium will be established in the end. In studies of ionic reactions two aspects are of interest, viz., the equilibrium state and, when there is no equilibrium, the reaction rates.

A very important parameter describing the equilibrium state of a reaction is its equilibrium constant, e.g., for reactions $\sum \bar{\nu}_j X_j = 0$ we have

$$K = \prod a_j^{\bar{\nu}_j} = \prod_p a_j^{\nu_j} / \prod_r a_j^{\nu_j}. \tag{10.97}$$

The equilibrium constant links the activities (and corresponding concentrations) of all components in the equilibrium state. For an arbitrary, given initial state the direction of the reaction is determined by the value of this constant.

In cases where the activities are not known or where the values are not sufficiently accurate, the component concentrations are used instead of activities $a_j = f_j c_j$. The parameter

$$K_c = \prod c_j^{\bar{\nu}_j} = K/K_f, \qquad \text{where } K_f = \prod f_j^{\bar{\nu}_j}, \tag{10.98}$$

is known as the formal equilibrium constant. When the concentrations are varied the values of K_j change less strongly than the values of f_j; we can assume to a first approximation, therefore, that the values of K_c are practically constant. In the remainder of the present section we shall use concentrations rather than activities, since this facilitates tie-up with the stoichiometric equations.

Instead of the equilibrium constants K themselves, often their exponents pK, defined as

$$pK \equiv -\log K \quad \text{or} \quad K \equiv 10^{-pK}, \tag{10.99}$$

are stated.

Ionic equilibria are dynamic; the reactions do not cease in the state of equilibrium but occur in both directions with identical rates given by the exchange rate v^0. Away from equilibrium the effective reaction rate v is given by the difference between the partial reaction rates in the two directions: $\overrightarrow{v} - \overleftarrow{v}$.

It had been assumed in the past that ionic reaction rates are very high, and that in ionic systems the equilibria are established practically instantaneously. We now know that in a number of cases the rates of such reactions are low; they may, in particular, be lower than the rates of electrochemical reactions occurring in the system.

10.9.1. Dissociation–Association Reactions

A very important reaction of this type is the dissociation of weak electrolytes. Equation (10.15) (see Section 10.2) describes the condition of equilibrium between the ions and undissociated molecules of a weak electrolyte. By solving this equation for the degree of dissociation α [Eq. (10.16)], we can calculate the concentration of ions for any given total electrolyte concentration c_k.

Sometimes one must calculate the degree of dissociation of a weak electrolyte in the presence of an excess of strong electrolyte having an ion in common with it. For the sake of simplicity, consider symmetric $z:z$ electrolytes where KA is weak (subscript 1) and MA is strong (subscript 2). From the conditions of the problem $c_2 \gg c_1$; hence, $c_A = c_2 + \alpha_1 c_1 \approx c_2$. Since electrolyte KA is weak, in the present case we also have $K_1 < c_2$. For the dissociation equilibrium of KA we have

$$K_1 = c_K c_A / c_{KA} = \alpha_1 c_1 (c_2 + \alpha_1 c_1) / (1 - \alpha_1) c_1 \approx \alpha_1 c_2 / (1 - \alpha_1),$$
$$(10.100)$$

whence

$$\alpha_1 = K_1 / (K_1 + c_2) \approx K_1 / c_2, \qquad (10.101)$$

and an analogous expression for c_K ($= \alpha_1 c_1$). Therefore, in the present case the degree of dissociation of the weak electrolyte is inversely proportional to the concentration of the strong electrolyte.

Consider another, more general case where both electrolytes are weak and their concentrations are comparable. Ion A^- is involved in the dissociation equilibria of both electrolytes, and its concentration is contained in the expressions for both dissociation equilibrium constants:

$$K_1 = c_K c_A / c_{KA} \quad \text{and} \quad K_2 = c_M c_A / c_{MA}. \qquad (10.102)$$

The two equilibria are independent, i.e., the components of one system have no influence on the equilibrium constant of the other system. Another three obvious stoichiometric relations connect the component concentrations:

$$c_K + c_{KA} = c_1, \qquad c_M + c_{MA} = c_2, \qquad c_A = c_K + c_M. \qquad (10.103)$$

Therefore, for given values c_1 and c_2 of the total concentrations, the five unknown concentrations c_K, c_M, c_A, c_{KA}, and c_{MA} are interconnected by five relations. The system of equations is readily solved by successive elimination of unknowns.

Complex formation often involves the successive addition of ligands L to a central ion M [see, e.g., Eq. (10.95)]. Each individual step is characterized by values of association constants or stability constants of the corresponding complexes. For instance, for formation of the complex ML_k from complex ML_{k-1} and ligand L,

$$K_k = c_{ML_k}/c_{ML_{k-1}}c_L. \qquad (10.104)$$

Consider in more detail a process where complex formation occurs in two steps. It is obvious that the combined concentrations of the three complex forms M, ML, and ML_2 are equal to the initial (given) concentration, c_M^0, of substance M, i.e., $c_M^0 = c_M + c_{ML} + c_{ML_2}$. Their relative concentrations depend on the ligand's concentration c_L and on the association constants $K_1 = c_{ML}/c_M c_L$ and $K_2 = c_{ML_2}/c_{ML}c_L$. Solving these expressions for the unknown concentrations we find

$$c_M = c_M^0(1 + K_1 c_L + K_1 K_2 c_L^2)^{-1},$$
$$c_{ML} = K_1 c_L c_M, \qquad c_{ML_2} = K_1 K_2 c_L^2 c_M. \qquad (10.105)$$

Figure 10.18 shows the relative concentrations c_M/c_M^0, c_{ML}/c_M^0, and c_{ML_2}/c_M^0 as functions of $\log(K_1 c_L)$ for a ratio $K_1/K_2 = 10$. As expected, the equilibrium shifts in the direction of ML_2 formation as the ligand concentration is raised. The intermediate form ML is formed in the region of intermediate ligand concentrations around $c_L \approx \sqrt{10}/K_1 = (K_1/K_2)^{-1/2}$. The maximum concentration of this intermediate form will be higher and the interval of concentrations c_L where it exists will be wider the larger the difference between the equilibrium constants of formation (K_1) and of further conversion (K_2) of this compound ML.

10.9.2. Protolytic Reactions

In aqueous solutions, protolytic reactions involving proton transfer from species HA to species B,

$$HA + B \rightleftarrows BH^+ + A^-, \qquad (10.106)$$

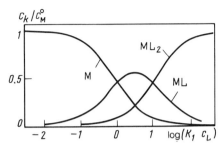

Fig. 10.18. Concentrations of the different complexes as functions of ligand concentration.

are very important (the species may carry charge different from that indicated in the equation).

In 1923 Brønsted gave a new definition for acids and bases according to which an acid is a substance able to give off a proton (a proton donor), and a base is a substance able to accept a proton (a proton acceptor). According to this definition, substances HA and HB^+ are acids, and substances A^- and B^- are bases. Thus, a protolytic reaction involves two acid–base systems, the system HA/A^- and the system HB^+/B.

A protolytic reaction is always invertible, and between the two systems an equilibrium is established for which

$$a_{BH^+} a_{A^-} / a_{HA} a_B = K, \qquad (10.107)$$

where K is the equilibrium constant of the protolytic reaction.

Different kinds of protolytic reactions are possible.

(i) The *"dissociation" of acids*

$$HA + H_2O \rightleftarrows H_3O^+ + A^-. \qquad (10.108)$$

As a result of this protolytic reaction, the acid functions are transferred from HA molecules to H_3O^+ ions. Dissociation reactions (like other protolytic reactions) are principally invertible. However, in the case of strong acids (e.g., $HClO_4$) the protolytic reaction occurs practically completely from left to right. This means that the "acidity" or donor properties of the HA molecules are much stronger than those of the H_3O^+ ion, so that HA is a strong acid undergoing complete dissociation. However, this is a relative concept. When the acid $HClO_4$ is dissolved in another proton-containing solvent, e.g., glacial acetic acid, then a protolysis reaction analogous to (10.108) may occur only to a minor degree. Therefore, the acid $HClO_4$ is a weak acid relative to the acid $(CH_3COOH_2)^+$, and dissociates only partly. In aqueous solutions all strong acids are leveled in their strengths by complete conversion to the acid H_3O^+, but when other proton-containing solvents are used these acids can be differentiated and arranged in order of decreasing acid strengths.

(ii) The *"dissociation" of certain bases* such as

$$H_2O + NH_3 \rightleftarrows NH_4^+ + OH^- \qquad (10.109)$$

is analogous to protolytic acid dissociation: the base functions are transferred (in our example) from the NH_3 molecule to the OH^- ion. Organic bases containing the groups $-NH_2$ or $>NH$ dissociate according to the same scheme.

(iii) The *hydrolysis of salts of a weak acid and strong base* (or vice versa), e.g.,

$$H_2O + CH_3COO^- \rightleftarrows CH_3COOH + OH^-. \qquad (10.110)$$

(iv) The *stepwise dissociation of polybasic acids*

$$
\begin{aligned}
H_2A + H_2O &\overset{(1)}{\rightleftarrows} H_3O^+ + HA^-, \\
HA^- + H_2O &\overset{(2)}{\rightleftarrows} H_3O^+ + A^{2-}.
\end{aligned}
\qquad (10.111)
$$

The equilibrium concentrations of the individual species H_2A, HA^-, and A^{2-} are calculated as in the case of stepwise complex formation [see Eq. (10.105), taking into account that a dissociation constant is the reciprocal of an association constant].

(v) *Autoprotolytic reactions*, an example of which is the "dissociation" of water,

$$2H_2O \rightleftarrows H_3O^+ + OH^-. \qquad (10.112)$$

All the protolytic reactions described involve water molecules and H_3O^+ or OH^- ions, hence the system's equilibrium state (the relative concentrations of the individual components) will depend on the solution pH. On the other hand, when the relative concentrations of all other components are given, the pH value of the solution will be defined in this way.

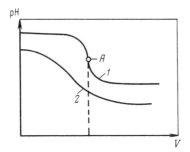

Fig. 10.19. Titration curves: (1) strong base titrated with a strong acid; (2) weak polyelectrolyte (base) titrated with a strong acid.

Consider a few other concepts related to protolytic ionic equilibria in aqueous electrolyte solutions.

Amphoteric electrolytes (ampholytes) can function as an acid or a base depending on solution pH. Thus, $Zn(OH)_2$ in alkaline solutions gives off protons to OH^- ions and is converted to a ZnO_2^{2-} ion, while in acidic solutions it adds protons and is converted to the ion Zn^{2+} aq. Sometimes the amphoteric properties are linked to the presence of different functional groups. In amino acids $H_2N-R-COOH$ for instance, carboxyl groups function as donors, and amino groups function as acceptors of protons. Within a certain range of intermediate pH values, internal autoprotolysis is possible for such compounds, viz., a proton is transferred from the acidic carboxyl to the basic amino group. This reaction yields a hybrid ion (zwitterion) $^+H_3N-R-COO^-$.

Titration curves. When base is added to an acid (or vice versa) the plot of pH against the volume of the added solution has the typical S-shape (Fig. 10.19, curve 1). Close to the equivalence point A, addition of a minor amount of acid or base leads to a large change in pH value. The buffer capacity of the solution (with respect to acid or base) is said to be very low in this case.

Buffer solutions are mixed solutions of a weak acid and its salt with a strong base (or of a weak base and its salt with a strong acid). Take for instance a solution containing CH_3COOH and CH_3COONa. The pH value of this solution is quite stable; upon addition of certain quantities of strong acid (e.g., HCl) most of the excess H_3O^+ ions will combine with anions to form undissociated acid molecules:

$$H_3O^+ + CH_3COO^- \rightleftarrows CH_3COOH + H_2O. \qquad (10.113)$$

Similarly, when strong base is added the excess OH^- ions will react with undissociated acid to form water and the corresponding salt. Under these conditions there is no accumulation of the H_3O^+ or OH^- ions added (only the relative concentrations of the salt and weak acid change), hence there is little change in solution pH, i.e., the buffer capacity of the solution is high.

Let us define more closely the concepts of pH and pOH of a solution. They are usually defined by the expressions

$$pH \equiv -\log a_{H_3O^+} \quad \text{and} \quad pOH \equiv -\log a_{OH^-}. \qquad (10.114)$$

Since the activities of individual ions are not known, the values of pH and pOH in acidic and alkaline solutions actually refer to mean ionic activities. In neutral (salt) solutions, i.e., in the pH interval between approximately 4 and 10, additional difficulties arise, since one cannot sufficiently accurately measure the mean ionic activities of ion combinations containing H_3O^+ or OH^- ions. For instance, when the emf method is used for the measurements, its accuracy will be insufficient owing to the boundaries with diffusion potentials which exist between electrolytes. Hence, for such solutions a special, conventional pH scale is used. In this scale, well-defined values of pH are assigned to five buffer solutions of given compositions:

Saturated potassium hydrogen tartrate solution   pH = 3.557

0.05 m potassium hydrogen phthalate solution   pH = 4.008

0.025 m KH_2PO_4 + 0.025 m Na_2HPO_4 solution   pH = 6.865

0.008695 m KH_2PO_4 + 0.03043 m Na_2HPO_4 solution   pH = 7.413

0.01 m sodium tetraborate solution   pH = 9.180

(all data for 25°C; data for temperatures between 0 and 96°C are listed in handbooks). These solutions are used to calibrate the electrodes to be used for measurements (e.g., hydrogen electrodes), and the pH values of test solutions are determined by interpolation.

10.9.3. Redox Reactions

Homogeneous redox reactions in solutions can formally be considered as analogs of protolytic reactions, except that an electron, rather than a proton, is the species transferred between particles, e.g.,

$$2Fe^{3+} + Sn^{2+} \rightleftarrows 2Fe^{2+} + Sn^{4+}. \tag{10.115}$$

For a redox reaction, two individual redox systems must be present in the solution, which in the above example are the systems Fe^{3+}/Fe^{2+} (the Fe^{3+} ions function as the electron acceptors) and Sn^{4+}/Sn^{2+} (the Sn^{2+} ions function as the electron donors). The relative strength of each redox system is determined by the redox potential that would be set up at a nonconsumable electrode present in this system. According to the Nernst equation, these potentials depend on the relative concentrations of the oxidized and reduced form of the substance involved. The direction of the reaction is determined from the condition that the system with the more positive value of redox potential functions as the electron acceptor (its oxidized form is reduced), while the system with the more negative redox potential functions as the electron donor. The reaction will continue until the relative concentrations of all components are such that the equilibrium redox potentials of the two systems are identical. In the example given, the standard potential of the Fe^{3+}/Fe^{2+} system ($E^0 = 0.783$ V) is much more positive than that of the Sn^{4+}/Sn^{2+} system ($E^0 = 0.15$ V), so that reaction (10.115) will go practically completely from the left to the right until one of the reactants, the Fe^{3+} or Sn^{2+} ions, is completely consumed.

Chapter 11
POLYELECTROLYTES AND NONAQUEOUS ELECTROLYTES

11.1. POLYELECTROLYTES

Polyelectrolytes is the name for polymeric compounds containing certain functional groups which in the presence of water will dissociate. At least one kind of the ions being formed has polymeric structure, i.e., is a polyion.

Many polyelectrolytes are derivatives of polyethylene with the general formula $[-CH_2-CHX-]_n$. The polyelectrolytes have different properties depending on the identity of functional group X. Examples are

X =	$-SO_3H$	Poly(ethylenesulfonic acid), a strong acid
	$-COOH$	Poly(acrylic acid), a weak acid
	$-C_5H_4NOH$	Poly(vinylpyridine), a weak base
	$-NR_3OH$	Poly(ethylenetriarylammonium hydroxide), a strong base

Polyelectrolytes can be derived, not only from organic but also from inorganic polymers, e.g., silica gel, aluminosilicates, or glass-forming compounds.

As in the case of the usual, nonpolymeric electrolytes, the dissociation of a polyacid is accomplished by transfer of protons to water molecules and the simultaneous formation of polyanions; the dissociation of a polybase yields polycations. In water the ions H_3O^+ and OH^- are formed as counterions.

When a strong nonpolymeric base, such as KOH, is added to a polyacid a system is formed which consists of polyanions and the K^+ ions as their counterions, and can be regarded as the dissociation product of a polymeric salt.

Polyelectrolytes may at the same time contain both acidic and basic functional groups. In this case they are amphoteric, and the effective charge of the polyions produced by dissociation depends on the solution's pH value. A typical example of such polyelectrolytes are the polymeric amino acids.

The polyelectrolytes can have linear structure or form a three-dimensional lattice by cross-linking of molecules having linear structure. The linear ones as a rule are water-soluble, those with three-dimensional structure are insoluble.

Polyelectrolytes which are insoluble, unlike those which are soluble, form independent phases which will not mix with the aqueous solution. These solid or liquid polymeric electrolytes are called *ion exchangers* (or, sometimes, ionites).

The ion exchangers will dissociate when in contact with water or aqueous solutions of simple electrolytes. The water penetrates into the space between the macromolecules and pushes them slightly apart, which causes swelling of the ion exchanger. The dissociation starts, not as a result of dissolution of the electrolyte, but as a result of its swelling when the water that has penetrated into the space between the molecules reaches the proton-donating or proton-accepting functional groups.

It is the special feature of ion exchangers that the polyions are fixed in a given phase and cannot move out into the surrounding solution. However, the simple counterions contained in this phase can freely exchange with ions of the same sign from the solution; an equilibrium distribution of the ions is established which is analogous to that described in Section 5.3. Ion exchangers representing polyanions which exchange cations are more specifically called cation exchangers, those representing polycations are called anion exchangers.

When the ion exchanger is in contact with an electrolyte solution rather than with pure water, not only counterions are taken up into the ion exchanger phase but also some number of coions, i.e., simple ions carrying charge of the same sign as the polyion.

In the case of soluble polyelectrolytes, high local concentrations of electric charges will arise even when their solutions are highly dilute, since multiply charged polyions are formed. These charges influence the properties of the macromolecules, for instance the mobility of the hydrocarbon chains. Moreover, in dilute solutions where the number of nearby counterions compensating the fixed polyion charges is low, the effective charge of the polyions is high and their mutual repulsion strong.

In the case of insoluble polyelectrolytes forming an independent electroneutral phase, a high concentration is found, not only for the fixed polyion charges but also for the counterions in the space between the molecules; these concentrations are as high as 5 or 6 M.

Because of the high concentration of charges, ion exchangers have a high electrostatic selectivity. In fact, according to Eq. (5.20), when the concentration of fixed charges, c_R, is high and the concentration of ions in the solution outside is low, the counterion concentration will practically constitute the charge equivalent of c_R, and the coion concentration will practically be zero. Superimposed upon the electrostatic selectivity, one often finds chemical selectivity (values of σ_{MA} which are different from zero) caused by differences in the chemical interactions between the polymeric moiety and different kinds of ion. In some cases chemical selectivity may arise for steric reasons, viz., when the distances between the molecules are not big enough for the passage of large, hydrated ions.

In solutions of a weak nonpolymeric electrolyte, any molecule has the same dissociation probability. In polyelectrolytes the polyion's charge increases as dissociation continues, and this actually interferes with further dissociation. Formally this implies that the dissociation constant of a polyelectrolyte is not constant but gradually decreases with increasing degree of dissociation. For this reason the titration curve of a weak polyelectrolyte (Fig. 10.19, curve 2) is much flatter than that of a weak nonpolymeric electrolyte.

Ion exchange can be employed to accomplish changes in ionic composition of a solution, particularly for the removal of undesired impurities and for the deionization of water. Nowadays ion-exchange materials are widely applied in different technical fields. Their commercial forms are pellets which are packed into special exchange columns, or membranes which exhibit selectivity.

11.2. NONAQUEOUS ELECTROLYTE SOLUTIONS

Nonaqueous electrolyte solutions are analogous to aqueous solutions; they, too, are systems with a liquid solvent and a solute or solutes dissociating and forming solvated ions.

The special features of water as a solvent are its high polarity (ε = 78.5), which promotes dissocation of dissolved electrolytes and hydration of the ions, and its protolytic reactivity. When considering these features, we can group the nonaqueous solvents as follows:

(i) protic, which are analogous to water, have polar character, and are involved in protolytic reactions;
(ii) aprotic polar (with values of $\varepsilon \geq 15$);
(iii) aprotic with low or zero polarity (values $\varepsilon \leq 15$).

The third group of solvents comprises the hydrocarbons and their halogen derivatives. They are not of interest for electrochemistry, since the solubilities and dissociation of salts, acids, and bases in them are low. Systems with protic or aprotic polar solvents are practically used and have been widely investigated.

In the following, we shall write the protic solvents conventionally as SH, where H is the proton and S is the solvent "residue". They include many compounds with –OH and >NH groups, also some others (Table 11.1). All these solvents are polar, have high values of ε (larger than 20, as a rule), and high solvating power, and many electrolytes have high degrees of dissociation when dissolved in them. The protic solvents distinctly tend to form hydrogen bonds, and all of them are involved in protolytic reactions. They may, like water, undergo autoprotolysis:

$$2S\text{H} \rightleftarrows S\text{H}_2 + S^-, \qquad (11.1)$$

which yields lyonium and lyate ions, $S\text{H}_2^+$ and S^-, as the analogs of hydroxonium and hydroxyl ions, H_3O^+ and OH^- (solvent "dissociation").

Of greatest interest among the protic solvents are liquid ammonia (where solutions with a very low freezing point can be prepared) and anhydrous (glacial) acetic acid (which has a high proton-donating power).

Aprotic polar solvents such as those listed in Table 11.1 are widely used in electrochemistry. In solutions with such solvents the alkali metals are stable, and will not dissolve under hydrogen evolution (by discharge of the proton donors) as they do in water or other protic solvents. These solvents find use in new kinds of chemical power sources (batteries) with lithium electrodes having a high energy density.

Compounds with the >CO group, particularly esters and ketones, are the most important aprotic polar solvents. The solubilities of most simple salts in these solvents are low, but salts with large (even complex) anions such as $LiClO_4$, $LiAlCl_4$, or $LiAsF_6$ have solubilities as high as 1 to 2 M.

CHAPTER 11 • POLYELECTROLYTES & NONAQUEOUS ELECTROLYTES 271

TABLE 11.1. Physical Properties of Certain Solvents at 25°C

Solvent	Formula	ε	η, MPa $\cdot$ s	d, kg/dm^3	t_f, °C	t_b, °C
Protic						
water	H_2O	78.5	0.89	0.997	0	100
ammonia (−20°C, 0.2 MPa)	NH_3	20	0.26	0.67	−78	−33
acetic acid	CH_3COOH	6.2	1.13	1.05	16.7	−118.1
methanol	CH_3OH	32.6	0.547	0.792	−97.9	64.5
Aprotic						
acetonitrile	CH_3CN	36	0.345	0.786	−45.7	81.6
dimethylformamide	$(CH_3)_2NCHO$	37	0.796	0.944	−61	153
propylene carbonate	$CH_3-CH\overset{\displaystyle CH_2-O-C=O}{-\!\!\!\!-\!\!\!\!-O}$	66.1	2.53	1.198	−49.2	242
γ-butyrolactone	$\begin{array}{c} CH_2-CH_2-O \\ \mid \quad\quad\quad \mid \\ CH_2-\quad\; C=O \end{array}$	39.1	1.75	1.125	−43.5	204
tetrahydrofuran	$\begin{array}{c} CH_2-CH_2-O \\ \mid \quad\quad\quad \mid \\ CH_2-\quad\quad CH_2 \end{array}$	7.4	0.46	0.880	−65	64

The solubilities of electrolytes and their degrees of dissociation in different solvents depend on the polarity of the solvent molecules, which also is the factor influencing the relative permittivity ε and the solvating power. The common tendency that the degree of dissociation and the conductivity of the solutions are higher in solvents with higher values of ε had already been noticed in the 1890s (Sir William Thomson [Lord Kelvin], Walther Nernst, Ivan A. Kablukov). However, this tendency is not in all cases unambiguous. Thus, the dipole moment of HCN molecules is 1.5 times that of water molecules, but the solubilities of most salts are lower in hydrocyanic acid than in water, which is due to the higher solvent reorganization energy required for solvation of the ions in HCN.

In some cases the degree of dissociation of a dissolved electrolyte depends, not only on the solvent's polarity or ε-value but on other properties as well. Thus, when dissolved in ethanol, hydrogen chloride dissociates practically completely, and behaves as a strong electrolyte. This is possible because of protolytic reactions yielding solvated protons $C_2H_5OH_2^+$. In nitrobenzene, though it has practically the same ε-value as ethanol, the degree of dissociation of HCl is low owing to the low proton-accepting power.

The ionic mobilities also depend on the solvent. In 1905/06 Paul Walden and Lev V. Pisarzhevskii established the rule according to which the product of limiting mobility of an ion and viscosity of the solution is approximately constant:

$$u_j^0 \eta \approx \text{const.} \tag{11.2}$$

This rule follows immediately from Stokes' law for the motion of spherical bodies in viscous fluids [Eq. (10.23)] when assuming constant radii. It is applicable in particular for the change in ionic mobility which occurs in a particular solvent when the temperature is varied. Between solvents it remains valid when the electrolytes have poorly solvated ions, such as $N(C_2H_5)_4I$. For other electrolytes we find rather significant departures from this rule. These are due in particular to the different degrees of solvation found for the ions in different solvents, and hence their different effective radii.

In aqueous electrolyte solutions the molar conductivities of the electrolyte, Λ, and of individual ions, λ_j, always increase with decreasing solute concentration [cf. Eq. (10.18) for solutions of weak electrolytes, and Eq. (10.21) for solutions of strong electrolytes]. In nonaqueous solutions even this rule fails, and in some cases maxima and minima appear in the plots of Λ against c (Fig. (11.1)). This tendency becomes stronger

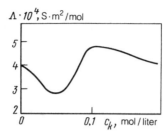

Fig. 11.1. Molar conductivities of $AgNO_3$ in pyridine as a function of solution concentration.

in solvents with low permittivity. This anomalous behavior of the non-aqueous solutions can be explained in terms of the different equilibria for ionic association (ion pairs or triplets) and complex formation. It is for the same reason that concentration changes often cause a drastic change in transport numbers of individual ions, which in some cases even assume values of less than zero or more than unity.

In aqueous solutions we see enhanced mobility and conductivity of the hydrogen ions, which is caused by additional proton transfer along chains of water molecules linked by hydrogen bonds (see Section 10.6.4). Solutions with nonaqueous, proton-containing solvents (e.g., in ammonia) sometimes also exhibit enhanced hydrogen ion mobility. In solutions with aprotic solvents which contain H^+ ions (for instance, after HCl addition), the mobility of these ions does not differ from that of other ions, i.e., the solvated protons present in the solution follow the usual migration mechanism.

11.3. IONICALLY CONDUCTING MELTS

Many electrochemical devices and plants (chemical power sources, electrolyzers, and others) contain electrolytes which are melts of various metal halides (particularly chlorides), also nitrates, carbonates, and certain other salts with melting points between 150 and 1500°C. The salt melts can be single-component (neat) or multicomponent, i.e., consist of mixtures of several salts (for their lower melting points in the eutectic region).

Melts are highly valuable as electrolytes, since processes can be realized in them at high temperatures which would be too slow at ordinary temperatures or which yield products which are unstable in aqueous solutions (e.g., electrolytic production of the alkali metals).

A special class of ionically conducting melts are the oxide-based systems (usually mixtures of a metal oxide and a nonmetal oxide, e.g., CaO and SiO_2) with melting points between 1200 and 2500°C. Such melts are often formed in the high-temperature processes of metallurgy.

The ionic conductivities of most solid crystalline salts and oxides are extremely low (an exception are the so-called solid electrolytes, which will be discussed in Section 11.4). The ions are rigidly held in the crystal lattices of these compounds and cannot move under the effect of applied electric fields. When melting, the ionic crystals break down forming free ions; the conductivities rise drastically and discontinuously, in some cases up to values of over 100 S/m, i.e., values higher than those of the most highly conducting electrolyte solutions.

When crystals with covalent bonds (e.g., $AlCl_3$ or $TiCl_4$) melt, the melt conductivity remains low, e.g., below 0.1 S/m, which implies that the degree of dissociation of the covalent bonds after melting is low. The covalent crystals also differ from the ionic crystals by their much lower melting points. The differences between these two types of crystal are rather pronounced, while there are few crystalline solids with intermediate properties.

A typical special feature of the melts of ionic crystals (ionic liquids) are their high concentrations of free ions, of about $25\ M$. Because of the short interionic distances, considerable electrostatic forces act between the ions, so that melts exhibit pronounced tendencies for the formation of different ionic aggregates: ion pairs, triplets, complex ions, etc.

Another special feature of ionic liquids is the lack of a foreign ("inert") molecular medium, and particularly a solvent, between the ions. Hence, they lack ion–molecule and the many kinds of nonelectrostatic interactions.

The results of X-ray studies show that upon melting, the ionic crystals retain some short-range order; an anion is more likely to be found in the immediate vicinity of a cation, and vice versa. The interionic distances do not increase upon melting, they rather decrease somewhat. Yet the volume of a salt markedly increases upon melting, usually by 10 to 20%. This indicates that melts contain a rather large number of voids (holes). These holes constantly form and perish owing to fluctuation phenomena attending the kinetic–molecular motion of the ions. Their mean size is between one and two interionic distances; they are uniformly distributed throughout the entire liquid volume.

TABLE 11.2. Conductivities σ, Cationic Transport Numbers t_+, Molar Conductivities of the Electrolyte (Λ) and Cation (λ_+), and Activation Energies of Conduction, A_σ, for a Number of Melts and Aqueous KCl Solution

Compound	T, °C melt.	T, °C meas.	σ, S/m	t_+	$\Lambda \cdot 10^4$, S·m^2/mol	$\lambda_+ \cdot 10^4$, S·m^2/mol	A_σ, kJ/mol
LiCl	614	620	587	0.75	183	137.2	5.2
NaCl	800	850	375	0.62	113.5	82.8	9.2
KCl	790	800	219	0.62	103.5	64.2	15
RbCl	715	715	117	0.58	94	54.2	16.7
CsCl	646	660	114	0.64	66.7	42.7	15
MgCl$_2$	712	730	105	0.48	28.8	13.8	15.9
CaCl$_2$	772	800	200	0.48	64	30.7	19.1
SrCl$_2$	873	900	198	0.26	55.7	14.5	18.1
PbCl$_2$	501	550	169	0.24	53	12.7	7.2
0.1 M KCl (aq)	—	18	1.12	0.494	112	55.3	11.3

Table 11.2 lists the conductivities, transport numbers t_+, and molar conductivities of the electrolyte ($\Lambda = \sigma/c_k$) and ions ($\lambda_+ = t_+\Lambda$) for a number of melts as well as for 0.1 M KCl solution. Melt conductivities are high, but the ionic mobilities are much lower in ionic liquids than in aqueous solutions; the high concentrations of the ions evidently give rise to difficulties in their mutual displacement.

It is now thought that the holes present in the melts are decisive for the conduction in melts. When an electric field is applied the ion nearest to a hole (in the direction of migration) will jump into the hole, leave a hole in its own former place, and thus the next ion can jump into this hole, etc. Ionic migration thus is not a smooth motion in a viscous medium but, rather, a sequence of ion–hole transitions.

For the jump of an ion into a hole a certain energy barrier must be overcome with the activation energy A_σ. The rate of this process (or value of conductivity) is subject to temperature dependence according to the well-known Arrhenius equation (see Section 14.1)

$$\sigma = B \exp(-A_\sigma/RT). \tag{11.3}$$

Values of A_σ for a number of melts are listed in Table 11.2.

The conductivities of melts, in contrast to those of aqueous solutions, increase with decreasing crystal radius of the anions and cations, since the leveling effect of the solvation sheaths is absent and ion jumps are easier when the radius is small. In melts constituting mixtures of two salts, often positive or negative deviations from additivity are observed for the values of conductivity (and also for many other properties). These deviations arise for two reasons, a change in hole size and the formation of new kinds of mixed ionic aggregates.

In a number of general properties such as viscosity and thermal conductivity, melts differ little from solutions. Their surface tensions are two to three times higher than those of aqueous solutions. It will be shown in Section 12.3 that this leads to poorer wetting of many solids, including important electrode materials such as carbon and graphite, by the ionic liquids.

Diffusion of ions can be observed in multicomponent systems where concentration gradients can arise. In individual melts, self-diffusion of ions can be studied with the aid of radiotracers. While the mobilities of ions are lower in melts, the diffusion coefficients are of the same order of magnitude as in aqueous solutions, viz., about 10^{-9} m^2/s. Thus, for melts the Einstein relation (4.6) is not applicable. This can be explained in terms of an appreciable contribution of ion pairs to diffusional transport; since these pairs are uncharged they do not carry current, so that values of ionic mobility calculated from diffusion coefficients will be high.

Equilibrium electrode potentials are readily established when metal electrodes are in contact with melts. However, two difficulties arise in attempts to measure them: suitable, sufficiently corrosion-resistant reference electrodes must be selected, and marked diffusion potentials develop at interfaces between different melts.

Experience shows that the potentials of metal electrodes in melts of their own salts (i.e., the activities of the cations) depend on the nature of the anions. However, the variation in the values of activity in melts is not very pronounced. This is due to the relatively small spread of interionic distances found in different melts (their entire volume is filled up with ions of similar size), as compared to the spread found in aqueous solutions. For this reason the electrostatic forces between the ions (which are very significant) do not greatly differ between different melts.

During electrolysis there is no change in composition of an individual melt close to the electrode surfaces, only its quantity (volume) will change. The resulting void space is filled again by flow of the entire liquid melt mass. This flow replaces the diffusional transport of ions customarily associated with aqueous solutions. This has particular consequences for the method used to measure ionic transport numbers; rather than determining the concentration changes in the melt layers near the electrodes by Hittorf's method, the volume changes occurring in these regions must be studied in the case of melts.

Because of the high temperatures, electrochemical reactions in melts as a rule are fast and involve little polarization. For such reactions the exchange current densities are as high as 10^4 to 10^5 A/m^2. Therefore, reactivities in melts (and also in high-temperature systems with solid electrolytes) are usually determined, not by kinetic but by thermodynamic features of the system.

Electrochemical processes in melts are often attended by side reactions and phenomena complicating the primary process. This is true, in particular, for the technically very important class of reactions in which a number of metals (calcium, barium, and others) are won electrometallurgically from molten salts. In many of these processes the metal which is deposited (sometimes in a highly disperse state) is found to interact with the corrosive melt, e.g., in a reaction such as

$$Ca + CaCl_2 \rightarrow 2CaCl (= Ca \cdot CaCl_2), \qquad (11.4)$$

producing valence-unsaturated (often colored) compounds. These compounds when forming at the cathode will dissolve in the melt, and can be reoxidized when reaching the anode (in our example, to $CaCl_2$). These processes markedly depress the current yields of metal. They are equivalent to the transport of metal (calcium) atoms or electrons through the melt from the cathode to the anode. Their intensity depends on melt composition, and can be lowered by the addition of salts which can form various ionic aggregates or complexes with the primary salt.

The melts as a rule have a strong corrosive effect, not only on the reaction products, but also on the various metallic and nonmetallic structural materials used to build the cells and reactors.

At high current densities sometimes the so-called "anode effect" occurs in melts during electrolysis: a gas skin is formed at the electrode surface, and there is intense sparking and a drastic increase in voltage. This effect depends on the anode material and on the melt anions, but its reasons are not fully understood. An important reason is insufficient wetting of the electrode surface by the melt, which causes "sticking" of gas bubbles to the surface.

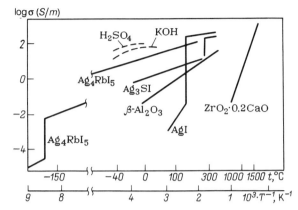

Fig. 11.2. Conductivities of a number of solid electrolytes as functions of temperature (dashed lines: the conductivities of 4 M H_2SO_4 and 8 M KOH solution).

11.4. SOLID ELECTROLYTES

11.4.1. Ionic Semiconductors

The conductivity of solid salts and oxides was first investigated by Michael Faraday in 1833. It was not yet known at that time that the nature of conduction in solid salts is different from that in metals. A number of fundamental studies were performed between 1914 and 1927 by Carl Tubandt in Germany and from 1924 onwards by Abram F. Ioffe and coworkers in Russia. These studies demonstrated that a mechanism of ionic migration in the lattice over macroscopic distances is involved. It was shown that during current flow in such a solid electrolyte, electrochemical changes obeying Faraday's laws occur at the metal/electrolyte interface.

In some cases (particularly at elevated temperatures) mixed electronic and ionic conduction is observed in solid salts. Typical materials with purely ionic conduction are the halides and sulfides of a number of metals, viz., $AgBr$, Ag_2S, $PbCl_2$, $CuCl_2$, and many others.

The conductivity, σ, of such materials is usually low at room temperature. The values of σ strongly increase with temperature (Fig. 11.2). The temperature dependence of conductivity can be described by Eq. (11.3). The appreciable temperature dependence is matched by the corresponding, high values of the activation energy of conduction, A_σ.

The conductivities of ionic crystals strongly depend on their purity. Impurities in the crystals markedly raise the values of σ, particularly at lower temperatures when the intrinsic conductivity of the pure material is still low.

All these features: low values of σ, a strong temperature dependence, and the effect of impurities, are reminiscent of the behavior of p- and n-type semiconductors. By analogy, we can consider these compounds as ionic semiconductors with intrinsic or impurity-type conduction.

As a rule (though not always), ionic semiconductors have unipolar conduction due to ions of one sign. Thus, in compounds AgBr, $PbCl_2$, and others the cation transport number t_+ is close to unity. In the mixed oxide $ZrO_2 \cdot nY_2O_3$ pure O^{2-} anion conduction ($t_- = 1$) is observed.

In an ideal ionic crystal all ions are rigidly held in the lattice sites where they only perform thermal vibratory motion. Transfer of an ion between sites under the effect of electrostatic fields (migration) or concentration gradients (diffusion) is not possible in such a crystal. Initially, therefore, the phenomenon of ionic conduction in solid ionic crystals was not understood.

Yakov I. Frenkel showed in 1926 that ideal crystals could not exist at temperatures above the absolute zero. Part of the ions leave their sites under the effect of thermal vibrations and are accommodated in the interstitial space leaving vacancies at the sites formerly taken up. Such point defects have been named Frenkel defects. These ideas were developed further by Walter Schottky in 1935, who pointed out that defects will also arise when individual ions or ion pairs are removed from the bulk lattice and brought, e.g., to the crystal surface; such defects have been named Schottky defects. Both types of point defect are in thermal equilibrium with the remainder of the crystal, and have the character of fluctuations; they spontaneously appear and disappear. Their concentrations drastically increase with temperature, i.e., their formation is associated with a high activation energy. The defect concentrations can be calculated statistically. For NaCl crystals close to the melting point it is about 10^{-3} M.

The point defects are decisive for conduction in solid ionic crystals. Ionic migration occurs in the form of relay-type jumps of the ions into the nearest vacancies (along the field). The relation between conductivity σ and the vacancy concentration is unambiguous, so that this concentration can also be determined from conductivity data.

In addition to the thermal vacancies, impurity-related vacancies will develop in ionic crystals. When the impurity ions bear charge different from the ions of like charge which are the crystal's main constituents, part of the lattice sites must remain vacant in order to preserve electroneutrality. Such impurity-type defects depend little on temperature, and their major effects are apparent at low temperatures when few thermal vacancies exist.

Because of the low concentration of thermal vacancies, pure ionic semiconductors have low conductivities, between 10^{-10} and 10^{-2} S/m. In the impurity-type ionic semiconductors the conductivities are sometimes higher; the best-known examples are the solid electrolytes on the basis of zirconium dioxide: $ZrO_2 \cdot nMO_x$ where $MO_x = CaO$, Y_2O_3, and others. The number of negative ions O^{2-} must decrease when a certain number of di- or trivalent cations are incorporated into the ZrO_2 lattice instead of the Zr^{4+} ions. For this reason oxygen vacancies are formed and oxygen ion conduction arises. This conduction is of practical importance only at high temperatures. Solid electrolytes $ZrO_2 \cdot 0.11Y_2O_3$ have a conductivity of about 12 S/m at 1000°C.

11.4.2. Ionic Conductors

It had been discovered long ago that the character of conduction in AgI changes drastically at temperatures above 147°C, when β- and λ-AgI change into α-AgI. At the phase transition temperature the conductivity, σ, increases discontinuously by almost four orders of magnitude (from 10^{-2} to 10^2 S/m). At temperatures above 147°C, the activation energy is very low and the conductivity increases little with temperature, in contrast to its behavior at lower temperatures (see Fig. 11.2).

Starting in the 1960s many compounds with such properties were discovered, i.e., with high conductivities and low temperature coefficients of conductivity. Some of them are double salts with silver iodide ($nAgI \cdot mMX$) or other silver halides where MX has, either the cation or the anion in common with the silver halide. The best-known example is $RbAg_4I_5$ (= $4AgI \cdot RbI$), where this sort of conduction arises already at -155°C and is preserved up to temperatures above 200°C. At 25°C this compound has a conductivity of 26 S/m, i.e., the same value as found for 7% KOH solution. Another example is Ag_3SI, which above 235°C forms an α-phase with a conductivity of 100 S/m.

The same conduction type is found for another class of compounds, the so-called sodium polyaluminates or β-aluminas $Na_2O \cdot nAl_2O_3$, where n has values between 3 and 11. Polycrystalline samples of these materials have a room-temperature conductivity of about 0.5 S/m, but at 300°C the conductivity is about 10 S/m.

Because of the high values of conductivity which, in individual cases, are found already at room temperature, such compounds are often called superionic conductors or ionic superconductors; but these designations are unfounded, and a more correct designation is "solid ionic conductors."

Strictly unipolar conduction is typical for all solid ionic conductors; in the silver double salts, conduction is due to silver ion migration, while in the sodium polyaluminates, conduction is due to sodium ion migration.

The discovery of the various ionic conductors has elicited strong interest, since they can be used in chemical power sources and other devices. It could be shown after numerous studies that the high ionic mobilities in these compounds are the result of particular lattice structures. In such lattices, the immobile ions of one kind (most often the anions) are fixed at their lattice sites and form a rather rigid, nondeformable sublattice. The sublattice of the other ions (most often the cations), to the contrary, is disordered: the cations are not bound to particular sites but can occupy any of a large number of equally probable sites. Since at any particular time an ion physically occupies just one site the other available sites function as the vacancies for the ion's motion. The differences between sites and interstitials are obliterated here, and a peculiar, highly mobile cation fluid is formed. The conductivity in the crystal will be anisotropic, i.e., depend on the direction in space, when the vacancies have a particular spatial configuration relative to the rigid anionic sublattice. As in the case of ionic semiconductors, this state is characteristic for certain structures, and can exist only within the temperature range where the particular crystal structure is stable.

Chapter 12
STRUCTURE AND PROPERTIES
OF SURFACE LAYERS

12.1. GENERAL CONCEPTS

12.1.1. Surface Layers and Interphases

Macroscopically, the region of contact between an electrode and an electrolyte is a two-dimensional surface separating the two phases. Microscopically, the same region is structured in a complex way; close to the boundary, surface layers of a certain thickness develop which differ from the principal phases in their properties. In a surface layer, particles are surrounded by other particles in an asymmetric fashion, and the forces acting on them do not balance. This gives rise to concentration changes relative to the values found in the bulk phase, it also leads to changes in the energy state of the individual particles and of the layer as a whole.

Figure 12.1 schematically shows the surface-layer structure of an aqueous electrolyte solution at the boundary formed with a gas phase. Despite their thermal motion, water molecules near the surface adopt an orientation in space. Ions close to the surface are less hydrated than those in the bulk solution, hence their sojourn in the surface layer is energetically disadvantageous, and they are pushed away from it into the bulk solution.

In the surface layers of metals an analogous, nonuniform distribution of the electrons is observed (see Fig. 2.8).

When two condensed phases (e.g., a metal and an electrolyte solution) are brought in contact, the surface-layer properties of each phase will change under the effect of the other phase. The set of two surface layers existing at the junction of two condensed phases is called the interphase.

283

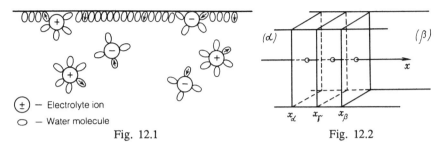

+ — Electrolyte ion
○ — Water molecule

Fig. 12.1   Fig. 12.2

Fig. 12.1. Surface-layer structure of an aqueous electrolyte solution (only those water molecules are shown which are oriented in the surface layer or in the first solvation sheath of ions; arrows point to the positive end of dipoles).

Fig. 12.2. Coordinates of interphase boundaries and Gibbs surface.

12.1.2. Surface Excesses

The thickness of the surface layer is usually measured in nanometers or fractions of a nanometer. There is a smooth change in properties as one moves from the surface into the interior of a phase, hence one cannot exactly define the limits of the surface layer or its thickness. This gives rise to difficulties in the layer's quantitative characterization.

These difficulties can be circumvented as follows. The outer limits of the interphase (Fig. 12.2, the planes with coordinates x_α and x_β) are deliberately positioned beyond the region affected by the superficial perturbations. In the interphase itself a conditional dividing plane, called the Gibbs surface (with the coordinate x_γ), is selected. The state of the real layer (with superscript s) is compared with that of an idealized layer having the same thickness but properties which, in each of the phases, remain unchanged right up to the Gibbs surface, i.e., a layer without changes (superscript id). The concept of *surface excesses* (with a superscript σ) of substances or energy in the real layer relative to the ideal layer is introduced, e.g.,

$$G^{(\sigma)} \equiv G^{(s)} - G^{(id)}. \tag{12.1}$$

Surface excesses are usually referred to unit surface area of the dividing plane S (surface excess densities).

The number of moles, n_j, of component j in the real interphase is given by

$$S \int_{x_\alpha}^{x_\beta} c_j^{(s)} \, dx,$$

and that in the ideal layer is given by

$$n_j^{(id)} = S[c_j^{(\alpha)}(x_\gamma - x_\alpha) + c_j^{(\beta)}(x_\beta - x_\gamma)].$$

We thus obtain an equation for the surface excess, $\Gamma_j \equiv n_j^{(\sigma)}/S$ (units: mol/m^2), of a substance per unit surface area:

$$\Gamma_j = \int_{x_\alpha}^{x_\beta} c_j^{(s)} \, dx - c_j^{(\alpha)}(x_\gamma - x_\alpha) - c_j^{(\beta)}(x_\beta - x_\alpha). \qquad (12.2)$$

The parameter $n_j^{(s)}$ always has positive values, while parameters $n_j^{(\sigma)}$ and Γ_j can have positive or negative values. Parameter Γ_j is called the *Gibbs adsorption* or *Gibbs surface excess*.

The adsorption value given by Γ_j is independent of the position selected for the outer limits of the interphase; even when the limits are pushed too far into a phase, then in the corresponding regions $c_j^{(s)}$ is equal to $c_j^{(\alpha)}$ (or $c_j^{(\beta)}$), and the value of Γ_j remains unchanged.

However, the value of Γ_j depends on the position of the Gibbs surface. By common convention, this position is so selected that for one of the components (with the index $j = 0$), the value of Γ_j defined by Eq. (12.2) will become zero. The solvent is chosen in this capacity when one of the phases in contact is a solution. Once the position of the Gibbs surface has been fixed one can unambiguously determine the Gibbs surface excesses of the other components. The adsorption of a component j is thus defined relative to the component $j = 0$ (*relative Gibbs surface excess* $\Gamma_{j(0)}$).

The position of the Gibbs surface thus selected is maintained when calculating the surface excesses of other parameters, such as that of the Gibbs energy $G^{(\sigma)}$.

Orientation of the water molecules at the interface between a metal and an aqueous solution causes some positive adsorption of these molecules, i.e., $c_{H_2O}^{(s)} > c_{H_2O}^{(\alpha)}$. But since by definition $\Gamma_{H_2O} = 0$, this implies that the Gibbs surface is inside the metal a small distance away from the physical metal/solution interface.

More commonly used is another definition of Gibbs surfaces excesses, according to which Γ_j is equal to the amount of substance j which must be added to the system (with a constant amount of the substance $j = 0$) so that the composition of the bulk phases will remain unchanged when the interface area is increased by unity. This definition can also be used when chemical reactions take place in the surface layer. In the case discussed here, the two definitions coincide.

The set of surface excesses of all components is sometimes called the surface phase (in contrast to the real surface layer or interphase).

12.1.3. Electrical Structure of Interphases

The interphase as a whole is electroneutral when an electrode is in contact with an electrolyte. However, *electric double layers* (edl) with a characteristic potential distribution are formed in the interphase because of a nonuniform distribution of the charged particles.

Two kinds of edl are distinguished: superficial and interfacial edl.

Superficial edl are wholly located within the surface layer of a single phase (e.g., the edl caused by a nonuniform electron distribution in the metal, the edl caused by orientation of the dipolar water molecules in the solution, the edl caused by specific adsorption of ions, etc.). The potential drops developing in these cases (the potential inside the phase relative to a point just outside) is called the surface potential $\chi^{(k)}$ of the given phase k.

Interfacial edl have their two parts in different phases, viz., the "inner" layer with the charge density $Q_{S,M}$ in the metal (on account of an excess or deficit of electrons in the surface layer), and the "outer" layer of counterions with the charge density $Q_{S,E} = -Q_{S,M}$ in the solution (an excess of cations or anions); the potential drop caused by this double layer is called the interfacial potential Φ.

The concept of edl formation at electrode surfaces was put forward by Hermann von Helmholtz in 1853. For a long time only the interfacial edl were taken into account. The considerable importance of various kinds of superficial edl was pointed out by Alexander N. Frumkin in 1919.

Each kind of edl and the potential drop produced by it contribute to the total Galvani potential, φ_G, at the interface considered:

$$\varphi_G^{(M,E)} = \Phi^{(M,E)} + \chi^{(M)} - \chi^{(E)} \tag{12.3}$$

(here the surface potential $\chi^{(E)}$ of the electrolyte solution can be the sum of several components, viz., those due to adsorption, $\chi_{ads}^{(E)}$, to orientation, $\chi_{or}^{(E)}$, etc.).

Adsorption and orientation in surface layers are phenomena evolving by their own specific laws, while the equilibrium value of the Galvani potential depends only on the bulk properties of the phases, according to

Eq. (3.34). The question arises, then, how the sum of all independent components on the right-hand side of Eq. (12.3) can always be constant, and equal to the given value of φ_G. The explanation must be sought in the regulating function assumed by the interfacial edl in the process of equilibration: between the phases (the electrode and the solution), charges will be transferred in one direction until the interfacial potential Φ has assumed a value satisfying Eq. (12.3). The interfacial potential will change in appropriate fashion when there has been, for whatever reason, a change in the adsorption or orientation components of the surface potentials.

The formation of any kind of edl implies the development of strong electrostatic fields in the interphase. The distance between the two sides of an edl as a rule is less than 1 nm, and the potential differences can attain several volts. Hence the field strengths E within edl can be higher than 10^9 V/m.

12.2. ADSORPTION

12.2.1. Kinds of Adsorption

Adsorption of a component j can be characterized by the distribution curve of its concentration $c_j^{(s)}$ in a section of the surface or interfacial layer. These curves may differ in their character (Fig. 12.3). The curve goes through a maximum when adsorption is positive (curve 1). Often a component accumulating in the surface layer is part of only one of the phases [e.g., phase (β)] but does not exist in the other phase. In this case the adsorption of this component (the adsorbate) on the surface of the other phase [the adsorbent, (α)] is discussed. In the limiting case, all adsorbed particles are packed right against the adsorbent surface, and the distribution curve has a high, narrow maximum (curve 2) corresponding to particle positions in the first layer from the adsorbent surface. This limiting case is called monolayer adsorption, in which the formal value of concentration $c_j^{(s)}$ in the adsorbed layer is high, and the value of Γ_j practically coincides with the amount of substance per unit area, $n_j^{(s)}/S$, which we shall label A_j and call real adsorption.

In other cases several layers of the adsorbing component can form on the adsorbent surface, and the distribution curve of adsorbate concentration has a distinct plateau (curve 3).

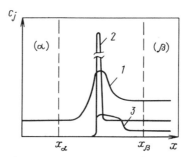

Fig. 12.3. Distribution curves of component concentrations in the surface layer.

When the component j can exist in both phases (curve 1) it will undergo redistribution after the phases have come into contact, and in particular some of it will be transferred into the interior of the phase where none of it had existed previously. In this case the term *absorption* (or bulk uptake) of the component is used.

When the two phases in contact are condensed phases and the entire volume is taken up by incompressible substances, positive adsorption of one component must be attended by negative adsorption (desorption) of other components. This phenomenon is called adsorptive displacement.

In the case of monolayer adsorption, a limiting adsorption value exists which is attained when the surface is completely covered by particles of a given substance, i.e., at full monolayer coverage. The limiting adsorption value A_j^0 depends on the effective surface area S_j taken up by one particle: $A_j^0 = 1/(S_j N_A)$ (N_A is the Avogadro constant). This parameter characterizes the number of sites which can be occupied by adsorbed particles on a given surface.

A convenient parameter for quantitative estimates of adsorption which is of the monolayer type is the degree of surface coverage defined by the relation

$$\theta \equiv A_j/A_j^0 \qquad (0 \le \theta \le 1). \tag{12.4}$$

In any particular case, the adsorption value depends on the properties of both the adsorbate (its "adsorbability") and the adsorbent (its "adsorptive power"). Substances with enhanced adsorbability are called surface-active. By convention, adsorption is regarded as insignificant when $\theta < 0.1$, and significant when $\theta > 0.5$.

12.2.2. Adsorption Energy

Adsorption of any component in the interphase is attended by a change in Gibbs energy and in enthalpy of the system, i.e., work $\bar{w}_j$ is performed and heat $\bar{q}_j$ is evolved. The adsorption energy is the algebraic sum of all energy effects associated with: (a) the formation of adsorbent–adsorbate bonds, (b) the desorption of other components, (c) bond breaking (e.g., in the dissociation or dehydration of adsorbate molecules), and (d) other kinds of system reorganization.

The (unbalanced) forces acting in surface layers are diverse in their intensities and character. Physical (van der Waals) forces between molecules are weak and give rise to slight energy effects (up to 20 kJ/mol). These forces decrease slowly with increasing distance (i.e., they operate within a relatively wide region) and are responsible for weak multilayer adsorption.

Unlike the van der Waals forces, chemical forces rapidly decay with increasing distance and are highly localized; they represent the chemical interaction between surface atoms (or ions) of the adsorbent and the adsorbate particles. These forces are strong; the energies of the bonds formed can be as high as 400 to 500 kJ/mol. Hence, in these cases adsorption is of the monolayer type, and it is often attended by chemical changes such as adsorbate dissociation. Thus, molecular hydrogen when adsorbed on a bare platinum surface will dissociate into atoms, and each adsorbed atom is covalently bonded to one surface atom of the platinum lattice:

$$2Pt^{(s)} + H_2 \rightarrow 2Pt^{(s)} - H_{ads}. \tag{12.5}$$

Adsorption due to the operation of chemical forces is called chemisorption.

12.2.3. Adsorption Isotherms

The adsorption of a component j in a given system depends on temperature T and on the component's concentration, $c_{V,j}$, in the bulk phase. The overall adsorption equation can be written as $A_j = f(T, c_{V,j})$. The relation between adsorption and the adsorbate's bulk concentration (or pressure, in the case of gases) at constant temperature is called the adsorption isotherm, while the relation between adsorption and temperature at constant concentration is called the adsorption isobar. From the shape of the adsorption isotherms, the adsorption behavior can be interpreted.

In the case of monolayer adsorption, the isotherms are usually written in the form of $\theta_j = f(c_{V,j})$. (Subscript j will be dropped in what follows.)

Henry isotherm. In the simplest case the degree of surface coverage is proportional to the bulk concentration:

$$\theta = Bc_V. \tag{12.6}$$

An analogous law was established in 1803 by William Henry for the solubilities of gases in water, hence this expression is called the Henry isotherm. The adsorption coefficient B (units: m^3/mol) depends on the heat of adsorption: $B = B^0 \exp(\bar{q}/RT)$. The Henry isotherm is valid for low surface coverages (e.g., at $\theta < 0.1$).

Langmuir isotherm. At higher values of θ when the number of free sites on the surface diminishes, one often observes relations of the form of

$$\theta = Bc_V/(1 + Bc_V) \quad \text{or} \quad \theta/(1 - \theta) = Bc_V. \tag{12.7}$$

At low values of the bulk concentration ($Bc_V \ll 1$) the degree of surface coverage is proportional to this concentration, but at high values it tends toward a limit of unity.

This equation was derived by Irving Langmuir in 1916 with three basic assumptions: (a) the number of adsorption sites is limited, and the value of adsorption cannot exceed A^0; (b) the surface is homogeneous: all adsorption sites have the same heat of adsorption and hence, the same coefficient B; and (c) no interaction forces exist between the adsorbed particles. The rate of adsorption is proportional to the bulk concentration and to the fraction, $(1-\theta)$, of vacant sites on the surface: $v_a = k_a c_V(1-\theta)$, while the rate of desorption is proportional to the fraction of occupied sites: $v_d = k_d\theta$. In the steady state these two rates are equal. With the notation of $k_a/k_d = B$ we obtain Eq. (12.7).

Temkin isotherm. Numerous departures of the experimental data from the Langmuir isotherm can be explained in terms of insufficient arguments for the second and third assumption. In 1939 Mikhail I. Temkin[1] examined the case of so-called uniform surface inhomogeneity where the heats of adsorption associated with different sites are different and range

---

[1]Pronounced Tyomkin.

from a maximum value $\bar{q}_0$ to a minimum value $\bar{q}_1$ (B varies between the values B_0 and B_1), while the different values of heat of adsorption are equally probable in this interval. The surface can be characterized by a dimensionless inhomogeneity factor $f \equiv (\bar{q}_0 - \bar{q}_1)/RT$. It follows from the solution obtained by Temkin that for $f \geq 5$ in the range of intermediate degrees of coverage ($0.2 \leq \theta \leq 0.8$)

$$\theta = g + (1/f)\ln c_V \quad \text{or} \quad \exp(f\theta) = B_0 c_V, \tag{12.8}$$

where $g \equiv (1/f)\ln B_0$. This equation is called the logarithmic Temkin isotherm.

Frumkin isotherm. In 1928 Frumkin derived an equation for interaction between the adsorbed particles. Mutual attraction leads to an increase in the heat of adsorption, while repulsion leads to a decrease. Quantitatively, these effects depend on the degree of surface coverage and can be written as $\bar{q}_\theta = \bar{q}_0 - f_{int}RT\theta$, where f_{int} is an interaction factor which has positive values when there is repulsion, and negative values when there is attraction between the particles (often the attraction constant $a \equiv -f_{int}/2$ is used instead of factor f_{int}). As a result an isotherm of the form of

$$[\theta/(1 - \theta)]\exp(f_{int}\theta) = B_0 c_V \tag{12.9}$$

is obtained. At values of $f_{int} \geq 5$ and intermediate degrees of coverage ($0.2 \leq \theta \leq 0.8$) this equation practically coincides with the equation of the Temkin isotherm. This implies that strong repulsive forces have the same effect as uniform surface inhomogeneity. Hence from the shape of an experimental adsorption isotherm one cannot deduce the reasons for departure from the Langmuir isotherm.

12.2.4. Adsorption in Electrochemical Systems

Adsorption phenomena at electrode/electrolyte interfaces have a number of characteristic special features.

(a) **Influence of electrode potential on adsorption.** Quantitative adsorption values usually depend on electrode potential. Substances which are strongly adsorbed in one region of potentials are often displaced from the interphase in another region of potentials.

(b) **Interaction forces.** There are two kinds of ionic adsorption from solutions onto electrode surfaces: electrostatic, under the effect of the charge on the metal surface, and specific, under the effect of chemical (nonelectrostatic) forces. Specifically adsorbing ions are called surface-active. Specific adsorption is more pronounced with anions.

(c) **Solvent adsorption.** The adsorption of solvent molecules is manifest in their orientation and ordered arrangement at the interface. Since in cases where a single solvent is employed, practically all of the surface is covered by these solvent molecules, the adsorption of other components is possible only upon desorption of part of the solvent molecules.

(d) **Reactant adsorption.** In many cases adsorption of a reactant is one of the first steps in an electrochemical reaction, and precedes charge transfer.

(e) **Adsorption of the electrochemical reaction products** is very common. The example of the adsorption of molecular hydrogen on platinum had been given earlier. Hydrogen adsorption is possible on the platinum electrode in aqueous solutions, even when there is no molecular hydrogen in the initial system; at potentials more negative than 0.3 V (rhe), the electrochemical reaction

$$Pt^{(s)} + H^+ + e^- \rightleftarrows Pt^{(s)} - H_{ads} \qquad (12.10)$$

proceeds, which yields hydrogen atoms adsorbed on the surface. This is a transient process ending when a certain (potential-dependent) value of θ is attained. In exactly the same way, at more positive potentials electrochemical oxygen adsorption

$$Pt^{(s)} + H_2O \rightleftarrows Pt^{(s)} - O_{ads} + 2H^+ + 2e^- \qquad (12.11)$$

proceeds on platinum and many other metals.

12.3. EXCESS SURFACE ENERGY. WETTING

The excess Gibbs energy of the surface layer of phase (α) at an interface with gas or vacuum, when referred to unit surface area of the phase, is called the *excess surface energy* (ese) of this phase, and designated as $\sigma^{(\alpha,G)}$ (units: J/m^2):

$$\sigma^{(\alpha,G)} \equiv G^{(\sigma)}/S = [G^{(s)} - G^{(id)}]/S. \qquad (12.12)$$

For the interphase between the two condensed phases (α) and (β), ese $\sigma^{(\alpha,\beta)}$ can be defined in a similar way. For instance, in the case of liquids

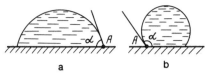

Fig. 12.4. Wetting angles at solids with surfaces lyophilic (a) and lyophobic (b) relative to the liquid.

one can define ese for interfaces formed with gas ($\sigma^{(L,G)}$), with another immiscible liquid ($\sigma^{(L,L)}$), and with a solid ($\sigma^{(L,S)}$).

By definition, ese is the work that must be performed to form unit area of new surface while preserving all other parameters. For stable interfaces the values of σ are always positive.

Because of ese, and because of the real forces acting on the particles in the surface layer, there is a tendency for spontaneous contraction of the interfacial area. In practice such a contraction can be realized only in the case of liquid phases where the particles can freely move relative to each other. Hence, liquids tend to assume spherical shape with a minimum ratio of surface area to volume (liquid drops), at least when the volume is small and gravitation does not interfere, and when they are not in contact with solid surfaces.

The tendencies of liquid surfaces to contract had in the past been regarded as the result of forces acting along the surface and causing it to contract (like an elastic film enveloping the liquid). This was the origin of the term "*surface tension*" still used today. The surface tension is stated as a force acting over unit length (units: N/m); for liquids, it numerically coincides with the value of ese stated in the units of J/m^2 (1 J = 1 N · m). The concept of surface tension is convenient for visualization of certain phenomena. However, it can merely conditionally convey the essence of the phenomena, since the real forces act in a direction perpendicular to the surface and not along the surface.

The *wetting of solid surfaces* by a liquid depends directly on the magnitude of ese. Consider a drop of liquid placed on the smooth surface of a solid (Fig. 12.4); the ambient air acts as the third phase (gas phase). The drop surface forms a certain angle α, the wetting angle, with the solid surface (this angle is measured in the liquid phase). It is assumed by convention that a surface is wetted by the liquid when $\alpha < 90°$ (Fig. 12.4a), and not wetted when $\alpha > 90°$ (Fig. 12.4b). Angles $\alpha \approx 0°$ correspond to complete wetting of the surface, in which case the drop spreads along the surface forming a thin liquid film.

Surfaces which are wetted by a liquid are called lyophilic with respect to this liquid, those which are not are called lyophobic. When the liquid is water or an aqueous solution, the terms used for the surface are hydrophilic and hydrophobic.

The wetting angle depends on three ese values: $\sigma^{(L,G)}$, $\sigma^{(L,S)}$, and $\sigma^{(S,G)}$. The system is in equilibrium when the overall Gibbs surface energy (the sum of the products of surface area of each interface times the corresponding ese value) has its minimum value. This implies that interfaces with high ese values will contract at the expense of those with lower ese values.

A quantitative relation can be derived by adducing the concept of tangential surface forces. Along the three-phase boundary, which constitutes a line (indicated in Fig. 12.4 as point A), the "forces" act in three directions. At equilibrium $\sigma^{(S,G)} = \sigma^{(S,L)} + \sigma^{(L,G)} \cos \alpha$, hence

$$\cos \alpha = [\sigma^{(S,G)} - \sigma^{(S,L)}]/\sigma^{(L,G)}. \tag{12.13}$$

It is a typical special feature of electrochemical systems that the ese values of interfaces between two conducting phases depend on the potential difference between them. The excess charge densities in the surface layer of each of the phases change with potential. The coulombic repulsion forces acting between excess charges of like sign along the surface, counteract the tendency of the surface to contract, and reduce the ese. Therefore, the ese will be lower the higher the charge density is in the interphase. The quantitative relation between these parameters is a very general thermodynamic one, and will be considered in Section 12.4. Wetting is influenced by electrode potential for the same reason, since it directly depends on ese.

12.4. THERMODYNAMICS OF SURFACE PHENOMENA

12.4.1. The Gibbs Equation

At constant temperature and pressure, the excess Gibbs energy of the surface layer depends on surface area S and on the composition of the layer, i.e., on the excess amounts $n_j^{(\sigma)}$ of the components. When there are changes in surface area or composition (which are sufficiently small, so that the accompanying changes in parameters σ and $\mu_j^{(\sigma)}$ can be disregarded) we have

$$dG^{(\sigma)} = [\partial G^{(\sigma)}/\partial S]_{n_j}\, dS + \sum [\partial G^{(\sigma)}/\partial n_j^{(\sigma)}]_{S,n_l}\, dn_j^{(\sigma)}$$
$$= \sigma dS + \sum \mu_j^{(\sigma)}\, dn_j^{(\sigma)}. \tag{12.14}$$

Parameter $G^{(\sigma)}$ is an additive function of surface area S and of the amounts of the components:

$$G^{(\sigma)} = \sigma S + \sum n_j^{(\sigma)} \mu_j^{(\sigma)}. \tag{12.15}$$

Subtracting Eq. (12.14) from the exact differential of Eq. (12.15) we find

$$S d\sigma + \sum n_j^{(\sigma)}\, d\mu_j^{(\sigma)} = 0. \tag{12.16}$$

This equation for the surface excesses is the surface analog of the Gibbs–Duhem equation (3.16) for bulk phases.

Thermodynamic discussions of surface-layer properties rely on the assumption of adsorption equilibrium, i.e., on the assumption that for each component the chemical potential in the surface layer is equal to that in the bulk phase:

$$\mu_j^{(\sigma)} = \mu_j^{(V)}. \tag{12.17}$$

When substituting into Eq. (12.16) the bulk value of chemical potential (simply written μ_j in what follows), and dividing all terms into S, we obtain the *Gibbs adsorption equation*, which is a very important equation of surface-layer thermodynamics:

$$-d\sigma = \sum \Gamma_j d\mu_j = RT \sum \Gamma_j d\ln a_j. \tag{12.18}$$

When the system is ideal and adsorption is possible for only one component, the Gibbs equation can be written as

$$-d\sigma = \Gamma RT d\ln c. \tag{12.19}$$

Thus, in the case of positive adsorption the value of σ decreases with increasing bulk concentration, but in the case of negative adsorption it increases.

For the adsorption of two components, e.g., the solvent (subscript 0) and a solute (subscript 1), the equation becomes

$$-d\sigma = \Gamma_0 d\mu_0 + \Gamma_1 d\mu_1. \qquad (12.20)$$

This equation cannot be used to calculate absolute values of adsorption, e.g., from the derivative $(\partial\sigma/\partial\mu_1)_{\mu_0}$, since the values of μ_0 and μ_1 are interrelated. We cannot alter μ_1 while keeping μ_0 constant (or vice versa). Therefore, if we alter μ_1 (by changing the concentration of component 1) we have

$$-d\sigma/d\mu_1 = \Gamma_1 + \Gamma_0(d\mu_0/d\mu_1). \qquad (12.21)$$

Using the Gibbs–Duhem equation (3.16) linking $d\mu_0$ and $d\mu_1$, we can modify Eq. (12.21):

$$-d\sigma/d\mu_1 = \Gamma_1 - \Gamma_0(n_1/n_0). \qquad (12.22)$$

It follows that thermodynamically the adsorption of a substance 1 can be calculated only when the concept of adsorption itself is defined, as described in Section 12.1.2, i.e., when it is assumed that adsorption of the other component (the solvent) is zero.

12.4.2. The General Equation of Electrocapillarity

When the Gibbs equation is used for an electrode/electrolyte interface, the charged species (electrons, ions) are characterized by their electrochemical potentials, while the interface is regarded as electroneutral, i.e., the surface density, $Q_{S,M}$, of excess charges in the metal caused by positive or negative adsorption of electrons ($Q_{S,M} = -F\Gamma_e$), is regarded as equal in size but opposite in sign to the charge density, $Q_{S,E}$ ($= F\sum z_j\Gamma_j$), in the solution's surface layer:

$$Q_{S,M} = -Q_{S,E} = -F\sum_j z_j\Gamma_j \qquad (12.23)$$

(in what follows, surface charge densities in the metal or electrolyte solution will often be called surface charge, for the sake of brevity).

The Gibbs equation for metal/electrolyte interfaces is of the form of

$$-d\sigma = \Gamma_e d\tilde{\mu}_e + \sum \Gamma_j d\tilde{\mu}_j \qquad (12.24)$$

(for uncharged species $\tilde{\mu}_j = \mu_j$). But the equation cannot be used in this form, since it contains parameters $d\tilde{\mu}_e$ and $d\tilde{\mu}_j$ for the electrons and ions which cannot be determined experimentally.

For charged species we have $d\tilde{\mu}_j = d\mu_j + z_j F d\psi$ when taking into account Eq. (3.17). For the electrons in the metal (which have a constant concentration) $d\mu_e = 0$. The Galvani potential at the interface considered is given by $\varphi_G = \psi^{(M)} - \psi^{(E)}$. Using these relations as well as equality (12.23), we can transform Eq. (12.24) to

$$-d\sigma = Q_{S,M} d\varphi_G + \sum \Gamma_j d\mu_j. \qquad (12.25)$$

The equation obtained can be used when the electrode potential can be varied independently of solution composition, i.e., when the electrode is ideally polarizable. For practical calculations we must change from the Galvani potentials, which cannot be determined experimentally, to the values of electrode potential which can be measured: $E = \varphi_G - \varphi_{RE} + \varphi_d +$ const (where φ_{RE} is the Galvani potential associated with the reference electrode, and φ_d is the diffusion potential between the working solution and the solution of the reference electrode). Two cases are considered:

(i) Constant reference electrode (e.g., a silver–silver chloride electrode in 0.1 m KCl solution) and working solutions sufficiently dilute so that φ_d will remain practically constant when their concentration is varied. In this case $dE = d\varphi_G$ and

$$-d\sigma = Q_{S,M} dE + \sum \Gamma_j d\mu_j. \qquad (12.26)$$

(ii) A solution containing only two kinds of ion (with the charge numbers z_+ and z_-) and a reference electrode in the same solution. In this case there is no diffusion potential. The reference electrode can be reversible, either with respect to the cation (then we conventionally write φ_+ and E_+ for the Galvani potential associated with the reference electrode and for the electrode potential of the working electrode) or with respect to the anion (the corresponding designations are φ_- and E_-). Consider, for the sake of definition, the latter case. For the reference electrode it follows from Eq. (2.11) that $d\varphi_- = (1/z_- F)d\mu_-$. The differential of electrode potential is given by $dE_- = d\varphi_G - d\varphi_-$, and the surface charge density of the metal by $Q_{S,M} = -F(z_+\Gamma_+ + z_-\Gamma_-)$. Substituting these expressions into Eq. (12.25) and allowing for the fact that $z_+/z_- = -\tau_-/\tau_+$ and [according to Eq. (3.24)] $\tau_+ d\mu_+ + \tau_- d\mu_- = RT(\tau_+ + \tau_-)d\ln a_\pm$, we obtain after simple transformations

$$-d\sigma = Q_{S,\mathrm{M}}dE_- + [1 + (\tau_-/\tau_+)]RT\Gamma_+ d\ln a_\pm. \qquad (12.27)$$

Analogously, for the case where a reference electrode which is reversible with respect to the cation is used, we obtain

$$-d\sigma = Q_{S,\mathrm{M}}dE_+ + [1 + (\tau_+/\tau_-)]RT\Gamma_- d\ln a_\pm. \qquad (12.28)$$

Equations (12.26) to (12.28) are different forms of the *general equation of electrocapillarity*. The parameters contained in them can be determined experimentally. These equations can be used to calculate one set of parameters when experimental values for another set of parameters are available. For instance, the electrode's surface charge density at any potential can be determined from the slope of a plot of ese against potential obtained by measurements in solutions of a particular, constant composition (in such a solution evidently $dE = dE_- = dE_+$):

$$Q_{S,\mathrm{M}} = -(\partial\sigma/\partial E)_{\mu_j}. \qquad (12.29)$$

This equation was first obtained by Gabriel Lippmann in 1875.

Cation adsorption in binary solutions can be determined, according to Eq. (12.27), from the relation between ese and the $a_\pm$-values:

$$\Gamma_+ = -(1/RT)[\tau_+/(\tau_+ + \tau_-)](\partial\sigma/\partial\ln a_\pm)_{E_-}, \qquad (12.30)$$

while anion adsorption can be determined, either from the analogous expression for Γ_- or from the experimental data for cation adsorption Γ_+ and surface charge $Q_{S,\mathrm{M}}$ [cf. Eq. (12.23)].

The Lippmann equation is of basic importance for electrochemistry. It shows that surface charge $Q_{S,\mathrm{M}}$ can be calculated thermodynamically from data obtained when measuring ese.

The plot of ese against potential is called the *electrocapillary curve* (ecc). Typical curves measured at a mercury electrode in NaF solutions of different concentration are shown in Fig. 12.5. Also shown in this figure is a plot of $Q_{S,\mathrm{M}}$-values against potential calculated via Eq. (12.29).

Electrocapillary curves have a maximum. At this point, according to Eq. (12.29), the surface charge $Q_{S,\mathrm{M}} = 0$. The potential, E_{zc}, of the maximum is called the *point of zero charge* (pzc). The metal surface is positively charged at potentials more positive than the pzc, and it is negatively charged at potentials more negative than the pzc. The point of zero

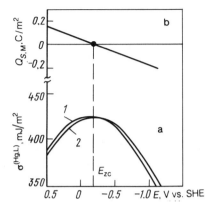

Fig. 12.5. (a) Electrocapillary curve of mercury in NaF solutions: (1) 0.01 M; (2) 0.1 M. (b) A plot of surface charge density vs. potential calculated from curve 2.

charge is a characteristic parameter for any electrode/electrolyte interface. The concept of pzc is of exceptional importance in electrochemistry.

Knowing the charge density $Q_{S,M}$ one can calculate the interfacial potential Φ contained in Eq. (12.3). This is insufficient however, for a calculation of the total Galvani potential, since other terms in this equation cannot be determined experimentally.

For metals where an electrochemical adsorption of components is possible [e.g., adsorption of hydrogen according to reaction (12.10)], the concept of surface charge becomes more complex; we then must distinguish free and total surface charge. These aspects will be considered in more detail in Section 12.8.6 in the instance of the platinum electrode.

12.5. STRUCTURE OF THE ELECTRIC DOUBLE LAYER

Consider in more detail the edl structure at the interface between a metal electrode and an aqueous electrolyte solution, as well as the influence of electrode potential, concentration of the electrolyte solution, and nature of the electrolyte (presence of surface active ions etc.) on edl structure.

It had been shown in Section 12.1.3 that edl structures at electrode surfaces are complex and include a number of components. Some of them, like a nonuniform electron distribution in the metal's surface layer and the layer of oriented dipolar water molecules in the solution surface layer adjacent to the electrode, depend on the above parameters to only a minor extent. Usually the contribution of these layers is regarded as constant, and it is only in individual cases that we must take into account any change in these surface potentials, $\delta\chi_e^{(M)}$ and $\delta\chi_{or}^{(E)}$, which occurs as a result of changes in the experimental conditions.

Changes in the parameters listed above primarily influence the interfacial edl, i.e., the excess charge densities $Q_{S,M} = -Q_{S,E}$ and the distribution of electrostatically adsorbed ions in the solution region next to the electrode, as well as that part of the solution's superficial edl which is associated with specific adsorption of solution ions. These two kinds of edl together are called the *ionic* edl. Many electrochemical properties of electrodes are determined precisely by the structure of the ionic edl and its changes during an experiment.

Because of mutual repulsion forces and also because of attraction forces arising on the other side of the edl, the excess charges in the metal are always tightly "packed" against the interface. The excess charges in the solution, i.e., the ions, are subject to thermal motion, and despite the electrostatic attraction can roam some small distance away from the surface.

A convenient parameter characterizing the charge distribution in the edl is the electrode's differential capacitance $C \equiv dQ_{S,M}/dE$ (units: F/m^2). Evidently, the value of capacitance will be lower the larger the (mean) distance of the excess ions from the surface (i.e., the thicker the edl). The value of differential capacitance can be found by differentiating the relation between $Q_{S,M}$ and potential (or by twofold differentiation of the electrocapillary curve). Much more accurate are values of capacitance measured directly (Section 12.6). A typical value of edl capacitance referred to true surface area is about 0.2 F/m^2.

The differential capacitance depends on potential and, in dilute solutions, on electrolyte concentration (Fig. 12.6). At low electrolyte concentrations a capacitance minimum is seen near the point of zero charge. The capacitance increases with increasing distance from the pzc, and at appreciable positive and negative potentials it tends toward certain limiting values. The capacitance in the region of the minimum increases with increasing concentration, and the minimum disappears when certain values of concentration are attained.

The surface charge density, $Q_{S,M}$, can be found, not only by differentiating the ecc but also by integrating the C vs. E curves from the pzc to a given potential E (Fig. 12.6). When the $Q_{S,M}$ vs. E curves cal-

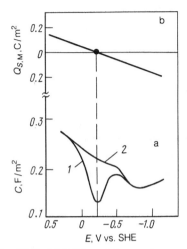

Fig. 12.6. (a) Differential capacitance of a mercury electrode as function of potential in NaF solutions: (1) 0.01 M; (2) 0.1 M. (b) A plot of surface charge density vs. potential calculated from curve 2.

culated from ecc coincide with those calculated from capacitance curves, this is evidence for the mutual compatibility of results obtained by these two methods of measurement. However, calculations associated with an integration of experimental data usually are more accurate than those associated with a differentiation. Still, the values of E_{zc} needed for integrating the C vs. E curves are taken from the positions of the ecc maxima.

Consider the distribution of electrostatic potential in the ionic edl as a function of distance x from the surface. By convention we locate the point of reference in the solution interior, i.e., shall assume that $\psi = 0$ when $x \to \infty$. The potential at $x = 0$ will be designated as ψ_0. Parameter ψ_0 represents the total potential drop across the ionic edl, and its sign corresponds to that of $Q_{S,M}$. When specific adsorption of ions is absent, the value of ψ_0 coincides with that of the interfacial potential Φ. When specific adsorption is present it also includes the adsorption potential $\chi_{ads}^{(E)}$ due to specific adsorption of ions of one sign and electrostatic adsorption of the corresponding number of counterions [this potential is counted in a direction opposite to that of the interfacial potential or of the metal's surface potential, see Eq. (12.3)].

At the pzc in the absence of specific adsorption of ions $\psi_0 = 0$. The values of ψ_0 at other potentials correspond to the difference $E - E_{zc}$, with minor corrections for the changes in surface potentials $\delta\chi$:

$$E - E_{zc} = \psi_0 + \delta\chi_e^{(M)} - \delta\chi_{or}^{(E)} \approx \psi_0; \qquad (12.31)$$

the values of $\delta\chi$ are usually small.

Since a number of electrochemical phenomena depend on the potential of the ionic edl, ψ_0, Lev I. Antropov between 1946 and 1951 suggested when comparing the properties of different electrodes to use the so-called reduced scale of electrode potentials ($E - E_{zc}$) where the potential of each electrode is reckoned against its own pzc.

Different structural models of the ionic edl have been suggested in order to describe the electrical properties of interfaces.

12.5.1. The Helmholtz Model

Following the concepts of Helmholtz, the edl has a rigid structure, and all excess charges on the solution side are packed against the interface. Thus, the edl is likened to a capacitor with plates separated by a distance δ, which is that of the closest approach of an ion's center to the surface. The edl capacitance depends on δ and on the value of ε for the medium between the plates. Adopting a value of δ which is $(1-2) \cdot 10^{-10}$ m, and a value of $\varepsilon = 4.5$ (the water molecules in the layer between the plates are oriented, and the value of ε is much lower than that in the bulk solution), we obtain $C \approx 0.2$–0.4 F/m^2, which corresponds to the observed values. However, this model has a defect, in that the values of capacitance calculated depend, neither on concentration nor on potential, which is at variance with experience (the model disregards thermal motion of the ions).

12.5.2. The Gouy–Chapman Model

Thermal motion of the ions in the edl was included in the theories developed independently by Georges Gouy in France (1910) and David L. Chapman in England (1913).

The combined effects of the electrostatic forces and of the thermal motion in the solution near the electrode surface, give rise to a diffuse distribution of the excess ions, and a *diffuse edl part* or diffuse layer is formed.

The distributions of charges and potential are calculated as in the Debye–Hückel theory of ion–ion interaction outlined in Section 10.8, except that now the "central charge" is not a point or spherical charge but a plane, i.e., a charged surface. (The Gouy–Chapman theory was created about 10 years earlier than the Debye–Hückel theory, and probably had great influence on its development.)

The Poisson equation for the flat, one-dimensional problem is

$$d^2\psi/dx^2 = -(1/\varepsilon_0\varepsilon)Q_V(x), \tag{12.32}$$

where $Q_V(x)$ is the volume density of excess charges.

The total excess charge, $Q_{S,E}$, in the solution per unit surface area is determined by the expression

$$Q_{S,E} = \int_0^\infty Q_V(x)\,dx. \tag{12.33}$$

Integrating the Poisson equation over x and allowing for Eq. (12.33), as well as for the fact that in the interior of the solution $(d\psi/dx)_{x\to\infty} = 0$, we find

$$Q_{S,M} = -Q_{S,E} = -\varepsilon_0\varepsilon(d\psi/dx)_{x=0} \tag{12.34}$$

(the minus sign indicates that for $Q_{S,M} > 0$ we should have $d\psi/dx < 0$).

The volume charge density depends on the ion distribution, which obeys the Boltzmann equation:

$$Q_V(x) = F\sum z_j c_j(x) = F\sum z_j c_{V,j}\exp(-z_j F\psi/RT). \tag{12.35}$$

Substituting this expression into Eq. (12.32) we obtain the basic differential equation for the potential distribution:

$$d^2\psi/dx^2 = -(F/\varepsilon_0\varepsilon)\sum z_j c_{V,j}\exp(-z_j F\psi/RT). \tag{12.36}$$

In this one-dimensional flat case, the Laplace operator [the left-hand part of Eqs. (12.32) or (12.36)] is simpler than in the problem with spherical symmetry arising when deriving the Debye–Hückel limiting law [the left-hand part of Eq. (10.64)]. Therefore, the differential equation (12.36) can be solved even without the simplification (of replacing the exponential factors by two terms of their series expansions) that would reduce its accuracy. We shall employ the mathematical identity

$$(d\psi/dx)^2 \equiv 2\int (d^2\psi/dx^2)\,d\psi, \qquad (12.37)$$

which is readily verified by differentiating both sides with respect to x. We substitute into it Eq. (12.36) and integrate it over ψ from the current value of ψ to $\psi = 0$ (at $x \to \infty$). We determine the integration constant from the condition that for $\psi = 0$ we also have $(d\psi/dx) = 0$. As a result we have

$$(d\psi/dx)^2 = (2RT/\varepsilon_0\varepsilon)\sum c_{V,j}[\exp(-z_j F\psi/RT) - 1]. \qquad (12.38)$$

As a simplification, we shall consider the binary solution of a $z : z$ electrolyte. In this case

$$(d\psi/dx)^2 = (2RTc_V/\varepsilon_0\varepsilon)[\exp(-zF\psi/RT) + \exp(zF\psi/RT) - 2]$$
$$= (2RTc_V/\varepsilon_0\varepsilon)[2\sinh(zF\psi/2RT)]^2. \qquad (12.39)$$

Parameters ψ and $d\psi/dx$ differ in sign, hence

$$d\psi/dx = -2(2RTc_V/\varepsilon_0\varepsilon)^{1/2}\sinh(zF\psi/2RT). \qquad (12.40)$$

When integrating this equation (which is readily accomplished for small and large values of $zF\psi/2RT$ separately) we can find the distribution of potential $\psi(x)$ relative to distance x. At low values of ψ this distribution is exponential [cf. the analogous Eq. (10.72)].

It is the major aim of diffuse edl theory to establish the relation between surface charge $Q_{S,M}$ and potential ψ_0 at the point $x = 0$ (the total potential drop across this layer). Substituting the value of the derivative $d\psi/dx$ for $x = 0$ into Eq. (12.34) we find for surface charge density:

$$Q_{S,M} = 2A\sqrt{c_V}\sinh(zF\psi_0/2RT), \qquad (12.41)$$

where $A \equiv (2RT\varepsilon_0\varepsilon)^{1/2}$. This function yields an expression for the very important parameter characterizing the degree of diffuseness of the edl, viz., the (specific) differential capacitance of the edl,

$$C \equiv dQ_{S,M}/d\psi_0 = (zFA\sqrt{c_V}/RT)\cosh(zF\psi_0/2RT). \qquad (12.42)$$

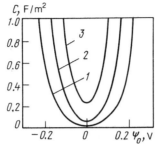

Fig. 12.7. Potential dependence of differential capacitance calculated from Gouy–Chapman theory for $z_+ = |z_-| = 1$ and different concentrations: (1) 10^{-4}; (2) 10^{-3}; (3) 10^{-2} M.

The higher the degree of diffuseness (or, the larger the mean distance x) the lower will be the differential capacitance.

Figure 12.7 shows plots of differential capacitance against the potential ψ_0 calculated for different concentrations c_V with Eq. (12.42), while assuming that $z = 1$ and that in the diffuse layer $\varepsilon = 78$. Unlike the Helmholtz model, the Gouy–Chapman model provides a qualitatively correct description of the capacitance minimum near the pzc and of the capacitance rise which occurs with increasing values of concentration and $|\psi_0|$.

However, for concentrations higher than 10^{-2} M and for potentials $|\psi_0| > 0.05$ V, the calculated capacitance values are much higher than the experimental ones (see Fig. 12.6). The discrepancies can be explained by the fact that the Gouy–Chapman model, unlike that of Helmholtz, disregards the ions' own size and allows the centers of the ions to come however close to the physical surface.

12.5.3. The Models of Stern and Grahame

In 1924 Otto Stern (Germany) suggested a new version of edl theory where the true size of the ions and the possibility of specific (nonelectrostatic) adsorption of the ions are taken into account. In 1947 David C. Grahame (USA) considerably improved this theory by eliminating a number of defects that had been present in earlier versions.

By combining the ideas of Helmholtz and of Gouy and Chapman, Stern introduced the notion of a plane of closest approach of the ions to the surface, the so-called *Helmholtz plane* with the coordinate x_H (which depends on the ionic radius) and the potential ψ_H.

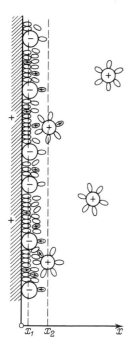

Fig. 12.8. Electric double-layer structure according to Grahame's model (see Fig. 12.1).

The charges on the solution side of the edl can be divided into two parts. One part ($Q_{S,H}$) is located at this plane and constitutes the compact or Helmholtz part of the edl. The other part ($Q_{S,d}$), which is under the effect of thermal motion, constitutes the diffuse part of the edl. Specifically adsorbed ions are located at the Helmholtz plane and determine the value of $Q_{S,H}$. However, according to the equations of Stern's theory, some nonzero value of $Q_{S,H}$ is retained even in the absence of chemical forces, which is physically difficult to explain.

Grahame introduced the idea that electrostatic and chemical adsorption are different in character. In the former, the adsorption forces are weak, the ions are not deformed during adsorption and continue to participate in thermal motion. Their distance of closest approach to the electrode surface is called the *outer Helmholtz plane* (coordinate x_2, potential ψ_2, charge of the diffuse edl part $Q_{S,2}$). When the more intense (and localized) chemical forces are operative the ions are deformed, undergo partial dehydration, and lose mobility. The centers of the specifically adsorbed ions constituting the charge $Q_{S,1}$ are at the *inner Helmholtz plane* with the potential ψ_1 and the coordinate x_1 ($x_1 < x_2$, Fig. 12.8).

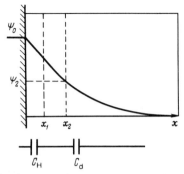

Fig. 12.9. Potential as a function of coordinate x within the electric double layer in the absence of specific adsorption of ions (Grahame's model).

Since the edl as a whole is electroneutral we have

$$Q_{S,M} = -(Q_{S,1} + Q_{S,2}).\qquad(12.43)$$

Consider first the situation when specific adsorption of ions is absent. In this case the ions cannot penetrate to the inner Helmholtz plane, the charge density $Q_{S,1}$ is zero, and Eq. (12.43) becomes

$$Q_{S,M} = -Q_{S,2}.\qquad(12.44)$$

Since no charges exist in the compact edl part ($0 \le x \le x_2$), the value of $d\psi/dx$ will be constant and the potential will vary linearly from ψ_0 to ψ_2 (Fig. 12.9). According to Eq. (12.34),

$$Q_{S,M}[= -\varepsilon_0\varepsilon(d\psi/dx)] = C_H(\psi_0 - \psi_2),\qquad(12.45)$$

where $C_H \equiv \varepsilon_0\varepsilon/x_2$ is the Helmholtz layer capacitance.

This capacitance depends on the nature of the ions on the solution side of the edl, but not on their concentration. When the metal is negatively charged the solution side is formed by cations, and $C_H \approx 0.2$ F/m². But when the metal is positively charged the solution side is formed by anions (for which the distances x_2 are smaller than for the cations), and the values of C_H can be as high as 0.4 F/m². Thus, in passing through the pzc the value of C_H changes discontinuously; but in each of the regions on the two sides of the pzc it is little potential-dependent.

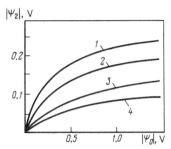

Fig. 12.10. Plots of ψ_2 vs. ψ_0 in NaF solutions of different concentrations: (1) 0.001; (2) 0.01; (3) 0.1; (4) 1 M.

The charge of the diffuse edl part can be described by the equations of Gouy–Chapman theory, but then the integration over variable x is from x_2 to ∞ rather than from 0 to ∞. In particular,

$$-Q_{S,2}[= Q_{S,M}] = 2A\sqrt{c_V}\,\sinh(zF\psi_2/RT). \qquad (12.46)$$

Combining the last two equations we finally obtain

$$Q_{S,M} = C_H(\psi_0 - \psi_2) = 2A\sqrt{c_V}\,\sinh(zF\psi_2/2RT). \qquad (12.47)$$

An important parameter of this combined edl is the potential ψ_2. Its dependence on concentration c_V and on the total potential ψ_0 (or total charge $Q_{S,M}$) can be found with the aid of Eq. (12.47) (Fig. 12.10). At sufficiently large values of $|\psi_0|$ this dependence can be written as

$$|\psi_2| = \text{const} + (2RT/F)\ln|\psi_0| - (RT/F)\ln c_V. \qquad (12.48)$$

By definition, the total differential capacitance, C, of the edl is given by $dQ_{S,M}/d\psi_0$. We shall introduce the notion of a capacitance of the diffuse edl part, $C_d \equiv |dQ_{S,2}/d\psi_2|$. It can be calculated via Eq. (12.42), except that ψ_2 must be used instead of ψ_0.

To establish the connection between the three kinds of capacitance, C, C_H, and C_d, let us differentiate Eq. (12.45) with respect to $Q_{S,M}$. It follows from the result obtained that in the case being discussed, when $-Q_{S,2}$ can be replaced by $Q_{S,M}$,

$$1/C = 1/C_H + 1/C_d. \qquad (12.49)$$

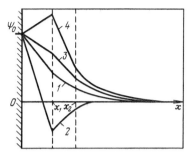

Fig. 12.11. Potential–distance plots in edl with specific adsorption of ions.

Therefore, electrically the edl can be likened to a network of two capacitances, C_H and C_d, connected in series (see Fig. 12.9). The total capacitance is lower than that of the smaller component. Considering that C_H depends little on potential while C_d varies as shown in Fig. 12.7, the relations between total capacitance C and potential obtained at different concentrations are in good agreement with the experimental data shown in Fig. 12.6. At very high concentrations and high $|\psi_0|$ the diffuse edl part is compressed, the capacitance C_d increases, and the total capacitance C tends toward the limiting value of C_H.

Specific adsorption occurs at the inner Helmholtz plane. Depending on the sign of charge of the specifically adsorbing ions and on adsorption strength, different types of potential distribution in the edl are possible (Fig. 12.11). When ions become adsorbed which carry charge opposite in sign to $Q_{S,M}$, the potential gradient in the inner Helmholtz layer between $x = 0$ and $x = x_1$ will increase in absolute value (curve 1). When specific adsorption is appreciable a change in the sign of surface charge is possible; this occurs when the charge density of opposite sign at the inner Helmholtz plane is larger than the charge density of the metal surface (curve 2). In this case the sign of ψ_1 is opposite to that of ψ_0, and the diffuse edl part must compensate this surplus charge $|Q_{S,1} - Q_{S,M}|$. When the specifically adsorbed ions have charge of the same sign the potential gradient in the Helmholtz layer decreases (curve 3), and it may even change sign (curve 4).

Specific adsorption of ions changes the value of E_{zc}, hence one distinguishes the notion of a point of zero charge, E_{zc}, in solutions of surface-inactive electrolytes, which depends on the metal, from that of a point of zero charge, $E_{zc,ads}$, in solutions of surface-active ions, which in addition depends on the nature and concentration of these ions. The difference between these quantities,

$$E_{zc,ads} - E_{zc} = -\chi_{ads}^{(E)}, \tag{12.50}$$

is the adsorption potential drop of the ions at the uncharged surface (this potential drop is opposite in sign to the change in E_{zc}-value).

Equation (12.31) remains valid when there is specific adsorption of ions, so long as the value of E_{zc} rather than that of $E_{zc,ads}$ is used in it.

In the presence of specific adsorption, Eq. (12.43) is valid, hence we cannot replace $-Q_{S,2}$ by $Q_{S,M}$, which implies that Eq. (12.49) is no longer obeyed.

Quantitative calculations are beset by a number of difficulties when specific adsorption occurs. Since the ions at the inner Helmholtz plane are localized (contrary to the ones at the outer Helmholtz plane, which are "smeared out"), it is not correct to define a unique potential ψ_1 for all points in this plane. Instead one must allow for a discrete distribution of charges and potential along the surface. As a rule, the micropotential $\psi_{1(0)}$ at a point of the inner Helmholtz plane which is not occupied, and to which an ion can be transferred from the solution, is lower (in absolute value) than the averaged macropotential ψ_1 which would be observed for delocalized charge. The effect of discreteness can be quantitatively characterized by the factor $\lambda \equiv \psi_{1(0)}/\psi_1$; this is unity for delocalized charge, and decreases with increasing discreteness-of-charge effect.

An important result of discreteness is weaker repulsive interaction between adsorbed ions of like charge. This leads to higher values of the adsorption coefficients and to a stronger dependence of adsorption on bulk concentration. It is these aspects, in particular, which can explain the anomalous pzc shift detected experimentally in 1939 by Oleg A. Esin and Boris F. Markov. According to the theory of electrocapillary curves, the pzc shift $dE_{zc}/d\ln c_j$ which occurs when varying the concentration of a specifically adsorbing ion j should be $RT/z_j F$. However, in halide solutions negative shifts as high as 100 mV (instead of 59 mV) were observed for a tenfold increase in concentration (the *Esin–Markov effect*). A more detailed analysis shows in fact, that the pzc shift will increase by a factor of $1/\lambda$ when the charge distribution at the inner Helmholtz plane is discrete.

12.6. METHODS FOR STUDYING ELECTRODE SURFACES

The methods that can be used to study surfaces appreciably depend on the nature of the electrode involved.

The electrodes most conveniently studied are the ideally polarizable ones; no steady-state electrochemical reactions occur at them, all charges supplied from outside accumulate in the edl and serve to shift the potential. Hence, one can study the potential dependence of various surface properties: the excess surface energy, edl capacitance, adsorption of components, etc. Electrodes which are not ideally polarizable have a much narrower range of potentials which can be realized (e.g., by variation of the current density). Moreover, reactions which take place often distort the electrode property being studied, for instance because of changes in the surface concentrations of components, so that such electrodes are difficult to study.

The typical example of an electrode which is ideally polarizable over a wide range of potentials is the mercury electrode in aqueous solutions of electrolytes such as KCl and K_2SO_4, the components of which will not react over a sufficiently wide range of potentials ("inert solution"). It is important for measurements of surface properties that throughout the region of ideal polarizability, both hydrogen and oxygen practically do not adsorb on mercury.

Another advantage of the mercury electrode is its liquid state ($t_f = -39°C$). The mercury surface is ideally smooth, the roughness factor is unity. Mercury surfaces are readily renewed. Electrocapillary curves can be measured highly accurately at the liquid metal. Certain alloys of mercury (amalgams), of gallium, and of other metals are liquids at room temperature, in addition to mercury. The surfaces of molten metals can also be studied at high temperatures.

Solid electrodes have rough surfaces, so that it is difficult to refer the parameters measured to the true surface area. The roughness factors of outwardly smooth, bright surfaces of solid metals can be three to five. Solid surfaces are composed of different faces of the metal crystallites, which often differ in their properties. Moreover, on solid surfaces one finds certain concentrations of structural defects (dislocations etc.). These two factors depend on the sample's prior history, and markedly influence the surface-layer properties.

The surface properties of a number of solid s- and p-metals (metals with incomplete s- or p-subshells), such as lead, tin, and cadmium, are largely analogous to those of mercury. But the d-metals (and in particular the metals of group VIII) are characterized by strong chemisorption of oxygen and hydrogen as well as of other substances.

Platinum is the typical example of a d-metal. Its surface has been studied in detail under various conditions. Smooth platinum electrodes are used, and also the electrodes with highly developed surface area which are obtained by electrochemical or chemical deposition of highly disperse platinum on a substrate of smooth platinum (platinized platinum) or other metal. The roughness factors of such electrodes can be as high as 10^4, and the true surface area of the platinum deposit (platinum black) is as high as $100 \text{ m}^2/\text{g}$. For such highly disperse electrodes, true surface areas can be measured sufficiently accurately by low-temperature argon or krypton adsorption (the Brunauer–Emmett–Teller or BET method).

Measurement of electrocapillary curves at mercury. The simplest device for measuring ecc at mercury and other liquid metals is Gouy's capillary electrometer (Fig. 12.12). Under the effect of a mercury column of height h, mercury is forced into the slightly conical capillary K. In the capillary, the mercury meniscus is in contact with electrolyte solution E. The radius of the mercury meniscus is practically equal to the capillary radius r_K at that point. The meniscus exerts a capillary pressure $p_K = 2\sigma^{(\text{Hg,E})}/r_K$ [see Eq. (15.1)] directed upwards which is balanced by the pressure $p_{\text{Hg}} = h\rho_{\text{Hg}}g$ of the mercury column (g is the acceleration of gravity); hence

$$\sigma^{(\text{Hg,E})} = h\rho_{\text{Hg}}r_K g/2. \qquad (12.51)$$

When potential is applied the meniscus moves owing to the resulting change in surface tension; when $\sigma^{(\text{Hg,E})}$ decreases it descends into the narrower part of the capillary, and when this quantity increases it rises into the wider part of the capillary. By varying the height of the mercury column during the measurements (which is done by raising or lowering the reservoir R) one returns the meniscus to its original position, which is checked with the aid of a horizontal microscope M. Thus, the meniscus radius remains constant and the value of $\sigma^{(\text{Hg,E})}$ measured is proportional to mercury column height. For calibration of the instrument it will suffice to make a single measurement under conditions where $\sigma^{(\text{Hg,E})}$ is known (in solutions of surface-inactive electrolytes at the pzc at $18°C$ it has a value of 426.7 mJ/m^2).

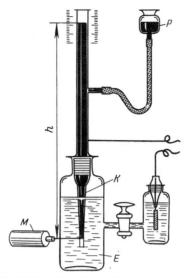

Fig. 12.12. Gouy's capillary electrometer.

Measurement of electrocapillary curves at solid metals. Several indirect methods exist which can be used to measure the excess surface energy of a solid electrode as a function of potential and other factors, but the accuracy of such measurements is lower than at liquid electrodes.

Changes in $\sigma^{(M,E)}$ can be roughly estimated by measuring the wetting angles α and using Eq. (12.13) for the calculation. Another method relies on the changes in hardness of metals which occur under polarization. Hardness is the resistance of a material against destructive forces. Destruction (grinding) of a material enlarges its surface area, hence the work of destruction is related to ese. In a method proposed by Peter A. Rehbinder and Evgenia K. Wenström (1945) one determines the rate of damping of the oscillations of a pendulum which rests on the test sample with its prismatic or spherical bearing made of hard material. The harder the sample the lower will be the damping rate. The hardness of a number of metals was obtained by this method as a function of potential, but quantitative estimates of ese are not possible.

In 1966 Aleksandr Y. Gokhshtein suggested a new method, the estance method, where the horizontal part of an L-shaped electrode touches the solution surface while the vertical part is attached to a sensitive piezoelectric crystal. Alternating current is passed through the electrode, which produces potential oscillations and the corresponding ese oscillations. This gives rise to very slight periodic deformations (bending) of the electrode which can be measured piezoelectrically. For s- and p-metals the amplitudes of the mechanical oscillations to a first approximation are proportional to the derivative $d\sigma^{(M,E)}/dE$, i.e., to the surface charge density in the edl. By performing the measurements at different potentials one can find the pzc and also the variation of ese with potential, but one cannot find absolute ese values. For d-metals more complicated functions are obtained.

Measurement of charging curves. In Section 9.4 the principles of transient polarization measurements in the presence of electrochemical reactions were described. When an electrode is ideally polarizable all of the current through it is nonfaradaic (charging current), and depends on the properties of the electrode surface:

$$i = i_{ch} \equiv dQ_{S,M}/dt = CdE/dt. \tag{12.52}$$

The method of galvanostatic charging curves was developed by Frumkin and Aleksandr I. Shlygin in 1935 for studies of platinized platinum electrodes. First hydrogen is passed through a cell with the test electrode and an inert solution; under these conditions a layer of adsorbed hydrogen atoms is formed on the surface [reaction (12.5)] and the equilibrium hydrogen potential is established. Then nitrogen or argon are passed through the cell in order to eliminate excess molecular hydrogen, which produces a minor shift (20–30 mV) of electrode potential in the positive direction, but the major quantity of the adsorbed atomic hydrogen remains on the surface. Next an anodic current of constant strength is made to pass the electrode (i_{ch} = const). The dependence of potential on time t or on the amount of charge $Q_{ext} = i_{ch}t$ that has passed the external circuit (the values of Q_{ext} are always referred to unit surface area of the electrode) is shown in Fig. 12.13. From the slope of the curve one can determine the electrode's capacitance C. In this method there is no need for complex equipment; it is very convenient for samples with large true surface areas (highly disperse deposits, powders, etc.).

Toward the end of the 1950s a number of workers developed methods to record potentiodynamic charging curves, which proved to be convenient for electrodes with smooth surfaces but require complex equipment. Figure 12.14 shows a typical voltammogram measured at a smooth platinum electrode. Such i_{ch} vs. E curves are the differential forms of the

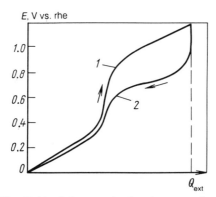

Fig. 12.13. Galvanostatic charging curve for a platinized platinum electrode in 0.1 M H_2SO_4 solution: (1) anodic scan; (2) cathodic scan.

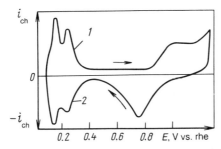

Fig. 12.14. Potentiodynamic charging curve for a smooth platinum electrode in 0.1 M H_2SO_4 solution: (1) anodic scan; (2) cathodic scan.

galvanostatic Q_{ext} vs. E curves. They reveal certain details which are less noticeable in the integral curves.

Ac techniques are widely used to investigate the capacitance and other surface properties of electrodes with small true surface areas. Their chief advantage is the possibility to apply them in the case of electrodes passing some faradaic current. It was shown in Section 9.5 that in this case the electrode's capacitance can be determined by extrapolating results obtained at different ac frequencies to the region of high frequencies. This extrapolation can be used for electrodes where electrode reactions occur which have standard rate constants, k_0^0, of up to 10^{-2} m/s.

Adsorption measurements. Methods exist by which adsorption of solution components can be measured directly. At high ratios of electrode surface area to solution volume (e.g., at platinized platinum electrodes in cells of small volume) and low solution concentrations, the adsorption of individual components leads to an appreciable concentration decrease in the bulk solution, which can be ascertained by analytical methods.

Radiotracer techniques have become very popular. There are two versions; one can measure the loss of radioactivity (concentration) of the test substance in the solution or the gain in radioactivity of the electrode resulting from adsorption. In the latter case one must employ special precautions in order to exclude background interference from the solution containing the radioactive substance (Vladimir E. Kazarinov, 1966).

Optical techniques. A variety of optical methods have become very popular in recent years for electrode surface studies. The measurements are made, either after taking the electrode from the solution (*ex situ* measurements) or directly in the solution (*in situ* measurements). Optical measurements in solutions are difficult, but their results are not affected by secondary processes which may occur when the electrode is no longer polarized and has been withdrawn from the solution.

Optical measurements yield assorted information about the nature, thickness, properties of adsorbed layers on electrode surfaces as well as the electronic properties of the metal electrode's surface layer.

12.7. THE MERCURY ELECTRODE SURFACE

12.7.1. Chief Properties; Anion Adsorption

In inert solutions the mercury surface is ideally polarizable within the region between its equilibrium potential (of 0.28 V in 1 M KCl solution, 0.61 V in 1 M K_2SO_4 solution, etc.) and the potential where discharge of the alkali metal ions or cathodic hydrogen evolution become noticeable, which is about -1.6 V at pH 7 (in this section, all potentials are given relative to SHE). When plotting the results of measurements concerning the edl at mercury, the horizontal axis by tradition is that of increasing negative values of potential (see Figs. 12.5 and 12.6).

An important aspect is that of studying the effects of composition of the electrolyte solution on the ecc and capacitance curves. Identical curves are obtained for solutions of fluorides, sulfates, and certain other alkali

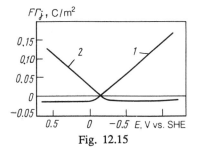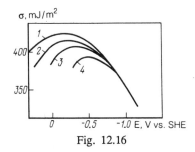

Fig. 12.15   Fig. 12.16

Fig. 12.15. Adsorption of cations (1) and anions (2) as functions of potential on a mercury electrode in 0.1 M NaF solution.

Fig. 12.16. Electrocapillary curves for a mercury electrode in 0.9 M solutions: (1) NaF; (2) NaCl; (3) NaBr; (4) NaI.

metal salts having identical concentrations. The pzc in such solutions is found at a potential of -0.193 V. The capacitance minimum found near the pzc disappears when the concentrations of these salts are raised (see Fig. 12.6); at the same time the ascending and descending branch of the ecc move somewhat closer together.

A constant potential of the electrocapillary maximum for different solution concentrations of a salt usually implies that at the pzc, ionic adsorption is minor. Figure 12.15 shows plots of the adsorption of cations and anions against potential for 0.1 M NaF solution (the values of adsorption are stated in electrical units of C/m^2, i.e., as $|z_j| F \Gamma_j$). When the mercury surface is negatively charged one finds electrostatic adsorption of cations (which form the solution side of the edl). When it is positively charged one finds a similar adsorption of anions. Ions are not adsorbed at the pzc or when the sign of surface charge is the same as that of the ions considered. Because of the effect of partial dehydration mentioned in Section 12.1, actually slightly negative values are found for Γ_j. Thus, no other, specific (nonelectrostatic) forces of interaction exist between these ions and the mercury surface; they are surface-inactive. Experience shows that surface activity is absent (or at least very low) in the case of ions F$^-$, OH$^-$, SO$_4^{2-}$, HCO$_3^-$, HPO$_4^{2-}$ and most inorganic cations. The halide ions Cl$^-$, Br$^-$, and I$^-$, a number of further anions, as well as cations Tl$^+$ and NR$_4^+$ are surface-active; they interact chemically with the mercury surface.

Specific adsorption of ions influences the shape of ecc and capacitance curves. Figure 12.16 shows the ecc for halide solutions. Anion adsorption depresses the ascending branch of the curves, which corresponds to positive surface charge. At negative potentials (starting from a potential about 0.2 V more negative than the pzc) the ecc for other halides practically coincide with the curve for fluoride, which implies that in this region the electrostatic repulsion forces prevail over the chemical attraction forces, and the anions are desorbed from the surface. The maximum value of ese decreases, and the pzc shifts to potentials more negative than the pzc of NaF solution. For 0.9 M NaI solution the value of $E_{zc,ads}$ is about -0.6 V, i.e., the pzc has shifted by more than 0.4 V. This shift is due to formation of an additional adsorptive edl consisting of specifically adsorbed anions at the inner Helmholtz plane and an equivalent number of cations diffusely distributed in the solution. Ionic adsorption at the pzc amounts to $1.4 \cdot 10^{-6}$ mol/m^2, which corresponds to a charge density of -0.14 C/m^2. As a result an additional component of the solution's surface potential arises, viz., the adsorption potential χ_{ads}. In NaI solutions, the mercury surface is positively charged at potentials between -0.2 and -0.6 V (in contrast to NaF solutions).

The effects of the anions, i.e., their specific adsorbabilities, increase in the order of $F^- < Cl^- < Br^- < I^-$. This trend is due to the fact that the solvation energy decreases with increasing crystal radius as one goes from F^- to I^-, and the transfer of the ions to the inner Helmholtz plane is facilitated accordingly.

The opposite picture is seen for surface-active cations, e.g., $[N(C_4H_9)_4]^+$; the descending branch of the ecc is depressed, and the pzc shifts in the positive direction.

Anion adsorption also influences the shape of the capacitance curves. In the region of the pzc and at positive surface charge the capacitance increases to values of 0.6–0.8 F/m^2 as edl thickness drops to a value of x_1. The capacitance minimum in dilute solutions is distorted, and its position no longer coincides with that of the pzc.

12.7.2. Adsorption of Organic Substances

Many neutral organic substances containing functional groups (e.g., the alcohols) are surface-active and adsorb on the mercury electrode. Figures 12.17a and 12.17b show how the ecc and capacitance curves change when n-butyl alcohol is added to the solution. It follows from these curves that adsorption of the alcohol molecules (which lowers the ese) occurs in a region around the pzc. The organic particles become desorbed when the potential is made much more positive or negative, and then the curves co-

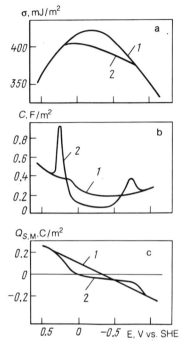

Fig. 12.17. Influence of the adsorption of organic substances: (a) on the electrocapillary curve, (b) on the capacitance curve, and (c) on the plot of surface charge against potential: (1) 0.1 M H_2SO_4 solution; (2) the same, with 0.1 M C_4H_9OH.

incide with those of the base-electrolyte solution. The potential of the pzc is shifted in the positive direction. It appears that the alcohol molecules which are dipolar have an orientation at the metal surface which is with the negative ends (the –OH groups) away from the surface. In other cases, e.g., in phenol adsorption, the pzc shifts in the negative direction, which indicates that the adsorbing molecules have the opposite orientation.

The values of edl capacitance are strongly depressed in the region where the organic substances are adsorbed, which indicates that their molecules are wedged in between the metal surface and the solution side of the edl. On one hand this leads to larger distances x_2, on the other hand the value of ε in the compact edl part decreases.

A typical feature are the sharp capacitance peaks appearing at the limits of the adsorption region of organic molecules, viz., the so-called adsorption or desorption peaks. They facilitate an accurate determination of the adsorption region. They arise because of the drastic change in distance x_2 which occurs over a very narrow potential interval during the adsorption or desorption of the organic substance, i.e., when the metal surface changes from covered to free or vice versa. In this narrow region, therefore, there is a drastic change in edl charge density, which is equivalent to a high capacitance value (Figs. 12.17b and 12.17c).

Desorption of the organic molecules at potentials where $|Q_{S,M}|$ is large, is due to a phenomenon known in electrostatics; viz., in any charged electrostatic capacitor, forces are operative which tend (when this is possible) to replace the medium with low ε-value by one with higher ε-value. Therefore, regardless of any chemical interaction of the organic molecules with the surface, they are electrostatically expelled from the edl at a certain value of $|Q_{S,M}|$, and replaced by water molecules.

12.8. THE PLATINUM ELECTRODE SURFACE

12.8.1. Surface Charge

The surface of platinum electrodes can be studied conveniently in inert solutions at potentials between 0 and 1.7 V (rhe). At more negative potentials cathodic hydrogen evolution starts, while at more positive potentials oxygen is evolved anodically.

It will be shown below that the values of $Q_{S,M}$ for platinum electrodes cannot be found with the aid of electrical measurements, since even within the potential range defined a current imposed from outside is consumed, not only for edl charging but also for other processes. The most unambiguous way of determining $Q_{S,M}$ is by measuring the adsorption of surface-inactive ions on the solution side of the edl, e.g., with the aid of radiotracers.

Figure 12.18 shows plots of the adsorption of cations and anions against potential for a platinized platinum electrode in acidified Na_2SO_4 solution (pH $\approx$ 3) as well as the values of $Q_{S,M}$ calculated from these adsorption values. We can see from the figure that in this solution the pzc of platinum is +0.34 V vs. rhe, which is +0.16 V vs. SHE. The edl capacitance of platinum at potentials between 0.1 and 0.7 V is 0.4 to 0.7 F/m^2. At more positive potentials the slopes of the $Q_{S,M}$ vs. E

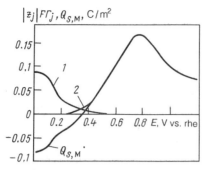

Fig. 12.18. Potential dependence of cation adsorption (1), anion adsorption (2), and total surface charge for a platinized platinum electrode in 10^{-3} M $H_2SO_4 + 3 \cdot 10^{-3}$ M Na_2SO_4 solution.

plots decrease, and may even assume negative values. The reasons for the anomalous shape of the curves found in this region are discussed below.

12.8.2. Charging Curves; Hydrogen Adsorption

The galvanostatic and potentiodynamic charging curves of platinum electrodes (see Figs. 12.13 and 12.14) shift approximately 60 mV in the negative direction when the solution pH is raised by one unit. This implies that when potentials E_r which refer to the equilibrium potential of a hydrogen electrode in the same solution (rhe) are used, these curves remain practically at the same place within a wide range of solution pH. Hence, we shall use this scale while analyzing these curves.

The galvanostatic charging curves distinctly reveal three linear sections with different slopes; in the first and third section the slopes correspond to capacitances of 4 to 7 F/m^2, in the second section the slope corresponds to a capacitance of 0.4 to 0.7 F/m^2. The capacitance of the second section is close to the value of edl capacitance determined from the slope of the $Q_{S,M}$ vs. E curve, hence this section is called the double-layer region. The anodic current which passes through the electrode when its potential is at points within the other two sections is consumed, not only for edl charging but also for the electrochemical oxidation and desorption of adsorbed atomic hydrogen [in the first section, reaction (12.10)] or for electrochemical oxygen adsorption [in the third section, reaction (12.11)]. In the case of cathodic currents the same processes occur in the opposite directions. Because of these processes, the total electrode capacitance (the so-called pseudocapacitance) is much higher than edl capacitance. These sections with high capacitance values correspond to sections with high currents in the potentiodynamic charging curves.

The region of atomic hydrogen adsorption stretches from 0 to 0.30–0.35 V. Hydrogen adsorption on platinum is a reversible or equilibrium process. The charging curves measured with low current densities in the hydrogen adsorption region, either in the cathodic (hydrogen deposition) or anodic (hydrogen stripping) direction, practically coincide. Because of its equilibrium character, hydrogen adsorption can be analyzed thermodynamically.

Let us define more closely the concept of Gibbs hydrogen adsorption Γ_i for an electrode at which the electrochemical reaction (12.10) can proceed. Consider an isolated (open-circuit) platinum electrode in acidified Na_2SO_4 solution; hydrogen and then argon have been bubbled in advance through the solution. When the electrode's surface area is increased by unity a number A_H of hydrogen atoms must be supplied to the surface in order to keep it in its former condition. Moreover, a certain number of hydrogen atoms or ions are needed for reaction (12.10) to produce an amount of surface charge $Q_{S,M}$. When $Q_{S,M} < 0$ one must supply a number $-Q_{S,M}/F$ of H atoms which, upon ionization, yield the required number of electrons in the metal; when $Q_{S,M} > 0$ one must supply a number $Q_{S,M}/F$ of H^+ ions which, by combining with electrons (i.e., by giving off their charge to the surface), are converted to H atoms and thus lower the number of atoms which have to be supplied. This number which, by definition, is equal to the Gibbs adsorption of hydrogen atoms, can be written in terms of the equation

$$\Gamma_H = A_H - (Q_{S,M}/F). \tag{12.53}$$

For negatively charged surfaces $\Gamma_H > A_H$, and the converse for positively charged surfaces.

Consider now the same electrode in an electrical circuit. Its initial condition is characterized by the values of parameters A_H and $Q_{S,M}$. When an external charge Q_{ext} is supplied to the electrode via this circuit it will be consumed, both for a change of surface charge, $\Delta Q_{S,M}$, and for a change in the number of adsorbed atoms, $-\Delta A_H/F$, through ionization (when charge Q_{ext} is positive the value of A_H will decrease):

$$Q_{ext} = \Delta Q_{S,M} - F\Delta A_H = -F\Delta\Gamma_H. \tag{12.54}$$

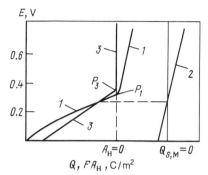

Fig. 12.19. Potential dependence of the values of Q_{ext} (1), $Q_{S,M}$ (2), and A_H (3) for a platinized platinum electrode.

We can see here that the external charge is equivalent to the change in Gibbs hydrogen adsorption.

When recording charging curves we actually determine the external charge Q_{ext}. When calculating the true amount A_H of adsorbed hydrogen we must take into account the charge $Q_{S,M}$ which has been determined, e.g., by adsorption measurements. Figure 12.19 shows experimental plots of Q_{ext} against E (curve 1) and of $Q_{S,M}$ against E (curve 2). Curve 3 is a plot of A_H against potential. Since from Eq. (12.54) one can only determine ΔA_H but not the absolute values of A_H, it has been assumed in plotting this curve that at the end of the hydrogen section in the charging curve (point P_1 in curve 1 or point P_3 in curve 3), hydrogen adsorption A_H is zero. Curve 3 intersects with curve 1 at the pzc.

The values of Q_{ext}, $Q_{S,M}$, and A_H refer to unit area of the true platinum surface. For absolute calculations one must first determine this area, e.g., by the BET method. Experience shows that $Q_{ext} \approx 2.2$ C/m^2 of true surface area are required in order to shift the potential from 0 to 0.35 V. Under these conditions the value of $Q_{S,M}$ changes from -0.1 to 0 C/m^2, i.e., by 0.10 C/m^2. It follows that the value of $A_H F$ at a potential of 0 V is about 2.1 C/m^2, while $A_H = 2.1 \cdot 10^{-5}$ mol/m^2 = $1.3 \cdot 10^{19}$ m^{-2}. This number actually coincides with the mean number of platinum atoms exposed on faces of the platinum lattice (the number varies slightly depending on the face index). Thus, at $E_r = 0$ V practically every platinum surface atom is bonded to an adsorbed hydrogen atom, i.e., the limiting surface coverage of $\theta_H = 1$ is attained. As the potential is made more positive, θ_H decreases until it attains a value of zero at 0.35 V.

To a first approximation it varies linearly with potential:

$$\theta_H = 1 - (1/0.35)E_r \tag{12.55}$$

(here E_r is stated in volts).

By using the value of Q_{ext} = 2.2 C/m^2 obtained one can readily determine the true surface area of any compact or disperse sample of platinum without adducing any other methods, simply by measuring the total amount of charge required to accomplish the said potential shift.

12.8.3. Hydrogen Adsorption Isotherm; Surface Inhomogeneity

It is not a trivial point that the θ_H vs. E_r curves are practically linear. In a reversible system the electrode potential can be linked to the activities (concentrations) of the potential-determining substances. In the system being discussed, this substance is atomic hydrogen. According to the Nernst equation we have: $E_r = \text{const} - (RT/F) \ln c_H$. It follows that the degree of coverage, θ_H, is linearly related to the logarithm of concentration c_H in the solution:

$$\theta_H = g + (1/f) \ln c_H, \tag{12.56}$$

i.e., by a logarithmic adsorption isotherm. This equation was derived in 1939 by Temkin, just for interpreting results obtained when measuring charging curves at platinum electrodes. We find, when comparing Eqs. (12.55) and (12.56), that the inhomogeneity factor f has a value of $0.35(F/RT) \approx 14$.

The shape of the isotherm arises because the heat of adsorption, $\bar{q}_\theta$, decreases with increasing degree of coverage of the electrode. Hydrogen ion discharge producing molecular hydrogen (gas) is possible, for thermodynamic reasons, only at values of potential more negative than 0 V. But since the intermediates (hydrogen atoms) are adsorbed, a discharge of hydrogen ions producing these atoms is possible even at more positive potentials because of the adsorption energy gained. When the electrode potential is moved from a value of +0.4 V (where the surface is known to be free of adsorbed hydrogen) in the negative direction, adsorption of the first numbers of hydrogen atoms will start at 0.35 V on the sites with the highest heat of adsorption, $\bar{q}_0$ (strongly bound hydrogen). As the potential is moved further to the negative side and θ_H increases the heat of adsorption gradually decreases, and its minimum value of $\bar{q}_1$ is attained near the potential of 0 V when $\theta_H \approx 1$. Thus, the total change in heat of adsorption is about 0.35 eV or, roughly, 34 kJ/mol.

The reasons for the change in hydrogen adsorption energy on platinum, or for inhomogeneity of the platinum surface, have not yet been established in an unambiguous way. The inhomogeneity may arise from different geometries of the adsorption sites (for instance, the exposed face of a metal crystallite) or by the influence of the hydrogen atoms adsorbed earlier. Accordingly, "geographic" or "biographic" and "induced" inhomogeneity are discussed for metal surfaces.

12.8.4. Adsorption of Other Substances

When charging curves are measured while the potential is shifted in the anodic direction, oxygen adsorption will start at a potential of about 0.75 V, i.e., long before the thermodynamic potential for the evolution of free oxygen (1.23 V) is attained. This indicates that the bond energy between platinum and adsorbed oxygen is high. The first amounts of oxygen evidently are adsorbed in the form of hydroxyl groups:

$$Pt + H_2O \rightleftarrows Pt\text{--}OH + H^+ + e^-. \tag{12.57}$$

At more positive potentials, bonds of the type of Pt=O or of more complex types can be formed. It follows from the amount of charge which is consumed when recording the charging curve, that formation of a monolayer of oxygen in the form of OH_{ads} is complete at a potential of about 1.15 V, and formation of a monolayer of oxygen in the form of O_{ads} is complete at a potential of about 1.5 V. Oxygen adsorption continues at more positive potentials, and at a potential of about 2.2 V the limiting value is attained which formally corresponds to degrees of coverage of 2.0–2.2 (for platinum, the so-called region of high anodic potentials, hap).

Oxygen adsorption which occurs at platinum at potentials more positive than 0.9–1.0 V is irreversible, in contrast to hydrogen adsorption. Oxygen can be removed from the surface by cathodic current, but the curves obtained in the anodic and cathodic scan do not coincide; cathodic oxygen desorption occurs within a narrower region of potentials, and these potentials are more negative than the region where the major amount of oxygen becomes adsorbed (see Figs. 12.13 and 12.14). Thus, immediately after adsorption there is a drastic increase in bond energy of the oxygen on the surface. The initial fast rise is followed by a further slow increase in bond energy, or "aging" of the adsorbed oxygen.

The adsorption of organic substances from solutions on platinum surfaces can be measured by different techniques. A platinum surface atom bonded to an organic particle loses its ability to adsorb hydrogen, hence the fraction of surface taken up by organic species can be estimated from the decrease in hydrogen adsorption noticed when recording cathodic charging curves. Sometimes a horizontal section of constant potential, which is due to anodic oxidation of adsorbed organic particles, appears in anodic charging curves at potentials of 0.6 to 0.8 V. The amount of particles adsorbed can be estimated from the length of this section.

Many organic species are adsorbed on platinum at potentials ranging approximately from 0.1 to 0.7 V (rhe). At more positive potentials they desorb because of oxidation or of displacement from the surface by adsorbing oxygen. The extent of adsorption of the organic substances can be considerable, and they can take up as many as 70% of the surface sites. The relation between θ_j and the bulk solution concentration often obeys the logarithmic Temkin isotherm with values of the factor f of 10–14, i.e., practically the same value as found in hydrogen adsorption. The adsorption of organic substances from solutions as a rule is irreversible; material already adsorbed will not desorb when the solution concentration is reduced (not even when it is reduced to zero). Often adsorption is attended by destruction of the organic molecule. For instance, in methanol adsorption on the platinum electrode three hydrogen atoms are split off from the CH_3OH molecule and become adsorbed on the surface, and adsorbed species of the type $\equiv COH$ are formed which are bonded to three platinum surface atoms.

The adsorbabilities of different anions on platinum increase in approximately the same order as on mercury. Hydrogen adsorption is not diminished when Cl^- or Br^- ions become adsorbed, but the shape of the charging curves changes somewhat: hydrogen adsorption and desorption occur within a narrower potential range which is between 0 V and approximately 0.25 V for Cl^- and between 0 V and approximately 0.2 V for Br^- (Fig. 12.20). Thus, there is a decrease in the maximum bond energy of hydrogen under the effect of these anions. The bond energy of oxygen on the surface is reduced in exactly the same way by the halide ions. However, anions as a rule become desorbed with increasing oxygen adsorption.

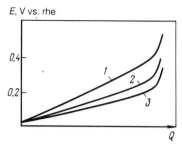

Fig. 12.20. Galvanostatic charging curves for a platinized platinum electrode in 1 M solutions of KF (1), KCl (2), and KBr (3).

This special ability of the adsorbed oxygen to displace other adsorbed substances from the surface is of great value for experimental work. The platinum electrode is held for a few seconds or minutes at potentials of 1.1 to 1.4 V (rhe) in order to rid the surface from impurities. Following this, if the potential is moved to a value of 0.4 V the oxygen will desorb and a clean surface is obtained. A highly sensitive indication for the degree of cleanliness attained by the surface is the shape of the potentiodynamic charging curves (see Fig. 12.14), which will be distorted by minute amounts of impurities.

12.8.5. Components of the Galvani Potential

We can see from the anomalous shape of the $Q_{S,M}$ vs. E curves recorded at positive potentials (see Fig. 12.18) that a change in potential gives rise, not only to a change in interfacial potential Φ of the ionic edl but also to changes in other components.

A major reason for this is the adsorbed oxygen appearing at potentials more positive than 0.8 V. The Pt–O bond is polar, and the center of the negative charges is closer to the oxygen. Therefore, an additional positive surface potential attaining values of 1 V and more arises at platinum. This potential is compensated by a decrease in interfacial potential, i.e., a decrease in charge density in the ionic edl (Fig. 12.21a). Even a charge reversal of the surface is possible, whereupon negative charges will appear on the surface in the region of positive potentials (Fig. 12.21b). In this case at a certain value of potential a new pzc is attained, which is that of the oxidized platinum surface.

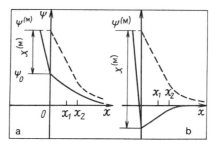

Fig. 12.21. Electric double-layer structure at the platinum electrode in the presence of adsorbed oxygen: (a) without surface charge reversal; (b) with charge reversal.

12.8.6. Free and Total Surface Charge; Partial Charge Transfer

We have seen that there are substantial differences in the surface properties of platinum and mercury electrodes. The latter are ideally polarizable; all of the charge supplied from outside accumulates in the edl, and merely produces a change in the value of surface charge $Q_{S,M}$ characterizing the condition of the system. Platinum electrodes (except for the double-layer region) are not ideally polarizable; electrochemical reactions (12.10) or (12.11) occur at their surfaces. It is true that these reactions are transient, and will cease when the adsorption of hydrogen or oxygen atoms has attained a certain level. Thus, even here any charge supplied from outside will merely produce a change in surface condition [according to Eq. (12.54), changes in A_H and A_O will occur in addition to the change in charge $Q_{S,M}$] but will not be consumed for the formation of reaction products escaping from the electrode surface. By convention, the transient electric current crossing the electrode is regarded as nonfaradaic in this case, even though it is associated in part with an electrochemical reaction.

The concept of ideal polarizability implying the total absence of charge transfer between the two edl sides was formulated in 1934 by Frederick Otto Koenig. But already in 1891 Max Planck had made the suggestion that the term "perfectly polarizable electrode" should be used when the surface condition of the electrode is uniquely defined by the amount of charge consumed. This concept is broader than that of ideal polarizability. It covers, not only ideally polarizable electrodes of the mercury type but also electrodes of the platinum type in the region of hydrogen adsorption. A platinum electrode in the region of oxygen adsorption is already outside this definition, since here the electrochemical reaction is irreversible, and owing to ageing of the adsorbed oxygen the surface condition is not uniquely defined by the amount of charge consumed. The concept, for exactly the same reasons, does not cover electrodes where phase layers of reaction products are formed on the surface because of a reaction.

Other types of adsorption processes are possible during which an electrode will be perfectly polarizable. Specific adsorption of ions is often attended by a partial transfer of their charge to the metal surface; for instance, in the specific adsorption of cations

$$M^{z+} + \lambda e^- \rightarrow M^{(z-\lambda)+}, \qquad (12.58)$$

where λ is the degree of charge transfer.

Complete charge transfer ($\lambda = z$) corresponds to the ion's complete discharge yielding an atom M adsorbed on the surface, i.e., an adatom. Partial charge transfer is possible, too, in the adsorption of neutral species when a (partly charged) adsorbed ion, i.e., an adion, is formed. The formation of hydrogen atoms adsorbed on platinum, which occurs as a result of hydrogen ion discharge, can be regarded as an example for almost complete charge transfer. The resulting Pt–H bonds are slightly polar, hence the degree of charge transfer still is somewhat different from unity.

The degree of charge transfer, λ, cannot be calculated thermodynamically directly from experimental data, and difficulties arise in defining the concept of surface charge $Q_{S,M}$. This parameter can be defined in terms of Gibbs adsorption of the ions [Eq. (12.23)], only in the case of ideally polarizable electrodes, i.e., when it is assumed that even partial charge transfer is lacking for the ions. Even the thermodynamic equation (12.29) can be used, only under the same assumption.

In 1970 Frumkin, Boris B. Damaskin, and Oleg A. Petrii introduced the concept of *total surface charge* $Q^0_{S,M}$ for the perfectly polarizable electrode. In contrast to "free" charge $Q_{S,M}$, this is a thermodynamic parameter and can be measured. The surface's total charge is defined (analogous to Gibbs adsorption) as the amount of electric charge that must be supplied to the electrode from outside in order to preserve the condition of the surface when its area is increased by unity.

Consider, as an example, a platinum electrode with adsorbed hydrogen for which the situation is such that, when the surface area is increased by unity, the system's condition is preserved by external supply of a charge $Q^0_{S,M}$ and of the required number of hydrogen ions, but not of hydrogen atoms (the electrode is kept in a medium that was purged with argon, i.e., is free of hydrogen). We shall assume that charge transfer in reaction (12.10) is complete, i.e., $\lambda = 1$. In this case part of the charge will be consumed for creating the free surface charge $Q_{S,M}$ of the new surface, and another (negative) part will be consumed for discharge of supplied hydrogen ions producing a layer of A_H adsorbed hydrogen atoms. Thus, when allowing for Eq. (12.53) we have

$$Q^0_{S,M} = Q_{S,M} - F A_H = -F\Gamma_H, \qquad (12.59)$$

i.e., the total surface charge corresponds to the Gibbs adsorption of hydrogen atoms, with the sign reversed. This simple relation between $Q^0_{S,M}$ and $Q_{S,M}$ is lost when $\lambda \neq 1$, and a charge transfer coefficient λ appears which is hard to determine from experiments.

It follows from Eq. (12.54) that when the electrode potential is varied (e.g., while recording the charging curve), the external charge supplied to the electrode or "withdrawn" from it will be equal to the change $Q^0_{S,M}$ in total surface charge.

For perfectly polarizable electrodes, precisely the quantity of total surface charge $Q^0_{S,M}$ rather than of free charge $Q_{S,M}$ appears in the Lippmann equation (12.29). Thus, total charge is a rigorously thermodynamic parameter. For ideally polarizable electrodes these two parameters coincide.

12.9. SURFACES OF OTHER ELECTRODES

It had been mentioned above that in their electrochemical surface properties, a number of metals (lead, tin, cadmium, and others) resemble mercury, while other metals of the platinum group resemble platinum itself. Within each of these groups, the trends of behavior observed coincide qualitatively, sometimes even semiquantitatively. Part of the differences between mercury and other s- or p-metals are due to their solid state. Among the platinum-group metals, palladium is exceptional, since strong bulk absorption of hydrogen is observed here in addition to surface adsorption, an effect which makes it hard to study the surface itself.

Surface studies are difficult in the case of many metal electrodes since their regions of ideal or perfect polarizability are very narrow, i.e., the potentials of anodic dissolution (or oxidation) of the metal and of cathodic hydrogen evolution are close together. In a number of cases such a region

TABLE 12.1. Potentials of Zero Charge of Certain Metals in Aqueous Solutions

Metal	Solution composition	E_{zc}, V (SHE)
Bismuth	0.002 M KF	−0.39
Gallium (liquid)	0.008 M HClO$_4$	−0.69
Iron	0.0006 M Na$_2$SO$_4$	−0.70
Gold	0.02 M NaF	+0.19
Indium	0.003 M NaF	−0.65
Cadmium	0.001 M KF	−0.75
Tin	0.001 M K$_2$SO$_4$	−0.38
Mercury	0.001 M NaF	−0.193
Lead	0.01 M NaF	−0.60
Silver (polycryst.)	0.002 M Na$_2$SO$_4$	−0.70
Silver [the (111) face]	0.001 M KF	−0.46
Silver [the (100) face]	0.002 M NaF	−0.61
Silver [the (110) face]	0.005 M NaF	−0.77
Antimony	0.002 M KClO$_4$	−0.15
Thallium	0.001 M NaF	−0.71

is not found at all. Often the qualitative results obtained when studying other electrodes are extended to such electrodes.

A very important characteristic of surface constitution for any metal is the position of its pzc. Table 12.1 reports values of the pzc for a number of metals. We can see that these values vary within rather wide limits.

The important difference between platinum-group metals and most other metals is the ability of the latter upon anodic polarization to form relatively thick superficial oxide or salt layers. Owing to their great practical value, these layers will be considered in more detail in Section 18.3.

In addition to metals, semiconductors with n- or p-type conduction are used as electrodes. The surface layers of semiconductors differ strongly from those of metals in their electrical constitution. The concentrations of free carriers are much lower in semiconductors (e.g., 10^{20} m^{-3} or $2 \cdot 10^{-7}$ mol/liter) than in metals (10^{28} m^{-3} or 20 mol/liter, in order of magnitude), therefore the excess electric charge in semiconductors (the electrode side of the edl) is not tightly packed against the surface but distributed diffusely within a surface layer 10^{-8}–10^{-6} m thick, i.e., extending over 10^2 to 10^4 atom layers. In this region a potential drop is generated by the resulting space charge. An appreciable part of the Galvani potential across the semiconductor/electrolyte interface is located in a relatively thick surface layer inside the semiconductor rather than on its

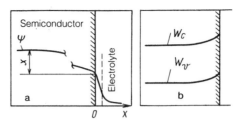

Fig. 12.22. (a) Potential distribution in the interphase at the semiconductor/electrolyte interface, and
(b) the bending of the energy levels at the top of the
valence band and bottom of the conduction band in
a semiconductor in the presence of surface states.

surface (Fig. 12.22a). The charge and potential distribution in this layer
are analogous to the same distributions in the diffuse edl part in very dilute electrolyte solutions, and can be described by the same equations (see
Section 12.5).

In impurity-type semiconductors where the concentrations of majority
and minority carriers are different, several possibilities exist for the formation of charge in the surface layer. For instance, in n-type semiconductors
negative charge arises by an accumulation of majority carriers (electrons),
the resulting layer is called an accumulation layer. Low positive charge
densities arise from the depletion of electrons (depletion layers). When the
electron concentration has dropped to zero, an excess of positive holes in
the surface layer can give rise to higher positive charge densities (inversion
layers).

Excess charge arises in the surface layer, not only as a result of charge
redistribution and the development of an interfacial potential when the
semiconductor comes in contact with another phase, but also as a result of
so-called *surface states* present in the semiconductor. This term describes
special sites at the surface where, owing to lattice discontinuities or adsorbed foreign material, electrons or holes can be bound. The density of
surface states is high in semiconductors (over 10^{17} m^{-2}). These states
give rise to a thin stratum in the edl compensating a space charge opposite
in sign which is present in a thicker surface layer.

The potential drop across the space charge region (in the surface layer)
can be regarded as the semiconductor's surface potential, $\chi \equiv \psi_V - \psi_{x=0}$,
where ψ_V is the electrostatic potential in the bulk of the semiconductor
[see Eq. (2.31)]. We should point out here that in semiconductor physics,
for the potential drop across the space charge region of the surface layer,

another parameter is used: $\varphi_{sc} \equiv \psi_{x=0} - \psi_V$, which differs in sign from the surface potential. When the semiconductor's surface layer is positively charged, the value of χ will be positive but φ_{sc} will be negative.

The potential drop in the semiconductor's surface layer leads to bending of the electronic energy levels, and particularly of the top of the valence band and bottom of the conduction band. Since electrons are negatively charged, a potential change which is in the negative direction as one approaches the surface (a positive surface potential), corresponds to an increase in energy, i.e., an upward bending of the levels (Fig. 12.22b).

The Galvani potential associated with the semiconductor electrode is the sum of surface potential in the semiconductor and interfacial potential at the semiconductor/electrolyte interface, $\varphi_G = \chi + \Phi$ (for the sake of simplicity, we include the potential drop across the diffuse edl part in Φ). Two limiting cases are possible when the electrode potential (the Galvani potential) changes: (i) under some conditions (high values of the Galvani potential, large number of surface states) χ is practically constant and independent of electrode potential; in this case any change in electrode potential basically will produce a change in interfacial potential: ΔE ($= \Delta \varphi_G$) $\approx \Delta \Phi$; (ii) under other conditions, a change in electrode potential basically will produce a change in surface potential: $\Delta E \approx \Delta \chi$. All intermediate cases are possible.

When the surface potential depends on electrode potential, a change in electrode potential will cause a change in the degree of bending of the band edges. At a certain potential the bending vanishes, and the energy levels flatten out. This value of potential is called the flat-band potential E_{fb}. At this potential the surface potential is zero, and the semiconductor's surface layer is free of space charge. This potential is the analog of the potential of zero charge of metal electrodes, and is very important in semiconductor electrochemistry.

The diffuse charge distribution in the semiconductor's surface layer leads to a drastically lower edl capacitance at the semiconductor/electrolyte interface. Typical values of capacitance for metal electrodes are 0.2–0.4 F/m^2, but for semiconductor electrodes these values are 2 to 3 orders of magnitude lower.

Owing to the change in charge density with potential, semiconductors also exhibit a potential dependence of capacitance. This dependence can be used for determining the flat-band potential.

12.10. TWO PROBLEMS IN ELECTROCHEMISTRY

12.10.1. The Volta Problem

During the 19th century opinions were divided as to where in galvanic cells of the type of

$$M_1 | E | M_2 | M_1 \qquad (12.60)$$

the interfacial potential differences responsible for the ocv value exhibited by the cell were located. Alessandro Volta thought that the potential difference resided wholly at the metal/metal junction and that Galvani potentials did not exist at metal/electrolyte interfaces ("physical emf theory"). According to Walther Nernst, to the contrary, the potential difference resided at the two metal/electrolyte interfaces where electrochemical reactions take place ("chemical emf theory"). The dependence of ocv on solution composition and its connection with the Gibbs reaction energy were arguments weighing heavily in favor of Nernst's ideas. However, experimental evidence exists which is in favor of the views of Volta, viz., the Volta potentials measured between any two metals in vacuum correlate with the ocv values observed when the same metals are immersed in an electrolyte solution; the metal that is more negative in vacuum as a rule is that which is more negative in the solution, and metal pairs with higher Volta potentials exhibit higher ocv values. The conflict which thus arises between these two viewpoints became known as the Volta problem in electrochemistry.

A solution to this problem was suggested in 1927 by Frumkin, as follows. In galvanic cells, interfacial potential differences develop across all interfaces. Each Galvani potential $\varphi_G^{(\beta,\alpha)}$ according to Eq. (12.3) can be written as the algebraic sum of three components, the interfacial potential $\Phi^{(\beta,\alpha)}$ and two surface potentials. Let the surface potential of phase (α) at its interface with vacuum be $\chi^{(\alpha/0)}$. At its interface with another condensed phase (γ), a somewhat different surface potential $\chi^{(\alpha/\gamma)}$ exists because of surface-layer interaction. Let the difference be $\delta\chi^{(\alpha/\gamma)}$ $[\equiv \chi^{(\alpha/\gamma)} - \chi^{(\alpha/0)}]$.

Using this partition of Galvani potentials into their components in Eq. (2.17) for the ocv, Frumkin obtained, after transformations,

$$\mathcal{E} = \Phi^{(1,2)} + \Phi^{(2,E)} - \Phi^{(1,E)} + [\delta\chi^{(1/2)} - \delta\chi^{(2/1)}]$$
$$+ [\delta\chi^{(2/E)} - \delta\chi^{(1/E)}] - [\delta\chi^{(E/2)} - \delta\chi^{(E/1)}]. \qquad (12.61)$$

According to Eqs. (2.34) and (12.3), the expression for the Volta potential can be written as

$$\varphi_V^{(1,2)} = \Phi^{(1,2)} + [\delta\chi^{(1/2)} - \delta\chi^{(2/1)}]. \qquad (12.62)$$

It will be assumed that the interactions between each of the metals (1) and (2) and the corresponding surface layers of the electrolyte solution are approximately identical, and also that specific adsorption of ions does not occur in the system being considered. In this case the values of the expressions in the last two square brackets in Eq. (12.61) become zero, and from (12.61) and (12.62) an important relation is obtained which links the ocv of galvanic cells with the Volta potential:

$$\mathcal{E} \approx \varphi_V^{(1,2)} + \Phi^{(2,E)} - \Phi^{(1,E)} \qquad (12.63)$$

This expression explains the qualitative agreement found to exist between the ocv values of galvanic cells and the Volta potentials of the corresponding metal pairs. But through terms $\Phi^{(1,E)}$ and $\Phi^{(2,E)}$, it also explains why ocv values depend on solution composition. All parameters of this equation can be measured experimentally.

Equation (12.63) yields another important result, viz., when both electrodes are at the potential of their respective pzc [and the values of $\Phi^{(M,E)}$ are zero], the cell voltage (which is the pzc potential difference between the two electrodes) will be equal to the Volta potential between the corresponding metals:

$$\Delta E_{zc}^{(1,2)} \approx \varphi_V^{(1,2)}. \qquad (12.64)$$

In a number of cases Eqs. (12.63) and (12.64) are in good agreement with experimental data. However, sometimes the quantitative agreement is not as good, and this can be attributed to the approximations made in connection with Eq. (12.63). Therefore, when comparing the calculations with experiment one can reach certain conclusions as to the way in which the surface potentials vary, i.e., as to the way a metal interacts with the electrolyte.

12.10.2. The Problem of Absolute Potential

Many attempts have been made by experiment or calculation to determine the absolute values of Galvani potentials at interfaces, and particularly across the electrode/electrolyte interface. Basically, if one knew the

Galvani potential between one metal and the associated electrolyte and that at the interface between the metal and another metal, one could then find the Galvani potentials for all other interfaces from the data of ocv measurements with other galvanic cells.

It had already been pointed out that there is not a single interface for which the Galvani potential can be either measured experimentally or calculated thermodynamically from indirect experimental data. The only way of determining it is through theoretical calculations based on nonthermodynamic models.

A frequent starting point for such calculations are the values of Volta potentials, $\varphi_V^{(M,E)}$, between electrodes and electrolytes (which can be measured by the same methods as Volta potentials between metals). According to Eq. (2.34),

$$\varphi_V^{(M,E)} = \varphi_G^{(M,E)} + \chi^{(E/0)} - \chi^{(M/0)}. \tag{12.65}$$

The problem of calculating Galvani potentials now reduces to that of calculating the surface potentials of the metal and solution.

Equation (2.32) can be used to calculate the metal's surface potential. The value of $w^{(M,0)}$, the electron work function, can be determined experimentally. The chemical potential of the electrons in the metal, $\mu_e^{(M)}$, can be calculated approximately from equations based on the models in modern theories of metals. The accuracy of such calculations is not very high. The surface potential of mercury determined in this way is roughly +2.2 V.

The surface potential of a solution can be calculated, according to Eq. (10.41), from the difference between the experimental real energy of solvation of one of the ions and the chemical energy of solvation of the same ion calculated from the theory of ion–dipole interaction. Such calculations lead to a value of +0.13 V for the surface potential of water. The positive sign indicates that in the surface layer, the water molecules are oriented with their negative ends away from the bulk (see Fig. 12.1).

It follows from these values that at the potential of the standard hydrogen electrode, the Galvani potential between a mercury electrode and the solution is about 1.6 V.

The accuracy of all such calculations is low.

Part 3
ELECTROCHEMICAL KINETICS

Chapter 13
MULTISTEP ELECTRODE REACTIONS

13.1. INTERMEDIATE REACTION STEPS

In Chapter 6 we considered the basic rules obeyed by simple electrode reactions occurring without the formation of intermediates. However, electrochemical reactions in which two or more electrons are transferred, more often than not follow a path involving a number of consecutive, simpler steps producing stable or unstable intermediates, i.e., they are multistep reactions.

The set of all intermediate steps is called the reaction pathway. A given reaction (involving the same reactants and products) may occur by a single pathway or by several parallel pathways. In the case of invertible reactions, the pathway followed in the reverse direction (for example, the cathodic) may or may not coincide with that of the forward direction (in this example, the anodic). For instance, the relatively simple anodic oxidation of divalent manganese ions which in acidic solutions yields tetravalent manganese ions: $Mn^{2+} \rightarrow Mn^{4+} + 2e^-$, can follow these two pathways:

$$Mn^{2+} \xrightarrow{-e^-} Mn^{3+} \xrightarrow{-e^-} Mn^{4+}, \qquad (I)$$

$$2[Mn^{2+} \xrightarrow{-e^-} Mn^{3+}]; \qquad 2Mn^{3+} \rightarrow Mn^{2+} + Mn^{4+}. \qquad (II)$$

The second pathway includes a step in which the trivalent manganese ions formed as intermediates disproportionate.

It is convenient to represent multistep electrode reactions involving one or more pathways in the form of tables such as this:

k	Step	l_k	$\mu_k(\text{I})$	$\mu_k(\text{II})$
1	$Mn^{2+} \rightarrow Mn^{3+} + e^-$	1	1	2
2	$Mn^{3+} \rightarrow Mn^{4+} + e^-$	1	1	—
3	$2Mn^{3+} \rightarrow Mn^{2+} + Mn^{4+}$	0	—	1

where l_k is the number of electrons in a step; μ_k is the *stoichiometric number of a step* in pathways I and II, which indicates how many times this step is repeated in an elementary reaction act, i.e., during the loss of two electrons from one Mn^{2+} ion. For instance, in a reaction following the second pathway the first step occurs twice ($\mu_1 = 2$), but one of the Mn^{2+} ions that had reacted is regenerated in the step that follows.

For electrochemical steps as a rule $l_k = 1$, though for the sake of generality we shall retain the symbol l_k in the equations. Purely chemical steps are also possible, in which electrons are not involved and $l_k = 0$, e.g., the third step in the above example. Obviously $\sum \mu_k l_k = n$, where n is the total number of electrons involved in one elementary reaction act (here $n = 2$).

Electrochemical steps are often denoted by the letter E (or e), and chemical steps by the letter C (or c). Thus, the first pathway in the above example can be said to follow an EE scheme, and the second an EC scheme.

Except for Section 13.7, the reactions considered below will occur by only a single pathway (both in the forward and reverse direction), and there will be no parallel path.

In multistep reactions, the number of particles of any intermediate B_k produced in unit time in one of the steps is equal to the number of particles reacting in the next step (in the steady state the concentrations of the intermediates remain unchanged). Hence, the rates of all intermediate steps are interrelated. Writing the rate v_k of an individual step as the number of elementary acts of this step which occur in unit time (in units of moles, i.e., divided by the Avogadro constant), and the rate v of the overall reaction as the number of elementary acts of the overall reaction which occur within the same time, we evidently have

$$v = v_1/\mu_1 = v_2/\mu_2 = \cdots = v_k/\mu_k = \cdots = v_z/\mu_z. \qquad (13.1)$$

Therefore, in the steady state the reduced rates v_k/μ_k are identical for all steps and equal to the rate of the overall reaction.

Each of the intermediate electrochemical or chemical steps is a reaction of its own, i.e., it has its own kinetic peculiarities and rules. Despite the fact that all steps occur with the same rate in the steady state, it is true that some steps occur readily, without kinetic limitations, and others to the contrary occur with limitations. Kinetic limitations that are present in electrochemical steps show up in the form of appreciable electrode polarization.

It is a very important task of electrochemical kinetics to establish the nature and kinetic parameters of the intermediate steps, as well as the way in which the kinetic parameters of the individual steps correlate with those of the overall reaction.

13.2. RATE-DETERMINING STEP

Consider the correlation between the kinetic parameters of the overall reaction and those of its individual steps in the instance of a very simple, invertible two-step chemical reaction:

$$A \overset{(1)}{\rightleftarrows} B \overset{(2)}{\rightleftarrows} D, \tag{13.2}$$

where B is the reaction intermediate and $\mu_1 = \mu_2 = 1$. We shall assume that both steps are first-order in the reactants, i.e.,

$$v_1 = k_1 c_A - k_{-1} c_B; \qquad v_2 = k_2 c_B - k_{-2} c_D. \tag{13.3}$$

In the steady state $v_1 = v_2 = v$. With the values of v_1 and v_2 from Eq. (13.3) we find

$$c_B = (k_1 c_A + k_{-2} c_D)/(k_{-1} + k_2) \tag{13.4}$$

for the steady concentration of intermediate B and

$$v = (k_1 k_2 c_A - k_{-1} k_{-2} c_D)/(k_{-1} + k_2) \tag{13.5}$$

for the rate of the overall reaction.

The direction of the reaction will depend on the relative concentrations of reactants and products; the reaction will go from left to right when

$k_1 k_2 c_A > k_{-1} k_{-2} c_D$, and in the opposite direction ($v < 0$) when the opposite inequality holds.

When we consider the overall reaction while disregarding the formation of intermediates, we can write its rate as

$$v = k_0 c_A - k_{-0} c_D. \tag{13.6}$$

By comparing Eqs. (13.5) and (13.6) we find the connection between the kinetic parameters of the overall reaction and those of the individual steps:

$$k_0 = k_1 k_2 / (k_{-1} + k_2); \qquad k_{-0} = k_{-1} k_{-2} / (k_{-1} + k_2). \tag{13.7}$$

A quantity of great importance in these equations is the ratio of parameters k_{-1} and k_2, i.e., of the constants in the expressions for the rate of reconversion of the intermediates to original reactants and for the rate of conversion of the intermediates to final products. In the particular case of $k_{-1} \ll k_2$ it follows from Eq. (13.5) that

$$v = k_1 c_A - (k_{-1} k_{-2} / k_2) c_D = k_1 c_A - k_{-1} c_{B(D)}^0. \tag{13.8}$$

Here $c_{B(D)}^0$ is the concentration of intermediate B that would be found under conditions of complete equilibrium between it and the reaction products ($k_2 c_{B(D)}^0 = k_{-2} c_D$):

$$c_{B(D)}^0 = (k_{-2} / k_2) c_D. \tag{13.9}$$

It is a very important conclusion following from Eq. (13.8) that in the case considered, the rate of the overall reaction is wholly determined by the kinetic parameters of the first step (k_1 and k_{-1}), while the second step influences this rate only through the equilibrium concentration of the intermediate B. We say, therefore, that the first step (with its low value of parameter k_{-1}) is the *rate-determining step* (rds) of this reaction. Sometimes the term "slow step" is used, which reflects the fact that under the conditions considered we have $\vec{v}_1 < \vec{v}_2$, but this term is not very fortunate inasmuch as the *effective* rates, v_1 and v_2, of the two steps actually are identical.

Analogously, when $k_{-1} \gg k_2$ we have

$$v = (k_1 k_2 / k_{-1}) c_A - k_{-2} c_D = k_2 c_{B(A)}^0 - k_{-2} c_D, \tag{13.10}$$

where

$$c_{B(A)}^0 = (k_1/k_{-1})c_A. \tag{13.11}$$

In this case, the overall reaction rate is determined by the parameters of the second step, i.e., this step is now rate-determining, and the concentration of the intermediate B is determined by the equilibrium of the first step. Generally, reactant A and product D will not be in chemical equilibrium since their concentrations, c_A and c_D, are arbitrarily defined. Hence, $c_{B(A)}^0$ and $c_{B(D)}^0$ will have different values; they will coincide only in the particular case of overall equilibrium between substances A and D which will be established at concentration ratios $c_D/c_A = k_1 k_2/k_{-1}k_{-2}$.

It is important to note that it is precisely the ratio of k_{-1} and k_2 that decides which step is rate-determining, and not the ratios of the parameters of both steps in the forward reaction (k_1 and k_2) or in the reverse reaction (k_{-1} and k_{-2}).

In chemical reactions, the kinetic parameters k_k and k_{-k} are constant for given conditions (of temperature etc.). Hence, the same step will be rate-determining in the forward and reverse direction of the reaction (providing the reaction pathways are the same in both directions).

The assumption had been made in deriving Eq. (13.4) that the concentration of the intermediate B is determined solely by the balance of rates of individual steps of the process. It is implied here that this intermediate cannot escape from the reaction zone by processes such as diffusion and evaporation.

13.3. TWO-STEP ELECTROCHEMICAL REACTIONS

13.3.1. Equilibrium Conditions

All the relations reported above are valid for simple two-step electrochemical reactions, when instead of rate constants k_k of the individual steps or of the reaction as a whole we use the corresponding kinetic parameters h_k. We shall assume, for the sake of definition, that in the electrochemical reaction substance A is the reducing, and substance D is the oxidizing agent, i.e., that in Eq. (13.2) the anodic reaction is that going from left to right.

Electrochemical reactions fundamentally differ from chemical reactions in that the kinetic parameters h_k are not constant (i.e., they are not rate "constants") but depend on electrode potential. In the typical case this dependence is described by Eq. (6.33). This dependence has an important consequence, viz., at given arbitrary values of the concentrations c_A and c_D, an equilibrium potential E_0 exists in the case of electrochemical reactions which is the potential at which substances A and D are in equilibrium with each other. At this point (E_0) the intermediate B is in common equilibrium with substances A and D. For this equilibrium concentration we obtain from Eqs. (13.9) and (13.11):

$$c_B^0 = (h_1^0/h_{-1}^0)c_A = (h_{-2}^0/h_2^0)c_D. \tag{13.12}$$

At the point of equilibrium, the exchange rate of the reaction as a whole is given by

$$v^0 = h_0^0 c_A = h_{-0}^0 c_D, \tag{13.13}$$

and the exchange rates of the individual steps are given by

$$v_1^0 = h_1^0 c_A = h_{-1}^0 c_B^0; \qquad v_2^0 = h_2^0 c_B^0 = h_{-2}^0 c_D. \tag{13.14}$$

With Eq. (13.7), this yields the following relation between the exchange rates:

$$1/v^0 = 1/v_1^0 + 1/v_2^0, \tag{13.15}$$

which is valid for electrochemical reactions at the equilibrium potential.

When the rate of the overall reaction is stated in electrical units, i.e., in terms of the current density (cd) $i \equiv nFv$, then it will be convenient to use the concept of partial current densities of the first and second step which are defined as $i_1 \equiv l_1 F v_1$ and $i_2 \equiv l_2 F v_2$. In the steady state $v = v_1 = v_2$ and $i = i_1 + i_2$. With these parameters, Eq. (13.15) becomes

$$n/i^0 = l_1/i_1^0 + l_2/i_2^0. \tag{13.16}$$

13.3.2. The General Kinetic Equation

The rates of an electrochemical reaction at potentials away from the equilibrium value are given by Eq. (13.5), which in this case can be written as

$$i = nF(h_1 h_2 c_A - h_{-1} h_{-2} c_D)/(h_{-1} + h_2). \tag{13.17}$$

Since parameters h_k depend on potential, the rds may not be the same in different regions of potential. Consider the two pairs of inequalities

$$\text{(1a) } h_1 h_2 c_A > h_{-1} h_{-2} c_D \quad \text{and} \quad \text{(1b) } h_1 h_2 c_A < h_{-1} h_{-2} c_D;$$
$$\text{(2a) } h_2 > h_{-1} \quad \text{and} \quad \text{(2b) } h_2 < h_{-1}. \tag{13.18}$$

Inequality (1a) is valid for anodic polarization, inequality (1b) for cathodic polarization ($i < 0$). In this case, the point of changeover is evidently the equilibrium potential E_0. However, the changeover from inequality (2a) to inequality (2b) generally occurs at another value of potential, that of the change in mechanism, E_{cm}. At potentials more positive than E_{cm} inequality (2a) holds, and step 1 is the rds; at potentials more negative than E_{cm}, step 2 is the rds.

At high anodic potentials when the electrode potential is more positive than both E_0 and E_{cm}, inequalities (1a) and (2a) hold and step 1 is the rds. The kinetic equation then is

$$i = nF h_1 c_A = nF k_1 c_A \exp(\beta_1 F E / RT) \tag{13.19}$$

(the reverse reaction can be disregarded). In exactly the same way, at high values of cathodic polarization when the electrode potential is more negative than both E_0 and E_{cm}, step 2 is the rds; it is the first step in the cathodic direction, i.e., of reaction (13.2) occurring from right to left. The kinetic equation now is

$$i \doteq -nF h_{-2} c_D = -nF k_{-2} c_D \exp(-\beta_2 F E / RT). \tag{13.20}$$

The behavior in the regions of moderate anodic or cathodic polarization depends on the relative positions of potentials E_{cm} and E_0, which in turn depend on the relative values of constants k_1 and k_{-2}. For E_{cm} which are more positive than E_0 (Fig. 13.1a), relation (13.20) for the cathodic cd remains valid at all values of cathodic polarization (except for the region of low values where the reverse reaction must be taken into account). At moderate values of anodic polarization, inequalities (1a) and (2b) are found to be valid at potentials more negative than E_{cm}, while step 2 becomes rate-determining, which is the second step along the reaction path. In this case [cf. Eq. (13.10)] we have

$$i = nF(h_1/h_{-1}) h_2 c_A \tag{13.21}$$

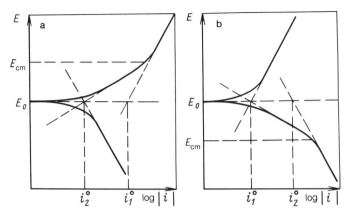

Fig. 13.1. Polarization curves for a two-step reaction.

or, when we decode the parameters h_k [Eq. (6.33)] and take into account that according to Eq. (6.20) $\beta_k + \beta_{-k} = l_k$,

$$i = nF(k_1/k_{-1})k_2 c_A \exp[(l_1 + \beta_2)FE/RT]. \tag{13.22}$$

However, for E_{cm} which are more negative than E_0 (Fig. 13.1b), relation (13.19) remains valid at all values of anodic polarization. At moderate values of cathodic polarization or potentials more positive than E_{cm}, step 1 will be rate-determining, which is the second step along the reaction path, and we find for the kinetic equation

$$i = -nF(h_{-2}/h_2)h_{-1}c_D = -nF(k_{-2}/k_2)k_{-1}c_D \exp[-(l_2+\beta_{-1})FE/RT]. \tag{13.23}$$

We can see here that a formal change occurs in the polarization equation at potential E_{cm}; in particular, transfer coefficient β_0 changes. When the two steps are one-electron steps ($l_k = 1$) and all transfer coefficients of the individual steps are close to 0.5, the value of β_0 will change from 0.5 in the region of very large polarization to 1.5 at lower values of polarization, on the other side of E_{cm}. Slope b' in the Tafel equation changes accordingly from 0.12 to 0.04 V, and the polarization curve exhibits two logarithmic sections (two Tafel slopes). Such breaks in the polarization curve (when plotted semilogarithmically) are a typical indication for multistep electrochemical reactions involving a change in the rds.

It follows from these kinetic equations that in reactions where the rds occurs after another step which is an equilibrium step, the kinetic coefficients k_0 and β_0 of the overall reaction are different from the corresponding coefficients of the rds; in fact, in the first case $k_0 \equiv (k_1/k_{-1})k_2$ and $\beta_0 \equiv l_1 + \beta_2$, and in the second case $k_{-0} \equiv (k_{-2}/k_2)k_{-1}$ and $\beta_0 \equiv l_2 + \beta_{-1}$. It is important to note that if the preceding equilibrium step is an electrochemical step ($l_1 \geq 1$), the transfer coefficient β_0 of the overall reaction will always be larger than unity. The sum of transfer coefficients in the forward and reverse direction of the overall reaction is given by

$$\beta_0 + \beta_{-0} = l_1 + \beta_2 + \beta_{-2} = l_1 + l_2 = n, \qquad (13.24)$$

i.e., relation (6.20) remains valid. However, when different steps are rate-determining under cathodic and anodic polarization the sum of transfer coefficients of these steps, $\beta_1 + \beta_{-2}$ or $\beta_2 + \beta_{-1}$, generally will be different from n, l_1, or l_2.

When the logarithmic sections of the polarization curves are extrapolated to the equilibrium potential E_0, they yield intercepts on the horizontal axis which represent the logarithms of exchange cd of the individual steps (up to factors n/l_k). By extrapolation of the section at high anodic potentials we obtain, according to Eq. (13.20), the exchange cd $(n/l_1)i_1^0$ of step 1, but by extrapolation of the section at high cathodic potentials we obtain, again according to Eq. (13.20), the exchange cd $(n/l_2)i_2^0$ of step 2. It follows that in both cases we obtain the exchange cd of the step which comes first in the reaction path. Analogously, when extrapolating the section in the region of less important polarization (between potentials E_{cm} and E_0) we obtain, according to Eqs. (13.22) or (13.24), the exchange cd of the step which comes second in the reaction path (see Fig. 13.1).

Thus, in the region of very high anodic or cathodic polarization the rds is always the first step in the reaction path. The transfer coefficient of the full reaction which is equal to that of this step, is always smaller than unity (for one-electron rds), while slope b' in the Tafel equation is always larger than 0.06 V. When the potential E_{cm} is outside the region of low polarization,[1] a section will appear in the polarization curve at intermediate values of anodic *or* cathodic polarization where the transfer coefficient is larger than unity and b' is smaller than 0.06 V. This indicates that in this region the step which is second in the reaction path is rate-determining.

A break in the polarization curve will not be observed when the kinetic parameters of the two steps (h_1 and h_{-2}, i_1^0 and i_2^0) are drastically different, and hence, potential E_{cm} is in the region of excessive anodic

---

[1] The region of low polarization is defined as that where the polarization curve cannot be represented as a straight line when plotted semilogarithmically.

or cathodic polarization where measurements become impossible or where the behavior observed is distorted by other phenomena (e.g., concentration polarization). For this reason two-step reactions often follow the behavior outlined in Chapter 6 for simple one-step reactions throughout the range where measurements can be made, and have the same rate-determining step in the forward and reverse direction (quasi-one-step reactions).

13.3.3. The Region of Low Polarization

For an analysis of the polarization curves at low values of polarization (low overpotentials), we shall use the general polarization equation

$$i = \frac{n i_1^0 i_2^0 (\gamma_1 \gamma_2 - \gamma_{-1} \gamma_{-2})}{l_2 \gamma_{-1} i_1^0 + l_1 \gamma_2 i_2^0}, \tag{13.25}$$

which follows from the kinetic equation (13.17) when h_k is replaced by $\gamma_k h_k^0$ and we take into account that $i_1^0 = l_1 F h_1^0 c_A = l_1 F h_{-1}^0 c_B$ and $i_2^0 = l_2 F h_2^0 c_B = l_2 F h_{-2}^0 c_D$.

At low values of polarization it will suffice to retain the first two terms of series expansions of the exponentials, i.e., to assume that $\gamma_k \equiv 1 + \beta_k F \Delta E / RT$ and $\gamma_{-k} \equiv 1 - \beta_{-k} F \Delta E / RT$. When substituting these values into Eq. (13.25) and taking into account that $\beta_1 + \beta_{-1} + \beta_2 + \beta_{-2} = l_1 + l_2 = n$, we find

$$i = \frac{n i_1^0 i_2^0 (nF/RT) \Delta E}{l_2 i_1^0 + l_1 i_2^0 + (l_1 \beta_2 i_2^0 - l_2 \beta_{-1} i_1^0)(nF/RT) \Delta E}. \tag{13.26}$$

This equation is of the general form of $i = K \Delta E / (M + N \Delta E)$ where K, M, and N are constants. Derivative $di/d\Delta E$ of this function has the value $KM/(M + N \Delta E)^2$; and in the particular case where $\Delta E = 0$ the derivative has the value K/M. Thus, when allowing for Eq. (13.17) we find for polarization resistance ρ, which for $\Delta E = 0$ is equal to $d\Delta E/di$:

$$\rho = (RT/nF)(l_2 i_1^0 + l_1 i_2^0)/n i_1^0 i_2^0 = (RT/nF)(1/i_0^0). \tag{13.27}$$

It follows that from the slope of the linear section in the polarization curve close to the equilibrium potential we can determine the exchange cd i_0^0 of the overall reaction.

Thus, in the case of two-step reactions, different methods of determining the exchange cd generally yield different results (in contrast to the case of simple reactions discussed earlier): extrapolation of the limiting anodic and cathodic sections of the semilogarithmic plots yields values i_1^0 and i_2^0, respectively, while the slope of the linear section in an ordinary plot of the polarization curve yields the value of i_0^0. It is typical for multistep reactions that the exchange cd determined by these methods differ.

The exchange cd determined by different methods will coincide only in the case of quasi-one-step reactions mentioned above. Thus, when the value of i_1^0 is so much higher than i_2^0 that the extreme anodic section cannot be measured and there is no break in the polarization curve, all three methods of determination lead to the same value of i_2^0. This implies that step 1 has no effect at all on the kinetics of the overall reaction and that its (high) exchange cd cannot be determined. The same conclusion holds in the opposite case of $i_1^0 \ll i_2^0$.

13.4. COMPLEX ELECTROCHEMICAL REACTIONS

In many electrochemical reactions the individual steps differ in their stoichiometric numbers, in contrast to what was found for reactions of the type of (13.2). A two-step reaction can generally be formulated as

$$\text{(step 1) } \nu_A A \rightleftarrows \nu_B B + l_1 e^- \tag{13.28}$$

$$\text{(step 2) } \nu_B' B \rightleftarrows \nu_D D + l_2 e^- \tag{13.29}$$

It follows from the steady-state condition for the numbers of particles B generated and consumed in the reaction that

$$\mu_1 \nu_B = \mu_2 \nu_B'. \tag{13.30}$$

where μ_1 and μ_2 are the stoichiometric numbers of steps 1 and 2. For the number of electrons the relation

$$\mu_1 l_1 + \mu_2 l_2 = n \tag{13.31}$$

is valid.

Consider the case of a quasi-one-step reaction for which step 1 is rate-determining at all potentials, while step 2 is in equilibrium. When using the Nernst equation (3.40) for this equilibrium we find

$$c_B^{\nu_B'} = K_2 c_D^{\nu_D} \exp(-l_2 FE/RT). \tag{13.32}$$

When a current flows the kinetic laws are determined by step 1. For the region where the anodic polarization is sufficiently high we obtain

$$i = nFk_1 c_A^{\nu_A} \exp(\beta_1 FE/RT) = i_1^0 \exp(\beta_1 F\Delta E) \qquad (13.33)$$

(here factor n/l_1 has conditionally been included into the value of i_1^0). For the region of cathodic polarization, and allowing for Eqs. (13.32) and (13.30), we find

$$-i = nFk_{-1} c_B^{\nu_B} \exp(-\beta_{-1} FE/RT)$$
$$= i_1^0 \exp[-(\beta_{-1} + \mu_2 l_2/\mu_1)F\Delta E/RT]. \qquad (13.34)$$

In the region of low polarization, when retaining only two terms in the series expansions of the exponentials we obtain

$$i = i_1^0(\beta_1 + \beta_{-1} + \mu_2 l_2/\mu_1)F\Delta E/RT \qquad (13.35)$$

and, when allowing for Eq. (13.31) and the relation $l_1 = \beta_1 + \beta_{-1}$ [cf. Eq. (6.24)], we obtain

$$i = i_1^0(1/\mu_1)nF\Delta E/RT. \qquad (13.36)$$

This equation is of interest since it contains the stoichiometric number, μ_1, of the rate-determining step. This number did not appear in the analogous Eq. (6.27), since in the example considered when deriving this equation it was unity. Using the exchange cd i_1^0 determined by extrapolation of the Tafel sections with Eq. (13.33) or (13.34), we can determine μ_1 from the slope of the linear section of the polarization curve and Eq. (13.36). In a number of cases this method can be used to find out which of the intermediate steps is rate-determining.

An analogous expression is obtained when step 2 is rate-determining. The stoichiometric number of the rds is often simply called the stoichiometric number of the reaction itself.

The general kinetic equations for electrode reactions with more than two steps are extremely complex and practically never used. The kinetic features of such reactions are usually examined separately in different regions of potential each having a particular step as the rds. When the exchange cd of one of the steps is much lower than the exchange cd of all other steps, the polarization characteristics of the multistep reaction will coincide with those of a simple one-step electrochemical reaction, practically throughout the region of measurements (as in the analogous case of two-step reactions).

13.5. REACTIONS WITH HOMOGENEOUS CHEMICAL STEPS

Consider an electrochemical reaction, of the type of Red $\rightleftarrows$ Ox $+ e^-$, which occurs under conditions when the chemical reaction (and equilibrium)

$$\nu_A A + \nu_j X_j \rightleftarrows \nu_{Red} Red + \nu_l X_l \qquad (13.37)$$

is possible between the component Red and a substance A in the bulk electrolyte; here X_j and X_l are other possible reaction components.

When the current is anodic, component Red is consumed and the equilibrium in the electrolyte close to the surface is disturbed; reaction (13.37) will start to proceed from left to right producing additional amounts of species Red. In this case the chemical precedes the electrochemical reaction. However, when the current is cathodic, substance Red is produced and the chemical reaction (13.37), now as a subsequent reaction, will occur from right to left. When component Ox rather than component Red is involved in the chemical reaction, this reaction will be the preceding reaction for cathodic currents, but otherwise all the results to be reported below remain valid.

Preceding and also subsequent homogeneous chemical reactions which occur in the bulk solution are very common. Examples are the dehydration (when only a nonhydrated form of the substance is involved in the electrochemical reaction), protonation (e.g., of the anions of organic acids), decay of complexes (in metal deposition from solutions of complex salts), etc.

Equilibrium between substances A and Red will be preserved during current flow, not only in the bulk solution but even near the electrode surface, when the chemical reaction has a high exchange rate. Therefore, a change in surface concentration of the substance Red which occurs as a result of the electrochemical reaction, will give rise to the corresponding change in equilibrium concentration of the substance A and to the development of concentration gradients with respect to both substances. When the current is anodic, reactant Red will be supplied from the bulk solution, both as such and in the form of particles of substance A which near the surface are converted to particles Red. The diffusion fluxes of the two substances have a common effect, and the total limiting diffusion cd i_l is the sum of two components, $i_{l,\mathrm{Red}}$ and $i_{l,\mathrm{A}}$. When calculating the surface concentration of substance Red in Eq. (4.17) we must use the combined limiting cd of both substances.

The situation is different when the chemical reaction is not very fast. In this case the equilibrium between substances Red and A in the solution layers near the electrode will be disturbed and the rate at which reactant Red is replenished on account of reaction (13.37) decreases. When the chemical reaction is very slow the limiting cd will approach the value $i_{l,\mathrm{Red}}$.

Consider the case when the equilibrium concentration of substance Red, and hence its limiting cd due to diffusion from the bulk solution, is low. In this case the reactant species Red can be supplied to the reaction zone only as a result of the chemical step. When the electrochemical step is sufficiently fast and activation polarization is low, the overall behavior of the reaction will be determined precisely by the special features of the chemical step; concentration polarization will be observed for the reaction at the electrode, not because of slow diffusion of the substance but because of a slow chemical step. We shall assume that the concentrations of substance A and of the reaction components are high enough so that they will remain practically unchanged when the chemical reaction proceeds. We shall assume, moreover, that reaction (13.37) follows first-order kinetics with respect to Red and A. We shall write c_0 for the equilibrium (bulk) concentration of substance Red, and we shall write c_S and c for the surface concentration and the instantaneous concentration (we shall not use subscripts Red in order to simplify the equations).

Under the assumptions made, the rate of the chemical step can be written as

$$v_{\mathrm{ch}} = k_{\mathrm{ch}}c_{\mathrm{A}} - k_{-\mathrm{ch}}c \qquad (13.38)$$

or, when we use the parameter of exchange rate, $v^0 = k_{\mathrm{ch}}c_{\mathrm{A}} = k_{-\mathrm{ch}}c_0$, as

$$v_{\mathrm{ch}} = v^0(1 - c/c_0) = k_{-\mathrm{ch}}(c_0 - c). \qquad (13.39)$$

Each of the particles of Red produced in the chemical reaction will, after some (mean) time t, have been reconverted to A. Hence, when the current is anodic, only those particles of Red will be involved in the electrochemical reaction which within their own lifetime can reach the electrode surface by diffusion. This is possible only for particles produced close to the surface, within a thin layer of electrolyte called the *reaction layer*. Let this layer have a thickness δ_r. As a result of the electrochemical reaction, the concentration of substance Red in the reaction layer will vary from a value c_0 at the outer boundary to the value c_S right next to the electrode; within the layer a concentration gradient and a diffusion flux toward the surface are set up.

It is a special feature of this diffusion situation that substance Red is produced by the chemical reaction, all along the diffusion path, i.e., sources of the substance are spatially distributed. For this reason the diffusion flux and the concentration gradient are not constant but increase (in absolute values) in the direction toward the surface. The incremental diffusion flux in a layer of thickness dx [$(dJ_d/dx)dx$ or $-D(d^2c/dx^2)dx$] should be equal to the rate, $v_{ch}dx$, of the chemical reaction in this layer. Hence, we have

$$d^2c/dx^2 = -v_{ch}/D = (k_{-ch}/D)(c - c_0). \qquad (13.40)$$

When the sign convention adopted for rate v_{ch} is taken into account this equation holds, both for anodic and for cathodic currents.

The method used for integrating Eq. (12.36) will again be used to integrate the differential Eq. (13.40) over the variable x. Then we obtain

$$(dc/dx)^2 = (k_{-ch}/D)(c^2 - 2cc_0) + K. \qquad (13.41)$$

Far from the surface ($x \to \infty$) the value of c tends toward c_0, while the value of (dc/dx) tends toward zero. Hence, we can determine the integration constant: $K = (k_{-ch}/D)c_0^2$. We finally have

$$dc/dx = \sqrt{k_{-ch}/D}\,(c_0 - c) \qquad (13.42)$$

[the sign of the root of Eq. (13.41) is selected so that for $c < c_0$ we have $dc/dx > 0$ and vice versa].

The current density is determined by the diffusion flux directly at the surface ($x = 0$) where substance Red has the concentration c_S:

$$i = nFD(dc/dx)_{x=0} = \pm i_r(1 - c_S/c_0), \qquad (13.43)$$

where

$$i_r \equiv nFc_0\sqrt{k_{-ch}D}. \qquad (13.44)$$

This equation links the current density to surface concentration. In the case discussed (where there is no activation polarization), the Nernst equation unequivocally links the electrode's polarization to the difference between surface and bulk concentration:

$$c_S/c_0 = \exp(-nF\Delta E/RT). \qquad (13.45)$$

For the polarization function we obtain, as a result,

$$i = \pm i_r[1 - \exp(-nF\Delta E/RT)]. \qquad (13.46)$$

When anodic polarization is appreciable ($\Delta E \gg 0$) the cd will tend toward the value i_r, and then remain unchanged when polarization increases further. Therefore, parameter i_r as defined by Eq. (13.44) is a limiting cd arising from the limited rate of a homogeneous chemical reaction when c_S drops to a value of zero; it is the *kinetic limiting current density*.

At low values of $|\Delta E|$ the exponential term in Eq. (13.46) can be replaced by the first two terms of the series expansion, and hence

$$i = i_r(nF/RT)\Delta E = (n^2 F^2/RT)c_0\sqrt{k_{-ch}D}\,\Delta E. \qquad (13.47)$$

Thus, at low values of polarization we again find proportionality between current density and polarization.

When cathodic polarization is appreciable ($\Delta E \ll 0$), Eq. (13.46) changes into

$$i = -i_r \exp(-nF\Delta E/RT). \qquad (13.48)$$

In this case the usual exponential dependence analogous to Eq. (6.50) is obtained.

All these equations differ from the corresponding equations for diffusion polarization, only in that the equilibrium concentration c_0 appears in them instead of bulk concentration c_V. Formally, diffusion can be regarded as a first-order reaction, the limiting diffusion flux being proportional to the first power of concentration.

The concentration distribution in the reaction layer can be found by integrating Eq. (13.42):

$$c(x) = c_0 - (c_0 - c_S)\exp[-\sqrt{(k_{-ch}/D)}\,x]. \qquad (13.49)$$

The concentration asymptotically approaches the value c_0 with increasing distance x, i.e., the reaction zone has no distinct boundary. Conventionally, thickness δ_r is defined just like the diffusion-layer thickness δ, i.e., by the condition that $c_0/\delta_r = (dc/dx)_{x=0}$ for zero surface concentration [cf. Eq. (4.48)]. Using Eq. (13.41) we find

$$\delta_r = \sqrt{D/k_{-ch}}. \qquad (13.50)$$

It can be seen here that, the larger the value of k_{-ch} the thinner will be the reaction layer, and the more readily will the particles avoid getting involved in the electrochemical reaction and instead participate in the reverse chemical reaction. However, because of the increase in concentration gradient, the flux to the surface and the current density will still increase.

The development of a kinetic limiting current is a characteristic of electrochemical reactions with a preceding chemical step. However, in contrast to limiting diffusion currents, these limiting currents do not depend on the intensity of electrolyte stirring. Thus, by examining the effect of stirring one can clearly determine the nature of the limiting current arising in the electrochemical system.

13.6. REACTIONS WITH MEDIATORS

One of the types of multistep electrochemical reactions with chemical steps are those involving mediators (transfer agents).

Often a dissolved oxidizing agent is electrochemically inactive, and at platinum or other nonconsumable electrodes the equilibrium redox potential is not set up; even at appreciable cathodic polarization of the electrode, the reduction reaction will either not occur at all or it will, but very slowly. Yet the same substance is readily reduced in a chemical way when reacting with other substances having reducing properties. This implies inhibition of the electrochemical step involving electron transfer from the electrode to the reacting species, but lack of inhibition of the chemical steps involving electron or hydrogen-atom transfer from other species.

The same situation is found in the oxidation of certain dissolved reducing agents; in many cases these reactions only occur by reaction with oxidizing agents, but not upon anodic polarization of an electrode. Such behavior is observed primarily in systems with organic reactants, more rarely in systems with inorganic reactants.

In systems of this type, the electrochemical reactions can be realized or greatly accelerated when small amounts of the components of another redox system are added to the solution. These components function as the auxiliary oxidizing or reducing intermediates of the primary reactants, i.e., as electron or hydrogen-atom transfer agents. When consumed they are regenerated at the electrode.

The oxidation of an anthracene suspension in sulfuric acid conducted in the presence of cerium salts can serve as an example of mediated oxidation. In the bulk solution the Ce^{4+} ions chemically oxidize anthracene to anthraquinone. The resulting Ce^{3+} ions are then reoxided at the anode to Ce^{4+}. Thus, the net result of the electrochemical reaction is the oxidation of anthracene, even though the electrochemical steps themselves only involve cerium ions, not anthracene. Since the cerium ions are continuously regenerated, a small amount will suffice to oxidize large amounts of anthracene.

In a similar fashion, chromium ions Cr^{2+} will reduce dissolved acetylene to ethylene, and then are regenerated at the cathode from the Cr^{3+} ions that were formed in the reaction. Or, at a platinum electrode in a solution of AsO_4^{3-} and AsO_3^{3-} ions, the equilibrium potential of this redox system is not established. After the addition of small amounts of iodine and iodide ions, an ionic reaction leads to equilibrium between the two redox systems, while the concentration ratio of the AsO_4^{3-} and AsO_3^{3-} ions which are present in excess remains practically unchanged. It is because of the iodine/iodide system that an overall equilibrium potential can be set up at the electrode; it practically coincides with the thermodynamic potential of the original arsenate/arsenite system.

Inorganic systems such as Br_2/Br^-, Ce^{4+}/Ce^{3+}, and Sn^{4+}/Sn^{2+} which have high electrochemical activity most often are used as the mediating redox systems. In a few cases organic redox systems are used, e.g., the quinone/hydroquinone system.

Mediating redox systems can be formed even without the addition of special reactants. In the electrochemical reduction of ethylene at platinum, a layer of adsorbed hydrogen atoms is formed in the first place on the electrode surface by the cathodic electrochemical reaction (12.10). These atoms chemically reduce the ethylene molecules. Hydrogen atoms consumed are continuously regenerated cathodically, and the reaction can continue. Similarly, the anodic oxidation of methanol at platinum occurs by chemical reaction of adsorbed methanol particles with –OH groups generated electrochemically on the electrode surface by reaction (12.57). In these two examples, the chemical reaction occurs not in the bulk solution, as in the earlier examples, but on the electrode surface. All these reactions have in common that the actual reducing or oxidizing agent is generated or regenerated during the reaction.

In the past it had been a popular belief that the electrochemical reduction of any inorganic or organic substance involves the primary electrochemical formation of a special, active form of hydrogen in the nascent state (*in statu nascendi*) and subsequent chemical reaction of this hydrogen with the substrate. However, for many reduction reactions a mechanism of direct electron transfer from the electrode to the substrate could be demonstrated. It is only in individual cases involving electrodes with superior hydrogen adsorption that the above mechanism with an intermediate formation of adsorbed atomic hydrogen is possible.

In other cases, to the contrary, certain substances may act as proton transfer agents in cathodic hydrogen evolution. Thus, in the presence of organic compounds containing –SH groups, hydrogen evolution at the mercury electrode is strongly accelerated, and we have the so-called catalytic hydrogen evolution at mercury. This acceleration arises from the cathodic reduction of –SH groups and simultaneous hydrogen evolution:

$$R–SH + e^- \rightarrow RS^- + H_{ads}(\rightarrow 1/2H_2). \tag{13.51}$$

The reactant R–SH consumed in this step is regenerated from the ions RS^- by their chemical reaction with the principal proton donors, which are H_2O molecules or H_3O^+ ions:

$$RS^- + H_2O \rightarrow R–SH + OH^-. \tag{13.52}$$

The system RSH/RS^- thus acts as a hydrogen transfer agent.

13.7. PARALLEL ELECTRODE REACTIONS

Current flow at electrode surfaces often involves several simultaneous electrochemical reactions which differ in character. For instance, upon cathodic polarization of an electrode in a mixed solution of lead and tin salt, lead and tin ions are discharged simultaneously, and from an acidic solution of zinc salt zinc is deposited, and at the same time hydrogen is evolved. Upon anodic polarization of a nonconsumable electrode in chloride solution, oxygen and chlorine are evolved in parallel reactions.

Different reactions (anodic and cathodic) can occur simultaneously at an electrode, even when there is no net current flow. In Section 2.5.1

we had mentioned the example of an iron electrode in $HCl + FeCl_2$ solution where anodic iron dissolution (2.24) and cathodic hydrogen evolution (2.25) occur simultaneously; these are the reactions of *spontaneous dissolution* of iron not requiring a net current.

The net (external or overall) current density at an electrode is the algebraic sum of the partial current densities of all reactions:

$$i = \sum i_m \qquad (13.53)$$

[here i_m denotes current densities, both of forward $(\overrightarrow{i})$ and back $(\overleftarrow{i})$ reactions]. In the particular case where the total current is zero we have $\sum i_m = 0$.

The *current yield* g_n is a useful parameter for the quantitative characterization of parallel reactions. This is the ratio of the partial cd, i_n, consumed in a given reaction n, to the total cd:

$$g_n = i_n/i = i_n \left/ \sum i_m \right. ; \qquad (13.54)$$

most often this parameter is used in connection with the desired (useful) reaction.

The *principle of independent electrochemical reactions* applies when several reactions occur simultaneously. It says that each reaction follows its own quantitative laws irrespective of other reactions. At a given potential, the rates of the different reactions are not at all interrelated, and at a given cd they are merely tied together by relation (13.53). This does not mean that the reactions have no influence at all on each other. One of the reactions may produce changes in the external conditions for other reactions, e.g., in the temperature or solution pH, the amount of impurities adsorbed on the electrode, etc. However, the form of the kinetic equation of each reaction is not affected by these changes. The principle of independent electrochemical reactions is quite general, and rarely violated (we shall discuss an instance of such a departure in Section 18.5.1).

All of the above remarks apply also to the case where a given reaction occurs along several parallel pathways. As a result of the principle of independence, the concept of a rate-determining step of the overall reaction becomes meaningless for such a reaction.

Consider in more detail the example mentioned, where a metal electrode dissolves anodically while hydrogen is evolved. This process is feasible when the equilibrium potential of metal dissolution and deposition (index 1) is more negative than the potential of the hydrogen reaction

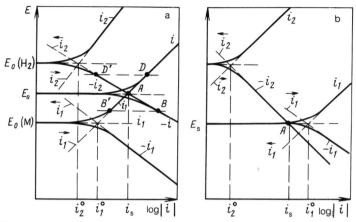

Fig. 13.2. Polarization curves for the partial current densities of the reactions involving the metal and hydrogen, and the polarization curves for the overall current density.

(index 2). In general, when the values of these equilibrium potentials are similar, both the partial anodic forward ($\vec{\imath}_1$, $\vec{\imath}_2$) and the partial cathodic reverse ($\overleftarrow{\imath}_1$, $\overleftarrow{\imath}_2$) reactions must be taken into account. The effective rate of metal dissolution, i_1, is given by $\vec{\imath}_1 - |\overleftarrow{\imath}_1|$, while the effective rate of cathodic hydrogen evolution, $|i_2|$, is given by $|\overleftarrow{\imath}_2| - \vec{\imath}_2$. At zero current the rates of these two reactions are identical, according to Eq. (13.53), i.e., the anodic conversion in one reaction is compensated, with respect to charge consumed, by cathodic conversion in the other reaction. Reactions of this type which are forced to proceed at identical rates are called *coupled reactions*. In the present example, the rate of the coupled reactions $i_s = i_1 = |i_2|$ is called the rate (or current density) of spontaneous metal dissolution.

Figure 13.2 shows anodic and cathodic polarization curves for the partial cd of dissolution ($\vec{\imath}_1$) and deposition ($\overleftarrow{\imath}_1$) of the metal and for the partial cd of ionization ($\vec{\imath}_2$) and evolution ($\overleftarrow{\imath}_2$) of hydrogen, as well as curves for the overall reaction current densities involving the metal (i_1) and the hydrogen (i_2). The spontaneous dissolution current density i_s evidently is determined by the point of intersection, A, of these combined curves.

The electrode's open-circuit potential (steady potential) E_s depends on the relative values of the exchange cd of both reactions and also on the slopes of the polarization curves. When the exchange cd and slopes are similar the open-circuit potential will have a value, the mixed (or "compromise") potential, which is intermediate between the two equilibrium potentials (Fig. 13.2a). However, when the exchange cd for one of the reactions is much higher than for the other, the open-circuit potential will practically coincide with the equilibrium potential of this reaction (Fig. 13.2b).

A relatively simple analytical expression linking the current density i_s and potential E_s to the kinetic parameters of the reactions can be obtained when the exchange cd of the reactions are comparable, while their equilibrium potentials strongly diverge. In this case the currents of metal deposition ($\overleftarrow{i}_1$) and hydrogen ionization ($\overrightarrow{i}_2$) can be neglected in the region of the open-circuit potential. It follows that at $E = E_s$ we have $i_s = \overrightarrow{i}_1 = |\overleftarrow{i}_2|$. Substituting kinetic equations of the type of (6.10) into these relations we obtain

$$i_s = 2Fk_1 \exp(\beta_1 F E_s/RT) = 2Fk_{-2}c_{H^+} \exp(-\beta_{-2}F E_s/RT).$$
(13.55)

Solving this equation we find the expression for the steady (or rest) potential E_s:

$$E_s = [RT/(\beta_{-2} + \beta_1)F]\ln(k_{-2}c_{H^+}/k_1).$$
(13.56)

In order to find the final expression for i_s, we must substitute this value of E_s into Eq. (13.55):

$$i_s = 2Fk_1^{\beta_{-2}/(\beta_{-2}+\beta_1)}(k_{-2}c_{H^+})^{\beta_1/(\beta_{-2}+\beta_1)}.$$
(13.57)

When more complex polarization functions are involved, and particularly when concentration polarization is superimposed, the values of E_s and i_s are preferably determined by graphical rather than analytical means.

When such a polyfunctional electrode is polarized the net current, i, will be given by $i_1 - |i_2|$. When the potential is made more negative the rate of cathodic hydrogen evolution will increase (Fig. 13.2b, point B), and the rate of anodic metal dissolution will decrease (point B'). This effect is known as *cathodic protection* of the metal. At potentials more negative than the metal's equilibrium potential its dissolution ceases completely. When the potential is made more positive the rate of anodic dissolution will increase (point D). However, at the same time the rate of cathodic hydrogen evolution will decrease (point D'), and the rate of spontaneous metal dissolution (the share of anodic dissolution not associated with the net current but with hydrogen evolution) will also decrease. This phenomenon is known as the *difference effect*.

Chapter 14
THE ELEMENTARY REACTION ACT

14.1. THE ENERGY OF ACTIVATION

14.1.1. Chemical Reactions

The rate constants, k_m, of most reactions increase with increasing temperature. This function is quantitatively described by the *Arrhenius equation* (Svante August Arrhenius, 1889),

$$k_m = B_m \exp(-A_m/RT), \tag{14.1}$$

where B_m (the preexponential factor) and A_m (the activation energy) are empirical parameters valid for a reaction m.

When the experimental data are plotted as $\ln k_m$ against T^{-1} they will fall onto a straight line. From its slope the value of A_m, and hence that of B_m, can be obtained. The value of A_m can also be determined from the derivative of Eq. (14.1),

$$A_m = RT^2(d \ln k_m/dT). \tag{14.2}$$

For an interpretation of activation energies one often uses potential energy–distance curves (Fig. 14.1). For a reaction X $\rightarrow$ Y, the potential energy (enthalpy) of the system of reacting particles is plotted on the vertical axis, and the conditional reaction pathway (the set of all intermediate states) is plotted as a distance λ (the reaction coordinate, which is not a true geometric distance) on the horizontal axis. In the initial state (point P_X) the system is stable and enthalpy H_X has a minimum value. The first stage of the reaction involves some change (activation) of the system, e.g., the stretching of chemical bonds, which needs additional

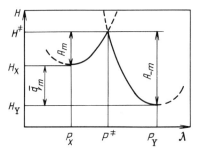

Fig. 14.1. Potential energy–distance curves for reactants and products in a chemical reaction.

energy input. The system's potential energy increases accordingly. The interaction between the activated particles forming new chemical bonds is attended by liberation of energy. Hence the potential energy after going through a maximum falls to a new minimum equilibrium value for the reaction products (point P_Y). The maximum of the curve (point $P^{\neq}$) can be regarded as the point of intersection of two partial curves (the dashed lines) for reactants X and products Y, respectively. The system's state in the maximum is called the transition state or activated state, ("activated complex"), and the enthalpy value in this point is written as $H^{\neq}$. The difference $\Delta H_m^{\neq} \equiv H^{\neq} - H_X$, which is the height of the potential-energy barrier, constitutes the activation energy (enthalpy) A_m of the forward reaction. For the reverse reaction Y → X the activation energy A_{-m} is given by $\Delta H_{-m}^{\neq} \equiv H^{\neq} - H_Y$.

The total energy effect of the reaction is given by $\bar{q}_m = -\Delta H_m \equiv H_X - H_Y$. When it is assumed that the forward and reverse reaction pass through the same transition state, then evidently

$$A_{-m} - A_m = \bar{q}_m = -\bar{q}_{-m}, \qquad (14.3)$$

i.e., the activation energies of the forward and reverse reaction are interrelated through the thermodynamic parameter $\bar{q}_m$.

Using Gibbs free energies rather than the enthalpies in constructing the potential energy–distance curves we will, accordingly, obtain the Gibbs free energy of activation $\Delta G^{\neq}$. The difference in values of this parameter between the forward and reverse reaction is the maximum work of reaction, $\bar{w}_m^0 \equiv -\Delta G_m$. Data for the enthalpies are more readily accessible than data for the Gibbs free energies, hence the potential energy–distance curves are usually constructed with enthalpies.

According to the theory of rate processes (Henry Eyring, Samuel Glasstone, and Keith J. Laidler, 1935 to 1940), reaction rate constants are determined by the expression

$$k_m = \varkappa_m (kT/h) \exp(-\Delta G_m^{\neq}/RT), \qquad (14.4)$$

where $k \ (\equiv R/N_A)$ is the Boltzmann constant, h is the Planck constant, and $\varkappa_m$ is the dimensionless transmission coefficient ($\varkappa_m \leq 1$).

This expression corresponds to the Arrhenius equation (14.1) and basically provides a possibility for calculating the preexponential factor B_m (a calculation of $\varkappa_m$ is in fact not easy). It also shows that in the Arrhenius equation it will be more correct to use parameter $\Delta G_m^{\neq}$ rather than $\Delta H_m^{\neq}$. However, since $\Delta G_m^{\neq} = \Delta H_m^{\neq} - T\Delta S_m^{\neq}$, it follows that the preexponential factor of Eq. (14.4) will contain an additional factor $\exp(\Delta S_m^{\neq}/R)$ reflecting the entropy of formation of the transition state when the enthalpy is used in this equation.

An important experimental rule for protolytic reactions was established by Johannes Nicolaus Brønsted in 1922 (it was later extended to other reactions). He showed that for a series of reactions of the same type the rate constants k_m and the equilibrium constants K_m are related simply as

$$k_m = \gamma_n K_m^{\alpha_n}, \qquad (14.5)$$

where γ_n and α_n are constants and $0 \leq \alpha_n \leq 1$ (here subscript n refers to the reaction type rather than to any specific reaction, i.e., the values of γ_n and α_n are constant for an entire series of reactions m of the same type n).

If two reactions differ in maximum work by a certain amount $\delta \bar{w}_m^0$ ($\equiv -\delta \Delta G_m^0$), it follows from the Brønsted relation [when taking into account the Arrhenius equation and the known relation between the equilibrium constant and the Gibbs standard free energy of reaction, viz., $K_m = \exp(-\Delta G_m^0/RT)$] that their activation energies will differ by a fraction of this work, with the opposite sign:

$$\delta A_m = -\alpha_n \delta \bar{w}_m^0. \qquad (14.6)$$

According to Eq. (14.3), the activation energies of the reverse reactions will also differ by a fraction of this work (but this time with the same

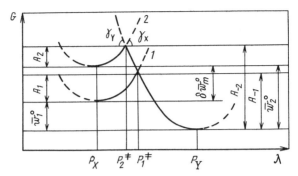

Fig. 14.2. Potential energy–distance curves for two reactions of the same type.

sign):

$$\delta A_{-m} = \alpha_{-n}\delta \bar{w}_m^0, \tag{14.7}$$

where between transfer coefficients α_n and α_{-n} the relation

$$\alpha_n + \alpha_{-n} = 1 \tag{14.8}$$

exists.

These relations can also be interpreted with the aid of potential energy–distance curves (Fig. 14.2). For reactions of a given type, the conditional reaction pathways and shapes of potential energy–distance curves are approximately identical in these diagrams. Here an increase in maximum work corresponds to a relative displacement of the curves along the vertical, viz., to an upward displacement of curve 1 for the reactants by an amount $\delta \bar{w}_m^0$ to the position of curve 2. We see that in this case the height of the energy barrier of the forward reaction actually decreases by a certain fraction α_n of the total displacement of the curve. The activation energy for the reverse reaction, in accordance with Eqs. (14.7) and (14.8), increases by a fraction $\alpha_{-n} = 1 - \alpha_n$. The values of these factors are related to the slopes ($\tan \gamma_j$) of the potential energy–distance curves of reactants X and products Y close to their point of intersection, as

$$\alpha_n = \tan \gamma_X / (\tan \gamma_X + \tan \gamma_Y). \tag{14.9}$$

When the potential energy–distance curves for the reactants and products are symmetric and have the same slope we have $\alpha_n = \alpha_{-n} = 0.5$.

14.1.2. Electrochemical Reactions

According to Eq. (14.2), the activation energy can be determined from the temperature dependence of the reaction rate constant. Since the overall rate constant h_m of an electrochemical reaction also depends on potential, it must be measured at constant values of the electrode's Galvani potential. However, as was shown in Section 3.6 the temperature coefficients of Galvani potentials cannot be determined. Hence, the conditions under which such a potential can be kept constant while the temperature is varied are not known, and the true activation energies of electrochemical reactions, and also the true values of factor B_m, cannot be measured.

For this reason and following a suggestion of Mikhail I. Temkin (1948), another conventional parameter is used in electrochemistry, viz., the real activation energy W_m described by Eq. (14.2), not at constant potential but at constant polarization of the electrode. These conditions are readily realized in the measurements (an electrode at zero current and the working electrode can be kept at the same temperature), and the real activation energy can be measured.

A more detailed analysis shows that the ideal and real activation energy are interrelated as

$$W_m = A_m \pm \alpha_m T \Delta S_m = A_m - (\pm \alpha_m \bar{q}_{\text{lat},m}), \tag{14.10}$$

where ΔS_m, the entropy change, and $\bar{q}_{\text{lat},m}$, the latent heat of the electrode reaction, are parameters that cannot be measured (the plus sign is valid for anodic, the minus sign for cathodic reactions).

In the Arrhenius equation, the real activation energy is combined with a real (measurable) preexponential factor. According to Eqs. (14.1) and (14.10), this factor differs from the true factor by the multiplicative entropy term $\exp(\pm \alpha_m \Delta S_m / R)$.

During the elementary act of an electrochemical reaction, charged particles cross the electrode/electrolyte interface, and the net charge on particles in the electrolyte changes by $\pm n$ [see Eq. (1.39)]. Hence, a term describing the change in electrostatic energy, $\pm n F \varphi_G$ or (to a constant term) $\pm n F E$, appears in the expression for the total Gibbs free energy of the reaction:

$$\bar{w}_m^0 \equiv -\Delta \bar{G}_m^0 = -(\Delta G_0^0)_m \pm nFE, \tag{14.11}$$

where $(\Delta G_0^0)_m$ is independent of potential, and the plus sign holds for anodic reactions.

Alexander N. Frumkin pointed out in 1932 that an electrochemical reaction occurring at different potentials can be regarded as an ideal set of chemical reactions of the same type, and suggested that the Brönsted relation be used to explain the potential dependence of electrochemical reaction rates. On the basis of Eqs. (14.6) and (14.11), the relation for the activation energy becomes

$$A_m = A_m^0 - (\pm \alpha_n n F E), \qquad (14.12)$$

where A_m^0 is the activation energy at the potential of the reference electrode ($E = 0$). Substituting (14.12) into the Arrhenius equation and using the notation of $\beta_m \equiv n\alpha_n$ for the electrochemical reaction m, we obtain the well-known relation between the rate constant of an electrochemical reaction, h_m, and potential:

$$h_m = k_m \exp(\pm \beta_m F E / RT), \qquad (14.13)$$

where the constant k_m which does not depend on potential (but depends on the reference electrode) includes the factor $\exp[-A_m^0/RT]$ in addition to the preexponential factor B_m.

In a diagram of the potential energy–distance curves for an anodic reaction, a potential change δE in the positive direction corresponds to an upward shift $nF\delta E$ of the curve for reactants relative to that for the products (or the equivalent downward shift of the curve for the products). The analogous shifts occur for a cathodic reaction when the potential is made more negative. In both cases the activation energies of these reactions decrease and the reactions themselves are accelerated. Such diagrams of potential energy–distance curves have first been used for an electrochemical reaction (cathodic hydrogen evolution) by Juro Horiuti and Michael Polanyi in 1935.

The popular polarization equations of the type of (6.6) for electrochemical reactions thus acquire some physical basis. However, quantum-mechanical concepts must be adduced for a theoretical calculation of the values of A_m^0 and B_m.

14.1.3. Nature of the Activated State

It had been assumed in the past that the main reason for development of an activated transition state with enhanced energy is a stretching of chemical bonds. Thus, in the model of Horiuti and Polanyi it was assumed that stretching of H^+-H_2O bonds, i.e., the "tearing away" of a proton from an hydroxonium ion and its approach to the electrode surface, is the initial stage in cathodic hydrogen evolution; the system's potential energy increases under these conditions. Some distance from the surface the proton is discharged, i.e., an electron is transferred to it and it is converted to a hydrogen atom. Under the effect of the new chemical bond between it and the electrode material, this atom then moves closer still to the surface (until reaching the equilibrium distance), and the potential energy decreases again.

This picture undoubtedly is valid for reactions involving complex molecules undergoing considerable structural change. However, according to current concepts the nature of the activated state is different. The actual act of transfer of a charged particle (electron, proton) is a quantum-mechanical event involving the fast tunneling of this particle slightly beneath the top of the potential-energy barrier. The condition required for such a transition is that of equal potential energies in the initial and final state in this transition. Hence, the energy of the original system must be raised to the level required for tunneling. Since charged species are involved in the reaction, their energy is determined by interaction with the polar solvent molecules. For particular orientations of these molecules the interaction energy assumes values higher than the average level. Because of the thermal motion of the solvent molecules and the fluctuations thus produced, a certain number of particles with the required orientation and energy exist at any specific time. After charge transfer the original orientation of the solvent molecules is restored. Therefore, the chief motive for attainment of an activated state is the need for reorganization (reorientation) of the polar solvent molecules. The reaction rate depends on the probability of finding such particles with a favorable orientation of the solvent molecules.

14.1.4. Activationless and Barrierless Reactions

Relation (14.12) cannot remain valid at all potentials. In an anodic reaction where the potential is highly positive (or in a cathodic reaction where it is highly negative), a potential E_{al} should be attained where A_m becomes zero. In the diagram of potential energy–distance curves (Fig. 14.3), curve 3 corresponds to the potential E_{al}; in the transition from the minimum in curve 3 to curve 4 no energy barrier must be overcome. Now the activation energy (which cannot be negative) ceases to depend

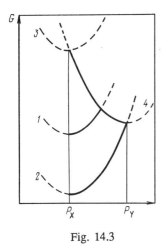

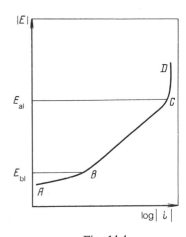

Fig. 14.3   Fig. 14.4

Fig. 14.3. Schematic potential energy–distance curves of reactants in (1) a normal, (2) a barrierless, and (3) an activationless reaction. (4) Potential energy–distance curve for the products.

Fig. 14.4. Schematic polarization curve with barrierless (section AB) and activationless (section CD) region.

on potential, which implies $\beta_m = 0$; rate constant h_m in this range of currents will not depend on potential, and the current will attain some limiting value (section CD in Fig. 14.4). Reactions occurring under such conditions are called *activationless* (hence a subscript "al").

It was shown by Lev I. Krishtalik in 1960 that another limiting case is possible in the opposite region of potentials (curves 2 and 4 in Fig. 14.3), when at potential E_{bl} the descending part of the plot of potential energy against the reaction coordinate vanishes. Starting from this potential the activation energy will be equal to the total energy change occurring during the reaction, i.e., a value of $\beta_m = 1$ is attained. For polarization curves plotted as E against $\ln i$, the slope changes from $RT/\beta_m F$ to RT/F, i.e., the Tafel coefficient b' decreases from 0.12 to 0.06 V (section AB in Fig. 14.4). Such reactions are called *barrierless* (hence suffix bl). We can see from Fig. 14.3 that the activationless region of the forward reaction corresponds to the barrierless region of the reverse reaction, and vice versa.

Activationless and barrierless regions cannot at all be realized in all reactions. Often E_{al} or E_{bl} are in regions of potentials where measurements are impossible or extremely difficult (for instance because of parallel reactions). The crossover to the barrierless region has been demonstrated experimentally for cathodic hydrogen and anodic chlorine evolution at certain electrodes. Clear-cut experimental evidence has not been obtained so far for limiting currents appearing as a result of an activationless reaction.

14.2. KINETIC INFLUENCE OF THE ELECTRIC DOUBLE LAYER

Reactant concentrations $c_{V,j}$ in the bulk solution as well as the Galvani potential between the electrode and the bulk solution (which is a constituent term in electrode potential E) appear in kinetic equations such as (6.8). However, the reacting particles are not those in the bulk solution but the ones close to the electrode surface, near the outer Helmholtz plane when there is no specific adsorption, and near the inner Helmholtz plane when there is specific adsorption. Both the particle concentrations and the potential differ between these regions and the bulk solution. It was first pointed out by Frumkin in 1933 that for this reason, the kinetics of electrochemical reactions should strongly depend on edl structure at the electrode surface.

Let ψ' be the potential at the point where the reacting particle had been prior to the reaction, or where a product particle just generated by the reaction would be. This potential (which is referred to the potential in the bulk solution) has a value similar to potentials ψ_2 or ψ_1, respectively, and gives rise to two effects important for the electrochemical reaction rates.

The first effect is that of a concentration change of the charged reactant particles in the reaction zone; this change is determined by Ludwig Boltzmann's distribution law:

$$c_{S,j} = c_{V,j} \exp(-z_j F \psi' / RT). \qquad (14.14)$$

Depending on the signs of parameters z_j and ψ', the concentration in the reaction zone can be higher or lower than the bulk concentration. There is no change in the concentration of uncharged particles.

The second effect is that of a change in the potential difference effectively influencing the reaction rate. The activation energy, by its physical meaning, should not be influenced by the full Galvani potential φ_G across the interface, but only by the potential difference $(\varphi_G - \psi')$ between the electrode and the reaction zone. Since the Galvani potential is one of the constituent parts of electrode potential E, the difference $(E - \psi')$ should be contained instead of E in Eq. (14.13):

$$h_m = k_m \exp[\pm \beta_m F(E - \psi')/RT].\qquad(14.15)$$

When substituting the new values of $c_{S,j}$ and h_m into the kinetic Eq. (6.10) for a simple first-order reaction we find

$$i = \pm n F k_m c_{V,j} \exp[-(z_j \pm \beta_m)F\psi'/RT] \exp(\pm\beta_m FE/RT)\quad(14.16)$$

(with the plus sign for anodic reactions).

The first exponential factor describes the influence of edl structure on reaction rate (the so-called ψ'-effect). Consider two examples. In the first example, hydrogen is evolved cathodically by H_3O^+ ion discharge at metals where this reaction occurs with high polarization, e.g., at mercury ($z_j = 1$, $\beta_m \approx 0.5$). In this case the reaction occurs at potentials much more negative than the pzc, and the value of ψ' is negative. It follows from Eq. (14.16) that when ψ' is made still more negative (or $|\psi'|$ is raised) the absolute value of current will increase, and vice versa. Therefore, when an excess of foreign electrolyte is added to a dilute solution of pure acid the reaction rate will decrease, at constant potential E, owing to the decrease in $|\psi'|$. For dilute solutions where $|\psi'|$ can attain values of more than 0.15 V, the reaction rate will decrease by one to two orders of magnitude.

In the second example, $S_2O_8^{2-}$ ions ($z_j = -2$) are reduced cathodically in dilute $Na_2S_2O_8$ solution at a mercury electrode:

$$S_2O_8^{2-} + 2e^- \rightarrow 2SO_4^{2-}.\qquad(14.17)$$

In the initial section of the curve, the current increases as usual with increasing polarization, and in the end attains the limiting (diffusion) value. In this region the mercury surface is positively charged. Then the potential moves through the pzc and attains the region where the surface charge and the values of ψ' are negative. The value of $|\psi'|$ increases more slowly with increasing polarization than the potential $|\psi_0|$ contained in E. However, since in Eq. (14.16) the factor $(z_j - \beta_m)$ in front of $\psi'F/RT$ has a value of about -2.5, the inhibiting effect of the ψ'-potential (the first exponential factor) prevails over the accelerating effect of the electric field (the second exponential factor). As a result, when the polarization increases the reaction rate decreases because of increasing repulsion of the anions by the negatively charged surface, and a distinct current drop appears in the curve (Fig. 14.5, curve 1). At still more negative potentials

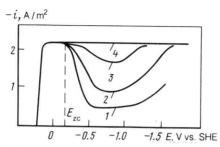

Fig. 14.5. Polarization curves for the reduction of $S_2O_8^{2-}$ ions at a rotating amalgamated silver electrode in $5 \cdot 10^{-4}$ M $Na_2S_2O_8$ solutions with different concentrations of Na_2SO_4: (1) 0; (2) 0.004; 3) 0.05; (4) 0.5 M.

the growth of $|\psi'|$ slows down, and the reaction rate resumes its rise. In the presence of base electrolyte the values of $|\psi'|$ are smaller, and the current drop is less pronounced (curves 2 to 4).

14.3. KINETIC INFLUENCE OF ADSORPTION

When considering how the adsorption of different substances on electrodes influences the kinetics of electrochemical reactions, we must distinguish two cases, that where components are adsorbed which are involved in the reaction, and that where incidental substances are adsorbed which are not involved in the reaction.

14.3.1. Reactant Adsorption

Often multistep reactions are encountered where a reactant j first becomes adsorbed on the electrode, then is converted electrochemically (or chemically) to a desorbing product. We shall consider the case where the electrochemical step involving adsorbed particles is rate-determining. With a homogeneous electrode surface and without interaction forces between the adsorbed particles, the assumption can be made that the rate of this step is proportional, not to the bulk concentration $c_{V,j}$ but to the surface concentration A_j or to the degree of surface coverage θ_j. Under these assumptions the adsorption follows the Langmuir isotherm (12.7), hence

$$i = nFh_m\theta_j = nFh_m[B_j c_{V,j}/(1 + B_j c_{V,j})], \qquad (14.18)$$

i.e., the relation between the current and the bulk concentration is formally the same as the analogous adsorption function: at low concentrations direct proportionality, and at high concentrations a limiting value are found.

However, with an inhomogeneous electrode surface and adsorption energies which are different at different sites, the reaction rate constant k_m and the related parameter h_m will also assume different values for different sites. In this case the idea that the reaction rate might be proportional to surface concentration is no longer correct. It was shown by Temkin that when the logarithmic adsorption isotherm (12.8) is obeyed the reaction rate will be an exponential function of the degree of surface coverage by the reactant:

$$i = nFh_m \exp(\gamma f\theta_j), \qquad (14.19)$$

where f is the surface inhomogeneity factor, and γ is a coefficient ($0 < \gamma < 1$); here the value of h_m will not depend on the surface segment chosen.

Substituting into this equation the expression for θ_j from the isotherm equation we find

$$i = nFh_m B_{0,j}^{\gamma} c_{V,j}^{\gamma}, \qquad (14.20)$$

i.e., the reaction rate is found to be proportional to fractional power of reactant bulk concentration (often $\gamma \approx 0.5$).

In a number of cases electrochemical reactions involving adsorbed substances exhibit special kinetic features. For instance, when the reactant bulk concentration is raised and the degree of coverage approaches unity (e.g., when $\theta_j > 0.9$), the reaction rate fails to tend toward a limiting value following a further concentration rise but instead starts to decrease. Sometimes this decrease is quite dramatic. This effect became known as the "effect of high coverages," and can be seen when several reactants are involved in the reaction. When the surface is almost completely covered by one of them, the others (whether adsorbing or not) will be displaced from the surface layer and cannot take part in the reaction.

Another example are the sometimes rather complex relations existing between potential and reaction rate. The electrode potential influences, not only the parameter h_m [see, e.g., Eq. (14.15)] but also the degree of surface coverage by reactant particles, i.e., the coefficients B_j in Eqs. (14.18) or (14.20). When a sharp drop in adsorption occurs with increasing electrode polarization (rising values of h_m), the monotonic relation between reaction rate and potential may break down and the current actually may decrease within a certain region while polarization increases.

14.3.2. ADSORPTION OF FOREIGN SUBSTANCES

Electrochemical reaction rates are also influenced by substances which, though not involved in the reaction, are readily adsorbed on the electrode surface (reaction products, accidental contaminants, or special additives).

Most often this influence comes about when the foreign species l by adsorbing on the electrode partly block the surface, depress the adsorption of reactant species j, and thus lower the reaction rate. On a homogeneous surface and with adsorption following the Langmuir isotherm, a factor $(1 - \theta_l)$ will appear in the kinetic equation which is the surface fraction free of foreign species l:

$$i = i_0(1 - \theta_l), \tag{14.21}$$

where i_0 is the reaction rate observed when species l are not present.

On inhomogeneous surfaces where adsorption obeys the Temkin isotherm, an exponential factor will appear in the kinetic equation:

$$i = i_0 \exp(-\gamma f \theta_l). \tag{14.22}$$

Adsorption of surface-active substances is attended by changes in edl structure and in the value of the ψ'-potential. Hence, the effects described in Section 14.2 will arise in addition. When surface-active cations $[NR_4]^+$ are added to an acidic solution, the ψ'-potential of the mercury electrode will move in the positive direction and cathodic hydrogen evolution at the mercury, according to Eq. (14.16), will slow down (Fig. 14.6, curve 2). When I^- ions are added the reaction rate to the contrary will increase (curve 3) owing to the negative shift of ψ'-potential. These effects disappear at potentials where the above ions become desorbed (at values of polarization of less than 0.6 V in the case of $[NR_4]^+$, and at values of polarization of over 0.9 V in the case of I^-).

In the adsorption of certain organic substances (α-naphthol, diphenylamine, and others) a strong inhibition of cathodic deposition is found for a number of metals (Mikhail A. Loshkarev, 1939). Under these conditions rather low limiting currents arise which are independent of potential up to

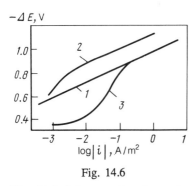

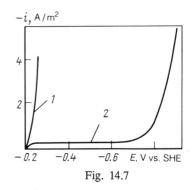

Fig. 14.6   Fig. 14.7

Fig. 14.6. Influence of surface-active ions $[N(C_4H_9)_4]^+$ (curve 2) and I^- (curve 3) on the polarization curve for hydrogen evolution at a mercury electrode in acidic solutions (curve 1 is that for the base electrolyte).

Fig. 14.7. Polarization curves for tin deposition from 0.125 M SnSO$_4$ solution: 1) without additive; 2) with 0.005 M α-naphthol and gelatin (1 g/liter).

the desorption potential of the organic substance (Fig. 14.7). This effect can be explained in terms of the difficulties encountered by the reactant metal ions when, in penetrating from the bulk solution to the electrode surface, they cross the adsorbed layer.

In certain cases an adsorbed foreign species l can directly influence the rate constant for conversion of the reactant species j. This can occur through changes in surface-layer properties of the electrode (e.g., changes in its electronic structure), through direct interaction between the reactant and foreign particles, or through other mechanisms, and can lead to lower or higher reaction rates.

14.4. INFLUENCE OF THE ELECTRODE MATERIAL

14.4.1. Electron Work Function

Consider a redox reaction of the type of (6.2) occurring at a non-consumable metal electrode. The electrode process includes a step where electrons are transferred between different phases: from the metal to an Ox species in the solution (in a cathodic reaction), or from a Red species in the solution to the metal (in an anodic reaction).

Often the claim is made that the rates of such reactions when occurring at different metals M will depend on the electron work function, $w_e^{(M,0)}$, and that the cathodic reaction rate should be higher (or the anodic reaction

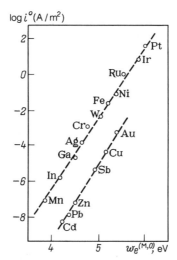

Fig. 14.8. Relation between the exchange current densities of hydrogen evolution and ionization at different metals and the electron work functions.

rate lower) the lower the work function, i.e., the more readily electrons can be extracted from the metal. In a number of cases a definite correlation between these parameters has actually been observed (John O'M. Bockris, 1965). The relation between work functions and exchange current densities for the evolution and ionization of hydrogen at metals is shown as an example in Fig. 14.8.

Generally, however, said claim is not valid. Tabulated electron work functions referring to vacuum [in which case the metal's surface potential $\chi^{(M)}$ must be overcome, see Eq. (2.32) in Section 2.6] are used when setting up empirical correlations such as that shown in Fig. 14.8. In electrochemical reactions, the electron is not extracted into vacuum; it is extracted into the electrolyte, and must overcome the full Galvani potential of the metal/electrolyte interface. The work function $w_e^{(M,E)}$ found at such an interface differs from the work function $w_e^{(M,0)}$ found in vacuum; it depends on the value of the Galvani potential. The work function is the difference between the electrochemical potentials of the electrons in two phases; therefore, we have (when we take into account the definition of Galvani potential)

$$w_e^{(M,E)} = \tilde{\mu}_e^{(E)} - \tilde{\mu}_e^{(M)} = [\mu_e^{(E)} - \mu_e^{(M)}] + Q^0 \varphi_G^{(M,E)} \qquad (14.23)$$

where $\mu_e^{(E)}$ is the chemical potential of solvated electrons in the electrolyte solution ($z_e = -1$); all values of $\tilde{\mu}_e$, μ_e, and $w_e^{(M,E)}$ refer to a single electron (see Section 2.6). For further details concerning solvated electrons see Section 21.1.

The work function will be higher the higher the value of $\varphi_G^{(M,E)}$, i.e., the more positive the metal's potential relative to the solution. At equal values of the Galvani potentials, the work functions of different metals still are different owing to differences in the values of $\mu_e^{(M)}$.

In electrochemical kinetics, electrode potentials E measured against a certain reference electrode M_R are used, rather than the Galvani potentials which cannot be measured experimentally. In particular, the rates of electrochemical reactions occurring at different electrodes are compared at identical values of electrode potential, rather than at identical values of the Galvani potential. Replacing $\varphi_G^{(M,E)}$ by E $[= \varphi_G^{(M,E)} - \varphi_G^{(M_R,E)} + \varphi_G^{(M_R,M)}]$ in Eq. (14.23) and taking into account that $Q^0\varphi_G^{(M_R,M)} = \mu_e^{(M)} - \mu_e^{(M_R)}$ we find, after transformations, that

$$w_e^{(M,E)} = w_e^0 + Q^0 E, \tag{14.24}$$

where w_e^0 is a constant which depends on the reference electrode and not on the nature of the nonconsumable working electrode M.

It is readily seen, in fact, that when the transition to electrode potentials E has been made, the term $\mu_e^{(M)}$ enters Eq. (14.23), both explicitly and as component part of the term $Q^0\varphi_G^{(M,M_R)}$, and hence cancels.

We thus reach the important conclusion that a metal's electron work function in solutions is independent of the nature of the metal when determined at the same value of electrode potential, i.e., it has identical values for all electrodes.

Electron work functions of metals in solutions can be determined by measurements of the current of electron photoemission into the solution. In an electrochemical system involving a given electrode, this current (I) depends, not only on the light's frequency ν (or quantum energy $h\nu$) but also on potential E. According to the quantum-mechanical theory of photoemission, this dependence is given by

$$I = A[h\nu - w_e^{(M,E)}]^{5/2} = A[h\nu - w_e^0 - Q^0 E]^{5/2} \tag{14.25}$$

(the "law of five halves"). Here A is a constant that depends on light intensity and on the experimental conditions, and the values w_e refer to a single electron rather than to one mole of electrons. In the measurements, light of a certain frequency ν is used and the photoemission currents are determined at different values of potential. Plots of $I^{0.4}$ against E are straight lines which are readily extrapolated graphically to $I = 0$. The potential of this point, the threshold potential, will be written as E_{thr}. At this point $h\nu = w_e^{(M,E)} = w_e^0 + Q^0 E_{\text{thr}}$.

Experience shows that the electron photoemission currents of different metals attain zero values at the same potential (Fig. 14.9), i.e., the values of w_e^0 are identical for all metals. Thus, experience confirms the conclusion mentioned earlier. The value of w_e^0 calculated from photoemission data is 3.10 ± 0.005 eV when the standard hydrogen electrode is used as the reference electrode.

Electron emission can be observed as well in the dark at highly negative electrode potentials when, according to Eq. (14.24), the work function approaches values of zero. The values of potential, of about -2.8 to -3.1 V (SHE), which are required cannot be realized in aqueous solutions owing to strong cathodic hydrogen evolution. However, such potentials can be attained in a number of nonaqueous solvents, e.g., hexamethylphosphoric triamide. In solutions of neutral salt (LiCl) in this solvent, one observes a direct cathodic emission of electrons into the solution, rather than the usual electrochemical reaction. It can be seen from Fig. 14.10 that under these conditions the polarization curves recorded for different metals coincide (curve 1), and this constitutes proof for the conclusion that the electron work functions of these metals have identical values.

Figure 14.10 also shows polarization curves obtained when adding hydrogen chloride HCl to said solution (curves 2 to 4). In this case cathodic hydrogen evolution starts at more positive potentials; the rate of this reaction will of course depend on the electrode material (until the limiting current with respect to the H$^+$ ions is attained). In particular, it follows from these curves that the step of cathodic electron emission is slower (its potential is more negative) than hydrogen evolution and many other reducing reactions. Therefore, the occasional claim that electron emission into the solution yielding solvated electrons is an independent first step in the overall pathway of most cathodic reduction reactions, is seen to be unrealistic. Actually the transition of an electron across the phase boundary which occurs during reactions is intimately linked to simultaneous reduction (change of charge) of the reactant particle.

Since the electron work function $w_e^{(M,E)}$ in solution is independent of the metal, at a given value of potential, it cannot influence the rates of

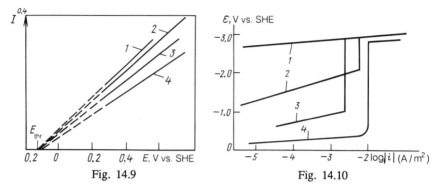

Fig. 14.9
Fig. 14.10

Fig. 14.9. Potential dependence of the electron photoemission currents at different electrodes in acidified Na_2SO_4 solution ($h\nu$ = 3.38 eV): (1) mercury; (2) 3% thallium amalgam; (3) 18% indium amalgam; (4) lead.

Fig. 14.10. Cathodic polarization curves recorded in 0.2 M LiCl solution in hexamethylphosphoric triamide without additives (1) and with the addition of 0.15–0.18 M HCl (2 to 4) at electrodes of: 1) Pt, Cu, Cd; 2) Cu; 3) Pt, and 4) Cd.

electrochemical reactions in a direct way. The correlation between rates and vacuum work functions $w_e^{(M,0)}$ which has been observed in many cases (see Fig. 14.8), can be explained by noting that reaction rates depend on special features of electronic structure of the metals, which in turn influence the work function values.

14.4.2. Bond Energies of Reactants on Electrode Surfaces

Electrochemical reactions conditionally can be divided into two groups: reactions not involving any adsorption of reactants, intermediates, or products of the reaction, and reactions involving adsorption of at least one of these substances. Experience shows that the rate constants of the first (small) group of reactions are practically independent of the electrode metal. A classical example of such reactions is the very simple cathodic reduction of peroxydisulfate ions (14.17). When this reaction is performed at different metals one can see some dependence of reaction rate on the metal involved, but this dependence only arises from differences in the pzc, and hence in the values of potentials ψ_0 and ψ' influencing the reaction kinetics. After correction for the ψ'-potential according to Eq. (14.16), the polarization curves recorded for this reaction at different

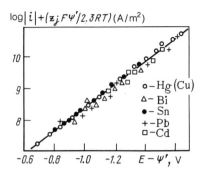

Fig. **14.11.** Polarization curves for the reduction of $S_2O_8^{2-}$ ions in $5 \cdot 10^{-4}$ M $Na_2S_2O_8 + 0.009$ M NaF solution at different metals.

metals merge practically completely (Fig. 14.11), which implies that the rate constants, k_m, of this reaction are independent of the electrode metal.

For the second (and much more populous) group of reactions one notices a strong dependence of reaction rate on the electrode material. This dependence is due primarily to changes in bond energies w_{M-X} between the adsorbed species and the electrode surface. When the reactants are adsorbed, their adsorption (degree of surface coverage) will increase with increasing bond energy, which according to Section 14.3 will lead to higher reaction rates. On the other hand, the same increase in bond energy is responsible for a decrease in the energy effect, $\bar{q}$ or $\bar{w}^0$, of subsequent conversion of the adsorbed reactant. According to Section 14.1, this will result in a higher activation energy of this step [cf. Eq. (14.6)], and the reaction rate will decrease; elimination of the species from the surface is more difficult. As a result of the opposing effects of these two factors, the plots of reaction rate against bond energy usually go through a maximum (Fig. 14.12). Such "bell-shaped' or "volcano"-type relations between the rates of catalytic reactions and bond energies have been described by Aleksei A. Balandin in 1955.

The activation energy will decrease with increasing bond energy when a reaction product (the product of the slow step) is adsorbed. However, increasing adsorption of a product has the effect that the number of surface sites available for the reaction will decrease, i.e., the reaction is made more difficult. As a result, again a curve with maximum is obtained for the plot of reaction rate against bond energy.

Functions of this kind have been studied in detail in the instance

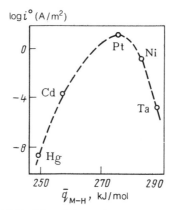

Fig. 14.12. Plot of the exchange current densities of hydrogen evolution and ionization at different metals against the bond energy between the metal and atomic hydrogen.

of cathodic hydrogen evolution, where the adsorption energy of atomic hydrogen as the intermediate has a strong effect on reaction rate.

14.4.3. The Special Features of Reactions at Semiconductor Electrodes

Electrochemical reactions at semiconductor electrodes have a number of special features relative to reactions at metal electrodes; these arise from the electronic structure found in the bulk and at the surface of semiconductors.

The electronic structure of metals is mainly a function, only of their chemical nature. That of semiconductors is also a function of other factors: acceptor- or donor-type impurities present in bulk, the character of surface states (which in turn is largely determined by surface pretreatment), the action of light, etc. Therefore, the electronic structure of semiconductors having a particular chemical composition can vary widely. This is part of the explanation for the appreciable scatter of experimental data obtained by different workers. For reproducible results one must clearly define all factors that may influence the state of the semiconductor.

Depending on the nature of the electrode and reaction, the carriers involved in an electrochemical reaction at a semiconductor electrode can be electrons from the conduction band (in the following to be called simply "electrons"), electrons from the valence band ("holes"), or both. The concentration of the minority carriers in semiconductors (electrons in p-type, and holes in n-type semiconductors) is always much less than that of the majority carriers, let alone the concentration of electrons in metals. Therefore, the specific features of reactions at semiconductor electrodes will be much more pronounced when the minority carriers are involved.

A typical feature of semiconductor electrodes is the space charge present in a relatively thick surface layer (see Section 12.9), which causes a potential drop across this layer, i.e., the appearance of a surface potential χ. This potential drop affects the rate of an electrochemical charge-transfer reaction in exactly the same way as the potential drop across the diffuse edl part (the ψ'-potential), viz., first through a change in carrier concentration in the surface layer, and secondly through a change in the effect of potential on the reaction's activation energy.

As an example, consider a simple anodic redox reaction involving electrons of the valence band, i.e., holes. The reaction equation can be written as

$$\text{Red} + nh^+ \rightarrow \text{Ox}, \qquad (14.26)$$

where h^+ is the symbol for holes.

We shall assume for the sake of simplicity that the total solution concentration is high enough for the influence of the ψ'-potential to be neglected. Other conditions being the same, the reaction rate will be proportional to the surface concentration of holes, c_{S,h^+}. We shall assume here that the relation between surface and bulk concentration of the holes is given by the Boltzmann distribution law (14.14) (an assumption which is not always justified). The activation energy of the reaction is influenced, not by the full Galvani potential φ_G of the interface, but only by the potential difference in the reaction zone between the semiconductor's outer surface and the solution, i.e., the potential difference $\Phi = \varphi_G - \chi$ between the phases. Allowing for these two factors we obtain an expression resembling Eq. (14.16) for the reaction rate:

$$i = nF k_m c_{V,\text{Red}} \exp[(1 - \beta_m)F\chi/RT] \exp(\beta_m FE/RT), \qquad (14.27)$$

where the bulk concentration of holes in the semiconductor, c_{V,h^+}, is contained in the value of rate constant k_m.

The form of the kinetic equation depends on the way in which the surface potential χ varies with electrode potential E. In the first of the cases discussed in Section 12.9, when the surface potential is practically

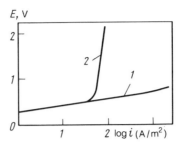

Fig. 14.13. Polarization curves for the anodic dissolution of p-type (1) and n-type (2) germanium in 0.1 M HCl solution.

constant the first factor in Eq. (14.27) will also be constant, and the potential dependence of reaction rate is governed by the second factor alone. The slope b of the polarization curve is $RT/\beta_m F$, i.e., has the same value as that found when the same reaction occurs at a metal electrode. In the second case a change in electrode potential E produces an equally large change in surface potential, i.e., $E \approx \chi + \text{const}$, while there is practically no change in interfacial potential. Then Eq. (14.27) changes into

$$i = nFk_m c_{V,\text{Red}} \exp(EF/RT),\qquad(14.28)$$

and the slope changes to RT/F, i.e., is about twice smaller than in the first case. In general, intermediate values of the slope are possible.

A typical feature of reactions involving the minority carriers are the limiting currents developing when the surface concentration of these carriers has dropped to zero and they must be supplied by slow diffusion from the bulk of the semiconductor. A reaction of this type, which has been studied in detail, is the anodic dissolution of germanium. Holes are involved in the first step of this reaction [Ge → Ge(II)], and electrons in the second [Ge(II) → Ge(IV)]. The overall reaction equation can be written as

$$\text{Ge} + 3\text{H}_2\text{O} + 2h^+ \rightarrow \text{H}_2\text{GeO}_3 + 4\text{H}^+ + 2e^-.\qquad(14.29)$$

It can be seen from Fig. 14.13 that the polarization curve for this reaction involving p-type germanium in 0.1 M HCl is the usual Tafel straight-line plot with a slope of about 0.12 V. For n-type germanium, where the hole concentration is low, the curve looks the same at low current densities. However, at current densities of about 50 A/m^2 we see a strong shift of potential in the positive direction, and a distinct limiting current is attained. Thus, here the first reaction step is inhibited by slow supply of holes to the reaction zone.

Under the effect of illumination, new phenomena arise at semiconductor electrodes, which will be discussed in Chapter 21.

Chapter 15
REACTIONS PRODUCING A NEW PHASE

15.1. INTERMEDIATE STAGES IN THE FORMATION OF NEW PHASES

In applied electrochemistry, reactions are very common in which a new phase is formed, viz., gas evolution, cathodic metal deposition, etc. They have a number of special features relative to reactions in which a new phase is not formed, and in which the products remain part of the electrolyte phase.

The first step in reactions of the type to be considered here is the usual electrochemical step, which produces the primary product that has not yet separated out to form a new phase. In gas evolution, this is the step which produces gas molecules dissolved in the electrolyte (possibly forming a supersaturated solution). In cathodic metal deposition, this is the formation of metal atoms by discharge of the ions; these atoms are in an adsorbed state (so-called adatoms) on the substrate electrode and have not yet become part of a new metal phase. These steps follow the usual laws of electrochemical reactions described in earlier chapters, and are uniformly spread out over all segments of the electrode surface.

These primary electrochemical steps may take place at values of potential below the equilibrium potential of the basic reaction. Thus, in a solution not yet saturated with dissolved hydrogen, hydrogen molecules can form even at potentials more positive than the equilibrium potential of the hydrogen electrode at 1 atm of hydrogen pressure. On account of their energy of chemical interaction with the substrate, metal adatoms can be produced cathodically even at potentials more positive than the equilibrium potential of a given metal/electrolyte system. This process is called the *underpotential deposition of metals.*

Subsequent steps are the formation of nuclei of the new phase and the growth of these nuclei. These steps have two special features.

1. The nuclei and the elements of new phase generated from them (gas bubbles, metal crystallites) are macroscopic entities; their number on the surface is limited, i.e., they emerge, not at all surface sites but only at a limited number of these sites. Hence, the primary products should move (by bulk or surface diffusion) from where they had been produced to where a nucleus appears or grows.

2. The process as a whole is transient; nucleation is predominant initially, and nucleus growth is predominant subsequently. Growth of the nuclei usually continues until they have reached a certain mean size. After some time a quasi-steady state is attained, when the number of nuclei which cease to grow in unit time has become equal to the number of nuclei newly formed in unit time.

Any of the steps listed can be rate-determining: formation of the primary product, its bulk or surface diffusion, nucleation, or nucleus growth. Hence, a large variety of kinetic behavior is typical for reactions producing a new phase.

Two types of reactions producing a new phase can be distinguished: (a) those producing a noncrystalline phase (gas bubbles; liquid drops as, e.g., in the electrolytic deposition of mercury on substrates not forming amalgams), and (b) those producing a crystalline phase (cathodic metal deposition, anodic deposition of oxides or salts having low solubility).

Features common to these two reaction types are the above sequence of steps and, particularly, the step producing nuclei of small size, e.g., in the nanometer range. The excess surface energy (ese) contributes significantly to the energy of these highly disperse entities (with their high surface-to-volume ratio). The thermodynamic properties of highly disperse (extremely small) particles differ from those of larger ones.

When crystal structure is involved it gives rise to special features in the reactions, and makes their mechanisms more complex. Therefore, at first we shall consider the common behavior of reactions producing a new phase in the instance of gas evolution reactions (Section 15.2), then we shall discuss the special features linked to crystal structure (Section 15.3).

15.2. FORMATION OF GAS BUBBLES

15.2.1. Nucleation

Consider the idealized spherical bubble nucleus of gas j with the radius r_{nucl}, the surface area $S_{nucl} = 4\pi r_{nucl}^2$, and the volume $V_{nucl} = (4\pi/3)r_{nucl}^3$. The molar volume of the gas j is V_j. The number of moles, n_{nucl}, of gas in the nucleus will be V_{nucl}/V_j, the number of molecules, N_{nucl}, will be $(V_{nucl}/V_j)N_A$.

Because of the small size of the nucleus, the chemical potential, μ_j^{nucl}, of substance j (the gas) in it will be higher than that (μ_j) in a sufficiently large phase volume of the same substance. Let us calculate this quantity.

At the curved surface of the sphere, a force is acting which is directed toward the center of the sphere and tends to reduce its surface area. Hence, the gas pressure p_{nucl} in the nucleus will be higher than the pressure p_0 in the surrounding medium or in a sufficiently large gas volume existing under the same conditions (with a radius of curvature $r \to \infty$). An infinitely small displacement dr of the surface in the direction of the sphere's center is attended by a surface-area decrease dS ($= 8\pi r\, dr$) and a volume decrease dV ($= 4\pi r^2\, dr$). The work of compression of the nucleus is given by $(p_{nucl} - p_0)dV$. It should be equal to the energy gain, σdS, resulting from surface shrinkage, where σ is the ese of the gas/solution interface. Hence, we find

$$p_{nucl} - p_0 = \sigma(dS/dV) = 2\sigma/r_{nucl} \qquad (15.1)$$

(the *Laplace equation*, 1806). This equation is valid for any curved phase boundary, and also for concave ones (for which $p_{nucl} < p_0$ and the radius of curvature is conventionally regarded as negative). Parameter $p_c \equiv p_{nucl} - p_0$ is called the capillary pressure of this curved surface.

We know from thermodynamics that when the pressure changes at constant temperature we have

$$(d\mu_j/dp)_T = V_j. \qquad (15.2)$$

We shall integrate this equation between limits given by the pressures p_0 and p_{nucl}:

$$\Delta\mu_j^{nucl} \equiv \mu_j^{nucl} - \mu_j = (p_{nucl} - p_0)V_j. \qquad (15.3)$$

Using Eq. (15.1) we finally find

$$\Delta\mu_j^{\text{nucl}} = 2\sigma V_j / r_{\text{nucl}} \qquad (15.4)$$

[the *Thomson (Kelvin) equation*, 1870].

Two conditions must be fulfilled for spontaneous nucleation: (a) the chemical potential of the primary product should be no less than μ_j^{nucl}, and (b) conditions enabling the "encounter" of N_{nucl} particles of the primary product should exist.

The first condition implies that the concentration, c_j^{nucl}, of the primary products in the nucleation zone should be higher than the equilibrium concentration c_j. Allowing for Eq. (3.13), we can define the required degree of supersaturation by the relation

$$\Delta\mu_j^{\text{nucl}} = RT \ln(c_j^{\text{nucl}}/c_j). \qquad (15.5)$$

It follows from Eqs. (15.4) and (15.5) that the required degree of supersaturation will be higher the smaller the size of the nuclei.

When this supersaturation exists the nucleation rate will be proportional to the probability, P_{nucl}, of formation of a favorable configuration of particles of the primary product. According to the Boltzmann law, this probability is determined by the work w_{nucl} of formation of a single nucleus:

$$P_{\text{nucl}} = B \exp(-w_{\text{nucl}}/kT), \qquad (15.6)$$

where B is a normalizing factor and k is the Boltzmann constant.

Let us find the work w_{nucl} of nucleation. At the concentration c_j^{nucl}, the nucleus is in equilibrium with the solution (the chemical potentials in the nucleus and in the supersaturated solution are the same, and particles of j are freely exchanged), but the work of nucleation in this solution is not zero. The total excess surface energy, U_{nucl}, is given by $S_{\text{nucl}}\sigma$ or $4\pi r_{\text{nucl}}^2\sigma$. Nucleation from a solution not supersaturated would require exactly the same work. But the work is less in the case of supersaturated solution, since formation of this solution was already attended by expenditure of a certain amount of work U_{ss}. Since the nucleus contains n_{nucl} moles of substance j, this work in accordance with what was said above will be determined by the expression

$$U_{\text{ss}} = n_{\text{nucl}}\Delta\mu_j^{\text{nucl}} = (4\pi r_{\text{nucl}}^3/3V_j)(2\sigma V_j/r_{\text{nucl}}) = (2/3)S_{\text{nucl}}\sigma. \qquad (15.7)$$

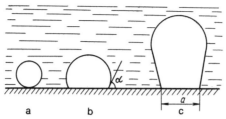

Fig. 15.1. Gas-bubble nuclei on an electrode with complete (a) and incomplete (b) wetting of the surface by the liquid, and a gas bubble at the moment of tearing away (c).

Thus, for the work of nucleation in supersaturated solution we obtain $w_{nucl} = U_{nucl} - U_{ss} = (1/3)S_{nucl}\sigma$ or, when replacing S_{nucl} (r_{nucl}) according to Eq. (15.4),

$$w_{nucl} = (16\pi/3)\sigma^3 V_j^2 / (\Delta\mu_j^{nucl})^2. \tag{15.8}$$

The smaller the nucleus (or higher the degree of supersaturation $\Delta\mu_j^{nucl}$) the smaller will be work w_{nucl} and the larger will be the probability of nucleation.

The above calculation is valid for a spherical nucleus forming in bulk solution or on an electrode surface completely wetted by the liquid electrolyte, where the wetting angle $\alpha \approx 0$ (Fig. 15.1a). The work of nucleation decreases markedly when wetting is incomplete (Fig. 15.1b), since the electrode/electrolyte contact area is smaller. The work also decreases when asperities, microcracks, and the like are present on the surface. Thus, Eq. (15.8) states merely the highest possible value of work w_{nucl}.

In an electrochemical system, gas supersaturation of the solution layer next to the electrode will produce a shift of equilibrium potential (as in diffusional concentration polarization). In the cathodic evolution of hydrogen (which is a reducing agent) the shift is in the negative direction, in the anodic evolution of chlorine (which is an oxidizing agent) it is in the positive direction. When this step is rate-determining and other causes of polarization do not exist, the value of electrode polarization will be related to solution supersaturation by

$$\pm\Delta E = \Delta\mu_j^{nucl}/nF = (RT/nF)\ln[c_j^{nucl}/c_j]. \tag{15.9}$$

With Eqs. (15.6) and (15.8) for the reaction rate and Eq. (15.9) for polarization, we obtain the following general form of polarization equation:

$$|i| = A \exp[-\gamma/(\Delta E)^2], \tag{15.10}$$

where A and γ are constants.

Thus, when plotted as $\ln |i|$ against $(\Delta E)^{-2}$ the experimental data should fall onto a straight line. Such a function is actually observed in a number of cases.

15.2.2. Nucleus Growth

After nucleation the degree of supersaturation of the solution in the immediate vicinity of the nucleus has fallen, and other nuclei can form only some distance away from the first nucleus. It follows that nucleus growth will occur (at least initially), not by the fusion of neighboring nuclei but by the direct addition of primary-product particles. For non-crystalline nuclei (bubbles or drops) no difficulties other than diffusional transport of particles to the nucleus are present at this stage. It is merely necessary that the chemical potential of these particles (or degree of supersaturation) be not inferior to the chemical potential in the nucleus itself, at the size attained. The requirements as to the needed degree of solution supersaturation diminish as the nucleus grows larger.

Another question that arises is the limiting size of the gas bubbles. As the bubble volume V_b increases the buoyancy force $V_b g \Delta \rho$ of the bubble increases (g is the acceleration of gravity, and $\Delta \rho$ is the density difference between the liquid and the gas). The bubble will tear away from the electrode surface as soon as this buoyancy force becomes larger than the force f_{ret} retaining the bubbles.

The retaining force depends on the "neck" perimeter πa along which the bubble is anchored on the surface (Fig. 15.1c), and on the wetting angle α; it can be formulated as $\pi a \sigma \sin \alpha$. It follows when the surface is readily wetted (α is small) that the retaining force, and hence the volume of the bubble tearing away, is considerably smaller than when the surface is poorly wetted. Figure 15.2 shows the relation between the wetting angle and the final bubble volume which was calculated and confirmed experimentally (Boris N. Kabanov, 1939).

The electrode's wetting angle depends on potential; it is largest at the pzc when $\sigma^{(S,L)}$ is largest [cf. Eq. (12.13)], and decreases with increasing distance from this point. This effect is the origin of a characteristic feature of hydrogen and oxygen evolution at nickel electrodes in the electrolysis of alkaline solutions. In these solutions, oxygen evolution occurs at potentials

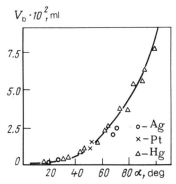

Fig. 15.2. Volumes of the departing gas bubbles as a function of wetting angle (the solid line is the calculated function).

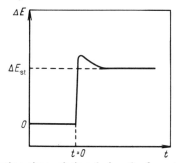

Fig. 15.3. Polarization–time relation during the formation of a new phase.

close to the pzc of nickel, hence, the oxygen bubbles are quite large. The potential of hydrogen evolution is far from the pzc, and the gas is evolved in the form of very fine bubbles forming a "milky cloud." This phenomenon provides the basis for technical degreasing of metal surfaces by strong cathodic or anodic polarization. The wetting of the surface by the aqueous solution increases with increasing distance from the pzc, the force with which oil droplets stick to the surface decreases, and they are carried away.

When a gas bubble has torn away, usually the small nucleus of a new bubble is left behind in its place. Therefore, in gas evolution an appreciable supersaturation is needed only for creating an initial set of nuclei, and subsequent processes require less supersaturation. Hence, in a galvanostatic transient the electrode's polarization will initially be higher, but will then fall to a lower, steady-state value (Fig. 15.3). Such a time dependence of polarization is typical for many processes involving formation of a new phase.

15.3. CRYSTAL PHASE FORMATION (METAL DEPOSITION)

15.3.1. The Initial Stage

Mercury electrodes can be used to study the kinetics of the initial step of cathodic metal ion discharge without complications due to subsequent steps. Here, the primary reaction product, metal atoms, do not form nuclei or crystallites but continue to exist as an amalgam or solution in mercury. We must remember however, that even the kinetics of the initial step depends on the electrode material; hence, the laws found for mercury cannot be used directly for other metals.

Experience shows that in the deposition of a number of metals (mercury, silver, lead, cadmium, and others), the rate of the initial reaction is high, and the associated polarization is low (not over 20 mV). For other metals (particularly of the iron group) high values of polarization are found. The strong inhibition of cathodic metal deposition which is found in the presence of a number of organic substances and which was described in Section 14.3 is observed also at mercury electrodes, i.e., it is associated precisely with the initial step of the process.

15.3.2. Nucleation

The formation of solid crystalline nuclei on a foreign substrate is basically subject to the same laws as the formation of noncrystalline nuclei. Specific features found in the case of crystalline nuclei are: (a) considerably higher ese values; (b) a faceted, rather than spherical, shape; (c) anisotropic properties, i.e., different ese values σ_i for different crystal faces i.

For any ideal crystal, a point can be found for which the distances h_i from these faces obey the relation

$$\sigma_1/h_1 = \sigma_2/h_2 = \cdots = \sigma_i/h_i = \text{const.} \qquad (15.11)$$

Hence, condition (15.4) for crystalline nuclei can be written as

$$\Delta\mu_j^{\text{nucl}} = 2\sigma_i V_j/h_i. \qquad (15.12)$$

In metal deposition, the primary products form adsorbates on the electrode surface, rather than a supersaturated solution. Their excess chemical potential $\Delta\mu_j^{\text{nucl}}$ is directly related to polarization and given by $nF\Delta E$.

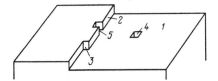

Fig. 15.4. Structure of a growing crystal face.

The total excess surface energy $U_{nucl} = \sum S_i \sigma_i$. Otherwise all the results described above remain valid.

15.3.3. Nucleus Growth

The basic difference between reactions producing a new crystalline phase and reactions producing a gas or liquid phase resides in the step of nucleus growth. Difficulties exist in the incorporation of primary reaction products into the lattice.

In metal deposition the nuclei practically grow by direct discharge of metal ions on them. Initially, elevated current densities are observed at the nuclei (which have a small area). Under these conditions concentration polarization can exist owing to slow metal ion supply.

The structure of growing crystal faces is inhomogeneous (Fig. 15.4). In addition to the lattice planes 1, it features steps 2 of a growing new two-dimensional metal layer (of atomic thickness), as well as kinks 3 formed by the one-dimensional row of metal atoms growing along the step. Lattice-plane holes 4 and edge vacancies 5 can develop when uniform nucleus growth is disrupted.

Metal ions are most readily discharged at the lattice plane (in positions 1). Discharge is hindered near steps 2 and kinks 3, since hydrated metal ions have difficulties in approaching the discharge site. On the other hand, the metal atoms are most readily built into the lattice in positions 3, where they interact with a larger number of neighbors than in positions 1. Hence, nucleus growth is associated with surface diffusion of adatoms (completely discharged) or adions (partly discharged) from positions of the type 1 to positions of the type 3 where they are added to the lattice. In this way, the rows along individual layer steps and the growing layers themselves are completed in succession.

After complete formation of each successive monolayer of atoms, the next layer should start to form. This requires two-dimensional nucleation by the union of several adatoms in a position 1. Like three-dimensional nucleation, two-dimensional nucleation requires some excess energy, i.e.,

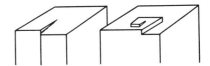

Fig. 15.5. Face growing on a lattice with a screw dislocation.

elevated electrode polarization. Introducing the concept of excess linear energy ρ of the one-dimensional face (of length L) of the nucleus, we can derive an expression for the work of formation of such a nucleus (in a way analogous to that used in Section 15.2.2). When the step of two-dimensional nucleation is rate-determining the polarization equation becomes, instead of (15.10),

$$|i| = A\exp[-\gamma/|\Delta E|]. \qquad (15.13)$$

When the growing crystal surface is nonideal and has defects, face growth can continue even without periodic two-dimensional nucleation. A crystal having the common, so-called screw dislocations in its lattice (Fig. 15.5) can serve as an example. A V-shaped step along which the atoms after discharge can be built into the lattice is formed on the surface of such a crystallite. The growing step continuously circles the face on top of the crystallite, and does not vanish as the face grows. Therefore, this spiral growth is a continuous process obviating intermediate two-dimensional nucleation. There are cases where in the presence of screw dislocations the growing faces are not monatomic but have appreciable thickness. For this reason spiral growth of crystallite faces can be followed visually (Rostislav Kaishev, 1954).

The polarization equations of the type of (15.10) or (15.13) contain the mean values of true current density. However, the rate-determining step is more often concentrated at just a few segments of the electrode; the true working area continuously changes and an exact determination of this area is practically impossible. This gives rise to difficulties in an interpretation of polarization data.

Part 4
SCIENTIFIC PRINCIPLES OF
APPLIED ELECTROCHEMISTRY

Chapter 16
MAIN AREAS OF APPLIED ELECTROCHEMISTRY

At all stages of the development of electrochemistry, an intimate connection existed between the development of theoretical concepts and the discovery of solutions for a practical application of electrochemical processes and phenomena.

In 1786 the Italian physiologist Luigi Galvani demonstrated in his remarkable experiments, that muscle contraction similar to that produced by discharge of a Leyden jar will occur when two different metals touch the exposed nerve in a frog. The phenomenon was correctly interpreted in 1794 by the Italian physicist Alessandro Volta, who showed that this "galvanic" effect originates from the contact established between the two metals and from the contact between these metals and the muscle tissue. In March 1800, Volta reported a device designed on the basis of the same phenomenon, which could produce "inexhaustible electric charge." Now known as the Volta pile, this was the first example of a practical electrochemical device, viz., a chemical power source.

The Volta pile was of extraordinary significance for developments in the science of electricity, since a continuous electric current, hitherto not known, could now be realized. Soon various properties and effects of the electric current were discovered, including many electrochemical processes. Already in May of the same year, 1800, William Nicholson and Sir Anthony Carlisle electrolyzed water producing hydrogen and oxygen. In 1803 processes of metal electrodeposition were discovered. In 1807 Sir Humphry Davy first isolated alkali metals by electrolyzing salt melts. All these processes constituted stimuli in the development of first theoretical notions concerning the mechanism of charge transport in solutions and melts (Theodor Freiherr von Grotthuss, Michael Faraday, and others).

The development of a variety of different chemical power sources taking place during the second half of the 19th century led to basic work in electrochemical thermodynamics (Josiah Willard Gibbs, Walther Nernst, Wilhelm Ostwald, and others). Likewise, electrochemical kinetic studies performed during the first half of the 20th century were intimately related to the advance of various electrochemical industries. The desire to further speed up the processes occurring in electrolyzers and power sources was responsible for the emergence of two new areas of theoretical electrochemistry in the years after 1950, viz., electrochemical macrokinetics (Section 17.3) and electrocatalysis (Section 19.6).

Thus, at all stages of the development of electrochemistry, theoretical investigations have been stimulated by the practical use of the different electrochemical phenomena and processes, and the theoretical concepts which were developed have in turn contributed significantly to the development of applied electrochemistry.

Today applied electrochemistry is of great value for the economy. Electrolytic processes are used to manufacture a number of valuable materials, chemical power sources (batteries) are widely used in many technical fields, and electrochemical methods are used to protect metals against corrosion, a phenomenon causing huge economic losses. It will suffice to mention that the world's electrochemical industries consume about 10% of all electrical energy generated today.

The various aspects of applied electrochemistry have been discussed in detail in numerous textbooks and monographs, both general and specialized. The present chapter only provides a brief outline characterizing the main areas of applied electrochemistry together with their economic and scientific value. Subsequent chapters will outline the particulars of certain electrochemical reactions and phenomena which are of interest for laboratory or manufacturing practice or for other fields of science.

Chemical power sources (batteries). During the first half of the 19th century batteries were the only source of electric current. In the 1860s a new source of electric power, the electric generator, was developed. However, even after this invention the batteries retained their value as autonomous power sources for portable devices and conveyances. Their importance soared early in the 20th century with the development of radios and the automobile, and again after 1950 with the development of rocketry, astronautics, and microelectronics. Worldwide today about 10 billion batteries of different types are produced every year, and are used in almost all areas of the economy. If all the batteries installed were turned on at once this would give an electric power comparable with the total of all power stations in the world (about 10^9 kW). The miniature cells in wrist watches work with a mean power of 10^{-5} W, while stationary storage batteries have up to 10^7 W. The mass of single chemical "power plants" can range from fractions of a gram to hundreds of tons.

A battery's source of electrical energy is the Gibbs energy of the current-generating electrochemical reaction. Among the chemical power sources we distinguish primary cells, storage batteries (secondary cells), and fuel cells. In the first two types reacting electrodes are used. Primary cells cease to be operable after their complete discharge (throw-away batteries), while secondary cells (storage batteries) can be recharged with current from an extraneous source sent through in the opposite direction. In fuel cells, which have nonconsumable electrodes, fresh amounts of the gaseous or liquid reactants are continuously supplied during discharge, and the reaction products are withdrawn, so that these devices can be discharged continuously for long periods of time.

In primary and storage batteries, metals are used as the active material of the negative electrodes: zinc, cadmium, iron, lead, etc., and solid compounds such as the oxides of manganese, lead, and nickel, copper chloride, etc. are used as the active material of the positive electrodes. Concentrated aqueous solutions of acids, alkalies, or salts, sometimes nonaqueous solutions, melts, or solid electrolytes are used as the electrolytes.

An example of primary cells is the popular zinc–manganese dioxide "dry" (or Leclanché) cell, which has an electrolyte on the basis of ammonium chloride. During discharge the (negative) zinc electrode is oxidized forming zinc hydroxide or complex zinc ions, while the manganese dioxide contained in the positive electrode is partly reduced. In a simplified way, the overall current-generating reaction can be written as

$$2MnO_2 + Zn + 2H_2O \rightleftarrows 2MnOOH + Zn(OH)_2. \quad (16.1)$$

A typical example for secondary batteries is the lead–acid battery. In the charged condition, the negatives in this storage battery contain spongy lead, and the positives contain lead dioxide PbO_2; the electrolyte is a sulfuric acid solution. At the electrodes and in the storage battery as a whole, the following reactions occur (from left to right during discharge, from right to left during charging):

$$(+) \ PbO_2 + 3H^+ + HSO_4^- + 2e^- \rightleftarrows PbSO_4 + 2H_2O, \qquad (16.2)$$

$$(-) \ Pb + HSO_4^- \rightleftarrows PbSO_4 + H^+ + 2e^-, \qquad (16.3)$$

$$(\text{battery}) \ PbO_2 + Pb + 2H_2SO_4 \rightleftarrows 2PbSO_4 + 2H_2O. \qquad (16.4)$$

In fuel cells, oxygen (pure or from the air) is the active material of the positive electrode, while hydrogen, and less frequently hydrazine, methanol, etc. are active materials of the negative electrode. The electrolytes used are concentrated solutions of alkalies or acids, also solid electrolytes and melts. The current-producing reaction in hydrogen–oxygen fuel cells is the electrochemical formation of water from hydrogen and oxygen:

$$2H_2 + O_2 \rightleftarrows 2H_2O. \qquad (16.5)$$

Fuel cells are of great interest, since they can be used to directly convert the chemical energy of natural fuels to electrical energy while bypassing the intermediate production of thermal energy and the thermodynamic limitations inherent in the further conversion of thermal to electric energy. Experimental fuel cell power plants of up to 5 MW have already been built, where the fuel used is hydrogen produced by coal gasification with steam (shift reaction) or by the steam reforming of natural gas or petroleum fractions. The overall efficiency of such plants is expected in the future to rise above that of current thermal power plants.

The characteristic of main interest in batteries is their discharge curve, i.e., the voltage $\mathcal{E}$ as a function of charge Q extracted or, when discharge occurs at constant current, as a function of time. The analogous charging curve is an additional characteristic of interest in the case of storage batteries. Typical charging and discharge curves of a lead–acid battery are shown in Fig. 16.1. The voltage decreases as discharge advances (the cell's total overvoltage increases). Discharge is continued to a defined, final voltage $\mathcal{E}_f$. The total amount of charge, Q_0, that can be extracted until this voltage is attained is called the discharge capacity of the particular battery. The product of capacity and average discharge voltage is the energy stored in the battery. Important performance figures of batteries are: the specific energy per unit mass or volume, the maximum specific power, the shelf life (for primary cells), the cycle life, which is the attainable number of charge–discharge cycles, and finally the energy efficiency, which is the ratio between the energy extracted on discharge

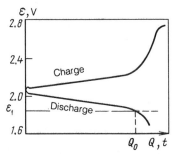

Fig. 16.1. Charging (1) and discharge (2) curves of a lead–acid battery.

and the energy invested for charging (in the case of storage batteries), the service life, the operational temperature range, the mechanical strength, the resistance to spilling of the electrolyte, and the like.

Electrolysis with nonconsumable electrodes. Among electrolytic processes used to produce materials, we customarily distinguish the ones in which electrodes are reacting, i.e., where the metal or other electrode material is involved in the reaction, from those with nonconsumable electrodes ("metalless electrolysis").

A very important industrial process with nonconsumable electrodes is the electrolysis of sodium chloride solution (brine) producing chlorine at the anode and sodium hydroxide NaOH (caustic soda) in the catholyte via the overall reaction

$$2Cl^- + 2H_2O \rightleftarrows Cl_2 + H_2 + 2OH^-; \qquad (16.6)$$

the third reaction product, hydrogen, is usually not utilized in chlor-alkali electrolysis.

Current annual world production of chlorine by electrolysis is over 30 million tons, that of alkali is 35 million tons, and it increases by 2 to 3% per year. This industry consumes about 100 billion kWh of electrical energy per year.

Three processes exist for chlor-alkali electrolysis: (a) the mercury-cell process, (b) the diaphragm-cell process, and (c) the membrane-cell process. The first two were developed in the 1880s, the third in the 1960s. They have in common that chlorine is electrolytically evolved at the anode by discharge of chloride ions. In the mercury-cell process, a mercury cathode is used at which, owing to the strong polarization of hydrogen evolution, sodium ions are discharged and sodium amalgam is formed. When the amalgam has attained a certain concentration it is pumped to another cell and decomposed with water:

$$2Na(Hg) + 2H_2O \rightarrow 2NaOH + H_2 + (Hg). \qquad (16.7)$$

This yields a solution of highly pure alkali (free of chloride ions), which can be used in the manufacture of synthetic fibers. The mercury which has been stripped of sodium is returned to the electrolyzer. The cost of chlorine is higher in the mercury-cell than in the diaphragm-cell process. In addition, the mercury-cell process is ecologically dangerous owing to the possible escape of mercury into the environment, hence, it has been increasingly discontinued in all countries. In the diaphragm-cell process a solid cathode (iron) is used where hydrogen is evolved [reaction (3.60b)]. Porous asbestos diaphragms are used to prevent mixing of the catholyte and anolyte, but owing to the finite permeability of these diaphragms, the alkaline solution which is produced near the cathode still contains important levels of chloride ions as an impurity. In the membrane-cell process, highly selective ion-exchange membranes are used which allow only the sodium ions to pass. Unfortunately, such membranes are still expensive.

For a long time synthetic graphite has been used as the anode material. However, the stability of graphite anodes is not very high; they undergo gradual surface oxidation and also erosion, and thus become thinner. Hence, the gap between the electrodes, and with it the ohmic voltage drop in the electrolyte, increases during operation. The graphite anodes have a service life of 6 to 24 months. The development of new electrode materials on the basis of RuO_2 and TiO_2 mixed oxides at the end of the 1950s was a significant advance. The titanium–ruthenium oxide anodes exhibit low polarization for chlorine evolution. Moreover, their chemical stability is high, so that sometimes they are also described as low-wear or "dimensionally stable" anodes (DSA^R).

The theoretical ocv of cells consisting of a hydrogen electrode (in alkaline solution) and a chlorine electrode is 2.17 V. Practical cell voltages of most modern electrolyzers are 3.3 to 3.8 V at a current density of 1 kA/m^2. With current yields of 95% for the anodic process, this corresponds to an energy consumption of 2600 to 3000 kWh per ton of chlorine.

When chlor-alkali electrolysis is conducted in an undivided cell with mild-steel cathode, the chlorine generated anodically will react with the alkali produced cathodically, and a solution of sodium hypochlorite NaClO is formed. Hypochlorite ions are readily oxidized at the anode to chlorate ions; this is the basis for electrolytic chlorate production. Perchlorates can also be obtained electrochemically.

Electrolytic hydrogen production [by the reverse reaction of (16.5)] is economically inferior to the large-scale chemical process based on the steam reforming of natural gas (methane). However, it is employed successfully in those cases when modest quantities of hydrogen are needed locally or when highly pure hydrogen is required. An appreciable increase in electrolytic hydrogen production is expected in the future as natural-gas resources diminish and the production of cheap electrical power in nuclear power plants increases. Water electrolyzers use alkaline electrolyte solution, which has the major advantage that iron resists corrosion in it, i.e., the electrodes and structural parts can be made from iron (steel). The theoretical ocv of the hydrogen–oxygen cell at 25°C is 1.229 V. Electrolysis is conducted with a cell voltage of 1.8 to 2.1 V at current densities of 2 to 4 kA/m^2. Improved designs where the working voltage is 1.6 to 1.7 V have been developed recently. The electrical energy consumption is 3.5 to 5 kWh per m^3 of hydrogen.

A process involving water electrolysis is the production of heavy water. During cathodic polarization the relative rates of deuterium discharge and evolution are lower than those of the normal hydrogen isotope. Hence, during electrolysis the solution is enriched in heavy water. When the process is performed repeatedly, water with a D_2O content of up to 99.7% can be produced.

Electrochemical methods are also widely used in manufacturing a variety of other inorganic and organic substances (Sections 18.4 and 18.5).

Electrometallurgy. In the electrolytic production of metals, both aqueous solutions (electrohydrometallurgy) and melts are used. Solutions involving nonaqueous solvents have also found uses in recent years. We distinguish electroextraction, which is the primary production of a metal from the materials obtained when processing and leaching the original ores, and electrorefining, which is the purification of metals by their anodic dissolution and subsequent cathodic redeposition. Zinc, cadmium, and manganese are the most important metals obtained by electroextraction from aqueous solutions; this route is also used to produce copper from poor oxide ores. Electrolysis in melts is used to produce aluminum and a number of the alkali and alkaline-earth metals (lithium, sodium, magnesium, calcium, and others), which cannot be produced from aqueous solutions, because they are unstable in water. Refining is widely used for the production of more highly pure copper, gold, nickel, lead, and other metals.

Electrolytic aluminum production is the most important process, both in volume and significance. World production is about 15 million tons per year, consuming about 240 billion kWh of electrical energy. Aluminum oxide (alumina) Al_2O_3 is subjected to electrolysis at a temperature of 950°C; to this end it is dissolved in molten cryolite Na_3AlF_6, with which it forms a eutectic melting at about 940°C. Carbon anodes which in the process are anodically oxidized to CO_2 are employed. The overall electrolysis reaction can be written as

$$2Al_2O_3 + 3C \rightarrow 4Al + 3CO_2. \tag{16.8}$$

The aluminum produced has a higher density than the melt and collects at the bottom of the cell. The anodes are replaced as they are consumed. The theoretical ocv for reaction (16.8) is about 1.2 V, and the theoretical consumption of electrical energy (assuming a 100% current yield) is 3.6 kWh/kg. However, the actual cell voltage is 4.2 to 4.5 V, and the current yields are 85 to 90%, which implies a practical energy consumption of 14 to 16 kWh/kg. The rather high voltage is required, because the side reaction of aluminum with CO_2 must be suppressed; this is achieved with a large gap between the electrodes (30 to 50 mm), which in turn implies ohmic voltage drops of more than 2 V in the electrolyte. The heat produced owing to these internal voltage losses serves to maintain the required working temperature of the cell. About 20% of the cost of aluminum metal are due to electrical energy consumption, and 45% are due to the production of sufficiently pure alumina as a starting material.

Electroplating (coating with metals). The electrolytic application of thin metal coatings to other metals evolved during the middle of the last century. Today it is one of the highest-volume chemical technologies; literally every metal-working plant has its own plating section.

Among the coatings we distinguish functional ones from those applied for protection and ornament. The latter have the chief purpose of protecting a base metal against corrosive and erosive attack by the medium surrounding it, and of giving its surface a particular appearance, such as luster, color, etc. Often coatings of nickel, chromium, and zinc are used here. Functional coatings are used for a variety of purposes: to produce reflecting surfaces, conducting paths (in printed circuits), magnetic layers, surfaces with specific friction parameters (sliding bearings), etc. Metal deposition is also used to join parts (electrochemical welding or brazing) and to restore the surfaces of worn parts.

During recent decades the demands as to the quality and properties of metal coatings have sharply increased. This is due, on one hand to the advances in microelectronics, and on the other hand to increasing uses of metal parts in corrosive environments.

Metal corrosion. Corrosive attack of metals is an electrochemical process causing huge losses to the economy. Corrosion (rusting) affects the metal structures of buildings and bridges, the equipment of chemical and metallurgical plants, river and sea vessels, underground pipelines and other structures. In the United States, for instance, corrosion-related losses approach a figure of 100 billion dollars per year, which is almost 5% of the gross national product. Direct losses attributable to corrosion include the expenditures for the replacement of individual parts, units, entire lines, or plants and for the various preventive and protective tasks (such as the application of coatings for corrosion protection). Indirect losses arise when corroded equipment leads to defective products which must be rejected, they also arise during downtime required for preventive maintenance or repair of equipment etc. About 30% of all steel and cast iron are lost because of corrosion. Part of this metal can be reprocessed as scrap, but about 10% are irrevocably lost.

The significance of corrosion protection has risen sharply in recent years for a number of reasons: (a) because of efforts to reduce the metal content of parts, e.g., by using thinner metallic support structures, (b) with the use of new types of equipment and processes involving expensive equipment operated under extreme conditions, such as nuclear reactors, jet and rocket engines, etc., (c) in connection with the development of products having extremely thin metal films, such as printed circuit boards, integrated circuits, etc.

Corrosion science has now developed into an independent branch of electrochemistry, having intimate connections with other fields of science, and particularly physical metallurgy. Its chief concern are the origin and mechanisms of the different forms of corrosive attack and the development of efficient ways to fight corrosion (Section 18.5).

Electrochemical methods of analysis. High accuracy and speed have been responsible for the popularity of electrochemical methods for quantitative and qualitative chemical analysis (Chapter 20). Electrochemical sensors of different types are used for the continuous or batch analysis of a number of substances in gaseous and liquid environments (e.g., for determining the concentration of carbon monoxide in air, of oxygen in biological fluids, etc.).

Electrochemical transducers. During the middle of this century a new field of applied electrochemistry emerged, where the objective was that of building functional devices converting nonelectrical signals (e.g., mechanical stimuli) into electrical signals or interconverting electrical signals of different types. These devices include electrochemically controlled resistors, discrete or continuous integrators, coulometers, pressure transducers, linear and angular accelerometers, seismic pickups, and many others. In a number of cases electrochemical transducers perform the same functions as semiconductor devices (diodes, transistors). Unlike semiconductor devices, the electrochemical devices are quite insensitive to high-frequency signals but, on the other hand, will function at low (below 20 Hz) and ultralow frequencies. This field is sometimes called chemotronics or molecular electronics.

Nonfaradaic processes. There is a large group of commercial processes where electrochemical phenomena other than reactions obeying the laws of Faraday are used. In many processes, so-called electrokinetic phenomena are used, i.e., the motion of a liquid relative to a solid wall which occurs when a tangential electric field is applied (Chapter 22). In recent decades various nonfaradaic processes of this type have found very wide commercial applications, for instance, in the electrophoretic painting of car bodies.

Photoelectrochemistry deals with the effects of electromagnetic radiation (e.g., visible or laser light) on electrochemical parameters and processes, and in particular on the open-circuit potentials of electrodes and on the nature and rate of the reactions occurring at the electrodes (Chapter 21). During the last decade there has been a drastic rise in the volume of work concerned with semiconductor photoelectrochemistry in connection with efforts to build electrochemical devices for the direct conversion of solar to electrical energy.

Bioelectrochemistry. Various electrochemical phenomena are of exceptional importance for essential biological processes such as the transport of ions across cell membranes, energy conversion at the molecular level, the transmission of nervous impulses, the perception of images in the eye, and many others (Chapter 23). Bioelectrochemistry is one of the youngest, but in the future undoubtedly one of the most important branches of electrochemistry.

Chapter 17
ELECTROCHEMICAL REACTORS

17.1. DESIGN PRINCIPLES

Electrochemical reactors (cells, tanks) are used for the practical realization of electrolysis or the electrochemical generation of electrical energy. In developing such reactors one must take into account the purpose of the reactor as well as the special features of the reactions employed in it.

Most common is the classical reactor type with plane-parallel electrodes in which positive and negative electrodes alternate and all electrodes having the same polarity are connected in parallel. Reactors in which the electrodes are concentric cylinders and convection of the liquid electrolyte can be realized by rotation of one of the electrodes are less common. In chemical power sources, occasionally the electrodes are in the form of two long ribbons with a separator in between, which are wound up as a double spiral.

The maximum values of electric power and unit output of electrochemical cells vary within wide limits. The total current load admitted by individual electrolyzers for the electrochemical production of various materials in plant or pilot installations (their capacity) is between 10 A and 200 kA, while the current loads that can be sustained by different types of battery (their current ratings) are between 10^{-5} A and 20 kA. Corresponding differences exist in the linear dimensions of the electrodes (between 5 mm and 3 m) as well as in the overall mass and size of the reactors.

In practice, groups (batteries) comprising a certain number of cells connected in series, and sometimes also in parallel, are usually used instead of single cells. The number of cells in series is determining for the

409

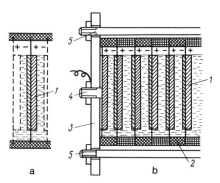

Fig. 17.1. (a) Bipolar electrode; (b) filter-press reactor. (1) Bipolar electrode; (2) gaskets; (3) end plate; (4) positive current collector; (5) tie bolts.

voltage. Large electrolysis plants commonly contain as many as 100 or more individual electrolysis cells, and power is supplied from rectifying equipment producing voltages of up to 400 V. Chemical power sources are often built in the form of batteries composed of individual primary or secondary cells so as to yield battery voltages of 9, 12, or 24 V and more.

For a series combination of the individual cells to groups, the positive lead from each cell is linked with an intercell bus to the negative lead of the following cell.

Another version of cell or group arrangement is that using *bipolar electrodes* (Fig. 17.1a). Here, combined electrodes are used: one side of the electrode functions as the positive electrode of one cell, the other side functions as the negative electrode of the next cell. The bipolar electrodes alternate with electrolyte compartments, and both must be carefully sealed along the periphery to prevent outflow of electrolyte, and provide a reliable separation of the electrolyte in neighboring compartments by the electrode plates. The separating plates also function as cell walls and intercell connectors, i.e., the current between neighboring cells merely crosses a thin wall having negligible resistance. This implies considerable savings in the size and mass of the reactor.

Several possibilities exist for the assembly of reactors with bipolar electrodes. In the sections of a filter-press design (Fig. 17.1b), the blocks of electrodes and separators forming electrolyte compartments are tightened with the aid of end plates and tie bolts. Sealing is realized with the aid of gaskets, which are compressed during tightening of the assembly. After sealing, the compartments are filled with electrolyte via special narrow channels in the gaskets or electrode edges, which also enable circulation of the electrolyte. In batteries one often finds a modular design where individual cells or groups of cells are sealed by a plastic lining.

Interelectrode gap. The relative electrolyte volume available per unit surface area of the electrodes is determined by the distance (gap) between the electrodes. This distance is between fractions of a millimeter and some 10 cm. The ohmic losses in the electrolyte increase with the distance between the electrodes. On the other hand, when the electrolyte volume is too small the reactant concentrations will rapidly change. Often the electrolyte volume in a reactor is increased by providing space for the electrolyte, not only between the electrodes, but also above or below the block of electrodes. Sometimes the electrolyte is pumped around in an external circuit including an additional electrolyte vessel.

Separators. Separators made of insulating material are almost always arranged between electrodes of different polarity. They have a variety of functions, ranging from a mechanical separation of the electrodes and prevention of accidental contact to the strictest separation of anolyte and catholyte (the electrolyte volumes at the anode and cathode) that is possible without upsetting the ionic conduction. Separators decisively influence reactor efficiency (for more details see Section 17.2).

Supply and withdrawal of reaction components. In the continued operation of a reactor with reacting solid electrodes (as in electrohydrometallurgy), provisions must be made for a periodic replacement of the electrodes. Reactants dissolved in the electrolyte are replenished and dissolved products withdrawn by appropriate adjustments of electrolyte composition, which may be continuous (e.g., when an external loop is operated) or batchwise. Gas-diffusion electrodes are used for the supply of gaseous reactants (Section 17.3). When gas is evolved in the reaction (as in water electrolysis) there is the danger of gas accumulation in the electrolyte, i.e., of large volumes of gas bubbles remaining in the space between the electrodes and producing a strong increase in electrolyte resistance. Various devices are used to prevent this, in particular the so-called gas lift, which is the natural vertical rise of the liquid electrolyte, caused by an ascending flow of the gas-filled (less dense) layer of electrolyte close to the electrode surface; the loop is completed by a descending flow of the liquid from which the gas bubbles have been separated. Copious gas evolution during the electrolysis of liquid solutions is associated with another problem, which is the loss of electrolyte in the form of mist consisting of very small droplets. Therefore, special spray separators are installed from which the liquid carried off with the gas is returned to the cell.

Scale factor. Large-volume reactors have a number of specific features. When the electrodes are of considerable height and the total current is large, ohmic losses will occur within the electrodes, leading to a nonuniform current-density distribution (Section 17.3). Special feeders of relatively large cross section are used to reduce the ohmic losses in electrodes. Nonuniformity of the current distribution can be made even worse when a stratification of the electrolyte solution occurs, i.e., when electrolyte layers of lower density which are depleted by the reaction will gather in the upper part of the reactor, and denser layers will accumulate in the lower part.

With increasing reactor size, the withdrawal of the heat evolved during the reaction becomes more difficult. This gives rise to a nonuniform temperature distribution within the reactor and thus, a differentiation of conditions for the electrochemical reaction in different reactor parts. In a number of cases the thermal conditions can be improved through cooled electrodes.

Selection of corrosion-resistant materials. The concentrated solutions of acids, alkalies, or salts, salt melts, and the like used as electrolytes in reactors as a rule are highly corrosive, particularly so at elevated temperatures. Hence, the design materials, both metallic and nonmetallic, should have a sufficiently high corrosion and chemical resistance. Low-alloy steels are a universal structural material for reactors with alkaline solutions, while for reactors with acidic solutions, high-alloy steels and other expensive materials must be used. Polymers, including the highly stable fluoropolymers such as PTFE, become more and more common as structural materials for reactors. Corrosion problems are of particular importance, of course, when materials for nonconsumable electrodes (and especially anodes) are selected, which must be sufficiently stable and at the same time catalytically active.

Mass- and volume-related figures of merit. All electrochemical reactors should be sufficiently compact in their design. For batteries which are used in mobile applications, the figures of prime importance are the mass or volume of the device per unit of stored energy, which are given in kg/kWh or dm³/kWh, or in the corresponding reciprocals representing the energy densities by mass or volume. For electrolyzers permanently installed in special rooms, the chief parameter is the productivity (e.g., the current load) per unit of floor space, since this is the parameter that determines the volume of the building required.

The productivity of modern electrolyzers per unit volume or unit of floor space as a rule is lower than that of chemical reactors with a similar purpose. This is due to the fact that in an electrochemical reactor the reactions occur only at the electrode surfaces, while in a chemical reactor they can occur in practically the full volume. Therefore, recent efforts go in a direction of designing new, more efficient electrochemical reactors.

With solid (and particularly polymeric) electrolytes which at the same time function as separators, one can appreciably reduce the distance between the electrodes, and hence, increase the electrode area per unit of reactor volume. Very compact equipment for water electrolysis which has no liquid electrolyte has been designed.

An appreciable increase in working area of the electrodes can be attained with porous electrodes (Section 17.3). Such electrodes are widely used in batteries, and in recent years they are also found in electrolyzers. Attempts are made to use particulate electrodes which consist of a rather thick bed of particulate electrode material, into which the auxiliary electrode is immersed together with a separator. Other efforts concern the so-called fluidized-bed reactors, where a finely divided electrode material is distributed over the full electrolyte volume by an ascending liquid or gas flow and continuously collides with special current collector electrodes.

The economic performance figures of electrochemical reactors depend on many factors. The *current yields* are a figure of particular significance. When expensive reactants are used (as, e.g., in the electrosynthesis of a number of organic substances) the *chemical yield* is another significant parameter; this is the ratio between the amount of product actually obtained and that which, according to the reaction equation, can theoretically be obtained from a given amount of reactant consumed. The analogous parameter for batteries is the degree of reactant utilization (the ratio between the capacity available on discharge and that theoretically calculated for the amount of active materials present in the battery).

17.2. SEPARATORS

17.2.1. Purpose, Types

The major functions of separators in electrochemical reactors are: (a) the mechanical separation of electrodes of different polarity and prevention of their contact (e.g., during vibrations) and of electronic conduction (short-circuiting) between them, (b) the prevention of dendrite growth and of the formation of metallic "bridges" between the electrodes during metal electrocrystallization, (c) the restriction of mixing of the anolyte and catholyte and prevention of transfer of solutes, colloidal or suspended particles between the electrodes (such separators are also known as diaphragms), (d) containment of the active material at the electrodes, prevention of its crumbling and shedding. While serving in all these functions the separator should not pose any significant resistance to ionic current flow in the electrolyte, and it should not interfere with the transport of those substances which are involved in the electrode reactions, or with the free access of electrolyte to all electrode segments. In some cases a separator also functions as a matrix material for the electrolyte, i.e., retains by capillary forces the electrolyte close to the electrode surface.

Plastic or ebonite rods, plastic cords 2 to 3 mm thick, separator sheets consisting of thin grids of perforated and corrugated PVC, polymer-fiber or fiber glass cloth, and other materials are used for a geometric separation of the electrodes. Such separators contain large pores or holes (between 0.1 and 5 mm). They cause little screening of the electrodes, but they also completely lack any capacity for electrolyte retention. Another group are the porous separators (diaphragms), which have pore radii between 20 nm and 100 μm. Sometimes these separators are arbitrarily divided into porous (with mean pore radii larger than 5 μm) and microporous (with smaller pore radii). They are used in electrolyzers and batteries. A special group of separators are swollen membranes, i.e., polymeric materials which interact with the aqueous or nonaqueous solutions. Upon entry of solvent molecules into the polymer structure the distance between individual macromolecules increases (the membrane swells), and migration of the solution ions in the membrane becomes possible. The pores in such membranes are not larger than 2 nm.

In porous separators the pore radii are large compared to the size of molecules. Hence, the interaction between the electrolyte and the pore walls has practically no qualitative effects on the ionic current through the separator; the transport numbers of the individual ions have the same values in the pores as in the bulk electrolyte. In swollen membranes the specific interaction between individual ions and macromolecules is very pronounced. Hence, these membranes often exhibit selectivity in the sense that different ions are affected differently in their migration. As a result, the transport numbers of the ions in the membrane differ from those in the electrolyte outside the membrane. In the limiting case, certain types of ion are completely arrested, and the membrane is called permselective (see Chapter 5).

Chemically inert and sufficiently stable materials are used to make the separators, e.g., the microfibrous chrysotile asbestos. Paper-making technology is used to prepare sheets (asbestos cardboard). Sometimes a mass containing asbestos fibers is applied to wire-gauze electrodes. More recently asbestos substitutes are introduced because of health risks in the asbestos industry. In batteries, porous diaphragms made of different synthetic resins (PVC etc.) are used. Cellophane (hydrated cellulose), polyethylene radiation-grafted with polyacrylic acid, and various types of ion-exchange resins are examples of materials for swelling membranes. Recently, a new kind of chemically and thermally highly stable ion-exchange membrane based on perfluorinated sulfonic-acid cation-exchange resins (of the Nafion^R type) has found successful use in chlor-alkali electrolysis and other applications.

17.2.2. Chief Parameters

Influence on electrolyte conductivity. In porous separators the ionic current passes through the liquid electrolyte present in the separator pores. Therefore, the electrolyte's resistance in the pores has to be calculated for known values of porosity of the separator and of conductivity, σ, of the free liquid electrolyte.

Such a calculation is highly complex in the general case. Consider the very simple model where a separator of thickness d has cylindrical pores of radius r which are parallel and completely electrolyte-filled (Fig. 17.2). Let l be the pore length and N the number of pores (all calculations refer to unit surface area of the separator). Ratio $\beta = l/d$ (where $\beta \equiv$ cosec $\alpha \geq 1$) characterizes the tilt of the pores, and is called the tortuosity factor of the pores. The total pore volume is given by $N\pi r^2 l$, the porosity by

$$\xi = N\pi r^2 l/d = N\pi r^2 \beta. \tag{17.1}$$

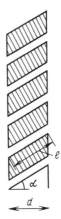

Fig. 17.2. Separator model with cylindrical pores.

The electrolyte resistance in the pores is given by

$$R_p = l/N\sigma\pi r^2. \qquad (17.2)$$

The ratio between R_p and the resistance, $R_E = d/\sigma$, of an electrolyte layer of the same thickness,

$$\varepsilon \equiv R_p/R_E = \beta/N\pi r^2, \qquad (17.3)$$

is the attenuation factor of conduction or coefficient of resistance rise (sometimes called the relative resistance). By definition $\varepsilon \geq 1$. Comparing Eqs. (17.1) and (17.3) we can see that

$$\varepsilon = \beta^2/\xi. \qquad (17.4)$$

Coefficient ε increases with increasing tortuosity and decreasing porosity. It is independent of pore radius when the values of these parameters are constant; the decrease in radius r which occurs while the total porosity is kept unchanged will be compensated by an increase in the number of pores, N.

Coefficient ε is independent of electrolyte conductivity. A quantity useful for practical calculations is the electrolyte's effective conductivity in the separator, which is related to ε as

$$\sigma_{\text{eff}} = \sigma/\varepsilon. \qquad (17.5)$$

Using this quantity one can abandon specific models of separator pore structure and calculate the overall resistance from the separator's overall geometry.

Real porous separators differ from the above model. It was found experimentally that in real porous systems the relation

$$\varepsilon \sim \xi^{-m} \tag{17.6}$$

holds (*Archie's law*), where m has values between 1.8 and 3.5, instead of $m = 1$ for the above model.

Coefficient ε should be low for separators in electrochemical reactors. It has values between 1.1 and 1.6 for simple separators, but for porous diaphragms and swollen membranes it has values between 2 and 10. The total porosity should be at least 50%, and the separator's pore space should be impregnated completely and sufficiently rapidly with the liquid electrolyte.

Diffusion through separators. Like current flow, the diffusion of dissolved components through separators will be delayed by decreasing porosity and increasing tortuosity. The attenuation factor of diffusion, ε_d (= D/D_{eff}), usually coincides with that of conduction.

Filtration of liquids. Depending on the specific electrochemical reactor type, the filtration rate of a liquid electrolyte through the separator should be, either high (in order to secure convective supply of substances) or very low (in order to prevent mixing of the anolyte and catholyte). The filtration rate which is attained under the effect of an external force Δp depends on porosity. For a separator model with cylindrical pores, the volume filtration rate can be calculated by Poiseuille's law:

$$\dot{V} = (N\pi r^4/8l\eta)\Delta p = (\xi r^2/8l\eta\beta)\Delta p, \tag{17.7}$$

where η is the liquid's viscosity.

It can be seen here that in contrast to conduction, liquid flow depends not only on the porosity and tortuosity but also on pore size. When other parameters remain unchanged, a decrease in pore radius by a factor of ten causes a decrease in liquid flow rate by a factor of 100.

17.3. MACROKINETICS OF ELECTROCHEMICAL PROCESSES

Electrochemical macrokinetics deals with the combined effects of polarization characteristics and of ohmic and diffusion factors on the current distribution and overall rate of electrochemical reactions in systems with distributed parameters. The term "macrokinetics" is used in order to conveniently distinguish these effects from effects arising at the molecular level.

17.3.1. Systems with Distributed Parameters

During electric current flow in a cell, the current is not always uniformly distributed over the full working area of the electrodes, i.e., the current densities are different at different surface segments. A nonuniform current distribution may arise for two reasons: differences in potential between different segments of the electrode surface (owing to differences in ohmic resistance between these segments and the auxiliary electrode or current collector) or differences in electrolyte composition (owing to differences in the diffusion conditions). In these cases it can also be said that different segments of the electrode are not equally accessible. Systems where such effects appear are called systems with distributed parameters. In electroplating, cells where the current distribution is uniform (nonuniform) are said to have good (poor) *throwing power*.

Consider a cell with one positive and two negative electrodes where the latter are at different distances, l_1 and l_2, from the former (Fig. 17.3). We shall assume for the sake of simplicity that polarization of the electrodes is proportional to current density, i.e., $\Delta E = \rho i$ (ρ is the combined polarization resistance of the positive and negative electrode). The voltages of the two halves of the cell, which are in parallel, are identical, hence, the sum of ohmic losses and polarization in the two halves should also be identical. The ohmic losses in the electrolyte are given by li/σ. Thus

$$\rho i_1 + l_1 i_1/\sigma = \rho i_2 + l_2 i_2/\sigma$$

or

$$i_1/i_2 = (\rho\sigma + l_2)/(\rho\sigma + l_1), \tag{17.8}$$

i.e., the current densities are not the same for the two negative electrodes.

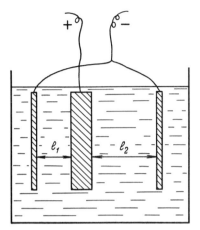

Fig. 17.3. Electrochemical cell with dissimilar distances between electrodes.

The current density distribution will be least uniform $(i_1/i_2 = l_2/l_1)$ when polarization resistance and (or) conductivity are low and $\rho\sigma \ll l$. Uniformity increases with increasing $\rho\sigma$, and the current density in the two halves becomes the same when $\rho\sigma \gg l$. Thus, the throwing power depends on parameter $\rho\sigma$; the higher this parameter the more uniform will be the current distribution. At high values of polarization ρ must be replaced by dE/di, and the value of this derivative is no longer a constant but decreases with increasing current density. Accordingly, the throwing power becomes poorer with increasing current density.

Consider as an example the current distribution in a cell with tall electrodes, where ohmic voltage losses arise in the current collector (Fig. 17.4a). Let b be the width, d the thickness, h the height of the current collector and σ the conductivity of the current-collector material. We shall assume that the electrode is working on both sides (i.e., its working area is given by $S = 2bh$) and that current collection at the upper edge is uniform over the full width of the current collector (its cross section is bd). The current density at height x will be designated as i_x. The total current in the current collector at this level, I_x, will evidently be given by

$$I_x = 2b \int_0^x i_x \, dx. \tag{17.9}$$

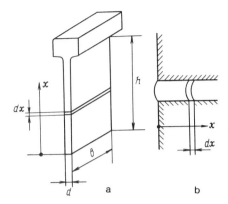

Fig. 17.4. Concerning the derivation of equations for nonuniform current distribution, (a) in a flat electrode, (b) in a cylindrical pore.

The ohmic potential drop along the section from x to $x + dx$ is given by

$$d\varphi_{\text{ohm}} = (dx/\sigma bd)I_x = (2dx/\sigma d)\int_0^x i_x\, dx,\qquad (17.10)$$

and the total potential drop in the current collector from height x to the upper edge of the electrode (at $x = h$), where the external lead is fastened, is given by

$$\varphi_{\text{ohm},x} = (2/\sigma d)\int_x^h dx \int_0^x i_x\, dx.\qquad (17.11)$$

In this example the current density distribution is nonuniform in the vertical, since at all heights x the sums of ohmic potential drops and polarization of the two electrodes must be identical. In the top parts of the electrodes, where the ohmic losses are minor, the current density will be highest, and it decreases toward the bottom. The current distribution will be more uniform the higher the polarization.

The most general form of a differential equation for the distribution of potential and current density in a system with ohmic losses is obtained when Eq. (17.10) is differentiated. Let s_V ($s_V = 2/d$) be the working surface area referred to unit volume (of the current collector). Considering that $dE = d\varphi_{\text{ohm}}$, owing to constancy of the sums of ohmic losses and polarization (which includes the appropriate signs), we find

$$d^2 E/dx^2 = (s_V/\sigma)i_x.\qquad (17.12)$$

Consider, in the instance of a cylindrical pore (Fig. 17.4b), the case where concentration gradients arise in the solution but ohmic potential drops can be neglected. With Π as the pore's perimeter, A as the pore's cross section, and S_p as the pore's surface area, it is evident that $s_V \equiv S_p/V = \Pi/A$. In a solution layer at distance x from the pore mouth, the concentration gradient of component j will be related to the current I_x crossing this layer as

$$I_x = (n/\bar{\nu}_j)FAD_j(dc_j/dx)_x. \tag{17.13}$$

The total current in layer x [by analogy with (17.9)] is given by

$$I_x = \Pi \int_0^x i_x \, dx. \tag{17.14}$$

Differentiating Eqs. (17.13) and (17.14) we find the general form of the differential equation in systems with concentration changes:

$$d^2c_j/dx^2 = (\bar{\nu}_j s_V/nFD_j)i_x. \tag{17.15}$$

For a solution of the differential Eqs. (17.12) and (17.15) and for a quantitative calculation of the current distribution, we must know how the current density depends on polarization at constant reactant concentrations, or on reactant concentrations at constant polarization. We must also formulate the boundary conditions. Examples of such calculations will be reported below.

17.3.2. Categories of Porous Electrodes

Porous electrodes have large true areas, S_p, of the inner surface as compared to their external geometric surface area S, i.e., large roughness factors $\gamma \equiv S_p/S$ (parameters γ and s_V are related as $\gamma = s_V d$). Using porous electrodes one can realize large currents at relatively low values of polarization.

In porous liquid-phase electrodes, all pores are filled with the liquid electrolyte (solution or melt). When part of the pores are gas-filled, the electrodes are called gas–liquid (when the gas is a reactant) or liquid–gas (when the gas is produced in the reaction). When the electrode is nonconsumable and chemically inert, its pore structure will remain unchanged during operation (or change very slowly on account of secondary ageing processes). The structure of an electrode which reacts changes continuously.

In the pores of the electrodes, practically no natural convection of the liquid takes place. Reactants dissolved in the liquid can be supplied in two ways from the external surface to the internal reaction zones (and reaction products transported away in the opposite direction): (a) by diffusion in the motionless liquid (*diffusion electrode*), and (b) with a liquid flow passing the porous electrode under the stimulus of an external force (*flow-through electrode*). In some cases, porous electrodes are working on one side only, viz., the front which is turned toward the auxiliary electrode while the other side (the back) is used for reactant supply. When the reactant is a gas, it is usually supplied to the reaction zone from the back of the electrode. Electrodes of this type are called gas-diffusion electrodes.

17.3.3. Liquid-Diffusion Electrodes

Let the net (overall) current density of the porous electrode be i. Under conditions of uniform work of the full internal surface area, the value of i would be γ times larger than the current density i_0 of a smooth electrode, working at the same value of polarization and in an electrolyte having the same composition. This case is rare in practice, and it is more common to find that i is smaller than its maximum value $i_{max} = \gamma i_0$. The ratio between these parameters,

$$h = i/i_{max} = i/\gamma i_0 \quad (h \leq 1), \tag{17.16}$$

is called the efficiency factor of the porous electrode.

Porous electrodes are systems with distributed parameters, and any loss of efficiency is due to the fact that different points within the electrode are not equally accessible to the electrode reaction. Concentration gradients and ohmic potential drops are possible in the electrolyte present in the pores. Hence, the local current density, $i_{S,x}$ (referred to the unit of true surface area), is different at different depths x of the porous electrode. It is largest close to the outer surface ($x = 0$), and falls with increasing depth inside the electrode.

Mathematical calculations of the current-density distribution in a direction normal to the electrode are rather difficult, hence, we shall limit ourselves to reviewing the simplest cases, in order to discuss the major qualitative trends. Consider the processes occurring in a porous electrode of thickness d operated unilaterally. The current density generated at depth x per unit volume will be designated as $i_{V,x}$, and it is obvious that $i_{V,x} = s_V i_{S,x}$.

We shall discuss the particular case where only ohmic potential drops are present, but concentration gradients are absent. The current-density distribution normal to the surface can be found by integrating the differential Eq. (17.12) with the boundary conditions

$$(\Delta E)_{x=0} = (\Delta E)_0; \qquad (dE/dx)_{x=d} = 0. \qquad (17.17)$$

The first condition implies that the electrode's polarization as a whole (which is measured at its front face where $x = 0$) is given, and has the value $(\Delta E)_0$. According to the second condition, a potential gradient does not exist close to the rear face, since here the total current flowing in the direction of the front face is low.

For a solution of Eq. (17.12) we must also know the dependence of current density on polarization. First we shall consider the simpler case of low values of polarization, when the linear function (6.7) with ρ as the kinetic parameter is valid.

Solving the differential equation for these conditions, we arrive at the following expression for the distribution of local current densities in the electrode in a direction normal to the surface:

$$i_{S,x} = i_0 \cosh[(d - x)/L_{\text{ohm}}]/\cosh(d/L_{\text{ohm}}), \qquad (17.18)$$

where

$$L_{\text{ohm}} \equiv \sqrt{\rho \sigma_{\text{eff}}/s_V}; \qquad i_0 = (\Delta E)_0/\rho. \qquad (17.19)$$

Curves showing the current densities as functions of x are presented for two values of electrode thickness in Fig. 17.5. Parameter L_{ohm} has the dimensions of length; it is called the characteristic length of the ohmic process. It approximately corresponds to the depth x at which the local current density has fallen by a factor of e (approximately 2.72). Therefore, this parameter can be used as a convenient characteristic of attenuation of the process inside the electrode.

The net current density can be computed by integrating the volume current density $i_{V,x}$ over electrode thickness:

$$i = s_V \int_0^d i_{S,x}\, dx = i_0 L_{\text{ohm}} s_V \tanh(d/L_{\text{ohm}}). \qquad (17.20)$$

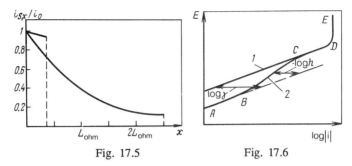

Fig. 17.5   Fig. 17.6

Fig. 17.5. Current density distribution inside a porous electrode [according to Eq. (17.18)] for two values of electrode thickness: $d_1 = 0.33 L_{ohm}$ and $d_2 = 2.5 L_{ohm}$.

Fig. 17.6. Schematic of the polarization curves for smooth (1) and porous (2) electrodes.

Substituting Eq. (17.20) into (17.16) we find

$$h = \tanh(d/L_{ohm})/(d/L_{ohm}). \qquad (17.21)$$

It can be seen here that for $d < L_{ohm}$ (a "thin" electrode) $h \to 1$, the electrode will work uniformly throughout, and the current only depends on electrochemical reaction kinetics (internal kinetic control of electrode operation). However, for $d > L_{ohm}$ (a "thick" electrode) h decreases (internal ohmic control of electrode operation). Owing to the ohmic potential drops, layers of the electrode deeper than $(2-3) L_{ohm}$ contribute practically nothing to the total current. At very low values of L_{ohm} (which are comparable with the dimensions of the structural elements in the electrode), the process is fully "pushed out" to the outer surface, and the inner segments of the electrodes are almost inoperative.

According to Eq. (17.18), the relative current density distribution, $i_{S,x}/i_0$, in a direction normal to the electrode surface is independent of the absolute values of current density. However, as the current density rises to values where polarization starts to depart from the linear function (6.7), the situation is perturbed and factors L_{ohm} and h start to decrease.

Figure 17.6 shows polarization curves for a smooth electrode and for a porous electrode with the roughness factor $\gamma = 100$. A system which at low currents works under internal kinetic control, i.e., with $h = 1$, was selected as the example. In this current range (segment AB in curve 2) the distance between the curves is constant and equal to the logarithm of the roughness factor. As the current increases, h begins to decrease owing to increasing influence of the ohmic factors, and the curves begin to merge.

The region of internal ohmic control is situated between points B and C. It is typical that in this region the slope of the polarization curve is double that found for a smooth electrode. As the current is raised further, the process is pushed out to the outer surface, and the porous electrode loses its advantages (external kinetic control, segment CD; external diffusion control, segment DE).

When only taking into account the concentration polarization in the pores but disregarding ohmic potential gradients, we must use an equation of the type of (17.15). Solving this equation for a first-order reaction ($i_{S,x} = nFh_m c_{j,x}$) leads to equations exactly like (17.18) for the distribution of the process inside the electrode, and like (17.20) for the total current. The rate of attenuation depends on the characteristic length of the diffusion process:

$$L_{\text{diff}} = \sqrt{nFD_{\text{eff}}/s_V h_m}. \tag{17.22}$$

Unlike the earlier case, L_{diff} here decreases with increasing current, not only at large but also at small values of polarization, since h_m depends on polarization. For $d \geq L_{\text{diff}}$ (and particularly when polarization is significant), the electrode will work under internal diffusion control.

17.3.4. Gas-Diffusion Electrodes

The solubilities of most gases in solutions are low, hence, for reactions involving gaseous reactants (and particularly for the reactions in fuel cells and other chemical power sources), usually gas-diffusion electrodes are employed, which contact the electrolyte with their front face, and the gas space with the other (rear) face (Fig. 17.7). The pore space of such electrodes is partly liquid-, partly gas-filled. The electrochemical reactions occur mainly beneath thin electrolyte films forming at the pore walls close to the places where gas and liquid pores meet, viz., near the so-called three-phase boundaries. These zones offer optimum conditions for the simultaneous access of gas (through the film) and dissolved reaction components (along the film). These zones must be made as large as possible in order to achieve high electrode efficiencies; this can be attained at certain volume proportions between gas and liquid pores. In addition these zones should be sufficiently uniformly distributed throughout the electrode.

Most electrode materials are hydrophilic and readily wetted by aqueous solutions. Two methods are used to create and maintain an optimum

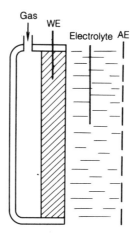

Fig. 17.7. Schematic of a gas-diffusion electrode. (1) Gas space; (2) electrolyte; (3) auxiliary electrode.

gas–solution ratio in the electrode. The first method employs a certain excess gas pressure in the gas space. This causes the liquid to be displaced from the wider pores, while in finer pores the liquid continues to be retained by capillary forces. The second method employs partial wetproofing of the electrode by the introduction of hydrophobic materials (for instance, fine PTFE particles). Then the electrolyte will only penetrate those pores in the hydrophilic electrode material where the concentration of hydrophobic particles is low.

The macrokinetics of processes in gas-diffusion electrodes is analogous to that in liquid-phase electrodes. In calculations one must take into account however, that the electric current and the solute species will only be carried through that part of pore space which is electrolyte-filled, while gas supply is mainly accomplished, not by diffusion through the liquid but by flow in the gas channels.

Chapter 18
REACTIONS INVOLVING METALS
OR OTHER SOLIDS

18.1. REACTING METAL ELECTRODES

Electrochemical systems with reacting metal electrodes are widely used in batteries, electrometallurgy, electroplating and other areas. Corrosion of metals is a typical example of processes occurring at reacting metal electrodes.

At most metals which are in contact with an electrolyte containing their own ions, the equilibrium potential of the metal's discharge and ionization reaction,

$$M \rightleftarrows M^{z+} + z_+ e^- \tag{18.1}$$

or, in the case of electrodes of the second kind, that of a reaction of the type of

$$M + z_+ A^- \rightleftarrows MA_{z_+} + z_+ e^-, \tag{18.2}$$

is relatively readily established (A^- is an anion, e.g., OH^-, Cl^-, $1/2\,SO_4^{2-}$, etc.).

When the metal is in contact with an electrolyte solution not containing its ions, its equilibrium potential theoretically will be strongly shifted in the negative direction. However, before long a certain amount of ions will accumulate close to the metal surface as a result of spontaneous dissolution of the metal. We may assume, provisionally, that the equilibrium potential of such an electrode corresponds to a concentration of ions of this metal of about 10^{-6} M. In the case of electrodes of the second kind, the solution is practically always saturated with metal ions, and their potential corresponds to the given anion concentration [an equation of the type of (3.50)].

When required, a metal's equilibrium potential can be altered by addition of complexing agents to the solution (see Eq. (3.53)].

Some metals are thermodynamically unstable in aqueous solutions, because their equilibrium potential is more negative than the potential of the reversible hydrogen electrode in the same solution. At such electrodes, anodic metal dissolution and cathodic hydrogen evolution can occur as coupled reactions, and their open-circuit potential (ocp) will be more positive than the equilibrium potential (see Section 13.7).

According to their exchange current densities i^0 (or standard rate constants k^0) and polarization in reactions involving their dissolution or deposition in solutions of simple salts (sulfates, chlorides, etc.), the metals can be provisionally divided into two groups: that of metals with high polarizability (low exchange current density), which includes chromium, manganese and the metals of the iron and platinum group, and that of metals with low polarizability (high exchange current density), which includes the others. In the second group, constants k^0 have values on the order of 10^{-4}–10^{-2} m/s, and exchange current densities i^0 between 1 and 100 kA/m^2. Up to current densities (cd) of 1 kA/m^2, these metals exhibit an activation polarization of no more than 20 mV. Metals of the first group have values of k^0 and i^0, which are two to four orders of magnitude lower, and polarization here can attain several tenths of a volt.

The reactions occurring at reacting metal electrodes are associated with structural changes, viz., lattice destruction or formation of the metal and, in certain cases, of other solid reaction components (oxides, salts, etc.). One should know the metal's original bulk and surface structure in order to analyze the influence of these structural changes.

Solid metals obtained upon solidification of the molten metal exhibit grain structure. They consist of fine crystallites randomly oriented in space. The size of the individual crystallites (grains) is between 10^{-7} m (fine-grained structure) and 10^{-3} m (coarse-grained structure). The crystal structure of the individual grains as a rule is not ideal. It contains different kinds of defect: vacant sites, interstitial atoms or ions and dislocations (lattice shearing or bending). Microcracks sometimes evolve in the zones between crystallites.

The metal surface is polycrystalline, and has a rather complex profile. Because of different crystallite orientations at the surface, different crystal faces are exposed, viz., smooth low-index faces and "stepped" high-index faces. Surface texture where a particular kind of face is predominant can develop in individual cases. Microcracks and different lattice defects (dislocations etc.) will also emerge at the surface.

Thus, the individual atoms (ions) at a metal surface exist under different geometric, and hence energy, conditions. Yet the surface particles as a rule retain a certain mobility. On account of their surface diffusion the original surface structure may change to a more stable state. When the surface is in contact with an electrolyte solution, this process is accelerated owing to the continued exchange occurring via the solution, viz., the dissolution of particles from the most active sites and their redeposition at other sites.

During mechanical working of the metal, both the bulk and surface structures are altered. Cutting, abrading, polishing and other treatments deform the surface layer and increase the number of defects in it. Under sufficient force a surface "milling" producing an almost amorphous surface layer up to 100 nm thick will occur. Beneath this layer the lattice distortion extends to depths attaining tenths of a millimeter ("work hardening" of the surface). A crystalline surface state can be restored by careful annealing or electrochemical polishing (Section 18.6).

It would be of great value for studies of different electrochemical phenomena if measurements could be made, not at polycrystalline surfaces but at particular faces of sufficiently large single crystals of a given metal. It is a rather difficult task, unfortunately, to produce the "pure" faces and work with them. In contact with solutions, segments or steps of other faces can appear on such a face owing to the exchange which occurs. Up to now, therefore, reliable data as to the electrochemical properties of individual faces or how they differ from those of polycrystalline surfaces are scarce.

Electrochemical processes involving metals, such as metal ion discharge and metal atom ionization, can be studied without the complications of structural changes when electrodes of the molten metal (at elevated temperatures and in nonaqueous electrolytes) or of the metal's liquid amalgam are used instead of the solid metal.

18.2. ANODIC METAL DISSOLUTION

Various electrochemical reactions are possible during anodic polarization of a metal electrode in aqueous solutions:

(a) anodic dissolution (oxidation) of the metal with the formation of soluble [reaction (18.1)] or insoluble [reaction (18.2)] products,

(b) the formation of adsorbed and phase oxide or salt layers (films) on the surface,

(c) the anodic oxidation of solution components, e.g., organic impurities,

(d) anodic oxygen evolution (also anodic chlorine evolution, in solutions containing chlorides).

Each of these reactions occurs in its own, typical potential range. Several reactions may occur in parallel. The oxidation of solution components and the evolution of oxygen and chlorine will be discussed in Chapter 19, the formation of surface layers in Section 18.3. In the present section we shall discuss anodic metal dissolution.

The shape of polarization curves for metals with low polarizability, mainly depends on concentration polarization. In the case of highly polarizable metals, where activation polarization can be measured sufficiently accurately, the polarization curve can usually be described by an equation of the type of (6.3), i.e., by a Tafel equation. For metals forming polyvalent ions, slope b' in this equation often has values between 30 and 60 mV.

Anodic dissolution reactions of metals typically have rates which strongly depend on solution composition, and particularly on the anion type and concentration (Yakov M. Kolotyrkin, 1955). The rates increase upon addition of surface-active anions. It follows, that the first step in anodic metal dissolution reactions is that of adsorption of an anion and chemical bond formation with a metal atom. This bonding facilitates subsequent steps in which the metal atom (ion) is torn from the lattice and solvated. The adsorption step may be associated with simultaneous surface migration of the dissolving atom to a more favorable position (e.g., from position 3 to position 1 in Fig. 15.4), where the formation of adsorption and solvation bonds is facilitated.

In iron dissolution in alkaline solutions, the rate is proportional to OH^- ion concentration, to a first approximation. The first step of this reaction is usually described as

$$Fe + OH^- \rightleftarrows Fe(OH)^-_{ads}. \tag{18.3}$$

This step is followed by steps involving electron transfer and the addition of further hydroxyl ions, which continues until the final reaction products are formed:

$$Fe(OH)^-_{ads} \xrightarrow[-e^-]{} Fe(OH)_{ads} \xrightarrow[-e^-]{+OH^-} Fe(OH)_2 \left[\xrightarrow[-H_2O]{+OH^-} HFeO^-_2 \right]. \quad (18.4)$$

Structural surface inhomogeneity influences the anodic dissolution process in the case of metals with appreciable activation polarization. Segments with perturbed structure, as a rule, dissolve more rapidly than ordered segments. In a number of cases this causes crystallites to break away from the electrode surface and form metal sludge.

In the case of metals of variable valency forming ions of different charge, peculiar effects are possible. Thus, the ions Cu^+ and Cu^{2+} between which the dismutation equilibrium (3.59) exists are formed in the case of copper. At equilibrium (and particularly at the electrode's equilibrium potential) the activities of these ions are interrelated through the equilibrium constant: $a^2_{Cu^+}/a_{Cu^{2+}} = K = 8.2 \cdot 10^{-7}$. When the electrode is anodically polarized and the potential moving in the positive direction, the relative amount of Cu^+ ions primarily formed increases. In the solution, the additional ions will undergo dismutation and yield Cu^{2+} ions as well as fine copper powder which deposits at the bottom of the cell in the form of sludge. Thus, formation of a Cu^{2+} ion via Cu^+ consumes two copper atoms, one of them ending up in the sludge, i.e., the current yield of copper dissolution (the relation between copper lost and charge consumed) has doubled.

A similar situation is found for certain metals not forming stable ions of variable valency. In anodic magnesium dissolution, Mg^+ ions are first formed. They do not undergo dismutation like the Cu^+ ions, but as a strong reducing agent react with water according to

$$2Mg^+ + 2H_2O \rightarrow 2Mg^{2+} + H_2 + 2OH^-. \quad (18.5)$$

In this case, one magnesium atom is consumed for the formation of one Mg^{2+} ion, but the current yield of magnesium dissolution again has doubled, because every other magnesium atom is spent, not for electron generation but for hydrogen evolution.

When the potential of the magnesium electrode is made more positive the rate of Mg^+ ion formation increases, and with it that of reaction (18.5). Therefore, the rate of hydrogen evolution increases with increasing anodic polarization of the magnesium, instead of falling off (see Section 13.7). This phenomenon has become known as the negative difference effect.

The phenomena occurring during the anodic dissolution of binary alloys of metals M and N (where M is the more electronegative metal) depend on the alloy type. In the case of heterogeneous alloys, the two components dissolve independently, each following its own behavior. At potentials intermediate between the equilibrium potentials of the two metals, metal M may dissolve selectively. When its concentration in the alloy is low, a pure surface of metal N may develop after some time. When its concentration is high, however, particles of metal N will lose their connection with the base, shed and form sludge. In the case of homogeneous alloys, both metals will dissolve at potentials intermediate between the equilibrium potentials of the alloy and of metal N. However, the N^{z+} ions formed can undergo discharge and redeposit, now as the pure metal rather than an alloy. Usually metal sludge is also formed under these conditions.

18.3. SURFACE-LAYER FORMATION

In the anodic polarization of metals, almost always surface layers of adsorbed oxygen are formed by reactions of the type of (12.11) occurring in parallel with anodic dissolution, and sometimes phase layers (films) of the metal's oxides or salts are also formed. Oxygen-containing layers often simply are produced upon contact of the metal with the solution (without anodic polarization) or with air (the air-oxidized surface state).

The first step of oxide-layer formation is oxygen adsorption (chemisorption). In the case of platinum, the process stops at this stage, and depending on the conditions an incomplete or complete monolayer of adsorbed oxygen is present on the platinum surface. In the case of other metals, layer formation continues. When its thickness δ has attained two to three atomic diameters, the layer is converted to an individual surface phase which is crystalline (more seldom amorphous) and has properties analogous to those of the corresponding bulk oxides.

The surface-phase layers will differ in character depending on the structures of metal and oxide. On certain metals (zinc, cadmium, magnesium, etc.), loose, highly porous layers are formed which can attain appreciable thicknesses. On other metals (aluminum, bismuth, titanium, etc.), compact layers with low or zero porosity are formed which are no thicker than 1 μm. In a number of cases (e.g., on iron), compact films are formed which have a distorted lattice owing to the influence of substrate metal structure and also of the effect of chemical surface forces.

The physicochemical and thermodynamic parameters of such films differ from those of the ordinary bulk oxides. Because of the internal stresses in the distorted lattice, such films are stable only when their thickness is insignificant, e.g., up to 3–5 nm.

As a rule, different kinds of oxide film will form simultaneously on metal electrodes, for instance, porous phase layers on top of adsorbed layers. Often ageing processes occur in the oxide layers, which produce time-dependent changes in the properties or even transitions between different forms.

Sufficiently strong cathodic polarization will reduce the oxide layers on many metals, but in certain cases (on titanium, tantalum, etc.) the oxide film cannot be removed by electrochemical means. A bare, nonoxidized metal surface can be obtained by thermal reduction under hydrogen, or by mechanical stripping of a metal surface layer in a medium (gas or solution) carefully freed of oxygen.

The anodic formation and cathodic reduction of an oxide layer can be studied by recording charging curves when other reactions (anodic metal dissolution or oxygen evolution, cathodic hydrogen evolution) do not occur at the surface. The anodic and cathodic scan of a galvanostatic charging curve are schematically shown in Fig. 18.1a. In the initial segment AB of the anodic curve, an adsorbed layer of oxygen atoms is formed; this segment is analogous to the oxygen region of the anodic charging curve for platinum (see Fig. 12.13). A phase oxide layer starts to form and grow when a certain potential has been reached. Initially, one usually sees an overshoot of the potential in the positive direction (segment BC), which is due to difficulties in the nucleation of new phase. After that, the potential as a rule remains almost unchanged; owing to polarization it is somewhat more positive than the equilibrium potential of this phase. At the end of layer growth (point D), the potential briskly shifts to a more positive value where another reaction starts, e.g., oxygen evolution. In the cathodic scan, following an initial overshoot of the potential in the negative direction (segment EF), the oxide layer is reduced, again at a practically constant value of potential, but this time somewhat more negative than the equilibrium potential.

In a potentiodynamic charging curve, narrow high current peaks are recorded in the regions where the oxide layer is formed and reduced (Fig. 18.1b).

Charging curves of a different type are recorded at certain metals: instead of the horizontal sections in the galvanostatic curves, sloping seg-

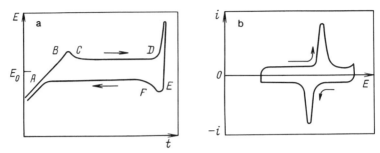

Fig. 18.1. Charging curves recorded when an adsorbed layer of oxygen adatoms or a phase oxide layer are formed: (a) galvanostatic; (b) potentiostatic.

ments arise, the hysteresis between the anodic and cathodic segments is larger, and the current peaks in potentiodynamic curves are more diffuse.

During anodic polarization, oxide layers may form even in solutions where they are soluble, particularly so in acidic solutions. In this case, the layers have a stationary thickness that depends on the equilibrium between the rates of electrochemical formation and chemical dissolution of the layer. Such layers vanish within a certain time after interruption of the current.

Oxide layers thinner than 0.4 μm are inaccessible to direct observation. Their thickness can be determined ellipsometrically; this method utilizes changes in optical polarization which occur during the reflection of plane-polarized light by the surface. The amount of oxygen contained in the layer can be determined from the length of the horizontal section in the cathodic galvanostatic charging curve. However, all these methods only yield average values, since possible unevenness of the layer is disregarded.

Many kinds of oxide layer have a certain, not very high electrical conductivity, of up to 10^{-4}–10^{-3} S/m. Conduction may be cationic (by M^{z+} ions) or anionic (by O^{2-} or OH^- ions), or of the mixed ionic and electronic type. Often charge transport occurs by a semiconductor hole-type mechanism, hence, oxides with ionic and ionic-hole conduction are distinguished (in the same sense as p-type and n-type conduction in the case of semiconductors, but here with anions or cations instead of the electrons, and the corresponding ionic vacancies instead of the electron holes). Electronic conduction is found for the oxide layers on iron-group metals and on chromium.

The behavior of metal electrodes with an oxidized surface depends on the properties of the oxide layers. Even a relatively small amount of chemisorbed oxygen will drastically alter the edl structure (see Fig. 12.21) and influence the adsorption of other substances. During current flow, porous layers will screen a significant fraction of the surface and interfere with reactant transport to and product transport away from the surface. Moreover, the ohmic voltage drop increases owing to the higher current density in pores. All these factors interfere with the electrochemical reactions, and particularly with a further increase in layer thickness.

In the case of electrodes with purely ionically conducting layers which are completely or almost completely nonporous, an electrochemical reaction is possible only at the inner surface of the layer (at the metal boundary). When conduction is cationic, an anodic current will cause metal ionization [and a cathodic current will cause metal ion discharge] at this boundary according to Eq. (18.1). Ions M^{z+} will migrate to [enter from] the layer's outer surface (the electrolyte boundary) where the reaction with the solution occurs, e.g.,

$$2M^{z+} + z_+H_2O \rightleftarrows M_2O_{z_+} + 2z_+H^+. \tag{18.6}$$

When conduction is anionic the O^{2-} ions produced [consumed] at the outer surface in the reaction

$$H_2O \rightleftarrows O^{2-}_{lattice} + 2H^+ \tag{18.7}$$

will migrate to [have migrated out from] the layer's inner surface where they undergo the electrochemical reaction [have formed in the reaction]

$$2M + z_+O^{2-} \rightleftarrows M_2O_{z_+} + 2z_+e^-. \tag{18.8}$$

The electrochemical overall reaction at the electrode can be written as

$$2M + z_+H_2O \rightleftarrows M_2O_{z_+} + 2z_+H^+ + 2z_+e^-, \tag{18.9}$$

i.e., the layer increases [decreases] in thickness during current flow.

When the layer has electronic in addition to ionic conductivity, the electrochemical reaction will be partly or completely pushed out to its outer surface. In addition, other electrochemical reactions involving the solution components, particularly anodic oxygen evolution, can occur on top of the layer.

In many cases the electrochemical reaction rates at such electrodes are governed by migration of the carriers through the layer. Their migration rates are proportional to the potential gradient, $\Delta\psi/\delta$, inside the layer, i.e., at a given electrode potential (given value of $\Delta\psi$) they are inversely proportional to layer thickness: $i = \text{const}/\delta$. When other reactions are absent, layer growth is proportional to current density: $d\delta/dt = \text{const} \cdot i$. Hence, $\delta(d\delta/dt) = \text{const}$, and after integration we have

$$\delta^2 = \text{const} \cdot t, \qquad (18.10)$$

i.e., layer thickness increases with the square root of time during electro-chemical oxide formation.

In very thin (nanometer) films, where the potential gradient may exceed 10^8 V/m, another mechanism of ion migration is observed, which involves periodic jumps of ions between equilibrium positions. In this case, the rate of migration is not proportional to the potential gradient but obeys the exponential law of

$$i = k \exp(\gamma \Delta\psi/\delta), \qquad (18.11)$$

with the corresponding change in the laws of film growth.

Nonporous layers which are practically nonconducting and completely insulate the electrode surface from the electrolyte solution will form during anodic polarization on certain metals (aluminum, titanium, etc.). Even when a high external voltage (e.g., 100 V) is applied, no anodic current will pass through the electrode. Such layers are employed in the production of electrolyte capacitors which are distinguished by high values of capacitance, since the layers are so thin.

In individual cases, anodic polarization of metals in electrolyte solutions will produce surface layers (adsorbed or phase) which instead of oxygen, contain the solution anions. Thus, anodic polarization of silver in chloride-containing solutions yields a surface layer of silver chloride, while the anodic polarization of lead in sulfuric acid solution yields a lead sulfate layer. Layers of sulfides, phosphates and other salts can be formed in the same way. In many respects the properties of such salt layers are analogous to those of the oxide layers.

Oxide and salt layers on metal electrodes are of great practical value. Electrodes with thick phase layers are used in batteries, while various kinds of thin layer will produce passivation of metals.

18.4. PASSIVATION OF ELECTRODES

"Passivation of an electrode with respect to a certain electrochemical reaction" is the term used for the strong hindrance experienced under certain conditions by the reaction which, under other conditions (in the electrode's active state), will occur without hindrance at this electrode. "Passivation of metals" implies the hindrance frequently observed with respect to anodic metal dissolution.

18.4.1. Passivation of Metals

Passivation of metals is very important in applied electrochemistry. It sharply retards the spontaneous dissolution of a number of metals when these are in contact with electrolyte solutions, i.e., raises their corrosion resistance. The passivation of metal anodes also interferes with the normal function of batteries and electrolyzers.

Passivation of metals is most distinctly displayed when, during anodic polarization, the potential is gradually made more positive. Figure 18.2a shows a typical anodic polarization curve recorded potentiodynamically with a slow linear potential scan. Segment AB corresponds to the region of ordinary anodic dissolution of nonpassivated (active) metal; the rate of the anodic reaction increases as the potential is made more positive. In segment BC, when potential E_{pass} and a certain critical current density i_{cr} have been reached, the reaction rate will drastically decrease to a new value i_{pass} when the potential is made more positive. Thus, for iron in 0.5 M H_2SO_4 solution at 25°C, the value of i_{cr} is about 2500 A/m^2, while that of i_{pass} is about 0.07 A/m^2, i.e., the current has decreased by four to five orders of magnitude. The value of i_{pass} remains almost constant within the wide range of potentials of section CD, which is the region of passivity of the metal. It is only when the potential is made much more positive (point D) that the current rises again, which is due to renewed acceleration of anodic metal dissolution or (and) to the start of oxygen evolution. In the former case, region DE is called the region of transpassivation of the metal; usually metal oxidation products of higher valency than in the active dissolution region AB are formed here.

When the polarization curve is recorded in the opposite (cathodic) direction, the electrode will regain its active state at a certain potential E_{act}. The activation potential E_{act} is sometimes called the Flade

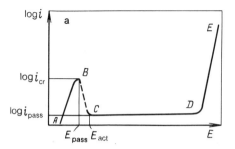 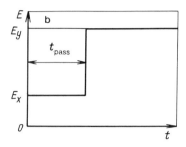

Fig. 18.2. (a) Potentiodynamic log i vs. E curve; (b) galvanostatic E vs. t curve for anodic metal dissolution and passivation.

potential (Friedrich Flade, 1911). The potentials of activation and passivation as a rule are slightly different.

The limits of transition region BC are not very distinct and depend on the experimental conditions. At high potential scan rates (short duration of the experiment) passivation will start later, i.e., potential E_{pass} will be somewhat more positive, and for a short time the currents may be higher than i_{cr}.

Passivation looks different when observed under galvanostatic conditions (Fig. 18.2b). The passive state will be attained after a certain time t_{pass} when an anodic current i_x which is higher than i_{cr} is applied to an active electrode. As the current is fixed by external conditions, the electrode potential at this point undergoes a discontinuous change from E_x to E_y, where transpassive dissolution of the metal or oxygen evolution starts. The passivation time t_{pass} will be shorter the higher the value of i_x. Often these parameters are interrelated as

$$(i_x - i_{cr})t_{pass}^{1/n} = \text{const}, \tag{18.12}$$

where constant n has values between one and two.

Under the effect of oxidizing agents, a metal may become passivated even when not anodically polarized by an external power source. In this case passivation is evident from the drastic decrease in the rate of spontaneous dissolution of the metal in the solution. The best-known example is that of iron passivation in concentrated nitric acid, which had been described by Mikhail V. Lomonosov as early as 1750. Passivation of the metal comes about under the effect of the oxidizing agent's positive redox potential.

For a number of metals the oxidizing action of air oxygen is sufficient to produce the passive state. In their air-oxidized state, metals such as tantalum, titanium, chromium, etc. are very stable in aqueous solutions.

The character of the passivation phenomena which can be observed at a given metal strongly depends on the composition of the electrolyte solution, and particularly on solution pH and the anions present. The passivating tendency as a rule increases with increasing pH, but sometimes it decreases again in concentrated alkali solutions. A number of anions, and particularly the halide ions Cl^- and Br^-, are strong activators.

In the process of passivation, metals usually are found only in one of the two extreme states, active or passive. The transition between these states occurs suddenly and discontinuously. The intermediate state in region BC can only be realized with special experimental precautions. It is in this sense, that passivation differs from the inhibition of electrochemical reactions observed during adsorption of a number of surface-active substances, where the degree of inhibition varies smoothly with the concentration of added material.

Practically all metals can be passivated. Even lithium, which is a highly active alkali metal, can be passivated in concentrated LiOH solution; this is the reason for its greatly reduced rate of reaction with water.

18.4.2. Different Kinds of Electrode Passivation

When metals become passivated their anodic dissolution rates are reduced, but the rates of other reactions will not necessarily be different. For instance, a platinum electrode is completely passivated with respect to its own anodic dissolution, but many electrochemical reactions involving dissolved and gaseous substances occur at it with great ease and speed over a wide range of potentials. Under certain conditions, though, other reactions also become passivated at this electrode. For instance, the ionization of molecular hydrogen [reaction (1.32) from right to left] is very fast at potentials of up to 0.2–0.3 V vs. rhe, but at more positive potentials this reaction begins to be hindered, and around 1.0 V its rate is greatly reduced (Fig. 18.3).

In the middle of the last century the Grenet primary cells were very popular. In these cells a negative zinc electrode was in contact with the acidic solution of a strong oxidizing agent, viz., with a mixture of sulfuric and chromic acid. Here, the zinc is relatively stable; its rate of reaction with the solution is low. However, under anodic polarization zinc

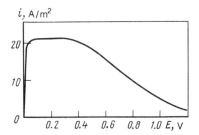

Fig. 18.3. Polarization curve for the ionization of molecular hydrogen at a rotating platinum electrode ($f = 16$ s^{-1}) in 0.5 M H$_2$SO$_4$
solution.

dissolution proceeds in the normal way, without any important hindrance.
Here we have an example where the zinc electrode is passivated, not
with respect to anodic metal dissolution but with respect to the cathodic
reactions coupled with it, viz., the reduction of chromic acid and the
evolution of hydrogen at zinc. For this reason the current of spontaneous
zinc dissolution is low. Since the exchange cd of the former reaction is
much higher than that of the other two reactions, the electrode's ocp will
practically coincide with the equilibrium potential of zinc discharge and
ionization. A similar situation is found in the primary cells with lithium
electrode which have been developed recently, and where the lithium
electrode is in contact with an electrolyte prepared with thionyl chloride
SOCl$_2$, another strong oxidizing agent.

Peculiar effects are seen during the passivation of magnesium and
aluminum in aqueous solutions. In acidic solutions, vigorous spontaneous dissolution is observed for magnesium, while in alkaline solutions
a compact oxide layer which completely passivates it is formed on its
surface. In neutral and weakly alkaline solutions (roughly between pH
of 6 and 11) a "semipassive" state is attained in which hydrogen evolution, and thus spontaneous magnesium corrosion, is strongly hindered,
while anodic magnesium dissolution is less difficult. As a result of such
passivation the steady-state ocp of magnesium is 0.6 to 1.1 V more positive than the thermodynamic value. The negative difference effect which
is seen in anodic magnesium dissolution (see Section 18.2) can, in part,
be attributed to partial mechanical disintegration of the passivating layer
during magnesium dissolution.

The behavior of aluminum in neutral and weakly alkaline solutions
resembles the behavior of magnesium, but the negative difference effect

is much less pronounced at aluminum. The steady-state potential of aluminum is approximately 1 V more positive than the thermodynamic value. Yet unlike magnesium, aluminum will not passivate in strongly alkaline solutions, but undergoes fast dissolution to soluble aluminates.

18.4.3. Origins and Mechanisms of Passivation

The suggestion that the passive state of metals is due to oxide layers present on their surface had been put forward as early as 1837 by Michael Faraday. In some form or other this concept has survived up to the present. The passive state is in fact brought about by factors (anodic polarization, the action of oxidizing agents) which promote the formation of a variety of layers on the surface.

Often passivation of metals has been attributed to a mechanical blocking of part of the surface by a chemically inert insulating layer. This mechanism is readily explained with the example of passivation of a lead electrode in sulfuric acid solution under galvanostatic anodic polarization (the process which occurs at the negative electrode of lead–acid batteries during discharge). A rather compact, porous layer of lead sulfate is formed on the electrode surface by reaction (16.3). After an initial overshoot a constant potential E_1 is attained at the electrode (Fig. 18.4); this is governed by the kinetics of lead sulfate crystallization. After some time, the potential begins to shift in the positive direction, slowly at first but then rapidly. This shift is due to the increase in true current density occurring in the pores of the layer when the total pore cross section decreases in the growing layer. The result is an increase in polarization of the electrode; the ohmic potential drop also increases in the pores. When galvanostatic polarization is continued the potential moves to a value E_2 where a new electrochemical reaction, the formation of lead dioxide PbO_2, begins. The oxidation of lead to lead sulfate completely ceases, even though an abundant supply of metallic lead remains. Passivation starts when the thickness of the lead sulfate layer is about 1 μm.

In corrosion studies of iron, the passivation of iron in acidic, neutral and alkaline solutions has been a topic of great interest. On passive iron one always finds a thin oxide film (3 to 5 nm) which in its structure and properties differs from known iron oxides (see Section 18.3). This film is nonporous and isolates the metal surface from the solution. It has marked electronic conductivity; during anodic polarization film growth continues on account of the migration of iron ions and electrons in the film.

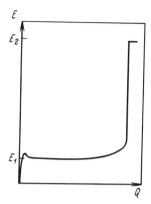

Fig. 18.4. Galvanostatic E vs. i curve for the
anodic polarization of lead in 5 M H_2SO_4.

However, in acidic solutions the film continuously dissolves from outside
with a certain constant rate. A stationary situation is attained in which the
rate of anodic oxidation of the passive metal, i_{pass} (the rate of film growth),
is equal to the rate of film dissolution. The slight potential dependence
of i_{pass}, or constant rate of migration of the ions in the film, implies that the
potential gradient $\Delta\psi/\delta$ in the film is practically constant. It follows that
film thickness δ should increase when the potential is made more positive
($\Delta\psi$ increases). The rate of film dissolution at constant potential decreases
with increasing solution pH; there is a corresponding decrease in the value
of i_{pass} and increase in steady-state film thickness. All these conclusions
concerning film thickness are confirmed by the results of measurements.

In the concepts described, passivation is attributed to the formation of
phase layers which mechanically screen the electrode surface and isolate it
from the solution, but the changes in surface properties of the metal have
been disregarded. Boris V. Érshler showed in 1942 that the passivation
of metallic platinum in hydrochloric acid solutions was due to minute
amounts of adsorbed oxygen atoms present on the platinum surface. At
a given potential, the rate of anodic platinum dissolution decreases with
increasing oxygen coverage, θ_O, of the electrode surface according to an
exponential law:

$$i = k\exp(-\gamma\theta_O). \tag{18.13}$$

Coefficient γ has a high value, of about 12. This implies that an increase
in θ_O by 0.06 (i.e., an increase in the number of sites taken up by oxygen

which amounts to 6% of the total number of sites) produces a 50% drop in reaction rate.

It had been shown in Section 12.8.5, that an important restructuring of the electric double layer occurs when an adsorbed oxygen layer appears on the surface of platinum; the metal's surface potential $\chi^{(Pt)}$ increases owing to the polar character of the Pt–O bonds, the interfacial potential Φ and potential ψ_0 of the ionic edl part decrease accordingly. The rate of anodic metal dissolution depends precisely on potential ψ_0, and hence, is strongly reduced by the edl rearrangement caused by adsorption. Therefore, in the case of platinum the passivating effect of the adsorbed oxygen layer is due, not to a protective mechanical action but to more subtle effects, viz., the redistributions of charges and potential drops in the surface layer.

Similar effects may exist at other metals. For instance, when the surface of an iron electrode is thermally reduced in hydrogen and then anodically polarized at a current density of 0.1 A/m^2 in 0.1 M NaOH solution, passivation already sets in after 1–2 min, i.e., after a charge flow of about 10 C/m^2. This amount of charge is much smaller than that required for formation of even a thin phase film. Since prior to the experiment, oxygen had been stripped from the surface, passivation can only be due to the adsorbed layer formed as a result of polarization.

Passivation phenomena on the whole are highly multifarious and complex. One must distinguish between the primal onset of the passive state and the secondary phenomena which arise when passivation has already occurred (i.e., as a result of passivation). It has been demonstrated for many systems by now that passivation is caused by adsorbed layers, and that the phase layers are formed when passivation has already been initiated. In other cases, passivation may be produced upon the formation of thin phase layers on the electrode surface. Relatively thick porous layers can form both before and after the start of passivation. Their effects, as a rule, amount to an increase in true current density and to higher concentration gradients in the solution layer next to the electrode. Therefore, they do not themselves passivate the electrode but are conducive to the onset of a passive state having different origins.

18.5. CORROSION OF METALS

Corrosion (from Latin *corrodere*, gnaw to pieces) of metals is the spontaneous chemical (oxidative) destruction of metals under the effect of their environment. Most often it follows an electrochemical mechanism, where anodic dissolution (oxidation) of the metal and cathodic reduction of an oxidizing agent occur as coupled reactions. Sometimes a chemical mechanism is observed.

Corrosion is often enhanced by different extraneous effects. Stress corrosion cracking can occur under appreciable mechanical loads or internal stresses; corrosion fatigue develops under prolonged cyclic mechanical loads (i.e., loads alternating in sign). Other factors are fretting and cavitation in a liquid (impact of the liquid). Corrosion can also occur on account of electric currents (stray currents in soils).

Corrosion phenomena can be classified according to the kind of corroding medium acting on the metal. Corrosion in nonelectrolytic media is distinguished from that in electrolytic media. The former include dry hot gases, organic liquids (e.g., gasoline) and also molten metals. Electrolytic media are most diverse, and include ambient air (with moisture and other components), water (seawater, tap water) and aqueous solutions (acids, alkalies, salt solutions), moist soil (for underground pipelines, piles, etc.), melts and nonaqueous electrolyte solutions.

Corrosion phenomena can also be classified according to the visible aspects of corrosive attack (Fig. 18.5). This may be general (continuous), affecting all of the exposed surface of a metallic object, or localized. General corrosion can be uniform and nonuniform. Depending on the width and depth of the segments affected by localized corrosion we may speak of spot, pit (large or small), or subsurface corrosion. Often intercrystalline corrosion is encountered, which propagates in the zones between individual metal crystallites. Cracks develop between or in individual crystals in the case of stress corrosion cracking.

Certain types of corrosion are selective. Thus, corrosion cracking is mainly observed in the case of alloys, and only when these are in contact with particular media.

Almost all metals are subject to corrosion, an exception being the so-called noble metals (the platinum metals, gold, silver) which under ordinary conditions do not corrode. Corrosion of iron is the most prominent problem, since parts and structures consisting of iron and steels are so widely used.

Different parameters are used to characterize the corrosion rate: the loss of mass by the metal sample within a certain length of time (per unit area), the decrease in sample thickness, the equivalent electric current density, etc. For most metals undergoing uniform general corrosion, these

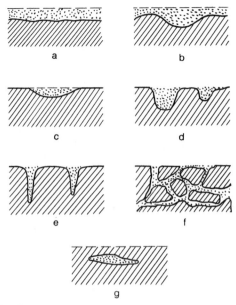

Fig. 18.5. Aspects of metallic corrosion: (a) uniform general; (b) nonuniform general; (c) localized (spots); (d) large pits; (e) small pits; (f) intercrystalline; (g) subsurface.

parameters in order of magnitude can be interrelated (while allowing for atomic masses and densities) as $1 \text{ g/m}^2\text{y} \approx 10^{-4} \text{ mm/y} \approx 10^{-4} \text{ A/m}^2$.

18.5.1. Mechanisms of Corrosion Processes

The analysis of corrosion processes comprises, examining the special features in a given metal's anodic dissolution, establishing the nature of the cathodic reaction (which is coupled with metal dissolution) and defining in greater detail the loci of the anodic and cathodic partial reaction.

In corrosion, the equilibrium potential of reduction of the oxidizing agent is always more positive than that of dissolution of the metal (at the given solution composition). The main cathodic reactions in metal corrosion are hydrogen evolution and the reduction of dissolved oxygen. It is only in special cases when the corresponding reactants are available that chlorine, nitric acid, or other oxidizing agents will be reduced. Hydrogen evolution occurs at much more negative potentials than oxygen reduction. Hence, corrosion coupled with hydrogen evolution can be observed

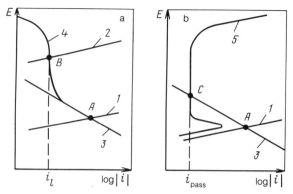

Fig. 18.6. Schematic polarization curves for the spontaneous dissolution: (a) of active metals; (b) of passivated metals. (1, 2) Anodic curves for active metals; (3) cathodic curve for hydrogen evolution; (4) cathodic curve for air-oxygen reduction; (5) anodic curve for the passivated metal.

only for metals having sufficiently negative equilibrium potentials: the alkali and alkaline-earth metals, aluminum, magnesium, zinc, iron, etc., and is encountered predominantly in acidic and alkaline media. Oxygen-depolarized corrosion occurs in contact with air, most often in neutral solutions (atmospheric corrosion, waterline corrosion in seawater, etc.).

It had been pointed out in Section 13.7 that the rate of corrosion (spontaneous dissolution) of metals depends on the shape and position, both of the anodic polarization curve for metal dissolution and of the corresponding cathodic curve, and is determined by the point of intersection of these curves. The anodic curves 1 and 2 in Fig. 18.6a are for metals with a more negative and more positive potential, respectively. For the former, the state of the system corresponds to point A; the corrosion current is high owing to the high rate of hydrogen evolution. For the latter, the potential is in a region where no hydrogen is evolved. Oxygen reduction is the only possible cathodic reaction. Because of the limited solubility of oxygen in water, this reaction occurs with concentration polarization (limiting current density i_l), which imposes a limit on the overall rate of the process: diffusion-controlled corrosion (point B). Oxygen-depolarized corrosion mainly occurs when the liquid film on the metal is thin, so that oxygen access to the electrode is sufficiently fast. In solutions, the rate of oxygen-depolarized corrosion depends on stirring intensity.

Passivation of the metal and the associated sharp decline of its anodic dissolution rate, have a strong effect on corrosion rates (curve 5 and the

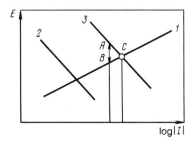

Fig. 18.7. Polarization curves: (1) anodic polarization of bare base metal; (2) cathodic polarization of bare base metal; (3) cathodic polarization in the presence of a foreign inclusion.

point of intersection C in Fig. 18.6b). Passivation is encountered more often under the effect of oxidizing agents, e.g., in the presence of oxygen.

When the metal surface is homogeneous, the anodic and cathodic partial reaction will be uniformly distributed over all surface segments; at different times both the anodic and the cathodic reaction will occur at each individual segment. The surface of a metal's liquid amalgam can be cited as an example of an ideally homogeneous surface. Rather good homogeneity is found as well on annealed surfaces of highly pure solid metals.

A nonuniform distribution of the reactions may arise when the metal's surface is inhomogeneous, and particularly when it contains inclusions of other metals. In many cases (e.g., zinc with iron inclusions), the polarization of hydrogen evolution is much lower at the inclusions than at the base metal, hence, hydrogen evolution at the inclusions will be faster (Fig. 18.7). Accordingly, the rate of the coupled anodic reaction (dissolution of the base metal) will also be faster. The electrode's ocp will become more positive under these conditions. At such surfaces, the cathodic reaction is concentrated at the inclusions, while the anodic reaction occurs at the base metal. This mechanism is reminiscent of the operation of shorted galvanic couples with spatially separated reactions: metal dissolves from one electrode, while hydrogen evolves at the other. Hence, such inclusions have been named *local cells* or microcells.

The idea that metal corrosion could be due to local-cell action was put forward in 1830 by Auguste Arthur de la Rive, and became very popular. An extreme view derived from this idea is the assertion that perfectly pure metals lacking all foreign inclusions will not corrode. However, it does not correspond to reality. It has been established long ago that even superpure metals with homogeneous surfaces are subject to corrosion, which is sometimes rather intense, because of coupled reactions occurring without spatial separation. Thus, local cells constitute an accelerating factor but are not the cause of metal corrosion.

The surface of the base metal is anodically polarized under the effect of local cells. For a graphical analysis of the phenomena, one must construct the polarization curves for the partial currents at the base metal, as well as the overall anodic I_a vs. E curve reflecting the effective rate of dissolution of this metal under anodic polarization. The rate of the cathodic process, I_c, at the inclusions is described by the corresponding cathodic polarization curve (since the surface areas of anodic and cathodic segments differ substantially, currents rather than current densities must be employed here). At open circuit the two rates are identical.

A current (the local cell's short-circuit current) flows between anodic and cathodic surface segments. When the solution's ohmic resistance, R, between these segments is low the resulting ohmic voltage drop can be neglected and the assumption made that the potentials are identical at the two types of surface segment. In this case the current, I_s, of spontaneous dissolution and also the electrode's potential will correspond to the point of intersection of the curves for I_a and I_c (point C in Fig. 18.7). However, in general, the ohmic voltage drop $I_s R$ in the solution cannot be neglected, and the current of spontaneous dissolution of the metal will have a value such that the corresponding potential difference, $E_a - E_c$, between the anodic and cathodic segments will be $I_s R$ (section AB).

Calculations of resistance R are associated with certain difficulties. For disk-shaped inclusions spaced far apart and having radii r_0, the resistance between the edge and center of a disk can be written, under certain assumptions, as

$$R = 2r_0 / \pi \sigma \qquad (18.14)$$

(σ is solution conductivity). In solutions where $\sigma \approx 10^{-3}$ S/m and with inclusions having a radius $r_0 \approx 1$ μm, the resistance is about 1 mΩ; it will have no effect even when the current at the cathodic segment is high, and the metal surface can be regarded as equipotential. For inclusions of large size or solutions with lower conductivities, the ohmic voltage drop as a rule must be taken into account.

The effects of impurities are less important for oxygen-depolarized than for hydrogen-depolarized corrosion, since the values of polarization for oxygen reduction found at different metals differ less strongly than those for hydrogen evolution.

When corrosion occurs under local-cell action, it may happen that the impurity concentration at the metal surface increases with advancing dissolution of the base metal, hence, the corrosion rate will increase.

In solutions of nonelectrolytes (e.g., of iodine in chloroform), metals are oxidized in a direct chemical reaction with the oxidizing agent. It has been shown recently that even in electrolyte solutions, metals may corrode by the chemical mechanism, which leads to kinetic laws differing from those described in Section 13.7 for coupled electrochemical reactions. In electrolyte solutions, the chemical mechanism can be visualized as the intimate local coupling of the two electrochemical process steps, i.e., these steps will occur simultaneously at the same surface site. This has an important effect on the energy requirements of the elementary reaction act. An isolated elementary act is attended by a change in charge of the particle in solution and, hence, by appreciable solvent reorganization, but when the anodic and cathodic act are intimately linked (e.g., when a hydrogen ion in the solution is replaced by a metal ion) the required degree of reorganization is much smaller. In this case the principle of independent electrochemical reactions (see Section 13.7) is no longer valid.

The chemical mechanism is of some importance in the corrosion of Fe, Cr, and Mn, particularly at elevated temperatures.

18.5.2. Corrosion Protection

Electrochemical and nonelectrochemical ways to protect metals against corrosion can be distinguished. The nonelectrochemical ways include dense protective films which isolate the metal against effects of the medium, and which may be paint, polymer, bitumen, enamel and the like. It is a general shortcoming of these coatings that when they are mechanically damaged they lose their protective action, and local corrosion activity arises.

Electrochemical ways of protection rest on different precepts: (a) electroplating of the corroding metal with a thin, protective layer of a more corrosion-resistant metal; (b) electrochemical oxidation of the surface or application of other types of surface layer; (c) control of polarization characteristics of the corroding metal (the position and shape of its polarization curves); and (d) control of potential of the corroding metal.

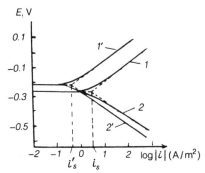

Fig. 18.8. Anodic (1, 1′) and cathodic (2, 2′) polarization curves for an iron electrode in pure 1 M HCl (1, 2) and in 1 M HCl with $5 \cdot 10^{-5}$ M $C_{14}H_{29}NC_5H_9Br$ (1′, 2′) (i_s and i_s' are the current densities of spontaneous dissolution).

The polarization characteristic of a corroding metal can be controlled by various additives to the solution, called *corrosion inhibitors*, which adsorb on the metal and lower the rates of the cathodic and (or) anodic reaction. Inhibitors are mainly used for acidic electrolyte solutions, sometimes also for neutral solutions. Various organic compounds with –OH, –SH, –NH$_2$, –COOH, etc. as the functional groups are used as inhibitors. The effects of an organic inhibitor, tetradecylpiperidinium bromide, on the polarization curves of hydrogen evolution and metal dissolution are shown in Fig. 18.8. This inhibitor markedly lowers the rate of both the anodic and cathodic process. However, since the effect on the anodic process is slightly stronger, the metal's ocp moves in the positive direction. The current of spontaneous dissolution decreases by about one order of magnitude when the inhibitor is present.

A condition for inhibitor action is its adsorption on the metal at the open-circuit potential. Neutral inhibitor molecules will not adsorb when this potential is far from the metal's point of zero charge (see Section 12.7.2). In this case, inhibitors forming ions are used: cations (e.g., from amino compounds) or anions (from compounds with sulfo groups), depending on the sign of surface charge. Inhibitor action is often greatly enhanced when mixtures of several substances are used.

When the metal is cathodically polarized its spontaneous dissolution rate will decrease. The potential of a base metal can be made more negative when this metal is electrically linked to another, more electronegative metal which is present in the same electrolyte medium. This produces a macroscopic galvanic couple where the base metal is cathodically polarized under the effect of the second metal. The latter, called protector, is anodically polarized and gradually consumed by anodic dissolution. This kind of metal protection is called cathodic protection. An example is galvanized iron, where zinc is functioning as a protective film and simultaneously, when the film has been damaged, as cathodic protector.

Sometimes anodic protection is used, in which case the metal's potential is made more positive. The rate of spontaneous dissolution will strongly decrease, rather than increase, when the metal's passivation potential is attained under these conditions. To make the potential more positive one only must accelerate a coupled cathodic reaction, which can be done by adding to the solution oxidizing agents readily undergoing cathodic reduction, e.g., chromate ions. The rate of cathodic hydrogen evolution can also be accelerated when minute amounts of platinum metals, which have a strong catalytic effect, are incorporated into the metal's surface layer (Nikon D. Tomashov, 1949).

The best-known way of lowering the corrosion of iron is by its alloying with chromium, nickel and other metals. The corrosion resistance of the corresponding stainless steels is due to the fact that chromium is readily passivated. This is a quality which is found even in alloys with relatively low chromium contents. Hence, stainless steels are practically always strongly passivated, and their spontaneous dissolution rates are very low.

18.6. ELECTROCHEMICAL METAL TREATMENTS

Different electrochemical surface treatments have found extensive use for the purposes of providing metal parts with particular properties, look and shape. This includes the application of superficial oxide or salt films (see Section 18.3), metal films (Section 18.7), and a number of methods exploiting the selective anodic dissolution of different segments of the metal surface. We shall briefly examine a few examples of the latter type.

Electrolytic etching (pickling) of metals (mainly ferrous) is used to rid the surface of the relatively thick oxide layers (scale, rust, etc.) prior to the application of various coatings. The treatment time is shorter and fewer chemicals are needed than in chemical etching. Etching can be anodic or cathodic. In anodic etching the dissolving metal and also the evolving oxygen bubbles mechanically lift the oxides from the surface.

The reaction is fast, hence, there is a danger of excessive etching. Cathodic etching is based on a partial electrochemical reduction of the oxides and on their mechanical removal from the surface by hydrogen bubbles. It is usually attended by hydrogenation of the metal. In both cases, electrolytes on the basis of sulfuric (more rarely hydrochloric) acid are used; the current densities are between 0.5 and 5 kA/m^2, the anodic etching times are between 1 and 5 min, the cathodic etching times are between 10 and 15 min. Etching of parts having complex profiles will be nonuniform owing to the low throwing power of the bath.

Electrochemical polishing (electropolishing) of metals is used in order to level and smoothen microroughness (of up to 1 μm) on metal surfaces, so that these will become mirror-smooth and more corrosion-resistant, and exhibit less friction. This process will not affect the macroscopic surface roughness, hence, sometimes a prior mechanical surface conditioning is required. Electropolishing is based on the selective anodic dissolution of raised points. Unlike mechanical polishing it will not produce any deformation of the metal's surface layer. Moreover, it is far less laborious and can be used for parts complex in shape. Electropolishing is used for the surface treatment of different steels, aluminum and silver, also nickel and copper coatings, etc.

Electropolishing is performed in concentrated mixtures of acids (sulfuric, phosphoric, chromic, etc.). Often organic acids and glycerol are added. It is somewhat inconvenient that almost all metals and alloys require their own solution composition. For electropolishing, intermediate and high current densities are used, between about 0.1 and 5 kA/m^2. Depending on current density, the process requires between 30 s and 20–30 min. Usually a metal layer 2–5 μm thick is removed under these conditions.

Different views exist as to the reasons for selective dissolution of the asperities. According to older concepts, convection of the liquid is hindered in the solution layers filling recesses, hence, reaction products will accumulate there and raise the concentration and viscosity in these layers. Both factors tend to lower the metal's anodic dissolution rate relative to that at raised points. According to other concepts, a surface condition close to passive arises during electropolishing. In this case, the conditions for passivation of the metal at raised points differ from those in recesses. Formation of a continuous passivating layer is difficult at raised points; the degree of passivation is lower, and dissolution will be faster.

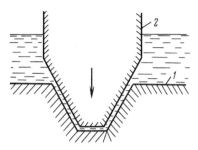

Fig. 18.9. Electrochemical machining of metals: (1) workpiece (anode); (2) tool (cathode).

The overall phenomenon of electrochemical polishing is highly complex, and a variety of factors contribute to the effect of surface leveling.

Electrochemical machining (ECM) of metals. This process rests on the selective local anodic dissolution of metal. It is used to give metal parts the required shape and size, to drill holes, create hollows, cut shaped slots, and fashion parts of complex pattern (e.g., the blades of gas turbines). It is an advantage of this method that it can also be used for hard metals (high-alloy steels and other alloys, metals in the quenched state, etc.).

Selective local dissolution of metals is attained under conditions set up so that the system's throwing power is very low. In particular, very small distances (0.05 to 0.2 mm) are maintained between the cathode and the anode section to be removed. The cathode's configuration is the reverse (negative) of the parts' desired shape (Fig. 18.9). The projecting part of the cathode (tool) is brought to working distance from the workpiece. As anode sections are machined away, the cathode is continuously advanced in the direction of the anode, so that the original distance is maintained; thus continuously new sections of the anode surface are engaged in the process.

Electrochemical machining is performed in concentrated solutions of salts, viz., alkali chlorides, sulfates, or nitrates. Very high current densities are used, viz., hundreds or thousands of kA/m^2 when referring to the surface area of the anodic working sections. At a current density of $100 \ kA/m^2$ the rate of iron dissolution is about 0.15 mm/min. This should also be the rate of advance of the cathode in the direction of the anode. High rates of solution flow through the working gap are used in order to eliminate the reaction products and heat evolved, e.g., flow rates of 10 m/sec.

The ECM accuracy depends on the extent to which it can be avoided that current flow spreads beyond the sections being machined. In order to reduce this spread, conditions are selected under which neighboring sections not to be machined are passivated. Under optimum conditions, parts can be machined to tolerances of 0.1 mm.

18.7. CATHODIC METAL DEPOSITION

Cathodic deposition (electrocrystallization) of metals is the basic process in electrometallurgy and electroplating.

Two kinds of metal deposition are distinguished: that on the same metal and that on a substrate (matrix) consisting of another metal. In the latter case a number of special features can arise.

Underpotential deposition of metal atoms: On account of the energy of interaction between a foreign substrate and the adsorbed metal atoms formed by discharge, cathodic discharge of a limited amount of metal ions producing adatoms is possible at potentials more positive than the equilibrium potential of the particular system, and also more positive than the potential of steady metal deposition.

Incorporation of metal: Metal atoms after their discharge, in certain cases, can penetrate into the substrate metal forming alloys or intermetallic compounds in the surface layer and down to a certain depth. This effect has been known for a long time in the discharge of metals at liquid mercury, where liquid or solid amalgams are formed. It has been shown in recent years that an analogous effect is present in metal ion discharge at many solid metals (Boris N. Kabanov, 1962).

Epitaxy: In many cases, the structure of the deposit will duplicate that of the substrate, when the crystallographic parameters of the metal being deposited are not overly different from those of the substrate metal (not by more than 15%). The influence of substrate structure can extend to several thousand atomic layers of the metal being deposited (which is equivalent to a deposit thickness of up to 1 μm). Epitaxy is pronounced in deposits of copper, silver and zinc.

An important problem in metal deposition is the quality of adhesion of the deposit to the substrate (both foreign and native). In electroplating, one of the major requirements is a high strength and reliability of adhesion. In electrohydrometallurgy and electroforming, metals are often deposited on a foreign matrix, and then separated from it; here, to the contrary, poor adhesion is required. The quality of adhesion—apart from other possible factors—depends on the structure and cleanliness of the substrate surface. Hence, prior to the start of deposition the substrate is carefully conditioned both by mechanical (grinding, polishing) and chemical (etching, degreasing) means.

After formation of a primary deposit layer on foreign substrates, further layer growth will follow the laws of metal deposition on the metal itself. But when the current is interrupted even briefly, the surface of the metal already deposited will become passivated, and when the current is turned back on, difficulties will again arise in the formation of first nuclei, exactly as at the start of deposition on a foreign substrate (see Section 15.3). This passivation is caused by the adsorption of organic additives or contaminants from the solution. Careful prepurification of the solution can prolong the delay with which this passivation will develop.

18.7.1. Polarization in Metal Deposition

In the steady state, polarization in the case of poorly polarizable metals mainly arises from limitations in the rate of reactant supply (concentration polarization) and from phenomena associated with crystallite growth. In the case of highly polarizable metals, an appreciable activation polarization which is associated with the step of metal ion discharge is found, and the contribution due to growth of the new phase is slight against this background. Experience shows that for such metals the values of polarization in electrocrystallization are similar to those in metal ion discharge at the mercury electrode (where a new phase is not formed).

Polarization in metal deposition is highly sensitive to solution composition. A number of complexing agents and surfactants when present in the solution will strongly enhance it. In electroplating, it is very common that metals are deposited from solutions of their complex salts, e.g., cyanide complexes. In such solutions, the metal exists in the form of different complex anions, of the type of $M(CN)_k^{(k-z+)-}$ with different values of k, which are in mutual equilibrium (see Section 10.9). The surface concentration of the anions will be lower than the bulk solution concentration, since cathodic metal deposition most often occurs at potentials where the surface charge and the ψ'-potential have negative values (see Section 14.2). This effect is more pronounced the higher the anion's charge. One can assume, therefore, that only the species having low negative charge $(k - z_+)Q^0$ will be discharged, even when their relative concentration is low.

The strong polarization observed in the discharge of these ions can have a number of reasons: a slow chemical step replenishing the supply of reacting species by the dissociation of complexes with higher values of k, slow ligand desorption from the surface after discharge of the ions, etc. Sometimes a drop in current is observed after crossing of the pzc; this is an effect of electrostatic repulsion of the anions from the negatively charged surface (see Section 14.5).

The influence of surfactants on polarization can also have a number of reasons: their effect on edl structure (change in ψ'-potential) or the formation of layers fairly strongly adsorbed on the electrode and hindering passage of the reacting species (see Section 14.3.2).

In almost all cases, typical peak values of enhanced polarization due to nucleation of the new phase are seen at the initial stage of the process.

18.7.2. Parallel Reactions

For many metal/solution systems, the equilibrium electrode potential is more negative than the equilibrium potential of the reversible hydrogen electrode in the same solution. However, even if it is not, polarization during cathodic metal deposition may push the electrode potential to a value more negative than that of the rhe. Then cathodic hydrogen evolution will occur in parallel with metal ion discharge (so long as the polarization of hydrogen evolution at the metal is not very high). A typical example is the cathodic deposition of zinc from acidic zinc sulfate solution, which is attended by copious hydrogen evolution. Under these conditions the current yields of metal deposition depend on many factors: the current density, solution composition, etc.

Parallel hydrogen evolution can have two effects.

(i) In contrast to metal ion discharge, hydrogen evolution according to reaction (3.60) causes a pH increase of the solution layer next to the cathode. At a certain value of pH in this layer, hydroxides or basic salts of the metal start to precipitate, which affects the mechanism of further metal deposition and also the structure and properties of the deposit produced.

(ii) In simultaneous hydrogen ion discharge, the resulting adsorbed hydrogen atoms are incorporated into the growing metal deposit. This alters the mechanical properties of the metal; hydrogen embrittlement occurs. This effect is pronounced in the iron-group metals, but less distinct in copper and zinc. It is not observed in cadmium or lead deposition.

When ions of different metals, M^{z+} and N^{z+}, are present together in the solution, simultaneous discharge of the ions and alloy formation between the two metals is possible. To this end the electrode potential must be more negative than that of the more electronegative metal, M. When the concentration of ions N^{z+} of the more electropositive metal is low, the electrode potential will shift to the said value as a result of concentration polarization. However, when the concentration of these ions is high, the equilibrium potentials of the two metals are highly different, and metal N is poorly polarizable, discharge of the more electronegative ions will be hindered. In this case the equilibrium potentials of the two metals can be moved closer together by suitably selected complexing agents. The electrolytic formation of metallic alloys is practiced in a number of technical fields.

The refining (purification) of metals rests on the separate dissolution and cathodic redeposition of different metals. In the anodic dissolution of the original metal sample, other more electronegative metals present as impurities dissolve together with the base metal. More electropositive impurity metals will not dissolve, but sediment as a sludge which can then be separated from the solution. During the ensuing cathodic deposition of base metal from the solution obtained, the electronegative impurity metals will not deposit at the potential selected but remain in solution. In this way a rather complete separation of the base metal from other metals, both more electropositive and more electronegative, is attained.

Highly pure metals can be obtained by prior fractional metal deposition at mercury and subsequent fractional redissolution from the resulting amalgam.

18.7.3. Structure of the Metal Deposits

In cathodic metal deposition as practiced in electrohydrometallurgy and electroplating, the main effort is toward producing compact deposits having low porosity. In other requirements which should be met by the deposits, these two areas of electrochemical technology are different. It is the principal task of electrohydrometallurgy to produce metal of a certain purity. However, in electroplating mechanical strength, hardness, freedom from internal stresses and certain optical properties of the films are required. Hence, in electroplating a large research effort is made in order to learn about the different factors influencing the structure of metallic coatings.

The structure of metallic deposits is determined primarily by the size, shape (faceting), type of arrangement and mutual orientation of the crystallites. Two factors may influence the orientation and spatial alignment of the microcrystals in electrocrystallization: the field direction (or direction of the electric current) and the nature of the substrate. The deposits are said to have texture when the crystallites are highly oriented in certain directions. Epitaxy implies that the lattice is altered under the influence of the substrate.

Different ways of a structural classification of deposits exist. In one system, the following structures are distinguished arbitrarily: (a) fine-crystalline deposits lacking orientation, (b) coarse-crystalline deposits poorly oriented, (c) compact textured deposits oriented in field direction (prismatic deposits), and (d) isolated crystals with a predominant orientation in field direction (friable deposits, dendrites).

The structure of metal deposits depends on a large number of factors: solution composition, the impurities present in the solution, the current density, surface pretreatment, etc.

A very general rule can be formulated: in metal deposition, higher activation polarization will favor the formation of fine-crystalline, compact deposits. Friable, coarse-crystalline deposits are formed in the deposition of poorly polarizable metals in solutions of simple salts, but relatively compact, fine-crystalline deposits are formed in the deposition of metals having high polarizability. A strong increase in polarization, and hence, the formation of fine-crystalline deposits, will result when various complexing agents or surfactants are added to salt solutions of poorly polarizable metals.

This situation is rather easy to explain. If the primary step of metal ion discharge is hindered, the appreciable electrode polarization associated with it will compensate the energetic difficulties of formation of new metal nuclei, and lead to the formation of more nuclei. Here, the overall charge is distributed over a large number of nuclei, and any individual nucleus will not undergo much further growth.

The formation of new nuclei and of a fine-crystalline deposit will also be promoted when a high concentration of the metal ions undergoing discharge is maintained in the solution layer next to the electrode. Therefore, concentration polarization will have effects opposite to those of activation polarization. Rather highly concentrated electrolyte solutions, vigorous stirring and other means are employed to reduce concentration polarization. Sometimes special electrolysis modes are employed for the same purposes, viz., currents which are intermittent, reversed (i.e., with periodic inverted, anodic pulses), or asymmetric (an ac component superimposed upon the dc).

Higher current densities and the associated increase in polarization also favor the formation of compact, fine-crystalline deposits. At very high cd approaching the limiting diffusion current, often the opposite effect is seen, viz., owing to depletion of reacting ions in the solution layer next to the electrode nonuniform crystal growth starts, i.e., an enhanced growth of projecting points of the deposit in the direction of the incoming flux of ions to be discharged. Then friable, spongy deposits, sometimes even distinct dendrites (branched or acicular treelike crystals) are formed. Electrolysis at currents close to the limiting cd is often used to produce metal powder, which spontaneously separates from the substrate. At high cd the coating quality may also deteriorate because of simultaneous hydrogen evolution, and because of the metal hydroxides and basic salts which are formed under these conditions owing to the attendant increase in solution pH.

Surfactants have very diverse effects on electrolytic metal deposition. Most often they drastically raise the polarization of the electrode, and hence, cause formation of fine-crystalline deposits. The surfactant effects strongly depend on the relative rates of new surface formation and adsorption of the surfactant on this surface, since in electrochemical deposition new sections of metal surface continuously form and grow. There are cases when the surfactants will adsorb on slowly growing faces (e.g., the lateral face of whiskers) but do not manage to become adsorbed on the rapidly growing front face (tip). This leads to an enhanced growth of dendritic crystals. In other cases, to the contrary, surfactants contribute to leveling of the surface profile and to the elimination of macroscopic roughness on the surface (brighteners).

It follows from the above that electrolyte solutions having a complex composition must be used in order to obtain high-quality metal coatings. As a rule, these solutions contain the following components: (a) salt of the metal being deposited, (b) a complexing agent, (c) base electrolyte (neutral salt, to increase the conductivity), (d) buffer agents helping to maintain the optimum value of solution pH, (e) additives to reduce anode passivation, (f) surfactants, etc.

18.8. REACTING NONMETAL ELECTRODES

In addition to metals, other substances which are solids and have at least some electronic conductivity can be used as reacting electrodes. During reaction, such a solid is converted to the solid phase of another substance (this is called a solid-state reaction), or soluble reaction products are formed. Reactions involving nonmetallic solids occur in batteries, where various oxides (MnO_2, PbO_2, NiOOH, Ag_2O, and others) and insoluble salts ($PbSO_4$, AgCl, and others) are widely used as electrode materials. These compounds are converted in an electrochemical reaction to the metal or to compounds of the metal in a different oxidation state.

Nonmetal electrodes are most often fabricated by pressing or rolling of the solid in the form of fine powder. For mechanical integrity of the electrodes, binders are added to the active mass. For higher electronic conductivity of the electrode and a better current distribution, conducting fillers are added (carbon black, graphite, metal powders). Electrodes of this type are porous and have a relatively high specific surface area. The porosity facilitates access of dissolved reactants (H^+ or OH^- ions and others) to the inner electrode layers.

The mechanism of solid-state reactions is not a simple one. In the reaction the lattice of one substance is destroyed, and that of another substance is freshly formed. Three types of solid-state reactions are distinguished.

Reactions via the solution phase. In the case of substances having at least some solubility (e.g., 10^{-6} mol/liter), the reaction often follows a scheme involving dissolved species:

$$A_S \rightleftarrows A_{sol} \overset{\pm ne^-}{\rightleftarrows} B_{sol} \rightleftarrows B_S. \tag{18.15}$$

This mechanism is followed in particular in the reactions of lead sulfate in the electrodes of lead–acid batteries [reactions (16.2) and (16.3)].

Reactions of this type can also occur when the conductivity of one of the phases is very low or practically zero. In these reactions, the sites of reactant lattice destruction and product lattice formation are spatially separated. During the reaction, dissolved species diffuse from the dissolution sites to the sites where they undergo further reaction and form the nuclei of the new phase. The length of the diffusion pathway in the solution depends on the degrees of dispersion of the original reactant and resulting product, and most often is between 10^{-5} and 10^{-3} m.

Topochemical reactions are those in which all steps, including that of nucleation of the new phase, occur exclusively at the interface between two solid phases, one being the reactant and the other the product. As the reaction proceeds, this interface gradually advances in the direction of the reactant. In electrochemical systems, topochemical reactions are possible only when the reactant or product is porous enough to enable access of reacting species from the solution to each individual reaction site. Few examples are known where an electrochemical reaction follows a truly topochemical mechanism.

Insertion reactions (reactions producing phases of varying composition). An example for insertion reactions is the cathodic reduction of manganese dioxide, which occurs during discharge of the positive electrodes in zinc–manganese dioxide cells. This reaction can be formulated as

$$MnO_2 + xH^+ + xe^- \rightarrow MnOOH_x. \qquad (18.16)$$

During the reaction protons, which have been produced from water molecules or from the hydroxonium (H_3O^+) ions of the solution, are inserted into the manganese dioxide lattice. At the same time an equivalent number of Mn^{4+} ions of the lattice are reduced to Mn^{3+} ions by the electrons arriving through the external circuit. Hence, the overall balance of positive and negative charges in the lattice remains unchanged. Even the lattice structure of manganese dioxide, up to a certain limit, will remain unchanged during proton insertion, though there may be some minor increase in the lattice parameters. Thus, the reaction product—Mn(III) oxide—accumulates without forming a new phase; the homogeneity of the original reactant phase is preserved. Hence, reactions of this type are also known as homogeneous solid-state reactions.

The degree of reduction in a phase of varying composition can be described by the ratio x between the number of Mn^{3+} ions and the total number of manganese ions in the lattice or, what amounts to the same thing, the ratio between the number of protons and the total number of manganese ions.

When the proton concentration x in the lattice has attained a certain limit the system becomes unstable, and a spontaneous phase change may occur, producing a new and independent MnOOH phase. The phase change itself follows a topochemical mechanism. Beyond this point, when reduction continues, the phase composition of reactant and product remains constant and only their relative amounts change (in a heterogeneous solid-state reaction). The phase change in manganese dioxide occurs at degrees of reduction, x, of 0.5 to 1 depending on the conditions.

The discharge of nickel oxide electrodes in alkaline storage batteries follows a similar mechanism, where protons are inserted into NiOOH and $Ni(OH)_2$ is formed. During charging, the reaction occurs in the opposite direction; protons are extracted from the nickel hydroxide lattice into the solution. In the case of manganese dioxide, the reaction can also be reversed within the homogeneous region. However, after onset of the phase change and formation of an independent MnOOH phase, it becomes practically noninvertible, and such an electrode can no longer be recharged.

It has been detected in recent years that a number of solids (certain oxides and other compounds of transition metals) can reversibly insert (intercalate), not just protons but other cations and anions as well, e.g.,

$$TiS_2 + xLi^+ + xe^- \rightleftarrows TiS_2Li_x. \tag{18.17}$$

These insertion compounds are of great interest for the design of new types of storage batteries, and particularly of batteries using lithium electrodes in nonaqeuous electrolytes.

That ions can be inserted into these compounds, is due to the fact that peculiar channels along which the foreign ions can move are present in their lattice structures. Certain compounds have a layered lattice structure where the gaps between planes are wide enough for ions from the solution to be inserted (they resemble graphite, which incidentally also has a disposition for ion insertion).

In heterogeneous solid-state reactions the electrode's equilibrium potential depends only on the nature of the two phases but not on their relative amounts. Hence, when the current is interrupted after partial reduction or oxidation, the electrode's potential is not a function of the oxidation state. However, when the solid-state reaction is homogeneous and a phase of varying composition is formed, there will be a smooth change in equilibrium potential as the degree of oxidation changes. Hence, the discharge curves of batteries in which homogeneous solid-state reactions take place (e.g., the reduction of manganese dioxide, curve 1 in Fig. 18.10) exhibit a gradually decreasing discharge voltage, while those of batteries in which

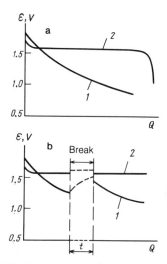

Fig. 18.10. Discharge curves of cells with nonmetal electrodes as the positive electrode: (a) continuous discharge; (b) interrupted discharge. (1) MnO_2; (2) Ag_2O. (A) Current interruption.

a heterogeneous process takes place (e.g., the reduction of silver oxide, curve 2) exhibit a discharge voltage which for a long time is constant.

A typical feature in homogeneous solid-state reactions is the slow diffusion of the inserted species in the host lattice. In any individual grain, therefore, a nonequilibrium concentration distribution of these species is found during the reaction; the concentration changes are more rapid at the surface than inside the grain. This gives rise to hysteresis effects. Thus, when a manganese dioxide electrode is partly discharged at a high current density and then the current is interrupted, the potential will be rather negative directly after the interruption. Then diffusional leveling of the concentration within the grains sets in, and the potential shifts in the positive direction until it reaches the equilibrium value for the actual oxidation state. This is the reason for the so-called "recovery" of zinc–manganese dioxide cells during breaks in discharge (Fig. 18.10b); in silver–zinc cells such an effect is not observed.

Chapter 19
REACTIONS AT NONCONSUMABLE ELECTRODES

19.1. HYDROGEN EVOLUTION AND IONIZATION

The equations of these reactions in aqueous solutions are:

$$2H^+ + 2e^- \rightleftarrows H_2, \qquad (19.1a)$$

$$2H_2O + 2e^- \rightleftarrows H_2 + 2OH^-. \qquad (19.1b)$$

Cathodic hydrogen evolution is one of the most common electrochemical reactions. It is the principal reaction in electrolytic hydrogen production, the auxiliary reaction in the production of many substances forming at the anode, such as chlorine, and a side reaction in many cathodic processes, particularly in electrohydrometallurgy. It is of considerable importance in the corrosion of metals. Its special characteristic is the fact that it can proceed in any aqueous solution; particular reactants need not be added. The reverse reaction, which is the anodic ionization of molecular hydrogen, is utilized in chemical power sources.

The equilibrium potential of the hydrogen electrode is established at electrodes of platinized platinum and other finely divided platinum-group metals, both in acidic and alkaline solutions. In the latter, it is also established at electrodes of finely divided nickel, and in the former, at electrodes of finely divided tungsten carbide WC. All these electrodes are sensitive to contamination; they will lose their activity even at low concentrations of an adsorbing impurity, so that their open-circuit potential (ocp) no longer corresponds to the equilibrium potential of the hydrogen reaction. In addition, tungsten carbide and nickel are sensitive to oxidizing action. At metals other than the above, the hydrogen potential is not established because of the low exchange current densities of the hydrogen reaction and also because of interference from spontaneous dissolution found for many metals (and particularly iron) in the region of the equilibrium hydrogen potential.

Most metals (other than the alkali and alkaline-earth metals) are corrosion-resistant when cathodically polarized to the potentials of hydrogen evolution, so that this reaction can be realized at many of them. It thus has been the subject of innumerable studies, and became the fundamental model in the development of current kinetic concepts for electrochemical reactions. Many of the principles presented in Chapters 6, 13, and 14 have been established, precisely in studies of this reaction.

In the past, elevated voltages in electrolysis cells (a cell overvoltage) had been attributed mainly to polarization of the hydrogen evolution reaction. Hence the term of "hydrogen overvoltage" became common for this kind of polarization.

The range of current densities encountered in cathodic hydrogen evolution is extremely wide. Corrosion of metals may become important at rates of hydrogen evolution which are of the order of 10^{-4}–10^{-3} A/m^2, while in industrial electrolyzers, where this reaction occurs at the cathode, the current densities attain values of 10^4 A/m^2 and more.

In alkaline solutions the equilibrium potential of the hydrogen reaction and the true potentials of hydrogen evolution (which include polarization) are moved to rather more negative values. Hence, the reaction is difficult to study: at many electrodes, discharge of the solution cations and an incorporation of the resulting metal atoms into the lattice of the electrode material (see Section 18.7) are possible in addition to hydrogen evolution. Owing to the energy of alloy formation, these additional processes will occur at potentials more positive than the equilibrium potential of the particular metal, and the contribution of metal ion discharge to the total current recorded during cathodic polarization may be large.

At intermediate values of pH, buffered solutions are used in order to avoid marked changes in solution acidity caused by the consumption of hydroxonium ions or formation of hydroxyl ions.

Figure 19.1 shows polarization curves for hydrogen evolution at electrodes of different metals in acidic electrolyte solutions. The results of polarization measurements are highly sensitive to the experimental conditions, and in particular to the degree of solution and electrode surface purification; for this reason marked differences exist between the data re-

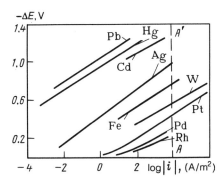

Fig. 19.1. Polarization curves for hydrogen evolution at different metals in acidic solutions.

ported by different workers. The curves shown still provide the correct picture of the common features.

It can be seen from Fig. 19.1 that in semilogarithmic plots of ΔE against $\log|i|$ the polarization characteristics are linear, i.e., obey the Tafel equation (6.3). Slopes b' practically coincide for most metals, and have values of 0.11 to 0.13 V. However, the absolute values of polarization recorded for a given current density (cd) vary within wide limits. The values of the exchange cd (intersections of the extrapolated polarization curves with the horizontal axis) and of constant a in the Tafel equation differ accordingly. Traditionally, the current densities are stated in A/cm^2 rather than A/m^2 when this equation is used; hence, the values of constant a reported in the literature refer to a current density of 1 A/cm^2 = 10^4 A/m^2 (intersections with the line AA').

The highest values of polarization are observed for the s- and p-metals: lead, mercury, and certain others ("metals with high hydrogen overvoltage"), the lowest are observed for transition metals (d-metals), and particularly for the platinum-group metals ("metals with low hydrogen overvoltage"). The constants a for lead and rhodium differ by 1.25 V, according to the data of Fig. 19.1. When allowing for the slopes of the polarization curves, this corresponds to a change in reaction rate at any given value of potential by a factor of about 10^{11}; the exchange cd varies between 10^{-14} A/m^2 for lead and $3 \cdot 10^{-3}$ A/m^2 for rhodium. Such a strong variation in reaction rate between different metal electrodes is particularly typical for cathodic hydrogen evolution.

In aqueous solutions, approximately one atom of deuterium, D, is present for every 7000 atoms of the ordinary hydrogen isotope (protium, H). In the evolution of "heavy" hydrogen HD, the polarization is approximately 0.1 V higher than in the evolution of ordinary hydrogen H_2. Hence during electrolysis the gas will be richer in protium, and the residual solution will be richer in deuterium. The relative degree of enrichment is called the *separation factor* (S) of the hydrogen isotopes,

$$S = (D/H)^L/(D/H)^{(G)} \qquad (19.2)$$

(here H and D are the corresponding atom fractions of protium and deuterium).

The values of S depend on the material and surface state of the electrode, the solution composition, and other factors, and at 75°C vary between 3–4 for mercury and platinized platinum, and 7–8 for iron and graphite, i.e., are unrelated to the overall value of electrode polarization. Measured values of S are somewhat reduced by isotope exchange between the gas and liquid:

$$H_2 \text{ (G)} + \text{HDO (L)} \rightleftarrows \text{HD (G)} + H_2O \text{ (L)}, \qquad (19.3)$$

which interferes with the isotope separation and is catalytically accelerated to a different extent by different metal surfaces. The higher values of S found at lower temperatures and higher current densities can be explained in terms of a lower influence of this reaction.

Unlike the cathodic reaction, anodic oxidation (ionization) of molecular hydrogen can only be studied at a few electrode materials, which include the platinum group metals, tungsten carbide, and in alkaline solutions nickel. Other metals either are not sufficiently stable in the appropriate range of potentials or prove to be inactive toward this reaction. For the materials mentioned, it can be realized only over a relatively narrow range of potentials. Adsorbed or phase-oxide layers interfering with the reaction form on the surface at positive potentials. Hence, as the polarization is raised the anodic current first will increase, then decrease (see Fig. 18.3), i.e., the electrode becomes passive. In the cases of nickel and of tungsten carbide, these changes may prove to be irreversible; the oxide layer cannot be completely reduced after a return to more negative potentials, and the activity remains low.

19.1.1. Mechanism of the Hydrogen Evolution Reaction

Cathodic hydrogen evolution is a complex two-electron reaction occurring through several consecutive, simpler intermediate steps. Each of these steps is sometimes referred to with the name of the scientist who had suggested that it was rate-determining for the overall reaction; the steps are:

A. the *discharge* (Volmer) step,

$$H_3O^+ + e^- \rightleftarrows H_{ads} + H_2O; \qquad (19.4)$$

B. the *recombination* (Tafel) step,

$$2H_{ads} \rightleftarrows H_2; \qquad (19.5)$$

C. the *electrochemical desorption* (Heyrovský) step,

$$H_3O^+ + H_{ads} + e^- \rightleftarrows H_2 + H_2O. \qquad (19.6)$$

These three steps can combine in different ways to three possible reaction pathways, I, II, and III. We shall represent them as a table analogous to that reported in Section 13.1:

Step k	l_k	$\mu_k(I)$	$\mu_k(II)$	$\mu_k(III)$
A	1	2	1	—
B	0	1	—	−1*
C	1	—	1	2

Note. The minus sign signifies that in cathodic hydrogen evolution following the third pathway, step B follows Eq. (19.5) from right to left.

It is important, not only to establish the true reaction pathway followed under different experimental conditions, but also to decide which of the two consecutive steps is rate-determining in a given pathway.

It had been thought in the past that cathodic hydrogen evolution follows pathway I constituting the sequence of a discharge and a recombination step. In 1905 Julius Tafel put forward the opinion that the recombination step is rate-determining while the discharge step is in equilibrium (the recombination theory of hydrogen overvoltage). The physical basis of this theory is the fact that the recombination of two atoms will not occur instantaneously (in every collision) but with some finite, low rate. This is due to the considerable energy liberated in the union of the two atoms; this energy is concentrated in the resulting molecule and causes it to redissociate into atoms. At different catalysts, and particularly at metals, recombination is accelerated by the elimination of part of the excess energy by the catalyst surface. A strong argument in favor of the recombination theory were the parallels existing between the catalytic activities of the various metals in the surface recombination of adsorbed hydrogen atoms in the gas phase and their activities in cathodic hydrogen evolution. The rates of both reactions increase in the order of Pb < Zn < Ag < Fe < Pt.

Slow recombination of hydrogen atoms has the effect that the surface concentration (degree of surface coverage) of the hydrogen atoms under cathodic current, $\theta_{H,i}$, is higher than that at zero current, $\theta_{H,0}$. According to recombination theory, polarization is precisely due to these changes, because step A is in equilibrium. Assuming that the activity of an adsorbed hydrogen atom is proportional to its surface concentration we have

$$-\Delta E = (RT/F)\ln(\theta_{H,i}/\theta_{H,0}). \tag{19.7}$$

On a homogeneous surface, the rate of the atom recombination step will be proportional to the square of the atom surface concentrations:

$$v_B = k_B \theta_{H,i}^2. \tag{19.8}$$

In a reaction following the first pathway, this rate will be proportional to the total current i: $-i = 2Fv_B$ (regardless of which step is the slow step). Combining the last two equations we obtain

$$-\Delta E = \text{const} + (RT/2F)\ln|i|. \tag{19.9}$$

Thus, the recombination theory provided the first theoretical interpretation of the linear relation between polarization and the logarithm of current density that had been established experimentally. It is true, though, that the preexponential factor in Eq. (19.9) [$2.303(RT/2F) \approx 0.03$ V] is four times smaller than the experimental values of slope b'; but it has been shown in later work that factors closer to the experimental values can be obtained when an inhomogeneous surface is assumed.

A basic defect of these ideas is their failure to provide an explanation of the substantial effects of solution composition, and in particular of the pH value, on the rate of the electrochemical reaction. Since hydrogen ions are not involved in the recombination step, the rate of this step according to Eq. (19.8) should not depend on solution pH. Yet in many cases the rate of hydrogen evolution at constant potential is proportional to the hydrogen ion concentration in solution.

The idea that the discharge of different ions could be slow was suggested already at the end of the 19th century. The slowness was attributed to the appreciable energy needed to break up the complexes (including those with solvent molecules) formed by the ions in solution. According to current concepts, another contribution to the activation energy arises from the need for reorganization of the solvent molecules close to the ions undergoing discharge.

In 1930 Max Volmer and Tibor Erdey-Grúz used the concept of a slow discharge step for cathodic hydrogen evolution (slow discharge theory). According to these ideas, the potential dependence of electrochemical reaction rate constants is described by Eq. (14.13). Since hydrogen ions are involved in the slow step A, the reaction rate will be proportional to their concentration. Thus, the overall kinetic equation can be written as

$$-i = -i_A = Fv_A = Fk_A c_{H^+} \exp(-\beta_A FE/RT). \qquad (19.10)$$

This equation explains the experimental reaction rates, both as a function of potential (under the plausible assumption that $\beta_A \approx 0.5$) and as a function of solution pH.

The idea that the adsorbed hydrogen atoms which have been produced by discharge can be eliminated from the electrode surface by recombination is faced with difficulties no matter which of the steps is the slow step. The degree of surface coverage, $\theta_{H,i}$, by adsorbed atoms can increase only up to a limiting value of unity. According to Eq. (19.8), at this point the limiting recombination rate is attained, and thus the limiting rate of the overall reaction, which cannot be faster than any of its consecutive steps. However, for hydrogen evolution such a limiting current is not observed experimentally (in the absence of concentration polarization with respect to H_3O^+ ions). Another difficulty are the improbably high values obtained for constant k_B in Eq. (19.8) in the case of metals where the values of $\theta_{H,i}$ are known to be low (e.g., mercury).

These difficulties can be avoided when it is assumed that the reaction follows pathway II, i.e., that the adsorbed hydrogen atoms are eliminated from the surface via an electrochemical desorption step. This step had been suggested first by Jaroslav Heyrovský in 1925.

For electrodes with homogeneous surfaces, the rate of this step can be described as

$$-i_C = Fv_C = Fk_C c_{H^+} \theta_{H,i} \exp(-\beta_C F/RT). \qquad (19.11)$$

The steady-state coverage, $\theta_{H,i}$, of the surface by hydrogen atoms can be found from the balance of their rates of formation and elimination. Discharge of the ions occurs only at sites free of adsorbed hydrogen. Hence the reaction rate will be proportional to the fraction of free surface when considerable amounts of hydrogen are present on the surface, and it can in brief be written as $Fh_A c_{H^+}(1 - \theta_{H,i})$. The rate of electrochemical desorption can be written, similarly, as $Fh_C c_{H^+}\theta_{H,i}$. In the steady state these two rates will be identical, which implies that

$$\theta_{H,i} = h_A/(h_A + h_C), \tag{19.12}$$

i.e., the degree of coverage always remains below unity, and the overall current is not limited by the attainment of complete surface coverage.

For a reaction following the second pathway, the total current is given as the sum of currents of the two steps: $i = i_A + i_C = 2i_A = 2i_C$. Hence

$$-i = 2Fh_C c_{H^+}\theta_{H,i} = 2Fc_{H^+}h_A h_C/(h_A + h_C). \tag{19.13}$$

The rates of the individual steps of cathodic hydrogen evolution at different metals are a complex function of bond energy $\bar{q}_{M-H}$ of the hydrogen adsorbed at the metal surface. For the discharge step, the rate first increases with increasing bond energy because of decreasing activation energy (see Section 14.4.2), then it goes through a maximum, and finally it decreases again owing to increasing hydrogen adsorption, i.e., decreasing fraction of free surface. A similar situation is found for the recombination and the electrochemical desorption step: the reaction first is accelerated with increasing bond energy owing to increasing surface concentration, but then it becomes slower owing to increasing activation energy of desorption.

A detailed analysis of these effects shows that in most cases the reaction follows the second pathway. For metals with low and intermediate bond energies (the s- and p-metals and some transition metals) slow discharge followed by electrochemical desorption is more likely. For the transition metals with high bond energy (nickel, tungsten, tantalum, etc.) slow electrochemical desorption is more likely. A reaction following the first pathway with a slow recombination step evidently is observed, only for activated electrodes of platinum metals at low values of polarization. Often conditions exist when different steps have comparable rate constants, i.e., when there is no clear-cut rate-determining step. No unambiguous evidence is available so far that the reaction may also follow the third pathway.

19.1.2. Influence of Solution Composition

Hydrogen can be evolved, not only as a result of discharge of hydroxonium ions H_3O^+ but also by discharge of other proton donors HA which may be present in the solution, including the water molecules themselves:

$$2HA + 2e^- \rightleftarrows H_2 + 2A^-, \tag{19.14}$$

where HA = H_3O^+, H_2O, NH_4^+, RSH, etc. It may then be asked what the relative contributions of reactions involving different proton donors are. As a general rule, in acidic solutions where the concentration of H_3O^+ ions is high, it will be precisely these ions which provide the main contribution to the overall current; but in alkaline solutions where their concentration is very low, hydrogen is mainly evolved by discharge of water molecules. In hydroxonium ion discharge, the reaction rate at a given potential E, according to Eqs. (19.10) and (19.11), will be proportional to the concentration of these ions, to a first approximation. It follows from these equations that $(\partial E/\partial \log c_{H^+})_i = 2.303 RT/\beta F$. The electrode potential shifts by about 0.12 V in the negative direction when the pH is raised by a unit, since $\beta \approx 0.5$ (Fig. 19.2, curve 1). In alkaline solutions where water molecules are discharged, the reaction rate at a given potential (or the potential at a given rate) is independent of solution pH. Since the equilibrium potential of the hydrogen electrode moves 0.06 V in the negative direction when the pH is raised by a unit (curve 3), polarization under these conditions will decrease. A plot of polarization against pH for constant current goes through a maximum (curve 2). In acidic solutions the slope of the curve is approximately 0.06 V, in alkaline solutions it is −0.06 V.

Other proton donors can display their influence, mainly in neutral solutions where the polarization values for the discharge of H_3O^+ ions and water molecules are highest.

The composition of the electrolyte solution influences the rate of hydrogen evolution, not only in a direct way, which is through the concentrations of species undergoing discharge, but also in an indirect way, through its effect on electric double-layer structure, and in particular through the value of the ψ'-potential. The general equation (14.16) which describes this effect becomes, for the reaction considered here (slow discharge in acidic solutions),

$$-i = Fk_A c_{H^+} \exp[-(1 - \beta_A)F\psi'/RT] \exp(-\beta_A FE/RT). \tag{19.15}$$

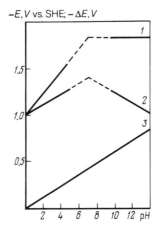

Fig. 19.2. The pH dependence of potential (1) and polarization (2) in cathodic hydrogen evolution at a mercury electrode (100 A/m²), and the pH dependence of equilibrium potential of the hydrogen electrode (3).

The ψ'-potential is negative for most metals in the region of cathodic hydrogen evolution. It depends, both on the total electrolyte concentration and on specifically adsorbing, surface-active ions present in the solution. Since the value of the ψ'-potential is close to that of the ψ_2-potential, Eq. (12.48) can be used to describe the concentration dependence.

Consider a few particular cases. The relation between polarization and solution pH at constant current which is reported in Fig. 19.2, refers to the case where there is no marked change in total solution concentration when the pH is varied (e.g., at high concentrations of a foreign or buffer electrolyte). When the measurements are made in pure acid solutions not containing a foreign electrolyte, the absolute values of the ψ'-potential will decrease with increasing acid concentration, i.e., this potential assumes less negative values. Putting function (12.48) into Eq. (19.15) we obtain

$$-i = F k'_A c_{H^+}^{\beta_A} \exp(-\beta_A F E / RT). \qquad (19.16)$$

It follows from this equation that in the case being considered, the reaction rate is proportional to a fractional power of hydrogen ion concentration. The value of $(\partial E / \partial \log c_{H^+})_i = 2.303 RT / F$, i.e., the potential shifts by 0.06 rather than 0.12 V in the negative direction when the solution pH is raised by a unit, while the value of polarization is independent of solution pH. When an excess of foreign electrolyte is added to the dilute solution of a pure acid, the reaction rate will decrease since the ψ'-potential shifts in the positive direction. The influence of surface-active ions (e.g., I^- or $[NH_4]^+$) on hydrogen evolution which arises through changes in ψ'-potential was described in Section 14.3.2 (see Fig. 14.6).

19.1.3. Ways of Influencing the Hydrogen Evolution Reaction

In industrial electrolyzers where hydrogen evolution occurs at the cathodes (as the principal or auxiliary reaction), any polarization of this reaction will raise the working voltage ("overvoltage"), and thus the electrical energy consumption, hence, it is a fundamental problem to reduce polarization to the largest possible extent. When hydrogen evolution is an undesirable side reaction (as, e.g., in metal deposition in electrohydrometallurgy), the problem to the contrary is that of raising the polarization of this reaction. This is even more important in corrosion processes attended by hydrogen evolution.

For lower values of polarization in hydrogen evolution one must first of all select the proper electrode materials. Metals and alloys of the iron group are most suitable under this aspect in alkaline solutions: iron, steel, binary and ternary alloys with nickel and cobalt. In certain cases, these materials can be used without noticeable corrosion, even in neutral and weakly acidic solutions, since during hydrogen evolution the solution next to the cathode surface becomes alkaline. In acidic solutions carbon or graphite cathodes are used most often. Electrodes containing platinum-group metals are used when it is necessary to lower the working voltage of the electrolyzers to the largest possible extent.

It is of great value in efforts to reduce polarization, to raise the true working surface area by using strongly roughened or even porous electrodes instead of smooth electrodes. Such electrodes find ever increasing use in industry.

Higher values of polarization in hydrogen evolution are attained mainly by the addition of different surfactants, both neutral and cationic, to the electrolyte. But other methods are used as well. For instance, in batteries the zinc electrodes are often amalgamated in order to reduce their corrosion (at amalgams, like at mercury, the rate of hydrogen evolution is low). For environmental reasons, though, the use of mercury in batteries should be minimized.

19.2. REACTIONS INVOLVING OXYGEN

The equations of the principal reaction in aqueous solutions are,

$$O_2 + 4H^+ + 4e^- \rightleftarrows 2H_2O, \qquad (19.17a)$$

$$O_2 + 2H_2O + 4e^- \rightleftarrows 4OH^-. \qquad (19.17b)$$

Anodic oxygen evolution is the principal reaction in electrolytic oxygen production, the auxiliary reaction in the production of many substances forming at the cathode, as for instance in electrohydrometallurgy, and a side reaction in a number of anodic processes, particularly in the production of chlorine and other oxidizing agents. Like hydrogen evolution, this reaction occurs in aqueous solutions without the addition of special reactants. The reverse reaction, which is the cathodic reduction of oxygen (also called the oxygen ionization reaction), is used in metal–air batteries and fuel cells. It is of considerable importance in the corrosion of metals.

Other reactions involving oxygen are those reaction steps in oxygen reduction which are of importance in their own right, viz., the formation of hydrogen peroxide as a relatively stable intermediate:

$$O_2 + 2H^+ + 2e^- \rightleftarrows H_2O_2, \qquad (19.18)$$

and its further cathodic reduction to water:

$$H_2O_2 + 2H^+ + 2e^- \rightarrow 2H_2O, \qquad (19.19)$$

as well as its electrochemical oxidation [the reverse of reaction (19.18)]. The reverse of reaction (19.19), i.e., the anodic oxidation of water to hydrogen peroxide, has not yet been achieved.

The standard electrode potential E_A^0 of reaction (19.17) calculated thermodynamically is 1.229 V (SHE) at 25°C. For reactions (19.18) and (19.19) these values are 0.682 and 1.776 V, respectively. The equilibrium potentials of all these reactions have the same pH dependence as the potential of the reversible hydrogen electrode; therefore, on the scale of potentials E_r (against the rhe), these equilibrium potentials are independent of pH. This picture is disturbed for reactions (19.18) and (19.19) in alkaline solutions because of the dissociation $H_2O_2 \rightleftarrows H^+ + HO_2^-$ (Fig. 19.3).

The relative values of the potentials given above indicate that hydrogen peroxide is thermodynamically unstable and may spontaneously

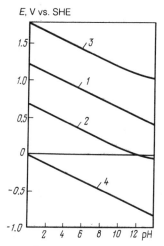

Fig. 19.3. The pH dependence of equilibrium potentials of various reactions: (1) reaction (19.17); (2) reaction (19.18); (3) reaction (19.19); (4) reaction (19.1).

decompose via the reaction

$$2H_2O_2 \rightarrow 2H_2O + O_2 \tag{19.20}$$

[compare the example of Cu^+ ions in Section 3.7.2]. Ordinarily this is a slow reaction, and hydrogen peroxide is relatively stable, but the reaction is very strongly catalytically accelerated by various solid materials, also by iron ions etc.

The oxygen reactions occur at potentials where most metal surfaces are covered by adsorbed or phase oxide layers. This is particularly true for oxygen evolution, which occurs at potentials of 1.5 to 2.2 V (rhe). At these potentials many metals either dissolve or are completely oxidized. In acidic solutions, oxygen evolution can be realized at electrodes of the platinum-group metals, the lead oxides, and the oxides of certain other metals. In alkaline solutions, electrodes of iron-group metals can also be used (at these potentials, their surfaces are practically completely oxidized).

For oxygen reduction, one can use electrodes made of carbon materials, and when the solutions are alkaline one can also use electrodes of silver and nickel, apart from those made of platinum metals.

At almost all electrodes, reaction (19.17) occurs with appreciable polarization, both in the cathodic and anodic direction; the exchange cd of this reaction is very low: 10^{-9}–10^{-6} A/m^2. For this reason the equilibrium potential of this reaction is not established at the electrodes. The ocp measured in oxygen are between 0.85 and 1.1 V (rhe), i.e., they are 0.15 to 0.4 V more negative than the equilibrium potential.

The equilibrium potential of reaction (19.18) is relatively readily established at mercury and graphite electrodes in alkaline solutions containing hydrogen peroxide.

19.2.1. Anodic Oxygen Evolution

The kinetics and mechanism of the oxygen evolution reaction, like those of hydrogen evolution, have been the subject of innumerable studies going back to the beginning of our century. However, progress has been modest. This is to be explained by the complexity of the reaction itself and by the possibility of simultaneous anodic reactions: the formation and growth of oxide layers, dissolution of the metal, and the oxidation of solution components. The experimental data obtained by different workers do not always coincide; there are effects of pretreatment of the electrode surface, time-dependent changes in the state of this surface, the impurities present in the solution, and other factors. An interpretation of the experimental data is further complicated by the fact that the cathodic reaction, as a rule, follows a pathway [e.g., that involving intermediate hydrogen peroxide formation via reaction (19.18)] which is not the same as that of the anodic reaction; hence it is not correct to compare the kinetic parameters of these reactions.

The polarization curves for the oxygen evolution reaction are more complex than those for hydrogen evolution. Usually several Tafel sections with different slopes are present. At intermediate cd their slope b' is very close to 0.12 V, but at low cd it sometimes falls to 0.06 V. At high cd higher slopes are found; at potentials above 2.2 V (rhe) new phenomena and processes are possible which will be considered in Section 19.5.

The polarization which is seen in oxygen evolution will of course depend on the nature of the electrode. It is interesting to note that in alkaline solutions much higher values of polarization are found at platinum or PbO_2 electrodes than at superficially oxidized iron-group metals. To a first approximation, the polarization at a given cd is independent of solution pH; this implies that the coefficient $-(\partial E/\partial pH)_i = 0.06$ V, like the analogous

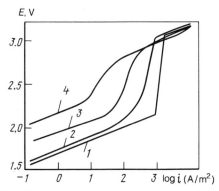

Fig. 19.4. Polarization curves for anodic oxygen evolution at a platinum electrode in perchloric acid solutions having different concentrations: (1) 1.34; (2) 3; (3) 5; (4) 9.8 M.

coefficient for the equilibrium potential. The reaction is also highly sensitive to the concentration of a number of solution components, particularly anions. Figure 19.4 shows an example, viz., polarization curves recorded in perchloric acid solutions having different concentrations.

The complexities of oxygen evolution, which is a four-electron reaction, are due to the fact that it is a multistep reaction and that oxygen adsorbed on the electrode is involved in it. As early as 1909 it was suggested by Fritz Foerster that oxygen is evolved as a result of decomposition of higher oxides forming as unstable intermediates. A large number of pathways have subsequently been suggested for this reaction. It is assumed in almost all of them, that the first step is the formation of additional adsorbed oxygen-containing species of the type of OH_{ads} or O_{ads} by discharge of water molecules or OH^- ions:

$$H_2O \underset{\xleftarrow{\hspace{1em}}}{\xrightarrow{-H^+,-e^-}} OH_{ads} \underset{\xleftarrow{\hspace{1em}}}{\xrightarrow{-H^+,-e^-}} O_{ads} \qquad (19.21)$$

These species are adsorbed, not only on the bare metal surface but also on surface sections already oxidized. They can undergo further reaction in a variety of steps, both purely chemical ones and steps of the electrochemical desorption type:

$$2O_{ads} \rightleftarrows O_2, \qquad (19.22)$$

$$O_{ads} + H_2O \rightleftarrows O_2 + 2H^+ + 2e^-, \qquad (19.23)$$

$$2[O_{ads} + MO_x \rightleftarrows MO_{x+1}] \rightleftarrows O_2 + 2MO_x \qquad (19.24)$$

(M is the electrode material), as well as many others including the analogous reactions involving OH_{ads}.

This great variety of pathways makes it difficult to decide which of the steps is the rate-determining step. It is most likely that at intermediate current densities the overall reaction rate is determined by the special kinetic features of step (19.21) producing the oxygen-containing species. The slopes of $b' \approx 0.12$ V observed experimentally are readily explained with the aid of this concept. Under different conditions one of the steps in which these species react further may be the slow step, or several of the consecutive steps may occur with similar kinetic parameters.

19.2.2. Cathodic Oxygen Reduction

Systematic studies of cathodic oxygen reduction, unlike those of its anodic evolution, were only started in the 1950s when required for the realization of fuel cells. The large polarization of this reaction is one of the major reasons why the efficiency of the fuel cells developed so far is not very high.

Two major pathways exist for this reaction, one bypassing hydrogen peroxide (first pathway) and the other involving intermediate peroxide formation via reaction (19.18) (second pathway). The peroxide formed is, either electrochemically reduced to water via reaction (19.19) or decomposed catalytically on the electrode surface via reaction (19.20), in which case half of the oxygen consumed to form it reemerges [in both cases the overall reaction corresponds to Eq. (19.17)].

The basic difference between the two pathways resides in the fate of the O–O bond in the original oxygen molecule, which in the first pathway is broken in the initial reaction step. The resulting oxygen atoms are adsorbed on the electrode, and then are reduced electrochemically:

$$O_2 \rightleftarrows 2 \left[O_{ads} \underset{\longleftarrow}{\overset{+2H^+, +2e^-}{\longrightarrow}} H_2O \right] \qquad (19.25)$$

[it cannot be ruled out that bond breaking occurs during addition of the first electron and (or) proton]. This pathway corresponds to the one followed in anodic oxygen evolution; but since the two reactions occur at different potentials, the nature and number of oxygen species on the surface may differ.

In the reaction following the second pathway, the O–O bond is not broken while the first two electrons are added; it is preserved in the H_2O_2 produced as an intermediate, and breaks in a later step, when the hydrogen peroxide is reduced or catalytically decomposed. An analog for this pathway does not exist in anodic oxygen evolution.

The second pathway is distinctly seen at mercury and graphite electrodes. These electrodes are quite inactive in the catalytic decomposition of H_2O_2. Moreover, at them the potential where the peroxide is further reduced is more negative than the potential where it is formed from oxygen. Hence, within a certain range of not too negative potentials the reaction can occur in such a way that the hydrogen peroxide formed accumulates in the solution.

At other electrodes the peroxide immediately undergoes further conversion, and practically none accumulates in the solution. In this case its intermediate formation can be detected and quantitatively examined with the rotating ring–disk electrode; at high rates of rotation of the electrode, the intermediate produced is rapidly carried away from the disk surface before it can react further, and can be ascertained at the ring electrode. According to these studies, intermediate H_2O_2 formation is not observed at a number of active metals, and particularly the platinum-group metals, i.e., the reaction apparently follows the first pathway. In other cases the reaction is found to occur in parallel following both of the pathways, while their relative importance depends on the experimental conditions, such as solution composition and electrode surface treatment. Solution impurities as a rule promote peroxide formation.

At mercury and graphite electrodes the kinetics of reactions (19.18) and (19.19) can be studied separately (in different regions of potential). It follows from the experimental data (Fig. 19.5) that in acidic solutions the slope $b' \approx 0.12$ V. The reaction rate is proportional to the oxygen partial pressure (its solution concentration). At a given current density the electrode potential is independent of solution pH, i.e., because of the shift of equilibrium potential the electrode's polarization decreases by 0.06 V when the pH is raised by a unit. These data indicate that the rate-determining step is addition of the first electron to the oxygen molecule:

$$O_2 + e^- \rightleftarrows O_2^-. \qquad (19.26)$$

The resulting unstable molecular ion O_2^- rapidly adds another electron and protons to yield hydrogen peroxide. In alkaline solutions the same

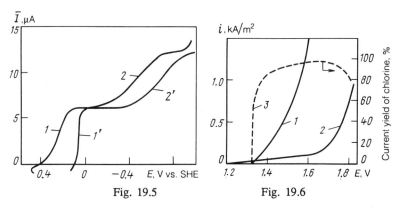

Fig. 19.5   Fig. 19.6

Fig. 19.5. Polarization curves for reactions (19.18) (1, 1′) and (19.19) (2, 2′) at the dropping mercury electrode in acidic (1, 2) and alkaline 1′, 2′) solutions.

Fig. 19.6. Polarization curves for anodic chlorine (1) and oxygen (2) evolution at a graphite electrode, and the current yields of chlorine as a function of potential (3).

pathway is followed, but owing to the much lower polarization the reaction becomes practically reversible ($b' = 0.03$ V); its rate then is determined by oxygen transport to the surface, and polarization is of the concentration type.

Polarization is much higher for the electrochemical reduction of hydrogen peroxide. The slope has the unusually high value of about 0.3 V. At a given current density the electrode potential in this reaction is again independent of solution pH. These, and certain other data, indicate that addition of the first electron to the peroxide molecule and simultaneous peroxide decomposition is the rate-determining step:

$$H_2O_2 + e^- \rightarrow OH_{ads} + OH^-. \qquad (19.27)$$

At the platinum electrode the individual steps of the four-electron reaction cannot be studied separately. Slope b' has its usual value of about 0.12 V, but the polarization, in contrast to what is seen at the mercury electrode, is practically independent of solution pH, i.e., the potential at a given current density shifts by 0.06 V in the negative direction when the pH is raised by a unit. It follows that the reaction rate depends on hydrogen ion concentration. The step in which an electron and a proton are transferred while the O–O bond is broken most probably is the rate-determining step.

19.3. REACTIONS INVOLVING CHLORINE AND OTHER HALOGENS

The principal reaction involving chlorine is

$$Cl_2 + 2e^- \rightleftarrows 2Cl^-. \tag{19.28}$$

The standard electrode potential of this reaction is 1.358 V vs. SHE; the equilibrium potential is independent of solution pH.

Anodic chlorine evolution by electrolysis of concentrated chloride solutions is used for the large-scale industrial production of chlorine. The cathodic reaction, which is the ionization of molecular chlorine, is used in certain types of batteries.

Because of the considerable corrosivity of chlorine toward most metals, anodic chlorine evolution can only be realized at a few electrode materials. In industry, mainly graphite had been used for these purposes in the past. Some oxide materials, manganese dioxide for instance, are stable as well. At present the titanium–ruthenium oxide anodes (DSA [trademark], see Chapter 16) are commonly used.

The mechanism of anodic chlorine evolution has been studied by many scientists. In many respects this reaction is reminiscent of hydrogen evolution. The analogous reaction pathways are possible. The most probable one is the second pathway in which the adsorbed chlorine atoms produced are eliminated by electrochemical desorption; but sometimes the first pathway is possible, too. As a rule the first step, which is discharge of the chloride ion, is the slow step.

One of the most important problems in chlorine production is that of reducing the rate of anodic oxygen evolution which occurs as a parallel reaction and is responsible, not only for a waste of electric power but also for contamination of the product. Figure 19.6 shows the relative positions of the polarization curves for anodic chlorine and oxygen evolution. Oxygen evolution starts earlier, since its equilibrium potential is more negative, but because of a higher polarizability in the oxygen reaction, at intermediate and high current densities chlorine evolution is predominant.

Higher current yields of chlorine can be obtained when raising the chloride concentration, which shifts the equilibrium potential of the chlorine electrode to more negative values and reduces the effect of concentration polarization. In practice, NaCl solutions having a concentration of about 5.3 M are used in electrolysis. It is very important in order to raise the current yields to maintain the optimum value of pH, which as a rule is between two and three. In diaphragm cells, the catholyte becomes strongly alkaline owing to hydrogen evolution. This produces a risk of OH^- ion diffusion through the diaphragm into the anode compartment, and thus of an increase in anolyte pH, which will accelerate the oxygen evolution reaction and lower the current yields of chlorine. To counteract this effect, the alkali being formed is continuously drained from the cathode compartment, and the chloride solution is introduced into the anode compartment. The (convective) filtration or solution flow through the diaphragm toward the cathode, which develops under these conditions, prevents a back diffusion of OH^- ions.

There are further side reactions occurring in chlorine evolution. Molecular chlorine reacts with aqueous solutions to hypochlorous acid HClO:

$$Cl_2 + H_2O \rightleftarrows HClO + HCl. \tag{19.29}$$

The concentration of the hypochlorous acid forming in acidic and weakly acidic solutions is low, but in alkaline solutions the equilibrium is shifted to the right, in the direction of hypochlorite formation. Since in chlorine cells two OH^- ions are produced for each Cl_2 molecule, this alkali will shift the equilibrium (19.29) completely to the right when electrolysis is conducted in a common solution (without a diaphragm, i.e., in an undivided cell), and no gaseous chlorine will be evolved. One thus can obtain dilute sodium or other hypochlorite solutions. At higher concentrations the ClO^- ions start to undergo anodic oxidation to chlorate ions ClO_3^-:

$$ClO^- + 2H_2O \rightleftarrows ClO_3^- + 4H^+ + 4e^-. \tag{19.30}$$

Chlorates also form by dismutation of the hypochlorite ions in the bulk solution:

$$3ClO^- \rightleftarrows ClO_3^- + 2Cl^-. \tag{19.31}$$

Current world chlorate production (about 700 kto per year) is based entirely on an electrochemical method where reactions (19.28) to (19.31) occur simultanously in undivided cells. A small amount of bichromate ions are added to the solution in order to reduce chlorate losses by rereduction at the cathode; these form a thin protective layer at the cathode which passivates the reduction of chlorate and hypochlorite ions.

Perchlorates are also produced electrochemically. The oxidation of chlorate to perchlorate ions occurs at a higher positive potential (above

2.0 V vs. SHE) than chloride ion oxidation. The current yield of perchlorate is lower when chloride ions are present in the solution, hence in perchlorate production concentrated pure chlorate solutions free of chlorides are used. Materials stable in this potential range are used as the anodes; primarily, these include smooth platinum, platinum on titanium, and also lead dioxide.

Methods have been developed for perchloric acid synthesis which involve the electrolysis of solutions containing hydrogen chloride or molecular chlorine. These processes occur at high anode potentials (2.8 to 3.0 V vs. SHE), when oxygen is evolved at the anode in parallel with perchloric acid formation. The current yields of perchloric acid will increase considerably when the reaction is conducted at low temperatures, e.g., $-20°C$.

Electrolytic fluorine production from molten eutectics of anhydrous hydrogen fluoride and potassium fluoride, with the compositions $KF \cdot HF$ ($t_f = 239°C$) or $KF \cdot 2HF$ ($t_f = 82°C$), is of great practical importance. The equation for anodic fluorine generation is the analog of reaction Eq. (19.28) from right to left. This reaction, which has been known for about 100 years, was developed commercially in the 1950s when fluorine began to be widely used for the fluorination of organic compounds and production of fluoropolymers. At present the total fluorine production is several tens of kto per year. Anodes of steel, copper, or of magnesium alloys are used for electrolysis. In the presence of fluorine, these metals quickly form a thin superficial fluoride layer protecting them against further corrosion in the highly aggressive medium. At the graphite cathodes, hydrogen is evolved. Hydrogen fluoride is added continuously to the melt in order to maintain the original HF/KF ratio. The cell voltage is rather high (8–12 V) because of the considerable ohmic losses and of high values of polarization of the electrodes. Often difficulties arise in the production because of poor wetting of the anodes by the melt and the development of so-called anode effects (see Section 11.3).

Unlike chlorine and fluorine, the free bromine and iodine are produced by chemical methods (reaction of chlorine with bromide or iodide solutions). Electrochemical methods are used to produce the salts of their oxygen-containing acids, the bromates and iodates, from the corresponding bromide and iodide solutions. These reactions are analogous to those in chlorate production [Eqs. (19.28) to (19.31)], and involve the intermediate formation of hypobromites and hypoiodites.

19.4. REACTIONS INVOLVING ORGANIC SUBSTANCES

Organic substances, like inorganic ones, can be anodically oxidized and (or) cathodically reduced. Such reactions are widely used for the synthesis of different organic compounds (electroorganic synthesis). Moreover, they are of importance for the qualitative and quantitative analysis of organic substances in solutions, as for instance in polarography.

Reactions involving organic substances have some special features. Many of these substances are poorly soluble in aqueous solutions. Sometimes their solubilities can be raised by adding to the solution the salts of aromatic sulfonic acids with cations of the type of $[NR_4]^+$ or alkali metal ions (Ralph Harper McKee, 1932). These salts have a salting-in effect on poorly soluble organic substances. In many cases solutions in mixed or nonaqueous solvents (e.g., methanol) are used. Suspensions of the organic substances in aqueous solutions are also useful for electrosynthesis.

The solutions of many organic substances in water or other solvents are not by themselves electrically conducting. Hence, substances providing conductivity as well as pH control are added to these solutions.

Many electrochemical reactions involving organic substances are invertible. Their efficiency strongly depends on the quality of the separators or diaphragms used to prevent the reaction products from reaching the auxiliary electrode.

Electrochemical reactors are inferior to chemical reactors in their productivity. Hence, electrosynthesis is used when chemical ways to synthesize a given substance are impractical or inefficient. It is an important advantage of electrosynthesis that special reactants (oxidizing or reducing agents) which would be hard to reextract from the products need not be added to the system. Moreover, the conditions under which electrosynthesis proceeds (the potential, current, etc.) are readily monitored and adjusted.

By now a large number of electrochemical reactions involving organic substances have been investigated, but so far few of them are commercially exploited.

Reactions involving organic compounds are highly diverse. Still, two basic types of mechanism can be distinguished, the electron–radical and the chemical mechanism. In the former, there is direct electron transfer between the electrode and an organic species yielding primarily the valence-unsaturated radicals $\dot{Q}H$ or the radical-ions Q^- or QH_2^+ [here and in the following, stable valence-saturated organic compounds will be written as Q and QH_2, RH and RR', etc.]:

$$Q \underset{\rightleftarrows}{+e^-} Q^- \underset{\rightleftarrows}{+H^+} \dot{Q}H \underset{\rightleftarrows}{+H^+} QH_2^+ \underset{\rightleftarrows}{+e^-} QH_2. \qquad (19.32)$$

Most often these radicals are unstable, and can exist only while adsorbed on the electrode, though in the case of polycyclic aromatic compounds (e.g., the derivatives of anthracene) they are more stable and can exist even in the solution. The radicals formed first can undergo a variety of chemical or electrochemical reactions. This reaction type is the analog of hydrogen evolution, where electron transfer as the first step produces an adsorbed hydrogen atom, which is a radical-type product.

The chemical mechanism rests on the effect of intervening redox systems (see Section 13.6). Here intermediate reactants, viz., H_{ads} species on a cathode surface, OH_{ads} species on an anode surface, or reducing and oxidizing agents in the solution layer next to the electrode, are first produced electrochemically from solution components. The further interaction of these reactants with the organic substance is purely chemical in character, e.g., following a reaction

$$Q + 2H_{ads} \rightleftarrows QH_2. \qquad (19.33)$$

Electrochemical equilibria between the oxidized and reduced form of an organic substance are observed in but a few cases. The best known instance is the equilibrium between quinone and hydroquinone (or their derivatives):

$$C_6H_4O_2 + 2H^+ + 2e^- \rightleftarrows C_6H_4(OH)_2. \qquad (19.34)$$

The equilibrium potential of this reaction is readily established at platinum electrodes; it is highly reproducible.

19.4.1. Reduction Reactions

For the cathodic reduction of organic substances, electrodes of two types are used: the platinum and the mercury type. Those of the first type (platinum metals, and in alkaline solutions nickel) exhibit low polarization in hydrogen evolution; their potential can be pushed in the negative direction, no further than to -0.3 V (rhe). Hydrogen readily adsorbs on these electrodes, which is favorable for reduction reactions following the chemical mechanism. Under the effect of certain impurities (poisons) they readily lose their activity. At electrodes of the mercury type (mercury itself, also lead, tin, cadmium, and others), much more negative potentials (up to -1.2 V vs. rhe, for instance) can be realized, since hydrogen evolution is slow. Here hydrogen adsorption is slight, hence reactions at electrodes of this type predominantly follow the electron–radical mechanism.

Unsaturated organic substances having double or triple bonds can be reduced (hydrogenated) only at electrodes of the platinum type, i.e., via the chemical mechanism. This process occurs with ease, but still the electrochemical method of hydrogenation is economically inferior to the common methods of catalytic hydrogenation of unsaturated compounds. If the compound in addition to a double bond has a functional group, its reduction owing to bond polarization is possible even at mercury-type metals, i.e., following an electron–radical mechanism. An example is the reduction of maleic to succinic acid:

$$COOHCH{=}CHCOOH + 2H^+ + 2e^- \rightarrow COOHCH_2CH_2COOH, \quad (19.35)$$

which will occur, both via the electron–radical and via the chemical mechanism.

At electrodes of the mercury type, electron–radical reactions can also be carried out when there is a deficit of proton donors in the solution. Then the radicals $\dot{Q}H$ formed initially, instead of adding a second electron and second proton will couple to a dimeric product:

$$2\dot{Q}H \rightarrow HQQH \qquad\qquad (19.36)$$

Reactions of this type are called electrochemical hydrodimerization. They are of great value for the synthesis of various bifunctional compounds.

A reaction which has found wide commercial use is the hydrodimerization of acrylonitrile to adiponitrile (the dinitrile of adipic acid):

$$2CH_2{=}CHCN + 2H^+ + 2e^- \rightarrow NC(CH_2)_4CN. \qquad (19.37)$$

Adiponitrile is readily hydrogenated catalytically to hexamethylenediamine, which is an important starting material for the production of nylons and other plastics. The electrochemical production of adiponitrile was started in the USA in 1965; at present its volume is about 200 kto per year. The reaction occurs at lead or cadmium cathodes with current densities of up to 2 kA/m^2 in phosphate buffer solutions of pH 8.5–9. Salts of tetrabutylammonium $[N(C_4H_9)_4]^+$ are added to the solution; this cation is specifically adsorbed on the cathode and displaces water molecules from the first solution layer at the surface. Therefore, the concentration of proton donors is drastically reduced in the reaction zone, and the reaction follows the scheme of (19.36) rather than that of (19.32), which would yield propionitrile.

Another group of reactions is that involving the reduction of different functional groups. A classical example is the reduction of nitrobenzene (and the analogous nitro compounds) which had been studied by Fritz Haber at the end of the 19th century. This is possible, both at electrodes of the platinum type and at electrodes of the mercury type. The main reaction sequence is

$$C_6H_5NO_2 \longrightarrow \underset{\text{(nitrosobenzene)}}{C_6H_5NO} \longrightarrow \underset{\text{(phenylhydroxylamine)}}{C_6H_5NHOH} \longrightarrow \underset{\text{(aniline)}}{C_6H_5NH_2} .$$
$$\text{(nitrobenzene)}$$

$$(19.38)$$

By appropriate selection of the electrode potential and other conditions one can control the reaction so that it will yield any one of these products. As a result of auxiliary chemical reactions, nitrosobenzene can be converted to azobenzene $C_6H_5N=NC_6H_5$ and azoxybenzene $C_6H_5N(=O)-NC_6H_5$, and the latter can be converted to hydrazobenzene $C_6H_5-NH-NH-C_6H_5$; phenylhydroxylamine can be converted to p-aminophenol $H_2NC_6H_4OH$ by internal rearrangement.

Another example for the reduction of functional groups is that of aromatic ketones. At the mercury electrode, a radical-type intermediate is formed as the primary product:

$$RR'C=O + H^+ + e^- \rightarrow RR'\dot{C}-OH. \qquad (19.39)$$

Depending on the electrode potential, this radical either adds a second electron and second proton to yield the corresponding secondary alcohol (A), or it dimerizes to the pinacol (B), a dihydric ditertiary alcohol:

$$\underset{A}{RR'CH-OH} \overset{+H^+,+e^-}{\longleftarrow} RR'\dot{C}-OH \longrightarrow \underset{B}{RR'COHCOHRR'}. \qquad (19.40)$$

19.4.2. Oxidation Reactions

Fundamentally, simple organic compounds can be oxidized all the way to CO_2 and water. A particular reaction that has been analyzed in detail is the anodic oxidation of methanol at platinum electrodes. Its first step is destructive chemisorption of the methanol molecules yielding radicals $:\dot{C}\text{-}OH$ bonded to three adsorption sites on the electrode surface:

$$CH_3OH \rightarrow :\dot{C}\text{-}OH + 3H_{ads}. \tag{19.41}$$

Subsequently this chemisorbed species is chemically oxidized by the oxygen-containing groups OH_{ads} which are formed in prior electrochemical steps (19.21):

$$:\dot{C}\text{-}OH \xrightarrow[-H_2O]{+OH_{ads}} :C{=}O \xrightarrow{+OH_{ads}} \dot{C}OOH \xrightarrow[-H_2O]{+OH_{ads}} CO_2. \tag{19.42}$$

The potential of the CH_3OH/CO_2 system has the thermodynamic value of 0.03 V (rhe). Numerous attempts have been made to use this reaction in fuel cells converting the chemical energy of methanol oxidation directly to electrical energy. However, even at the costly platinum electrodes the reaction is too slow, at tolerable values of polarization, for the design of adequate functional devices.

Reactions involving mild oxidation are of much greater interest in the electrosynthesis of various organic compounds. Thus, at gold electrodes in acidic solutions, olefins can be oxidized to aldehydes, acids, oxides, and other compounds. A good deal of work was invested in the oxidation of aromatic compounds (benzene, anthracene, etc.) to the corresponding quinones. To this end various mediating redox systems, e.g., the Ce^{4+}/Ce^{3+} system, are employed (see Section 13.6).

A reaction of practical importance is the oxidation of a carbohydrate aldehyde group to a carboxyl group. This is the basis for a process converting glucose to calcium gluconate, a substance of pharmaceutical interest. The oxidation reaction occurs at graphite electrodes in presence of the Br_2/Br^- redox system. Calcium salt is added to the solution in order to prevent a further oxidation of free gluconic acid.

Common reactions are the anodic substitution of hydrogen atoms in organic compounds by halide or other functional groups:

$$RH + X^- \rightarrow RX + H^+ + 2e^-, \tag{19.43}$$

where X^- is the nucleophile: a halide ion, CN^-, OH^-, RO^-, $RCOO^-$, or others. These reactions are conducted chiefly with aromatic compounds; in the case of aliphatic compounds, nonelectrochemical methods of substitution, as a rule, are more convenient.

As in reduction reactions, two possible mechanisms exist for substitution reactions: (a) electron–radical, involving the intermediate formation of radicals and their reaction with nucleophiles X^-:

$$RH \xrightarrow[-e^-,-H^+]{} \dot{R} \xrightarrow[-e^-,+X^-]{} RX, \qquad (19.44)$$

and (b) chemical, involving the prior discharge of species X^-:

$$2X^- \xrightarrow[-2e^-]{} X_2 \xrightarrow{+RH} RX + HX. \qquad (19.45)$$

In certain cases several hydrogen atoms can be replaced in the original organic compound.

Of great practical value is the thorough fluorination of hydrocarbons and other organic compounds, which can be used to produce valuable perfluorinated substances, e.g.,

$$C_2H_6 + 12F^- \rightarrow C_2F_6 + 6HF + 12e^-. \qquad (19.46)$$

These reactions will occur at graphite, nickel, or platinum electrodes in solutions of sodium or potassium fluoride in anhydrous liquid hydrogen fluoride. They have rather complicated mechanisms (the example shown involves 12 electrons!). In many cases the starting materials are broken down to simpler ones, while in aromatic compounds, benzene rings may be opened. As a result, mixtures of different reaction products having different chain lengths and different degrees of fluorination are obtained. The reaction mechanism of thorough fluorination has not been established in a definite way.

Another group of reactions involving anodic oxidation is the anodic dimerization occurring at high anodic potentials. These reactions will be considered in Section 19.5.

19.4.3. Reactions Involving Organometallic Compounds

It had been shown in the preceding two sections that the initial step in a number of cathodic and anodic reactions yields organic radicals which then undergo further oxidation, reduction, or dimerization. In some cases reactions of another type are possible, viz., reaction of the radical with the electrode metal yielding organometallic compounds which are then taken up by the solution. Such reactions can be used in the synthesis of these compounds.

The best known example is the electrosynthesis of tetraethyllead (TEL) $Pb(C_2H_5)_4$, which has been in wide use as an antiknock additive of gasoline, and still is in a number of countries. This substance is readily produced by reaction of ethyl radicals with the lead electrode:

$$Pb + 4\dot{C}_2H_5 \rightarrow Pb(C_2H_5)_4. \qquad (19.47)$$

Ethyl radicals can be produced in various ways, for instance by cathodic reduction of ethyl bromide:

$$C_2H_5Br + e^- \rightarrow \dot{C}_2H_5 + Br^-. \qquad (19.48)$$

The reaction proceeds in a solution of $[N(C_2H_5)_4]Br$ in acetonitrile at lead cathodes. The current yield is about 70%.

Anodic processes can also be used for tetraethyllead electrosynthesis. Here solutions of organometallic compounds are used, i.e., the overall reaction is replacement of the metal in these compounds by another metal, lead. One such process uses a melt of the compound $NaAl(C_2H_5)_4$, from which radicals $\dot{C}_2H_5$ are produced anodically. The process is highly efficient, but it is not easy to isolate the TEL produced from the melt. More convenient is a commercial process involving the anodic oxidation of the Grignard reagent C_2H_5MgCl:

$$C_2H_5MgCl \rightarrow \dot{C}_2H_5 + Mg^{2+} + Cl^- + e^-. \qquad (19.49)$$

The reaction occurs at lead anodes; at the cathode metallic magnesium is produced. An ether solution of the Grignard reagent is used to which ethyl chloride is added. The latter reacts with the magnesium that is formed, to regenerate part of the Grignard reagent consumed. The overall reaction in the electrolyzer follows the equation

$$2C_2H_5MgCl + 2C_2H_5Cl + Pb \rightarrow Pb(C_2H_5)_4 + 2MgCl_2. \qquad (19.50)$$

19.5. REACTIONS AT HIGH ANODIC POTENTIALS

A number of anodic reactions will proceed at platinum and some other electrodes at potentials between about 2.2 and 3.5 V vs. rhe, which is the so-called *region of high anodic potentials* (hap). According to the usual laws of electrochemical kinetics, rather vigorous oxygen evolution could be expected in this region in aqueous solutions, and decomposition (oxidation) of the solvent with evolution of the corresponding products in nonaqueous solutions. But under certain conditions these reactions are suppressed, and others proceed instead. The main reason for these effects are changes in the electrode's surface state.

It had been shown in Section 12.8 that as the potential of a platinum electrode is moved from 0.75 to 2.2 V the amount of oxygen adsorbed on it will increase up to a limiting value corresponding to degrees of surface coverage of 2 to 2.2. The properties of the "oxidized" platinum surface differ from those of "nonoxidized" surfaces. For instance, the surface potential changes upon oxidation, which in turn produces a change in surface charge density and in the pzc. The adsorption properties also change. There is a drastic increase in anion adsorption from the solution on account of chemical (specific) interaction forces. It had been shown in Section 12.8.4 that at potentials more positive than 0.8 V the organic substance is desorbed from the platinum surface because of its displacement by adsorbing oxygen (which can be used, in particular, to clean the surface from organic contaminants). However, at higher positive potentials, in the region which starts at about 1.6–1.7 V, adsorption of organic substances increases again, though now it occurs on the oxidized surface.

The enhanced adsorption of anions and other substances which occurs at increasingly positive potentials causes a gradual displacement of water (or other solvent) molecules from the electrolyte layer next to the electrode. This leads to a markedly slower increase in the rate of oxygen evolution from water molecules and facilitates a further change of potential in the positive direction. As a result, conditions arise which are favorable for reactions involving the adsorbed species themselves (Fig. 19.7). In particular, adsorbed anions are discharged forming adsorbed radicals:

$$A_{ads}^- \rightleftarrows \dot{A}_{ads} + e^-, \qquad (19.51)$$

which subsequently dimerize by recombination:

$$2\dot{A}_{ads} \rightleftarrows AA, \qquad (19.52)$$

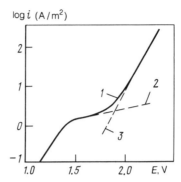

Fig. 19.7. Anodic polarization curves recorded at a platinum electrode in the region of high anodic potentials in the presence of acetate ions: (1) total current; (2) partial current of oxygen evolution; (3) partial current of oxidation of adsorbed species.

or via an electrochemical desorption step:

$$\dot{A}_{ads} + A_{ads}^- \rightleftarrows AA + e^- \tag{19.53}$$

(anodic dimerization).

In surface layers still containing a certain amount of water, reaction (19.21) may occur in parallel yielding OH_{ads} groups, so that two different radical-type intermediates can react:

$$\dot{A}_{ads} + \dot{O}H_{ads} \rightleftarrows AOH. \tag{19.54}$$

In concentrated sulfuric acid solutions at hap, the adsorbed HSO_4^- ions are converted, according to reaction (19.51), to $H\dot{S}O_4$ radicals which dimerize forming peroxydisulfuric (persulfuric) acid $H_2S_2O_8$. This acid is the intermediate for one of the commercialized methods of hydrogen peroxide production. The first efforts toward the electrosynthesis of peroxydisulfuric acid go back to 1878; commercial production started in 1908. The standard electrode potential of the overall reaction

$$2HSO_4^- \rightleftarrows S_2O_8^{2-} + 2H^+ + 2e^- \tag{19.55}$$

is 2.10 V. The process occurs in 5.5–6.5 M H_2SO_4 solution at smooth platinum electrodes using current densities of 5–10 kA/m^2 and anode potentials of 3.0–3.2 V. The current yield of the product is 70 to 75%. Parallel reactions are oxygen evolution and, by a reaction of the type of (19.54), the formation of peroxymonosulfuric acid H_2SO_5. Higher current yields can be achieved at a high current density and therefore, high polarization of the oxygen evolution reaction. The region of hap is not attained and peroxydisulfuric acid is not formed at platinized platinum, where the true current density and polarization are much lower. Sometimes, the process is conducted in ammonium sulfate solution and not in sulfuric acid, and a solution of ammonium peroxydisulfate is formed.

Many other peroxy compounds can analogously be produced in the region of hap, for instance sodium perborate $Na_2(BO_3)_2$ (from sodium metaborate $NaBO_2$) and peroxycarbonates. These compounds are used as stable oxidizing and bleaching agents.

Peculiar reactions occur during the oxidation of organic acid anions. Michael Faraday found in 1843 already that a hydrocarbon is evolved during the electrolysis of acetate solutions. These reactions were studied in 1849 by Hermann Kolbe, who showed that the acetate ions during electrolysis condense with the evolution of CO_2, i.e., the reaction (in modern notation)

$$2CH_3COO^- \rightarrow C_2H_6 + 2CO_2 + 2e^- \qquad (19.56)$$

(the *Kolbe reaction*) occurs. Other aliphatic acids, including substituted ones, react in an analogous way. The overall reaction can be written as

$$2\left[RCOO^- \xrightarrow[-e^-]{} RCO\dot{O} \xrightarrow[-CO_2]{} \dot{R} \right] \longrightarrow RR. \qquad (19.57)$$

Thus, the initially produced $RCO\dot{O}$ radicals are unstable, and before (or while) undergoing dimerization split up into simpler radicals and CO_2 molecules.

The anions of substituted acids will not always react according to this scheme. Some of them, and particularly the α-substituted acids, form unsaturated compounds. More complex reactions also occur when dicarboxylate ions $^-OOCRCOO^-$ are used. When one of the carboxyl groups in the dicarboxylic acid is esterified the Kolbe reaction proceeds without difficulties:

$$2R'OOCRCOO^- \rightarrow R'OOCRRCOOR' + 2CO_2 + 2e^-. \qquad (19.58)$$

Such reactions usually are called *Brown–Walker* (Alexander Crum Brown and James Walker, who first described them in 1891). They are of practical importance for the synthesis of higher carboxylic acids.

In the Soviet Union in the 1960s an industrial production of sebacic acid $HOOC(CH_2)_8COOH$ (an important intermediate for different plastics) was started up which involves the anodic condensation of monomethyl adipate $CH_3OOC(CH_2)_4COO^-$. Dimethyl sebacate is obtained via the scheme of (19.58), and then hydrolyzed in autoclaves to the final product. Methanol is used as a solvent in order to lower the rates of side reactions. The reaction occurs with current yields attaining 75%, and with chemical yields (degrees of utilization of the original adipate) of 82–84%. Upon introduction of this process it was no longer necessary to use castor oil, an expensive raw material which was needed to produce sebacic acid by the chemical process.

When several types of radical are present on the electrode surface they can interact in different ways. One of these reactions is that between organic radicals $\dot{R}$ with OH_{ads} groups yielding alcohols ROH according to the scheme of (19.54). Reactions of this type or, in the more general case, reactions yielding ethers by interaction of radicals $\dot{R}$ with groups produced by decomposition of other solvents (e.g., $CH_3\dot{O}$), are known as *Hofer–Moest reactions* (Hans Hofer and Martin Moest, 1902). In electrosynthesis these reactions are often undesirable, since they lower the yields of the wanted product.

Reactions involving a crossed anodic condensation are of practical interest when they combine organic radicals of different type. In solutions containing anions $RCOO^-$ and $R'COO^-$, condensation products of the type of RR' are formed together with the standard products RR and $R'R'$ by the reactions described.

Unsaturated organic compounds which are present in the solution while these reactions proceed, will also become adsorbed on the electrode and may act as acceptors for the radicals yielding addition products:

$$2RCOO^- + C_2H_4 \rightarrow RH_2CCH_2R + 2CO_2 + 2e^- \qquad (19.59)$$

(additive dimerization reactions). When dienes with two conjugated double bonds are used, the condensation products are unsaturated; for instance, with butadiene,

$$2RCOO^- + H_2C=CHCH=CH_2 \rightarrow RH_2CCH=CHCH_2R + 2CO_2 + 2e^-$$
$$(19.60)$$

(Mikhail Ya. Fioshin and Leonid A. Mirkind, 1962). In solutions containing more than one compound of the RCOOH type, asymmetric products of the crossed addition of different radicals may form together with the standard additive dimers.

19.6. ELECTROCATALYSIS

It had been pointed out in Section 19.1 that the rates of cathodic hydrogen evolution recorded at a given potential vary within wide limits for electrodes of different metals; between lead and rhodium they increase by a factor of 10^{11}. Similar, though less drastic, differences between the reaction rates at different metals can be noticed for other reactions at nonconsumable electrodes. This signifies that the so-called "inert" electrodes not merely supply or accept electrons but have strong catalytic effects on the reactions. It had been shown in Section 19.2 in the instance of cathodic oxygen ionization, that the nature of the electrode is also important for the reaction pathway (the intermediate formation of hydrogen peroxide). In complex reactions, e.g., those involving organic substances, the electrode material is decisive for selectivity, i.e., the general direction of the reaction as well as the nature and yields of the principal and secondary reaction products.

These two catalytic effects of the electrode material: that on the rate and that on the selectivity of many reactions, are of great importance in the practical use of such reactions in electrolyzers, batteries, etc. It is mandatory to select for each reaction the optimum material as the electrode and catalyst. The catalytic effects also require a theoretical foundation. All these problems are the subject of electrocatalysis, a new branch of electrochemistry which has seen an accelerated development since about 1960. One of the reasons for the efforts made in this field were the attempts to build highly efficient fuel cells.

Catalysts for electrochemical reactions must meet a number of requirements: they should be electronically conducting, corrosion-resistant under the reaction conditions, remain active in long-term use, etc.

Good and sufficiently universal catalysts for many electrochemical reactions are platinum and other metals of the platinum group. However, their general use is not feasible for economic reasons; they are expensive and scarce. They are often applied to inert supports in the form of very finely divided deposits (platinum black) in order to reduce the amount needed in an electrode and be utilized more efficiently. Stable platinum deposits having true surface areas of up to 100 m^2/g can be obtained on carbon supports. The highest possible specific surface area of platinum, which can be calculated for the case where a grain contains no more than 8–10 atoms and hence each atom is a surface atom, is 270 m^2/g. However, it was seen in the example of reactions occurring in the region of hap (Section 19.5) that the properties of finely divided catalysts may differ from those of smooth ones.

Other catalyst types are less universal than the platinum metals. In many cases they have insufficient chemical stability and cannot be used for this reason. The following are practically used as catalytic electrodes: metals (nickel and other iron-group metals, silver, gold, mercury), carbon materials (graphite, active carbons, glassy carbon, carbon black), oxides (the simple oxides of a number of metals, mixed oxides with spinel or perovskite structure), and solid compounds (tungsten carbide). It was shown in recent years that organic (metal-containing) complexes (phthalocyanines, porphyrins, even the polymeric materials produced by their heat treatment) can be used as catalysts in a number of reactions.

A unified theory that could explain the catalytic effects of given electrode materials on reaction rates and selectivities does not exist up to now. There can be no doubt that the activity depends, both on the bulk properties of the catalyst (its electronic structure) and on its surface state and structure. Numerous attempts have been recorded to establish correlations between the catalytic effects and individual catalyst properties. However, the functions found are specific and cannot be generalized. It can merely be asserted that catalytic action is linked to adsorption of the reaction components, viz., of the reactants, intermediates, and (or) products. Catalytic effects are not found in systems completely lacking adsorption phenomena (e.g., in the reduction of peroxydisulfate to sulfate ions considered in Section 14.4.2); here the reaction rate is independent of the electrode material.

During recent decades a number of interesting catalytic phenomena have been discovered in electrochemical reactions. One of these is the synergistic (greater-than-additive) effect seen in multicomponent systems. Thus, the catalytic activity of a mixed platinum–ruthenium deposit toward anodic methanol oxidation is several orders of magnitude higher than that of deposits of platinum or ruthenium alone. Similar effects are produced by minute amounts of a foreign metal (e.g., tin) present in the form of adatoms on the surface of a platinum catalyst.

It can be expected that more detailed studies of these and other catalytic effects will significantly contribute to an understanding of catalysis in the different electrochemical reactions.

Chapter 20
ELECTROCHEMICAL METHODS
OF ANALYSIS

Electrochemical phenomena and processes are useful for the quantitative and qualitative chemical analysis of different substances and media including liquids, gases, and solids. Electrochemical methods of analysis have first been described at the end of the 19th century. Since then there has been continuous progress and improvement.

The high accuracy of the electrochemical methods of analysis derives from the fact that they are based on highly exact laws, e.g., those of Faraday. The methods of electrochemical analysis are instrumental. It is very convenient that electrical signals are used for the perturbation: current, potential, etc., and that the result (the response) again is obtained as an electrical signal. This is the basis for the high speed and accuracy of the readings, for the extensive possibilities of automated recording of the results, as well as for automation of the entire analysis. Electrochemical methods of analysis are distinguished by their high sensitivity, selectivity (the possibility of analyzing certain substances in the presence of others), speed of the measurements, and other advantages. In many cases extremely small volumes, less than 1 ml, of the test solution will suffice for electrochemical analysis.

The following are the major groups of electrochemical methods for chemical analysis:

(i) *Conductometry*, which measures the electrical conductivity of the electrolyte solution being examined,

(ii) *Coulometry*, which measures the amount of charge Q consumed for the complete conversion (oxidation or reduction) of the substance being examined,

(iii) *Voltammetry*, which determines the steady-state or transient polarization characteristics of electrodes in reactions involving the substance being examined, and

(iv) *Potentiometry*, which measures the open-circuit equilibrium potential of an indicator electrode, for which the substance being examined is potential-determining.

Electrochemical methods are of importance in their own right for direct chemical analyses; but in addition, in a number of cases, they are ancillary to other methods of analysis, e.g., the titration of solutions. Volumetric titration is a convenient and exact method of quantitative chemical analysis. However, in titrations, difficulties often arise in an exact determination of the titration end point. In the titration of an acid with base (and vice versa), this point can be determined from the change of color of an added indicator, but in other kinds of titration such a possibility as a rule is not available. Therefore, many electrochemical methods have been developed for indicating the end of titration, and particularly those based on the changes in conductivity (*conductometric titration*), potential (*potentiometric titration*), or current (*amperometric titration*).

All the electrochemical methods listed are also valuable as research tools: conductivity measurements for problems in solution theory, voltammetric characteristics for the determination of electrochemical reaction mechanisms and kinetic parameters, potentiometry for the thermodynamic properties of different substances, etc.

20.1. CONDUCTOMETRY

Conductometry is a nonselective method of analysis; all kinds of mobile ion present in the solution (or other medium being examined) contribute to conductivity, and the contributions of the individual kinds cannot be distinguished in the values measured. Hence, conductometry is primarily useful when determining the concentrations in binary electrolyte solutions, e.g., for determining the solubilities of poorly soluble compounds. In the case of multicomponent systems, conductometry is used when the qualitative composition of the solution is known and invariant, as, for example, in the continuous or batch analysis of solutions in process streams. Conductometry can also be used to monitor the rinsing of deposits and materials. Straightforward conductometry is important in determining the total ion content (degree of mineralization) of natural waters and in the quality control of water, after purification or distillation.

Conductometric titration rests on the marked changes which occur near the titration end point in the relation between conductivity and the amount of titrant added (an extreme or inflection point). It is used in particular for the titration of acids with base (and vice versa) in colored and turbid solutions or solutions containing reducing and oxidizing agents, i.e., in those cases where the usual color change of acid–base indicators cannot be seen.

Conductometric analysis is performed, both in concentrated and dilute solutions. The accuracy depends on the system; in binary solutions it is as high as 0.1%, but in multicomponent systems it is much lower.

20.2. COULOMETRY

Coulometry can be regarded as an analog of titration where the substance being examined is quantitatively converted to a reaction product, not by the addition of titrant, but by a certain amount of electric charge Q. As in titration, the end point must be determined. To determine the end point during current flow, one combines coulometry with another of the electrochemical methods described, and accordingly is concerned with conductometric, potentiometric, or amperometric coulometry.

In coulometry, one must exactly define the amount of charge that was consumed at the electrode up to the moment when the end point signal appeared. In galvanostatic experiments (at constant current), the charge is defined as the product of current and the exactly measured time. However, in experiments with currents continuously changing in time, it is more convenient to use special coulometers, which are counters for the quantity of charge passed. Electrochemical coulometers are based on the laws of Faraday; with them the volume of gas or mercury liberated, which is proportional to charge, is measured. Electromechanical coulometers are also available.

For coulometric analysis, the substance being examined must react in 100% current yields, i.e., other (secondary) reactions must be entirely absent. In efforts to avoid side reactions, coulometry most often is performed potentiostatically (amperometrically), i.e., the electrode potential is kept constant during the experiment, and the current consumed at the electrode is measured. The current is highest at the start of the experiment; it decreases as the substance being examined is consumed. The coulometric end point is that where the current has become zero.

The drop in current which occurs in coulometric experiments may arise, not only from the decrease in bulk concentration of the substance being analyzed, but also from a decrease in its surface concentration caused by the development of concentration gradients (see Section 9.3.1). Low values of current density and strong solution stirring are used in order to avoid the interference of such effects. Thin-layer cells where the electrodes are very close together (tens of micrometers) and the parameter ratio S/V is high, are often used in order to shorten the experiments.

The galvanostatic version of coulometric analysis is used more rarely, even though the determination of charge Q is greatly facilitated by constancy of the current. Here, the end point is determined from the typical potential rise associated with the changeover to a different reaction. The shape of the potential–time curve is the same as that of the curve for transient concentration polarization under galvanostatic conditions (see Fig. 9.10), though in the present case the potential change should occur when the bulk concentration drops to zero, not when merely the surface concentration drops to zero.

Coulometric titration is the term used when the substance being examined cannot be oxidized or reduced directly at the electrode, but a mediator is produced by the electrochemical reaction which will then react with the substance, i.e., functions as the titrant. An example is the coulometric determination of arsenous acid in the presence of bromide ions, where bromine is produced as the oxidizing mediator. When titration is complete (the arsenous acid has been completely oxidized), free bromine begins to accumulate in the solution, and can be detected either amperometrically (by the current recorded at an additional, cathodically polarized indicator electrode) or potentiometrically (from the sharp change in potential of the indicator electrode). Otherwise coulometric titration does not differ from ordinary coulometry.

Direct coulometric methods cannot be used unless the reactions with the substance being analyzed occur with 100% current yields. In some cases electrogravimetry can be used for such reactions; here the substance being examined is electrochemically deposited on a suitable electrode, and then its mass is determined by weighing.

20.3. VOLTAMMETRY

In the voltammetric methods, one measures the characteristic parameters (half-wave potential, limiting currents, potentials or currents in minima or maxima, etc.) of steady-state, quasi-steady-state, and transient polarization curves.

20.3.1. Amperometry

In many cases the concentration of a substance can be determined by measuring its steady-state limiting diffusion current. This method can be used when the concentration of the substance being examined is not very low, and other substances able to react in the working potential range are not present in the solution.

An example of amperometric methods used for analytical purposes is the sensor proposed in 1953 by Leland C. Clark Jr. for determining the concentration of dissolved molecular oxygen in aqueous solutions (chiefly biological fluids). A schematic of the sensor is shown in Fig. 20.1. A cylindrical cap 1 houses the platinum or other indicator electrode 2, the cylindrical auxiliary electrode 3, and an electrolyte, e.g., KCl, solution 4. The internal solution is separated by the polymer membrane 5 from the external test solution 6. The oxygen contained in the test solution diffuses through the membrane into the internal solution and is reduced at the cathodically polarized indicator electrode. The reduction current is determined by the rate of oxygen diffusion through the membrane, which in turn depends (through Fick's law) on the concentration of dissolved oxygen in the test solution.

20.3.2. Improvements of Classical Polarography

Polarography is a particular form of voltammetry associated with the dropping mercury electrode (DME). Classical polarography is the measurement of quasi-steady-state polarization curves with linear potential scans (lps) applied to the DME sufficiently slowly so that within the lifetime, t_{dr}, of an individual drop the potential would not change by more than 3–5 mV.

The polarographic method can be used to qualitatively and quantitatively analyze a large group of solutes that can be reduced within the working potential range of the DME. It is an advantage of the method

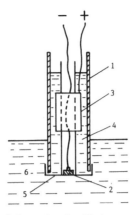

Fig. 20.1. Schematic of a Clark oxygen sensor.

that solutions with low concentrations of the test substances can be analyzed, approximately down to $(1-5) \cdot 10^{-5} \ M$. The volume of the solution sample needed for analysis can be as small as 1 ml or less. Hence, one can detect less than 0.01 mg of the substance being examined. The error limits of analysis are $\pm 2\%$ when appropriate conditions are maintained.

Because of these advantages, polarography became very popular immediately after its inception in 1922. Over the period from 1922 to 1960, several tens of thousands of papers concerned with the use and improvement of polarography were published. However, interest in this method markedly declined in the 1960s, mainly due to a drastic increase in the requirements to be met by methods of chemical analysis. With the production of new superpure materials and increasing awareness for ecological problems, it became necessary to develop much more sensitive methods of analysis able to detect the different impurities down to a level of $10^{-8} \ M$.

The major defects of classical polarography are as follows: (a) it is not possible to drastically increase the sensitivity owing to interference from DME charging currents; (b) the measuring time is long (between 3 and 10 min for a single solution sample); (c) substances which can be oxidized but not reduced cannot be analyzed by the method.

Numerous attempts have been made to overcome some of the defects listed. Electrical circuits for an automatic compensation of charging currents and a direct recording of the faradaic current are available in modern polarographs in order to reduce the influence of the charging currents. However, the accuracy of such compensation is limited, particularly at low reactant concentrations.

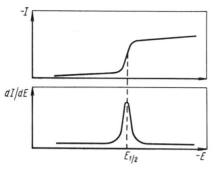

Fig. 20.2. Polarographic curves plotted as I against E and as dI/dE against E.

The sensitivity and selectivity can be raised when recording as a function of potential, not the current but its derivative with respect to potential. In this case, a curve with maximum is obtained (Fig. 20.2) instead of the polarographic wave. The potential of the maximum corresponds to the half-wave potential in an ordinary polarographic curve, while the height of the maximum is proportional to the concentration of the substance being examined. A signal proportional to the derivative of current with respect to potential can be formed in polarographs with the aid of relatively simple electric circuitry.

For higher signal-to-background ratios one can use the differences in time dependence of the faradaic and charging current, I_f and I_{ch}, during growth of the mercury drop. According to Eq. (9.37), the faradaic current increases with time as $t^{1/6}$, while [see Eq. (9.33)] the charging current falls as $t^{-1/3}$. Therefore, the ratio of I_f over I_{ch} is highest and measuring conditions most favorable at the end of drop life. In *tast polarography* (from German *tasten* to probe, to sample), the current is measured for only a brief time, 5 to 20 ms, prior to drop detachment, but not continuously (with averaging) during the drop's life. Using this method one can raise the sensitivity of the method by about an order of magnitude, and detect inpurities in concentrations down to $(1-5) \cdot 10^{-6}$ M.

In the past, polarographic measurements were made in two-electrode cells, and the potential of the DME was varied by changing the full cell voltage. In improved equipment, three-electrode cells and potentiodynamic lps circuits are used. In this way measurements can be performed even in poorly conducting (and particularly in nonaqueous) solutions, where considerable ohmic potential drops develop during current flow.

For reduced analysis times, fast-dropping electrodes producing as many as several tens of drops per second were proposed. In this case the scan rate can be increased to 0.1–0.2 V/s, and the measurement completed in a few tens of seconds. However, here the influence of the charging current increases markedly, and thus the sensitivity and measuring accuracy decrease.

20.3.3. Transient Voltammetric Techniques

Many versions of transient methods of voltammetric analysis using single or repetitive potential or current signals with different shapes and amplitude have been described. These versions have been developed with the basic aim of raising the method's sensitivity by increasing the ratio between the levels of useful signal (the faradaic current) and background (charging current etc.). Under transient diffusion conditions the faradaic currents are much higher than in the steady state (see Section 7.2). Charging currents arise for two reasons: (i) a change of potential and associated change in edl charge density, and (ii) a continuous increase in electrode area. In classical polarography only the second of these reasons is of practical importance, while at solid electrodes only the first reason is important.

In transient measurements one must record rapidly changing currents or potentials. In the past, cathode-ray oscilloscopes have been used for this purpose (at present, improved recording devices or computers are used as well), hence the term of "oscillographic polarography" (or "oscillographic voltammetry"). This term is unfortunate, since it only reflects the device used to record the results, rather than the essential features of the method used for the measurements.

Linear potential scan voltammetry. Voltammograms with characteristic current maxima are obtained (see Fig. 9.9) when lps which are not particularly slow are applied to an electrode. The potentials at which a maximum occurs depend on the nature of the reactant, while the associated current depends on its concentration. When several reactants are present in the solution, several maxima will appear in a curve.

According to Eqs. (9.8) or (9.12), the faradaic current is proportional to the square root of scan rate v. According to Eq. (9.13) however, the charging current is proportional to scan rate. Thus, ratio I_f/I_{ch} decreases with increasing v, and the measuring sensitivity falls. For this reason relatively low scan rates, of 20 to 50 mV/s, are used in measurements at solid electrodes.

Under optimum conditions lps voltammetry is an order of magnitude more sensitive than polarography, i.e., the detection limit is about 10^{-6} M. As in classical polarography, somewhat higher sensitivity and selectivity can be attained when using a differential version, i.e., when recording as a function of potential, not the current but its derivative with respect to potential.

Impressed-ac method. This method was proposed in 1941 by Jaroslav Heyrovský under the name of oscillographic polarography. Here, an adjustable sinusoidal alternating current is galvanodynamically applied to the electrode; its strength (amplitude) is selected so that the electrode potential is scanned through the full potential range required. In a solution not containing reacting substances (Fig. 20.3, curves 1), the potential–time relation is also sinusoidal (Fig. 20.3a). When a reacting substance is present (curves 2), steps or arrests appear in the E vs. t curve at the potential where the corresponding reaction starts. The lengths of the arrests correspond to the transition time t_{lim} after which the surface concentration of the reacting species has fallen to zero [see Section 7.2, Eq. (7.9)]. The higher the reactant concentration the longer the step will be. The changes in the curves can be seen more distinctly when plotting dE/dt against t (Fig. 20.3b). When the curve is plotted as dE/dt against E (the corresponding transformation is accomplished in the oscilloscope itself), closed curves are obtained which, when a reactant is not present, are ellipses (Fig. 20.3c). When a reactant is present, typical dents appear in the upper and lower half of the ellipse. The position of the dents relative to the axis of potentials is characteristic for the nature of the reactant, their height is characteristic for its concentration. When several reactants are present in the solution, several pairs of dents develop in the curve.

This method allows very rapid measurements to be made, but in its sensitivity and selectivity it is inferior to other transient voltammetric techniques.

Method of consecutive potentiostatic pulses. In this method, which was proposed in 1957 by Geoffrey C. Barker, a series of potentiostatic pulses of increasing amplitude (Fig. 20.4a) are applied to the electrode. Between pulses the electrode is at a potential where there is no reaction; during this time the concentration changes caused by a prior pulse will

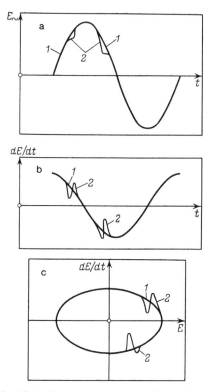

Fig. 20.3. Curves obtained when impressing an alternating current on the electrode, and plotted in different sets of coordinates: (1) base-electrolyte solution; (2) solution with reactant.

level off. The typical length of an individual pulse is 40 to 60 ms, that of the interval between pulses is 1 s. The current is measured during a brief period of time (5 to 15 ms) toward the end of the pulse. By this time the transient current due to the change of edl charge has practically fallen to zero, and the current being measured is basically faradaic. The diffusion conditions are transient because of the short duration of the pulses, and the current is many times higher than the steady-state current. This is a situation favorable for the determination of small amounts of impurities.

The method of potentiostatic pulses is often combined with the DME (so-called pulse polarography). In this case the pulse frequency should match the drop frequency, where each pulse is applied at a definite time during drop life.

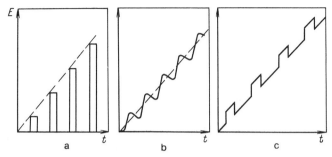

Fig. 20.4. (a) Sequence of potentiostatic pulses of increasing amplitude; (b) linear potential scan with superimposed sinusoidal signals; (c) linear potential scan with superimposed rectangular signals.

In Barker's method, large pulse amplitudes are used. Other versions of the potentiostatic pulse technique are square-wave and staircase voltammetry; here small-amplitude pulses are used.

Ac voltammetry. This term (which should give rise to confusion with the impressed-ac method considered earlier) combines a group of methods where periodic alternating potential signals of sinusoidal (Fig. 20.4b), rectangular (Fig. 20.4c), or other form are superimposed on a slow lps. The signal amplitude is low, usually 10–20 mV. The total current through the electrode is not measured, only its alternating component. Plots of this component against the dc potential (without its periodic excursions) pass through a maximum. As in other cases, the position of the maximum is characteristic for the nature of the reactant, and the height of the maximum is characteristic for its concentration.

Rather high charging currents cross the electrode when a variable potential component is applied. Therefore, in order to reduce the influence of these currents in the case of rectangular pulses, the measurements are made at a specific time after the potential change, when the charging current has drastically decreased. In the case of sinusoidal superimposed currents, one uses another device based on the fact that the alternating charging current (or capacitive current) and the alternating faradaic current have different phase shifts relative to the applied alternating voltage. The capacitive current component has a phase lead of 90° relative to the voltage, while for the faradaic current this lead is 45° or less, depending on the character of the reaction (see Section 9.5). Hence, one can use phase-sensitive instruments to measure the ac only in a particular phase. For instance, when measuring the current with a phase shift of 90° relative to voltage, the capacitive current will not appear at all, since at this point it goes through zero, yet a significant fraction of the faradaic current can still be measured at that time. When used in combination with the DME this method has been called *vector polarography*.

The methods of alternating current voltammetry are widely used for kinetic studies of different electrochemical reactions. The sensitivity for analytical purposes is about 10^{-7} M. It can be raised by about an order of magnitude when versions are used in which the ac signal is recorded, not at the fundamental frequency of the ac voltage, but at its second harmonic, or when still more complicated effects are used.

Stripping voltammetry (voltammetry with concentration).   This method differs from the others, not in the shape of the signal used, but in the principle of analysis.   At first the substance to be examined is completely deposited electrochemically on an inert electrode or substrate from a sample of the solution (concentration step). This method is most commonly used for metal ions which are cathodically deposited on a quiescent (not dropping!) mercury electrode or at platinum, gold, and similar electrodes. After complete extraction of the ions from the solution the electrode, with its thin layer of deposited metal, is subjected to anodic polarization with a linear potential scan. The position and height of the current maximum in the voltammogram characterize the nature and total amount (concentration) of the substance being examined. Several maxima will develop in the curve when several reactants are present. This method is extremely sensitive; in some cases metal impurities can be detected in concentrations of 10^{-9} M or even less. It must be remembered that at very low concentrations, a complete deposition of the ions from the solution will require considerable times of concentration, sometimes as long as 1 h.

20.4. POTENTIOMETRY

Potentiometry is suitable for the analysis of substances for which electrochemical equilibrium is established at a suitable indicator electrode at zero current. According to the Nernst equation (3.44), the potential of such an electrode depends on the activities of the potential-determining substances, i.e., this method determines activities rather than concentrations.

As an example, consider the potentiometric determination of concentration $c_N^{(x)}$ of ions N^{z+} in a solution E_x which is to be accomplished with the aid of an electrode of metal N. Using a simple cell with the reference electrode M_R/E_R:

$$M_R|E_R|E_x|N, \tag{20.1}$$

we find that the measured value of potential is distorted by the diffusion potential φ_d present between solutions E_R and E_x, which in most cases cannot be exactly calculated. For activity determinations we must also know the standard potential E^0 of the electrode used. The activity coefficients of the electrolyte must be known as a function of concentration in order to change over from activities to concentrations.

The problem can be simplified when the peculiar concentration cell

$$M_R|E_R|E_1|N|E_x|E_R|M_R \tag{20.2}$$

is used. The test solution is on the right-hand side of the cell, and solution E_1 with the exactly known concentration $c_N^{(1)}$ of ions N^{z+} (the standard solution) is on the left-hand side of the cell. This cell has a voltage of

$$\mathcal{E}_{(x)} = (RT/z_+F)\ln[\gamma_\pm^{(x)} c_N^{(x)}/\gamma_\pm^{(1)} c_N^{(1)}] + \Delta\varphi_d. \tag{20.3}$$

When a standard solution is used which has an ionic strength close to that of the test solution, the activity coefficients in the two solutions can be assumed to be approximately the same. If in addition the two solutions differ but slightly in their ionic compositions, the difference in diffusion potentials, $\Delta\varphi_d$, between the left- and right-hand part of the cell can be neglected. As a result Eq. (20.3) assumes the simpler form of

$$\mathcal{E}_{(x)} = (RT/z_+F)\ln[c_N^{(x)}/c_N^{(1)}], \tag{20.4}$$

offering the possibility to directly calculate concentrations $c_N^{(x)}$ from measured ocv values $\mathcal{E}_{(x)}$.

The concentrations of anions can be measured analogously when electrodes of the second kind are used.

It is a special feature of potentiometry that the response signal depends, not on the parameter (activity or concentration) which is measured but on its logarithm. On one hand this implies a poorer accuracy and sensitivity of the measurements; for instance, an error of 0.2 mV committed in determining the potential or ocv produces an error in the concentration value found, of 0.8% for $z_j = 1$, 1.6% for $z_j = 2$, and 2.4% for $z_j = 3$. On the other hand however, the accuracy does not depend on the absolute value of the concentration being measured. This is an important difference relative to methods of analysis where the limiting current or current of a maximum is determined, and where the relative error often increases drastically with decreasing concentration and decreasing faradaic current. Hence, potentiometry can be used over a much wider range of concentrations than the voltammetric methods, viz., from concentrated solutions down to solutions with a concentration of 10^{-5}–10^{-7} M of the ion being determined, or down to even lower concentrations when ionic equilibria exist (see Section 3.6.5). Thus, the solution pH can be determined over the pH range from -2 to 15. When dealing with wide concentration ranges of different ions it is convenient to use the concentration exponent $pJ = -\log c_j$, which is the analog of parameter pH for the hydrogen ions.

An important condition for potentiometry is high selectivity; the electrode's potential should respond only to the substance being examined, but not to other components in the solution. This condition greatly restricts the possibilities of the version of potentiometry described here when metal electrodes are used as the indicator electrodes. Thus, the solution should be free of ions of more electropositive metals, since they could deposit on the indicator electrode by cementation and impose their own potential. The solution should also be free of the components of other redox systems; in particular, dissolved air or air in the cell's gas space cannot be tolerated in many cases. Only corrosion-resistant materials can be used as electrodes. It is not possible at all with this method to determine alkali or alkaline-earth metal ions in aqueous solutions.

20.4.1. Ion-Selective Electrodes

Considerable progress was made in overcoming the difficulties pointed out above, when highly selective membranes started to be used in potentiometry instead of the metallic indicator electrodes. The first work involving thin glass membranes goes back to the beginning of the 20th century (Max Cremer, Fritz Haber). Work involving a variety of new membrane types was started in the 1960s; these studies dramatically expanded the potential applications of potentiometry.

Consider the same example, of determining the concentration of ions N^{z+}, but now while using a membrane having ideal permselectivity, i.e., a membrane which is permeable only for ions N^{z+}, but completely impermeable for all other ions present in the system. Let this membrane be designated as $\mu\{N^{z+}\}$. We use the cell

$$M_R|E_R|E_1|\mu\{N^{z+}\}|E_x|E_R|M_R, \qquad (20.5)$$

which is the analog of cell (20.2). At the interfaces between the two solutions and the membrane, equilibria are established for ions N^{z+}, and certain potential differences develop between the phases (see Sections 5.3 and 5.4). According to Eq. (5.13), for the left-hand side this potential difference can be written as

$$\varphi_G = \text{const} + (RT/z_+F)\ln[a_N^{(1)}/a_N^{(\mu)}]; \qquad (20.6)$$

the expression for the right-hand interface is analogous (and contains the same value of the constant). Hence, we have for the full cell's ocv (under the assumptions made above),

$$\mathcal{E}_{(x)} = (RT/z_+F)\ln[c_N^{(x)}/c_N^{(1)}], \qquad (20.7)$$

i.e., exactly the same ocv value is obtained as in the case of cell (20.2).

Thus, the behavior of the selective membrane $\mu\{N^{z+}\}$ is completely equivalent to that of an electrode of metal N. Hence, membranes of this type are called ion-selective electrodes, and in the particular case discussed, the membrane is called an N^{z+}-selective electrode. Sometimes the term is extended to the entire left half of cell (20.5), which in addition to the membrane contains the standard solution and the reference electrode.

The above conclusion is valid for ideally selective membranes. Real membranes in most cases have a limited selectivity. A quantitative criterion of membrane selectivity for an ion N^{z+} to be measured, relative to another ion M^{z+}, is the selectivity coefficient $\sigma_{M/N}$. The lower this coefficient the higher the selectivity will be for ions N^{z+} relative to ions M^{z+}. An electrolyte system with an imperfectly selective membrane can be described by the scheme (5.30). We assume, for the sake of simplicity, that ions N^{z+} and M^{z+} have the same charge. Then the membrane potential is determined by Eq. (5.31), and the equation for the full cell's ocv becomes

$$\mathcal{E}_{(x)} = (RT/z_+F)\ln[(c_N^{(x)} + \sigma_{M/N}c_M^{(x)}/c_N^{(1)}] + \varphi_d. \qquad (20.8)$$

A diffusion potential φ_d can develop in the membrane, since in the case being considered it contains two kinds of mobile ion. However, this potential is small.

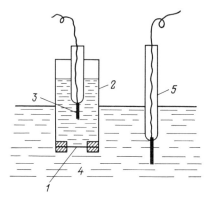

Fig. 20.5. Schematic of a cell with ion-selective electrode.

It can be seen from Eq. (20.8) that so long as concentration c_M^x is much lower than $\sigma_{M/N}^{-1} c_N^{(x)}$, the ions M^{z+} will have no influence on the readings of the N^{z+}-selective electrode. If for instance $\sigma_{M/N} \approx 10^{-6}$, this conclusion holds up to concentrations of ion M^{z+} which are four to five orders of magnitude higher than those of ions N^{z+}. The influence of the foreign ion must be taken into account, only at still less favorable concentration ratios.

Equation (20.8) had been derived in 1937 by Boris P. Nikol'skii for glass electrodes used to determine the hydrogen ion concentration, when the sodium ions contained in the glass membrane are the foreign ions.

Different cell designs exist for potentiometry with ion-selective electrodes. The design most often used is shown schematically in Fig. 20.5. Membrane 1 covers the lower part of a cylindrical cell. The cell contains the standard (internal) solution 2 and the reference electrode 3. The membrane is in contact with the test (external) solution 4 which contains a second reference electrode 5. The cell can be miniaturized as a capillary for analyses in extremely small solution volumes, e.g., the fluid inside an individual (physiological) cell.

It is the major task of practical ion-selective potentiometry to develop sufficiently selective and stable membranes. Different membranes are presently used: solid and liquid, organic and inorganic, homogeneous and heterogeneous.

An example of homogeneous solid membranes is the membrane which consists of a thin layer of single-crystal lanthanum fluoride LaF_3, which has unipolar fluoride ion conduction. This membrane can be used for the potentiometric determination of F^- ions in solutions at concentrations between approximately 10^{-6} and 1 M. The selectivity of this membrane is such that measurements can be made with a 1000-fold excess of most other kinds of ion. Fluoride-selective electrodes are widely used for the analysis of noxious fluorides in the air or in industrial fumes (following their prior absorption by aqueous solution). The sensitivity of the method is so high that as little as 0.1 μg/m^3 can be detected.

Another example is the membrane of silver sulfide Ag_2S (single-crystal or compressed powder), which can be used to determine very low concentrations of silver ions, down to 10^{-20} M (equilibrium concentration). Since the solubility product of silver sulfide is very low (about 10^{-51}), such a membrane can also be used to determine the concentration of sulfide ions in solutions, analogous to an electrode of the second kind.

In heterogeneous solid membranes, the active (selective) material is mixed in the form of a powder with a suitable binder, e.g., silicone rubber, and cast in the form of thin membranes. In this case, the difficulties associated with single crystal fabrication are avoided, and better stability is obtained than in membranes compressed from polycrystalline powder.

Unfortunately, solid materials with high ionic selectivity are rare. Many organic and inorganic ion exchangers are poorly selective with respect to different ions of like charge.

For this reason liquid membranes have become very popular. They consist of an interlayer of liquid that is immiscible with aqueous solutions (or test solutions in other solvents), and displays selectivity with respect to an individual ion or group of ions. Immiscibility of the working fluid with water is attained when aliphatic compounds with long hydrocarbon chains or aromatic compounds are used. Selectivity is attained through the specific chemical interaction between functional groups of these compounds or other, added substances and the given sort of ion.

With liquid membranes, a problem is mechanical stability of the interface between the test solution and the working fluid. Two basic designs exist for solving this problem. One can use a fine-porous matrix (in the form of a thin disk) made of a material which is not wetted by the aqueous solution, but is well wetted by the working fluid, and which strongly retains this fluid in its pores by capillary forces. The matrix can be replenished from a small reservoir inside the cell to compensate losses of working fluid arising from evaporation or a partial dissolution in the test solutions.

Alternatively, one can mix the working fluid with a film-forming polymer material, e.g., poly(vinyl chloride), and cast an elastic film from it.

An electrode widely used is the Ca^{2+}-selective electrode based on the liquid solution of calcium didecyl phosphate in octylphenyl phosphonate. Its selectivity rests on the pronounced tendency of Ca^{2+} ions to form strong complexes with polyphosphate ions. However, the selectivity deteriorates when another solvent, e.g., decanol, is used for the same salt, and now the membrane reacts to all divalent cations. In this way it can be used, in particular, to determine the total hardness of water.

In some cases the high selectivity toward a certain ion is due, not so much to a high energy of interaction between this ion and the membrane medium, but to geometric factors. For instance, the macrocyclic antibiotic valinomycin when dissolved in a suitable solvent has a very high selectivity with respect to potassium ions. Using membrane electrodes containing this substance, one can analyze potassium ions in the presence of a 10^4-fold excess of sodium ions. This selectivity can be explained in terms of the configuration of the valinomycin molecule; it contains a void just matching the potassium ions.

By now a large number of ion-selective substances are known. They are used for the potentiometric concentration determination of practically all kinds of cations and anions. Numerous publications and monographs of the last two decades are concerned with ion-selective electrodes.

20.4.2. The Glass Electrode

Glass electrodes were the first ion-selective electrodes to become known. For more than 60 years, they have been widely used for pH determinations in solutions, both in industry and in scientific research, and particularly in biology and medicine.

The conductivity of glass at room temperature is very low. It arises from the slight mobility of sodium ions in the glass. Special kinds of glass are used for glass electrodes in order to lower their resistance. In addition, they are made as thin as possible. Usually the electrodes are blown from glass tubing in the form of bulbs having a wall thickness of about 0.1 mm.

In its usual form, glass does not contain hydrogen ions. However, during the leaching of glass in aqueous solutions its surface layer is altered to a certain depth; water molecules enter, hydration processes and some swelling occur. Part of the sodium ions in the surface layer are

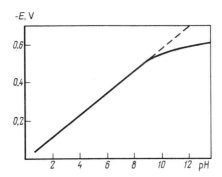

Fig. 20.6. Potential of a glass electrode as function of solution pH.

leached out and are replaced by hydrogen ions from the solution. Equilibrium is established between the hydrogen ions in the surface layer and those in solution; hence, an equilibrium potential difference between the phases arises. The special feature of this surface layer is its exceptionally high selectivity toward hydrogen ions (relative to sodium ions or other cations). For different kinds of glass, the selectivity coefficients for other ions relative to hydrogen ions are between 10^{-9} and 10^{-14}. For this reason the two sides of the membrane act as good hydrogen electrodes. In contrast to other kinds of ion-selective membranes, the current between the two surface layers is transported across the central part of the glass membrane by different ions, viz., the sodium ions.

Figure 20.6 shows the readings of a glass electrode [the measured values of $\mathcal{E}_{(x)}$ of a cell of the type of (20.5)] as a function of solution pH. In the range from acidic to neutral solutions, this curve perfectly obeys Eq. (20.7), i.e., the potential varies linearly by 0.06 V per unit of pH. However, in alkaline solutions the curve departs from this function ("alkali error of the glass electrode"); in strongly alkaline solutions the readings of the electrode are practically independent of solution pH. This is due to violation of the selectivity conditions. At a pH value of 10 and a sodium ion concentration in the solution of 0.1 M, the concentration ratio of sodium and hydrogen ions is 10^9. Under these conditions the electrode's potential is appreciably codetermined by the sodium ions. At still higher pH values, the membrane functions entirely as a sodium electrode. The shape of the curve shown in Fig. 20.6 is in full agreement with Nikol'skii's equation (20.8). The limiting value of solution pH that can be determined with a glass electrode is between 8 and 13, depending on the kind of glass used and on its selectivity coefficient.

When required one can, on the other hand, lower the selectivity of the glass toward hydrogen ions and use the glass electrode for determining the concentration of sodium ions. It will of course be necessary then, that the hydrogen ion concentration is at least 10^4 times lower than that of the sodium ions (the pH value should be four units higher than the pNa value).

Chapter 21
PHOTOELECTROCHEMISTRY

The effects of luminous (or other electromagnetic) radiation on the properties of electrodes and on electrochemical reactions are the subject of photoelectrochemistry.

Luminous radiation (light) can produce changes in the open-circuit potentials and also in the polarization characteristics of electrodes; at constant potential, the current may change (anodic or cathodic photocurrents appear), while at constant current, the electrode potential may change (photopotentials appear). It is an important special feature that electrochemical reactions may become possible which at the same potentials in the dark are thermodynamically prohibited, i.e., associated with an increase in Gibbs energy; under illumination, such reactions are possible because of the energy supplied from outside.

The first observations of photoelectrochemical phenomena were made by Alexandre Edmonde Becquerel in 1839. He used symmetric galvanic cells consisting of two identical metal electrodes in a dilute acid. When illuminating one of the electrodes he observed current flow in the closed electric circuit.

Studies of photoelectrochemical phenomena are of great theoretical value. With light as an additional energy factor, in particular, studies of the elementary act of electrochemical reactions are expedited. Photoelectrochemical phenomena on the other hand are of practical value. Today, it is a major practical task to build electrochemical devices for a direct conversion of luminous (solar) into electrical energy. Another task is that of lowering the consumption of electrical energy in the electrolytic production of hydrogen and other energy-rich substances. During the last decade numerous photoelectrochemical studies have been carried out in connection with these tasks.

Light in the visible and near ultraviolet part of the spectrum is primarily used for photoelectrochemical measurements. The energy of the discrete photons or light quanta, which is given by $h\nu$ (where h is the Planck constant and ν is the frequency of the light wave), and the light intensity or photon flux striking the object being examined, are used to characterize the light. For visible light the photon energies are between 1.63 eV (for the red limit, with a wavelength of $\lambda = 760$ nm) and 3.1 eV (for the violet limit, with $\lambda = 400$ nm). The energies are higher in the ultraviolet, e.g., 6.2 eV at $\lambda = 200$ nm. Electrochemical systems are less strongly influenced by photons having lower energies (in the infrared part of the spectrum). The usual light sources for laboratory work are mercury or xenon lamps, which are used in combination with monochromators as filters for light of a particular frequency. The cell walls (or special windows in the walls) should be transparent to the light used.

Lately, laser light is used in order to have higher intensities. An undesirable side effect at high intensities is heating of the solution layer next to the electrode. This effect can be reduced when intermittent (pulsed) light is used. Light pulses offer the additional possibility to examine aftereffects of the illumination, viz., the relaxation processes which occur when the system returns to its original condition.

The first step in a photoelectrochemical reaction is photon absorption (capture) by a substrate sensitive to light, and the change in electron energy (*photoexcitation*) occurring in the substrate as a result of photon absorption. This step is followed by other chemical or electrochemical reaction steps involving the activated substrate.

We distinguish two basic types of photoelectrochemical reaction:

a) photoexcitation of the electrode (of the electrons in its surface layer) and subsequent reaction of ordinary, nonactivated reactants at the electrode (Section 21.3), and

b) photoexcitation of reactant particles in the solution and their subsequent reaction at a nonactivated electrode (Section 21.4).

Electron photoemission from an electrode into an electrolyte solution, which yields solvated (hydrated) electrons, can be regarded as a particular case of reactions with photoexcitation of the electrode (Section 21.2).

It will be necessary to provide quantitative criteria for studies of photoelectrochemical phenomena. These phenomena are a function of photon energy, and appear only at energies above a particular *threshold energy* $h\nu_{thr}$ (or below a particular threshold wavelength, λ_{thr}, of the light). The threshold energy is an important characteristic of any given phenomenon.

For phenomena involving electrons crossing the phase boundary (photocurrents, electron photoemission), the *quantum yield* γ of the reaction is a criterion frequently employed. It is defined as the ratio between the number of electrons, N_e, that have crossed and the number of photons, N_{ph}, that had reached the reaction zone (or, in another definition, the number of photons actually absorbed by the substrate): $\gamma = N_e/N_{ph}$.

21.1. ENERGY LEVELS OF ELECTRONS IN ELECTRODES AND IN THE ELECTROLYTE

21.1.1. Electrochemical Potential of Electrons in Metals and Semiconductors

We shall at first amplify on the concepts of "Fermi energy" and "Fermi level" of the electrons in metals, previously mentioned in Section 1.3.1.

The Fermi energy W_F which can be calculated via Eq. (1.14) is reckoned from the energy of the valence-band bottom ("zero-point energy"), and gives the kinetic energy of the electrons at the highest occupied level of this band. This energy is equal to the chemical potential of the electrons.

The points of reference thus defined for the different metals, i.e., their valence-band bottoms, are all different and exist in different phases, so that it is difficult to compare the electron energies in these metals. It will be advantageous to always choose a point of reference that is present in the same phase. Our choice is a point m in vacuum just outside the conductor (the concept of "just outside the conductor" had been explained in Section 2.6.1). When the metal as a whole is uncharged, i.e., when there is no external field and the outer potential $\psi_{ex}^{(M)}$ is zero, a point in vacuum infinitely far from the metal surface can also be used as the point of reference, since the work required to transfer a charge from the point $x \to \infty$ to point m is zero.

The Fermi level (for which the symbol U_F is used in order to distinguish it from the Fermi energy) is the level of total energy of the energy-richest electrons relative to this new point of reference, and is given by the work that must be expended in transferring an electron from vacuum (from point m) to the highest occupied level of the valence band in the metal. In this transfer, work must be expended to overcome both electrostatic and chemical forces. The concept of "Fermi level" U_F completely coincides with that of "electrochemical potential of electrons in the metal" $\tilde{\mu}_e^{(M)}$ (relative to the given point of reference).

In electron emission from a metal into vacuum, primarily electrons from the highest occupied level are extracted. Therefore, the work function $w_e^{(M,0)}$ involved in this act, under the assumptions made, is equal to the Fermi level or electrochemical potential of electrons in the metal, but with an inverted sign [compare with Eq. (2.32)]:

$$w_e^{(M,0)} = -\tilde{\mu}_e^{(M)} \equiv -U_F^{(M)}. \qquad (21.1)$$

In the present chapter, all values of μ_j, $\tilde{\mu}_j$, w_e, also W_F and U_F, refer to a single electron (or single species) and are stated in electron volts, as in the earlier Sections 2.6.2 and 14.4.1. We recall that the values of work functions are always positive, hence, the values of $\tilde{\mu}_e^{(M)}$ and $U_F^{(M)}$ are always negative (electron transfer from vacuum into a metal is associated, not with an expenditure but with a gain of energy).

It had been pointed out in Section 1.3.1 that the Fermi level in intrinsic semiconductors is approximately in the middle between the energy levels of valence-band top and conduction-band bottom; in impurity-type semiconductors it is in the middle between one of these two levels and the energy level associated with the acceptor- or donor-type impurity (see Fig. 1.2). Therefore, the Fermi level is inside the band gap, and in the semiconductor there are no real electrons having this Fermi energy. However, the physical meaning is preserved; it can be shown that the work required to transfer an electron from vacuum into the semiconductor, corresponds exactly to the energy level thus defined, i.e., this level characterizes the electrochemical potential of electrons in the semiconductor.

Consider the more complicated case of a junction between two different metals α and β. Generally, they will have different values of the Fermi energy and work function. Between the two metals, a certain Volta potential $\varphi_V^{(\beta,\alpha)}$ will be set up. This implies that the outer potentials $\psi_{ex}^{(M)}$ at points a and b, which are just outside of the two metals, are different. However, it will be preferable to count the Fermi levels or electrochemical potentials from a common point of reference. This can be either the point a or the point b. Since these two points are located in the same phase, the potential difference between them (the Volta potential) can be measured. Hence, values counted from one of the points of reference are readily converted to the other point of reference when required.

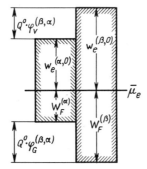

Fig. 21.1. Energy levels of the electrons in two metals (α) and (β) which are in contact.

Unlike the values of $\tilde{\mu}_e$, values of electron work function $w_e^{(M,0)}$ always refer to the work of electron transfer from the metal to "its own" point of reference. Hence, in this case, the relation established between these two parameters by Eq. (21.1) is disturbed.

The condition for electronic equilibrium between two phases is that of equal electrochemical potentials $\tilde{\mu}_e$ of the electrons in them [Eq. (2.10)]. In Fig. 21.1, the energies of the valence-band bottoms (or negative values of the Fermi energies) are plotted downward relative to this common level, in the direction of decreasing energies, while the values of the electron work functions are plotted upward. The difference in energy levels of the valence-band bottoms, i.e., the difference in chemical potentials of the electrons in the two metals, according to Eq. (2.11) corresponds to the work $Q^0\varphi_G^{(\beta,\alpha)}$ associated with overcoming the Galvani potential. On the other hand, the work function difference according to Eq. (2.35) corresponds to the work $Q^0\varphi_V^{(\beta,\alpha)}$ associated with overcoming the Volta potential between the two metals.

21.1.2. Electrochemical Potential of the Electrons in Solutions

Electrolyte solutions ordinarily do not contain free electrons. The concept of *electrochemical potential of the electrons in solution*, $\tilde{\mu}_e^{(E)}$, can still be used for those among the bound electrons which will participate in redox reactions in the solution.

Consider the equilibrium $Ox + ne^- \rightleftarrows Red$ in the solution. In equilibrium, the total change in Gibbs energy in the reaction is zero, hence, the condition for equilibrium can be formulated as

$$\tilde{\mu}_e^{(E)} = (1/n)[\tilde{\mu}_{Red}^{(E)} - \tilde{\mu}_{Ox}^{(E)}]. \tag{21.2}$$

The electrochemical potential of the electron depends on the redox properties of the system; the higher the value of $\tilde{\mu}_{Ox}^{(E)}$ (or: the stronger the oxidizing agent and the higher its electron affinity), the higher will be the energy gain in electron transfer from the point of reference to species Ox, i.e., the lower (or: more negative) will be the value of $\tilde{\mu}_e^{(E)}$. In the reaction considered, the species Ox and Red differ by n unit charges; therefore, the expression for electrochemical potential of the electrons in solution can also be written [when allowing for Eq. (3.17)] as

$$\tilde{\mu}_e^{(E)} = (1/n)[\mu_{Red}^{(E)} - \mu_{Ox}^{(E)}] - Q^0\psi^{(E)}. \tag{21.3}$$

When the solution in this redox system is in contact with a nonconsumable metal electrode (e.g., a platinum electrode), the equilibrium set up also implies equal electrochemical potentials, $\tilde{\mu}_e^{(M)}$ and $\tilde{\mu}_e^{(E)}$, of the electrons in the metal and electrolyte. According to Eq. (3.34) and taking into account that the electrode potential differs by a constant term from the metal/solution Galvani potential, we thus have an expression for the equilibrium potential of this electrode and, at the same time, for the equilibrium potential, E_{Redox}, of this redox system:

$$Q^0 E_{Redox} = (1/n)[\mu_{Red}^{(E)} - \mu_{Ox}^{(E)}] + \text{const}. \tag{21.4}$$

It follows from the last two equations that between the values of $\tilde{\mu}_e^{(E)}$ and the redox potential, the relation

$$\tilde{\mu}_e^{(E)} = A - Q^0 E_{Redox} \tag{21.5}$$

exists, i.e., the more positive the redox potential E_{Redox} the more negative will be the value of $\tilde{\mu}_e^{(E)}$, and vice versa. Constant A in this equation (which is the value of $\tilde{\mu}_e^{(E)}$ for $E_{Redox} = 0$) depends only on the reference electrode against which potential E_{Redox} has been measured, but not on the nature (metal or semiconductor) of the electrode which is in contact and equilibrium with the electrolyte solution.

When the electrochemical potential of the electrons in the metal is counted from the point of reference just outside the metal, then at the electrode potential $E = 0$ V it should also be equal to the value of A. At other potentials it will be determined by an equation analogous to (21.5):

$$\tilde{\mu}_e^{(M)} = A - Q^0 E. \qquad (21.6)$$

It is important to notice that this relation between the values of $\tilde{\mu}_e^{(M)}$ and E is preserved, even when the electrode is not in equilibrium with the solution: $\tilde{\mu}_e^{(M)} \neq \tilde{\mu}_e^{(E)}$, and in particular when it is ideally polarizable.

Constant A in Eqs. (21.5) and (21.6) is about -4.4 eV when the standard hydrogen electrode is used as the reference electrode. This value has been determined from experimental values for the electron work function of mercury in vacuum, $w_e^{(M,0)}$, which is 4.48 eV, and for the Volta potential, $\varphi_V^{(M,E)}$, between the solution and a mercury electrode polarized to $E = 0$ V (SHE), which is -0.07 V (the work of electron transfer is -0.07 eV). The sum of these two values, according to Eq. (2.35), corresponds to the solution's electron work function at this potential, i.e., to the value of constant A with an inverted sign.

Thus, an unambiguous correlation exists between the values of electrode potential (electrochemical scale) and the Fermi levels or values of electrochemical potential of the electrons defined as indicated (physical scale); see the symbols at the vertical axes in Fig. 21.2.

It had been pointed out in Section 14.4.1 that at electrode potentials more negative than approximately -2.8 V (SHE), free solvated electrons appear in the solution as a result of (dark) emission from the metal. At this potential the electrochemical potential of the electrons according to Eq. (21.6) is about -1.6 eV, which is at once the energy of electron hydration in electron transfer from vacuum into an aqueous phase.

The value of $\tilde{\mu}_e^{(M)}$ defined by Eq. (21.6) is sometimes called the "absolute electrode potential measured against vacuum". We must remember here that we are concerned with electrochemical potentials stated in electron volts, rather than with electrostatic potentials stated in volts. Hence, this absolute potential, which can be determined experimentally, is unrelated to the problem of determining absolute values of the Galvani potential that had been discussed in Section 12.10.2. Using absolute (electrochemical) potentials we can state the emf of a galvanic cell as the difference between two parameters,

$$\mathcal{E} = (1/Q^0)[\tilde{\mu}_e^{(M_2)} - \tilde{\mu}_e^{(M_1)}], \qquad (21.7)$$

each referring to just one of the electrodes; in Section 2.4, the same had been achieved, analogously, with the aid of electrode potentials. Thus, "absolute potentials" in a certain sense are equivalent to "electrode potentials", but differ from them insofar as they are not tied to any particular reference electrode.

21.1.3. Electron Transitions between Electrode and Solution

The elementary act of an electrochemical redox reaction is the transition of an electron from the electrode to the electrolyte or conversely. Such transitions obey the so-called Franck–Condon principle, which says that the electron transition probability is highest when the energies of the electron in the initial and final state are identical.

It follows from the Franck–Condon principle that in electrochemical redox reactions at metal electrodes, practically only the electrons residing at the highest occupied level of the metal's valence band are involved, i.e., the electrons at the Fermi level. At semiconductor electrodes, the electrons from the bottom of the conduction band or holes from the top of the valence band are involved in the reactions. Under equilibrium conditions, the electrochemical potential of these carriers is equal to the electrochemical potential of the electrons in the solution. Hence, mutual exchange of electrons (an exchange current) is realized between levels having the same energies.

When a net current flows and the electrode's polarization is ΔE, its Fermi level is shifted by $-Q^0 \Delta E$ relative to level $\tilde{\mu}_e^{(E)}$, down in the case of anodic polarization, and up in the case of cathodic polarization. The higher the electrode's catalytic activity toward a given reaction the lower will be the polarization at a given current density, and the smaller will be said shift.

21.2. ELECTRON PHOTOEMISSION INTO SOLUTIONS

Events of electron photoemission from a metal into an aqueous solution had first been documented in 1963 by Geoffrey C. Barker and Arthur W. Gardner on the basis of indirect experimental evidence. The formation of solvated electrons in nonaqueous solutions (e.g., following the dissolution of metallic sodium in liquid ammonia) had long been known, but it was only in the beginning of the 1950s that their existence in aqueous solutions was first thought possible. It is probably for this reason that even nowadays in aqueous solutions we more often find the term "solvated" than "hydrated" electrons.

When photons are absorbed in a metal the ensemble of electrons are excited and part of the electrons are lifted to higher energy levels. The excited state is preserved in the metal for only a short time, and the system rapidly returns to its original state. When the photon energy $h\nu$ is higher than the metal's electron work function, $w_e^{(M,E)}$, in the solution at a given potential, individual excited electrons can be emitted into the solution. Usually the quantum yields of this process are low, e.g., around 10^{-4}, and depend on the depth of the layer in which the photons are absorbed.

Photoemission produces "dry" (nonsolvated) electrons leaving the metal surface with a high initial velocity. An excess photon energy of 0.05 eV is enough to impart a velocity of about 10^5 m/s to the electron. Already in the first layers of solution the electrons are strongly decelerated and "thermalized", i.e., their kinetic energy is reduced to values typical for thermal motion at the given temperature. Then, at distances of 2–4 nm from the electrode surface, the electrons are solvated. The electron's hydration energy in aqueous solutions is about 1.5–1.6 eV.

The further fate of the solvated electrons depends on solution composition. When the solution contains no substances with which the solvated electrons could interact, they will accumulate near the electrode and rather rapidly form a negative space charge pushing the electrons now being emitted back to the electrode. Part of the electrons already emitted will be reabsorbed by the metal, i.e., are "oxidized", since the electrochemical potential of electrons in the metal is markedly lower than that of solvated electrons in the solution. A steady state is attained after about 10^{-4} s; at this time the rate of oxidation has become equal to the rate of emission, and the original, transient photoemission current (the electric current in the galvanic cell in which the illuminated electrode is the cathode) has fallen to zero.

Steady photoemission currents can be realized when acceptors (scavengers) for the solvated electrons are present in the solution. These are substances which will interact with the solvated electrons (are reduced by them) but are not themselves involved in any electrochemical reaction at the electrode surface. The acceptors most often used are nitrous oxide N_2O and also hydroxonium ions H_3O^+ (at potentials where these ions are not electrochemically reduced at the given electrode). A reaction with acceptors implies that the current of reoxidation of the solvated electrons becomes lower, and thus a steady photoemission current appears. Water molecules will practically not react with the solvated electrons, and hence do not function as acceptors.

Steady electron photoemission in the presence of acceptors is subject to certain laws of behavior enabling us to distinguish the photoemission currents from the cathodic currents of ordinary electrochemical reactions.

The basic law of electron photoemission in solutions which links the photoemission current with the light's frequency and with electrode potential is described by Eq. (14.25) ("the law of five halves"). This equation must be defined somewhat more closely. As in the case of electrochemical reactions (see Section 14.2), not the full electrode potential E as shown in Eq. (14.24) is affecting the metal's electron work function in the solution but only a part $(E - \psi')$ of this potential, which is the part associated with the potential difference between the electrode and a point in the solution which is just outside the electrode [compare with Eq. (14.15)]. Hence, the basic law of photoemission should more correctly be written as

$$I = A[h\nu - w_e^0 - Q^0(E - \psi')]^{5/2}. \tag{21.8}$$

Two typical features of the photoemission current follow from this law: (i) the linear relation between $I^{2/5}$ and the potential E (at least at potentials where the value of ψ' depends little on E), and (ii) the existence of a threshold frequency ν_{thr} which is independent of the electrode material but depends on electrode potential.

At a potential of $E = 0$ V (SHE) the work function has a value of 3.15 eV. Therefore, it is possible when using light having exactly this photon energy, to observe photoemission currents in the cathodic region starting from approximately the potential of the standard hydrogen electrode.

Another feature of the photoemission current is its unusual dependence on the acceptor concentration. It follows from theory and is confirmed by experiments that at low concentrations, the current is proportional to the square root of acceptor concentration: $I = c_{\text{acc}}^{1/2}$. At high concentrations, when practically all electrons which have been emitted react with the acceptor, the photoemission current no longer depends on the acceptor concentration.

Photoemission phenomena are of great value for a number of areas in electrochemistry. In particular, they can be used to study the kinetics and mechanism of electrochemical processes involving free radicals as intermediates.

Photoemission measurements can also be used to study electric double-layer structure at electrode surfaces. For instance, by measuring the photoemission currents in a dilute solution and, under identical conditions, in a concentrated solution (where we know that $\psi' = 0$), we can find the value of ψ' in the dilute solution by simple calculations using Eq. (21.8). We can also determine the point of zero charge of the metal involved. In fact, at the potential of zero charge (where always $\psi' = 0$) the photoemission current is independent of total concentration of the electrolyte solution, while it increases upon dilution of the solution at more positive potentials (the values of ψ' increase), and it decreases upon dilution at more negative potentials. Thus, curves recorded at different electrolyte concentrations and plotted as $I^{2/5}$ against E intersect at the pzc.

21.3. PHOTOEXCITATION OF SEMICONDUCTOR ELECTRODES

21.3.1. Behavior of Illuminated Semiconductor Electrodes

Semiconductor electrodes exhibit electron photoemission into the solution, like metal electrodes, but in addition they exhibit further photoelectrochemical effects due to excitation of the electrode under illumination.

The first observations in this area were made toward the middle of our century. At the end of the 1940s Vladimir I. Veselovskii studied the photoelectrochemical behavior of metals covered with oxide layers having semiconductor properties. In 1955 Walter H. Brattain and Charles G. B. Garrett published a paper in which they established the connection between the photoelectrochemical properties of single-crystal semiconductors and their electronic structure.

The behavior of a semiconductor depends on its own nature and also on that of the phase in contact with it. We shall examine n-type semiconductors as an example, which are often used and which contain electron-donating additives. These additives produce a strong decrease in work function. This implies that the electrochemical potential, $\tilde{\mu}^{(S)}$, of the electrons in them is rather high (just slightly negative). When contact is made with an electrolyte, the original value of $\tilde{\mu}_e$ in the semiconductor, as a rule, is higher than that of the redox system in the solution. For this reason, a partial transition of electrons into the solution starts to take place,

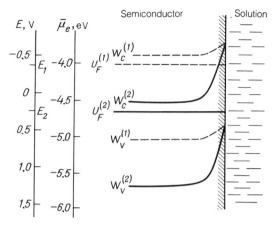

Fig. 21.2. Energy bands in an n-type semiconductor in contact with an electrolyte solution, for two values, E_1 and E_2, of electrode potential ($E_2 > E_2$).

the semiconductor becomes positively charged, its potential moves in the positive direction, and the values of $\bar{\mu}_e$ come closer together (in the case of semiconductors, complete electronic equilibrium with the solution is often not established, i.e., some difference between the $\bar{\mu}_e$-values remains).

The excess positive charge appearing in the semiconductor's surface layer (which constitutes a depletion of electrons in this layer) leads to an upward bending of the band edges (see Fig. 12.22).

As the semiconductor's potential is moved in the positive direction, the Fermi level, in accordance with Eq. (21.6), moves in the negative direction. The edges of valence and conduction band in the bulk semiconductor move together with it. The energies of the band edges at the surface itself, i.e., the degree of band bending, depend on the repartition of the total potential change ΔE between the surface and interfacial potential (see Section 12.9). Figure 21.2 schematically shows the positions of the band edges and Fermi level for two values of electrode potential $E_2 > E_1$ in the case where only the surface potential changes, while the interfacial potential remains practically constant (the band edges are "pinned to the surface").

When semiconductors are irradiated with photons of high energy, electron photoemission is possible, as in the case of metals. When the photon energy is lower than the electron work function in the solution, under given

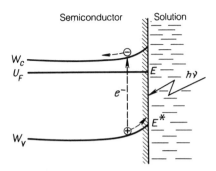

Fig. 21.3. Charge separation during photoexcitation of a semiconductor electrode.

conditions, but is still higher than the semiconductor's band gap W_g:

$$w_e^{(S,E)} > h\nu > W_g \qquad (21.9)$$

(index S stands for the semiconductor), then interband electron transitions can occur from the valence band to the conduction band above. The result of such a transition is the formation of a pair of free carriers, an electron in the conduction band and a positive hole in the valence band. This is an event equivalent to the electron transitions to higher energy levels in the valence band which occur in metals.

The difference between metals and semiconductors becomes apparent when the further fate of these excited charges is considered. In metals an excited electron will very quickly (within a time of the order of 10^{-14} s) return to its original level, and the photon's original energy is converted to thermal energy. Photoexcitation has no other consequences.

In semiconductors, which have a band gap, recombination of the excited carriers—return of the electrons from the conduction band to vacancies in the valence band—is greatly delayed, and the lifetime of the excited state is much longer than in metals. Moreover, in n-type semiconductors with band edges bent upward, excess electrons in the conduction band will be driven away from the surface into the semiconductor by the electrostatic field, while positive holes in the valence band will be pushed against the solution boundary (Fig. 21.3). The electrons and holes in the pairs produced are thus separated in space. This leads to an additional stabilization of the excited state, to the creation of some steady concentration of excess electrons in the conduction band inside the semiconductor, and to the creation of excess holes in the valence band at the semiconductor/solution interface.

These two groups of excited carriers are not in equilibrium with each other. Each of them corresponds to a particular value of electrochemical potential; we shall call these values $\tilde{\mu}_e^{(S)*}$ and $\tilde{\mu}_h^{(S)*}$. Often these levels are called the quasi-Fermi levels of excited electrons and holes. The quasi level of the electrons is located between the (dark) Fermi level and the bottom of the conduction band, while the quasi level of the holes is located between the Fermi level and the top of the valence band. The higher the relative concentration of excited carriers the closer to the corresponding band will be the quasi level. In n-type semiconductors, where the concentration of electrons in the conduction band is high even without illumination, the quasi level of the excited electrons is just slightly above the Fermi level, while the quasi level of the excited holes, $\tilde{\mu}_h^{(S)*}$, is located considerably lower than the Fermi level.

Because of the excess holes with an energy lower than the Fermi level which are present at the n-type semiconductor surface in contact with the solution, electron transitions from the solution to the semiconductor electrode are facilitated ("egress of holes from the electrode to the reacting species"), and anodic photocurrents arise. Such currents do not merely arise from an acceleration of reactions which, at the particular potential, will also occur in the dark. The electrochemical potential, $\tilde{\mu}_h^{(S)*}$, according to Eq. (21.6) corresponds to a more positive value of electrode potential (E^*) than that which actually exists (E). Hence, anodic reactions can occur at the electrode, even with redox systems having an equilibrium potential more positive than E (between E and E^*), i.e., reactions which are prohibited in the dark.

When the illuminated electrode is not potentiostated, the space charge in its surface layer will decrease under the effect of the holes accumulating at the surface itself; the degree of band bending will also decrease, and the electrode potential will move in the negative direction. It can be shown that the maximum value of this potential shift is equal to the original (dark) value of the semiconductor's surface potential in contact with the solution, i.e., to the original value of band bending.

The electrons produced in the conduction band as a result of illumination can participate in cathodic reactions. However, since in n-type semiconductors the quasi-Fermi level is just slightly above the Fermi level, the excited electrons participating in a cathodic reaction will practically not increase the energy effect of the reaction. Their concentration close to the actual surface is low, hence, it will be advantageous to link the n-type semiconductor electrode to another electrode which is metallic, and not illuminated, and to allow the cathodic reaction to occur at this electrode. It is necessary then, that the auxiliary metal electrode has good a catalytic activity toward the cathodic reaction.

Analogous effects are seen at p-type semiconductors at which cathodic reactions are accelerated when the electrode is illuminated. For heightened effects, one can combine in a single cell an n- and a p-type semiconductor, and allow the anodic reaction to occur at the former, and the cathodic reaction to occur at the latter.

21.3.2. Devices Based on the Photoexcitation of Semiconductor Electrodes

The phenomena listed can be used for the design of practical devices in which luminous (solar) energy is directly converted to electrical or chemical energy.

The devices for the production of electrical energy are galvanic cells in which the anodic reaction occurring at an illuminated electrode is fully compensated by the reverse (cathodic) reaction occurring at a dark electrode, i.e., there is no overall current-producing reaction, and no overall chemical change whatever occurs in the system. Cells of this type, where the electrodes are operated under different conditions but where the same electrode reaction occurs in different directions at the two electrodes, are called regenerative.

Figure 21.4 shows an example, viz., the energy diagram of a cell where n-type cadmium sulfide CdS is used as a photoanode, a metal that is corrosion-resistant and catalytically active is used as the (dark) cathode, and an alkaline solution with S^{2-} and S_2^{2-} ions between which the redox equilibrium $S_2^{2-} + 2e^- \rightleftarrows 2S^{2-}$ exists is used as the electrolyte. In this system, equilibrium is practically established, not only at the metal/solution but also at the semiconductor/solution interface. Hence, in the dark, the electrochemical potentials of the electrons in all three phases are identical.

The band edges are flattened when the anode is illuminated, the Fermi level rises, and the electrode potential shifts in the negative direction. As a result a potential difference ΔE_{ph} which amounts to about 0.6–0.8 V develops between the semiconductor and metal electrode. When the external circuit is closed over some load R, the electrons produced by illumination in the conduction band of the semiconductor electrode will flow through the external circuit to the metal electrode, where they are consumed in the cathodic reaction. Holes from the valence band of

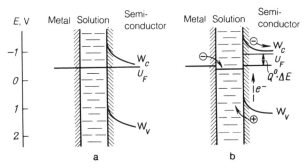

Fig. 21.4. Energy diagrams for a cell with CdS photoanode in an alkaline solution containing ions S^{2-} and S_2^{2-}, (a) in the dark, (b) under illumination.

the semiconductor electrode at the same time are directly absorbed by the anodic reaction. Therefore, a steady electrical current arises in the system, and the energy of this current can be utilized in the external circuit. In such devices, the solar-to-electrical energy conversion efficiency is as high as 5–10%. Unfortunately, their operating life is restricted by the low corrosion resistance of semiconductor electrodes.

Devices for the production of chemical energy differ from the ones described, in that different electrode reactions now occur at the cathode and anode. As the result of overall current-producing and current-consuming reactions, energy-rich products are generated (*photoelectrolysis*); their chemical energy can be utilized. Of greatest interest is the photoelectrolytic production of hydrogen. The first cells of this type were built in 1972 by Akira Fujishima and Kenichi Honda in Japan. Their photoanode was titanium dioxide TiO_2 (an n-type semiconductor), their cathode was platinized platinum. A porous separator was used in the cell in order to keep the anodic and cathodic reaction products apart.

The energy diagram of a cell of this type is shown schematically in Fig. 21.5. In a semiconductor having a sufficiently wide band gap (e.g., in strontium titanate $SrTiO_3$ we have $W_g = 3.2$ eV), the quasi level $\tilde{\mu}_h^{(S)*}$ of the holes is located appreciably below $\tilde{\mu}_e^{(E)}$ for the water/oxygen redox system, while the quasi level $\tilde{\mu}_e^{(S)*}$ of the electrons is located appreciably above $\tilde{\mu}_e^{(E)}$ for the water/hydrogen redox system. Hence, under illumination and with the external circuit closed, anodic oxygen evolution will readily occur at such an electrode (high electrode polarization and thus high current densities are possible), while cathodic hydrogen evolution consuming the electrons which have been generated in the semiconduc-

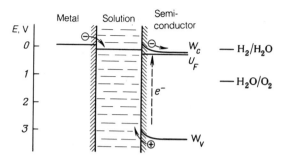

Fig. 21.5. Energy diagram for water photoelectrolysis with a TiO$_2$ anode; U_{ex} is the required minimum voltage of the external power source.

tor's conduction band and have traveled to the platinum electrode will occur at this electrode.

According to Eq. (21.9), only light of the ultraviolet part of the spectrum can be absorbed when the band gaps are so wide, and such devices are not suitable for solar energy conversion. Titanium dioxide has a slightly narrower band gap (about 3 eV). However, in this material the quasi level of the electrons is slightly below $\tilde{\mu}_e^{(E)}$ for the water/hydrogen redox system, so that hydrogen evolution is no longer possible. Other materials having lower values of W_g may have an even larger difference between these levels. This difficulty can be overcome when a power source of low voltage (e.g., 0.4 to 0.6 V) which will shift the potential of the platinum electrode in the negative direction is included in the external circuit. Using this device, one can electrolytically produce hydrogen and oxygen with cell voltages about four times lower than by electrolysis in the dark (the theoretical voltage for water electrolysis is 1.23 V, see Section 19.1).

The overall energy conversion efficiency (the ratio between the chemical energy of hydrogen and the solar energy striking the electrode, corrected for the energy consumed to sustain the additional voltage mentioned) is about 6%. Even in this case problems concerning stability of the semiconductor electrode arise.

Numerous studies aiming at finding more efficient and more stable semiconductor systems are still going on at the present time.

21.4. PHOTOEXCITATION OF REACTING SPECIES

The absorption of visible light of a particular frequency by atoms or molecules of a solute is perceived by our eyes as the color of this substance. Hence, only colored substances—natural or synthetic dyes—are excited by visible light. Under ultraviolet light, substances not absorbing in the visible part of the spectrum can also be excited.

Upon excitation of the molecule, one or several electrons are lifted to a higher energy level. When the lifetime of the photoexcited state of molecules j is sufficiently long (that is, longer than the time required for electronic transitions between particles), this state can be described in terms of a new value $\tilde{\mu}_j^{(E)*}$ of the electrochemical potential, which is higher than the value $\tilde{\mu}_j^{(E)}$ of nonexcited particles.

Often the primary photoexcited species are unstable and are converted (e.g., by chemical reaction with other solution components) to more stable secondary species which, as a rule, still have an electrochemical potential higher than the original, unexcited species. Sometimes a whole chain of such conversions may be involved.

An excited particle which is to become involved in the electrochemical reaction, must be sufficiently close to the electrode surface in order to diffuse to the surface within the lifetime of its excited state. It is better yet, when it is present on the surface as an adsorbate. Sometimes dyes are applied to the surface which are not themselves involved in any electrochemical reaction, but which when excited react with the solution to produce a soluble secondary substance which will react (sensitization of the electrode surface).

In the case of species involved in a redox reaction, any change in their electrochemical potential upon photoexcitation will also cause a change in electrochemical potential $\tilde{\mu}_e^{(E)}$ of the electrons in the solution. When the excited species is an oxidizing agent, the value of $\tilde{\mu}_e^{(E)}$ according to Eq. (21.2) will decrease; but when it is a reducing agent, the value of $\tilde{\mu}_e^{(E)}$ will increase. A corresponding change will also occur in the redox potential of a reaction involving photoexcited species. This change can be ascertained by measuring the open-circuit potential of a metallic indicator electrode in the solution containing the photoexcited species (when the excited species are sufficiently stable, the indicator electrode may even be outside the illuminated region).

A change in thermodynamic properties of the excited species has several consequences. For instance, cathodic reduction of an excited oxidizing agent is possible at less negative potentials than that of a nonexcited one. In particular, in polarographic measurements at the dropping mercury electrode this leads to a shift of half-wave potential in the positive direction (*photopolarography*, Hermann Berg, 1965). In electrochemical reactions, the apparent kinetic parameters may change.

When, in a symmetric galvanic cell suitably assembled, the region of the solution close to one of the electrodes is illuminated, a potential difference between the electrodes will develop owing to photoexcitation of dissolved species. An electric current is recorded in the system when the circuit is closed; the reaction of the photoexcited species at the first electrode (e.g., the cathodic reduction of an excited oxidizing agent) will be compensated by the reverse reaction at the second electrode. Thus, a cell of this type is regenerative, and luminous energy is converted in it to electrical energy.

The classical example of such a device is the cell where thionine dye is used. Thionine is the oxidizing agent in the reaction: $T + e^- + H^+ \rightleftarrows$ TH. Thionine itself is hard to reduce electrochemically. Therefore, the mediating redox system Fe^{3+}/Fe^{2+} is used which functions as an electron shuttle. The excited form of thionine, T^*, produced under illumination is readily reduced by divalent iron ions:

$$T^* + Fe^{2+} + H^+ \rightleftarrows TH + Fe^{3+}. \tag{21.10}$$

The thionine reduction product TH is anodically reoxidized to thionine, while the Fe^{3+} ions are cathodically rereduced to Fe^{2+} ions. Thus, the chemical composition of the system will not change during current flow. The potential difference between the electrodes which can be used to extract electrical energy is 0.2 to 0.4 V under current flow. The conversion factor of luminous to electrical energy is very low in such cells, viz., about 0.1%. This is due to the numerous side reactions which drastically lower the overall efficiency. Moreover, the stability of such systems is not high. Therefore, the chances for a practical use are not evident so far.

Chapter 22
ELECTROKINETIC PROCESSES

Electric double layers are formed in heterogeneous electrochemical systems at interfaces between the electrolyte solution and other conducting or nonconducting phases; this implies that charges of opposite sign accumulate at the surfaces of the adjacent phases. When an electric field is present in the solution phase which acts along such an interface, forces arise which produce (when this is possible) a relative motion of the phases in opposite directions. The associated phenomena historically came to be known as electrokinetic phenomena or electrokinetic processes. The term is not very fortunate, since a similar term, "electrochemical kinetics," commonly has a different meaning (see Part 3).

Four different electrokinetic processes are known. Two of them, *electroosmosis* and *electrophoresis*, were described in 1808 by Ferdinand Friedrich Reuss, professor at the University of Moscow. The schematic of a cell appropriate for realizing and studying electroosmosis is shown in Fig. 22.1a. An electrolyte solution in a U-shaped cell is divided in two parts by a porous diaphragm. Auxiliary electrodes are placed in each of the half cells in order to set up an electric field in the solution. Under the influence of this field, the solution starts to flow through the diaphragm in the direction of one of the electrodes. The flow continues until a hydrostatic pressure differential (height of liquid column) has been built up between the two cell parts which is such as to compensate the electroosmotic force.

Electrophoresis can be observed in solutions containing suspended matter (solid particles, liquid drops, gas bubbles) in a highly disperse state (Fig. 22.2a). Under the influence of an electric field, these particles start to be displaced in the direction of one of the electrodes. Often this movement

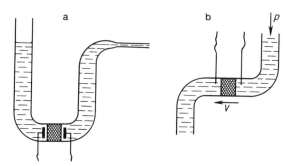

Fig. 22.1. Schematic design of cells for studying electroosmosis (a) and streaming potentials (b); the velocity of electroosmotic transport can be measured in terms of the rate of displacement of the meniscus in the capillary tube (in the right-hand part of the cell).

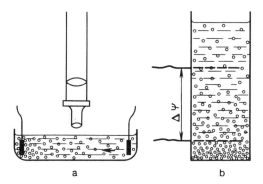

Fig. 22.2. Schematic design of cells for studing electrophoresis (a) and sedimentation potentials (b).

is toward the negative electrode or cathode; hence, electrophoresis has occasionally been called cataphoresis.

In 1859 Georg Hermann Quincke described a phenomenon which is the converse of electroosmosis: when an electrolyte solution is forced through a porous diaphragm by means of an external hydrostatic pressure P (Fig. 22.1b), a potential difference called the *streaming potential* arises between indicator electrodes placed on different sides of the diaphragm. Exactly in the same sense, in 1880 Friedrich Ernst Dorn described a phenomenon which is the converse of electrophoresis: during the sedimentation or floating up of suspended particles under the force of gravity (or, e.g., the centrifugal force in a centrifuge), a potential difference called the *sedimentation potential* arises between indicator electrodes set up in the top and bottom part of the solution (Fig. 22.2b).

All four processes have the same origin, since they are all based on the phenomenon of slip of the liquid along the surface of the other phase when a tangential electric field is present, or conversely, on the phenomenon that an electric field will arise during slip of the liquid.

The electrokinetic processes can actually be observed, only when one of the phases is highly disperse, i.e., with electrolyte in the fine capillaries of a porous solid in the cases of electroosmosis and streaming potentials, with finely divided particles in the cases of electrophoresis and sedimentation potentials (we are concerned here with degrees of dispersion where the particles retain the properties of an individual phase, not of particles molecularly dispersed such as individual molecules or ions). These processes are of great importance in particular for colloidal systems. Theoretically, electrokinetic processes should also occur in nondisperse systems, but then additional factors arise (vortex formation in the liquid, settling of the particles, etc.) which produce a strong distortion. Hence, electrokinetic processes can be regarded as one of the aspects of the electrochemistry of disperse heterogeneous systems.

Electrokinetic processes only develop in dilute electrolyte solutions. The second phase can be conducting or nonconducting. Processes involving insulators are of great importance, since they provide the only way of studying the structure and electrical properties of the surface layer of these materials when they are in contact with the solution. Hence, electrokinetic processes can also be discussed as one of the aspects of insulator electrochemistry.

Auxiliary electrodes are placed into in the solution in order to set up the electric field that is needed to produce electrophoresis or electroosmosis. Under these conditions an electric current passes through the solution and the external circuit; its value depends on the applied voltage and on solution conductivity. The lower this conductivity, the higher will be the electric field strength E (or ohmic voltage drop) in the solution that can be realized at a given value of current.

It is important to point out that electrokinetic transport processes have no direct connection with the electrochemical reactions occurring at the auxiliary electrodes during current flow, nor to current flow in the solution (even though as a rule the transport is proportional to current, see below). In electrophoresis, the finely divided particles move in the direction of one of the electrodes and, when adhesion is sufficiently strong, can deposit on it forming relatively thick layers. Transport of these particles and their subsequent deposition on the electrode do not obey the laws of Faraday [Eq. (1.44)], in contrast to the migration and electrochemical deposition of metal ions. Apparent transport numbers calculated via Eq. (10.24) are usually much larger than unity (they can be as high as tens of thousands). Solvent transport during electroosmosis is also much greater than that in the solvation sheaths of ions migrating in the electric field.

Transport processes of this type are called *nonfaradaic transport*. The nonfaradaic transport considered here is a steady-state process, in contrast to nonfaradaic currents which had been mentioned previously and which were due, e.g., to charging of the electric double layer.

Electrokinetic processes are of great practical significance, as will be discussed in Section 22.3.

22.1. ELECTROKINETIC POTENTIAL

22.1.1. The Metal/Solution Interface

The electrokinetic processes have electrostatic origins; they are linked to the charges present on both sides of the slip plane close to the phase boundary. The charge and potential distribution in the surface layer can be described by the relations and laws outlined in Chapter 12.

Consider in more detail the phenomena occurring at the interface between the solution and a small metallic particle involved in electrophoresis. Upon contact with the electrolyte solution, the metal acquires a certain value of electrode potential E, which may be the equilibrium value (with respect to ions of the same metal in the solution or to any redox system) or a nonequilibrium value. In accordance with this potential, there will exist a certain Galvani potential φ_G and, depending on the composition of the electrolyte solution (and particularly on any surface-active ions that may be present) a certain value of the interfacial potential Φ and certain type of ionic double-layer structure, particularly of its diffuse part.

As the particle moves relative to the electrolyte solution, the layer of water molecules which is directly adjacent to the particle surface is strongly bound and will be pulled along. The thickness of this bound layer is approximately one or two diameters of a water molecule. We shall write x_k for the x-coordinate of this layer's outer boundary, which is the slip plane. The electrostatic potential at this plane relative to the potential in the bulk solution is designated with the Greek letter ζ and called the *zeta potential* or *electrokinetic potential* of the interface discussed. This potential is a very important parameter characterizing the electrokinetic processes in this system.

It follows from the definition cited that the size of the zeta potential depends on the structure of the diffuse part of the ionic edl. At the outer limit of the Helmholtz layer (at $x = x_2$) the potential is ψ_2, in the notation adopted in Chapter 12. Beyond this point the potential asymptotically approaches zero with increasing distance from the surface. The slip plane in all likelihood is somewhat further away from the electrode than the outer Helmholtz layer. Hence, the value of ζ agrees in sign with the value of ψ_2 but is somewhat lower in absolute value.

When a tangential electric field is present, all charges which exist in the surface layers of the particles from the surface out to the slip plane will be pulled along by the particles, i.e., the charges $Q_S^{(M)}$ on the surface of the metal itself and the charges $\Delta Q_S^{(E)}$ of opposite sign in the solution in the region between x_1 or x_2 and x_k (here and in the following, Q_S will always be the symbol for surface charge density). The effective charge, $(Q_S^{(M)})_k$, of the moving particle [for which we can write: $(Q_S^{(M)})_k = Q_S^{(M)} - \Delta Q_S^{(E)}$] is compensated by charge of opposite sign in more distant regions of the diffuse edl part.

All factors influencing the potentials of the inner or outer Helmholtz plane will also influence the zeta potential. For instance, when owing to the adsorption of surface-active anions a positively charged metal surface will, at constant value of electrode potential, be converted to a negatively charged surface (see Fig. 12.11, curve 2), the zeta potential will also become negative. The zeta potential is zero around the point of zero charge, where an ionic edl is absent.

It also follows from what was said, that a zeta potential will be displayed only in dilute electrolyte solutions. This potential is very small in concentrated solutions where the diffuse edl part has collapsed against the metal surface. This is the explanation why electrokinetic processes will develop only in dilute electrolyte solutions.

22.1.2. The Insulator/Solution Interface

Insulators lack free charges (mobile electrons or ions). At interfaces with electrolyte solutions, steady-state electrochemical reactions involving charge transfer across the interface cannot occur. It would seem, for this reason, that there is no basis at this interface for the development of interfacial potentials.

Experience shows, however, that this is not so. A number of processes exist which will lead to charge accumulation in the surface layers of insulators. In the interaction with aqueous solutions a dissociation of acidic or basic functional groups existing on the insulator surface may occur, also a hydrolysis of saltlike groups. These processes can be reinforced by a partial interaction between the surface layer and the solvent and by swelling of this layer. Polar or charged surface groups may also arise upon rupture of chemical bonds directly at the surfaces of individual crystallites. An important source of electric charges at solid surfaces is the specific adsorption of solution ions.

The charges present on the insulator surface in contact with the solution give rise to an accumulation of ions of opposite sign in the solution layer next to the surface, and thus formation of an electric double layer. Since straightforward electrochemical measurements are not possible at insulator surfaces, the only way in which this edl can be quantitatively characterized is by measuring the values of the zeta potential in electrokinetic experiments (see Section 22.2).

Experiments show that in highly dilute electrolyte solutions (with concentrations between 10^{-5} and 10^{-7} M), the zeta potentials at insulators can be as high as ± 0.2 V. Often (but not always) the zeta potentials of solid insulators are negative. In the past this fact has been linked to Alfred Coehn's famous rule (1898), according to which the insulator with the lower permittivity ε during triboelectrification with another insulator will charge up negatively. However, the rule cannot be applied to conductors (the electrolyte solution). The negative surface charge of many insulators such as silicates (glasses) is due to the dissociation of acidic surface groups. Proton-accepting surfaces (metal oxides or hydroxides) will charge up positively. Experience shows that the specific adsorption of solution ions is very important for the charging of insulators. For instance, when colloidal silver iodide is prepared from solutions of potassium iodide and silver nitrate, the AgI particles will charge up negatively owing to anion adsorption when an excess of KI is present (relative to the stoichiometric requirement for AgI precipitation), but when an excess of $AgNO_3$ is present, they will charge up positively owing to cation adsorption.

The absolute value of the zeta potential decreases with increasing solution concentration. Sometimes the concentration dependence is nonmonotonic. Solutions containing polyvalent ions with charge opposite in sign

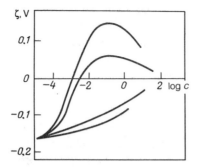

Fig. 22.3. Zeta potentials of a glass surface as functions of the concentrations of salt solutions with the cations (from bottom to top): K^+, Ca^{2+}, Al^{3+}, Th^{4+}.

to that of the surface have a particularly strong effect. Figure 22.3 shows plots of zeta potentials for a negatively charged glass surface against the concentrations of salt solutions having cations of different charge. We see that as the cation charge increases, the concentration required to depress the absolute value of ζ to a certain level decreases drastically. Charge reversal of the surface is observed at higher concentrations in the case of tri- and tetravalent cations. All these observations are in perfect agreement with the theories of edl structure (Chapter 12).

22.1.3. Electrochemical Properties of Colloidal Solutions

A solution is called colloidal when it contains particles having a size between approximately 5 and 200 nm, i.e., outside the limits where ordinary optical measurements are possible. The colloidal particles can be crystalline or constitute an amorphous agglomeration of individual molecules. The definition also includes nonaggregated large macromolecules, such as proteins. An arbitrary distinction is made between hydrophobic colloids (sols) and hydrophilic colloids (gels), which depends on the degree and type of interaction with the aqueous solvent.

Hydrophobic colloidal particles readily move in the liquid phase under the effect of thermal motion of the solvent molelcules (in this case the motion is called Brownian) or under the effect of an external electric field. The surfaces of such particles as a rule are charged (for the same reasons for which the surfaces of larger metal and insulator particles in contact with a solution are charged). As a result, an edl is formed and a certain value of the zeta potential developed.

A very important property of colloidal solutions is linked to this edl, viz., their stability against coagulation (coalescence of the particles and their precipitation as a separate phase). From the fact that they have a zeta potential, it can be concluded that the moving colloidal particles have some effective charge, $(Q_S)_k$. Because of this charge, electrostatic repulsion forces develop between the individual particles and prevent a close approach and coagulation of the particles moving about in the solution. The fact that the particles are charged and have a certain zeta potential also influences the rheological properties (viscosity) of the colloidal solutions.

As at other interfaces, the effective surface charge of colloidal particles depends on the total concentration and composition of the solution, and particularly on polyvalent or surface-active ions which may be present. When the zeta potential is reduced below a certain critical (absolute) value, which is approximately 25–30 mV, the colloidal solution becomes unstable.

22.2. BASIC EQUATIONS OF ELECTROKINETIC PROCESSES

The basic equations of electrokinetic processes establish the connection between zeta-potential values and the parameters of the electrokinetic processes, viz., the velocity of relative motion of the phases or the potential difference established in the solution as a result of such motion. These equations are predominantly used for calculating the zeta potentials of the various interfaces from parameters of the electrokinetic processes which have been found experimentally. The equations are used less often for the inverse problem, i.e., for calculating the velocity of motion from known values of the zeta potential.

The equations of the electrokinetic processes have been derived in 1903 by the Polish physicist Maryan Ritter von Smoluchowski, on the basis of ideas concerning the function of edl in these processes which had been developed by Hermann Ludwig Ferdinand von Helmholtz in 1879. Nowadays, these equations are often called the Helmholtz–Smoluchowski equations.

Electroosmosis. Consider the electroosmotic motion of a solution in a porous solid consisting of insulating material. We shall assume for the

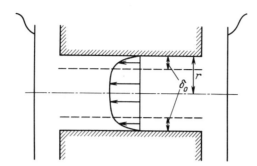

Fig. 22.4. Velocity distribution in electroosmotic solution flow in a cylindrical pipe (pore).

sake of simplicity that the pores are cylindrical and have a radius r. When an electric field having the field strength $\mathbf{E}$ is applied, the liquid will start to move under the influence of the electric force acting on the excess charges $(Q_S)_k$ present in a thin layer between the slip plane x_k and the outer limit of the edl in the diffuse edl part. We shall write δ_0 for the thickness of this layer (of slipping charges). Since the potential ψ of the solution approaches zero asymptotically, we cannot exactly define the coordinate of the outer limit of this layer. For this reason, we define thickness δ_0 in a manner analogous to diffusion-layer thickness in Chapter 4 (Fig. 4.7).

The charge density $(Q_S)_k$, the zeta potential and the thickness δ_0 are interrelated by the plate-capacitor relation [cf. Eq. (2.6)]

$$(Q_S)_k = \varepsilon_0 \varepsilon \zeta / \delta_0. \tag{22.1}$$

The electrical force acting on charge $(Q_S)_k$ is

$$f_e = \mathbf{E} \cdot (Q_S)_k = \varepsilon_0 \varepsilon \mathbf{E} \zeta / \delta_0. \tag{22.2}$$

The linear velocity of the liquid developing under the effect of this force is zero directly at the solid surface, and increases to some maximum value v at the distance $x = \delta_0$ from the surface. Solution regions further out lack the excess charges which could come under the effect of the external electric field, hence, there is no further increase in liquid velocity (Fig. 22.4). When the layer (δ_0) is much thinner than the capillary radius $(\delta_0 \ll r)$, the assumption can be made that the bulk of the solution moves with a uniform velocity v.

The velocity, v, of liquid motion can be found from the condition that the electric force should be compensated by the viscous force f_{fr}.

The latter is proportional to the velocity gradient in the layer of slipping charges:

$$f_{fr} = \eta \cdot v / \delta_0 \tag{22.3}$$

(η is the viscosity; for aqueous solutions $\eta \approx 1 \cdot 10^{-3}$ Pa $\cdot$ s). Equating the two forces, we find the desired equation linking the velocity of liquid motion and the zeta potential:

$$v = \varepsilon_0 \varepsilon \mathbf{E} \zeta / \eta. \tag{22.4}$$

At an electric field strength in the solution of, e.g., 100 V/m and a zeta potential of 0.05 V, the velocity of liquid motion will be $3.5 \cdot 10^{-6}$ m/s, which is of the same order of magnitude as the velocities of the ions in solutions.

The electric field also gives rise to a migration current density $\mathbf{E} \cdot \sigma$ [see Eq. (1.3)]. Hence, in Eq. (22.4) we can replace the parameter of field strength $\mathbf{E}$ by current density, which is more readily measured:

$$v = (\varepsilon_0 \varepsilon \zeta / \eta \cdot \sigma) \cdot i. \tag{22.5}$$

Sometimes, not the linear but the space velocity $\dot{V}$ of the liquid is used:

$$\dot{V} = v \cdot S = (\varepsilon_0 \varepsilon \zeta / \eta \cdot \sigma) \cdot I \tag{22.6}$$

(S is the total solution cross section in the porous solid).

A number of special features follow from the equations reported. The linear velocity of electroosmotic transport of the liquid is independent of the geometric parameters of the porous diaphragm (the size and number of pores, the thickness, etc.), and the space velocity only depends on S. At given values of the current, the transport rate increases with decreasing solution conductivity (increasing field strength).

Streaming potential. When the solution is forced through the porous solid under the effect of an external pressure P, the character of liquid motion in the cylindrical pores will be different from that in electroosmotic transport. Since the external pressure acts uniformly upon the full pore cross section, the velocity of the liquid will be highest in the center of the pore, and it will gradually decrease with decreasing distance from the pore walls (Fig. 22.5). The velocity distribution across the pore is quantitatively described by the Poiseuille equation

$$v_x = P[r^2 - (r - x)^2] / 4\eta \cdot l, \tag{22.7}$$

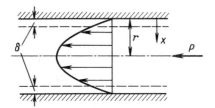

Fig. 22.5. Velocity distribution in the motion of a solution in a cylindrical pipe (pore) under the effect of the external force P.

where x is the distance from the pore wall (which has a maximum value of $x = r$) and l is pore length.

At the distance $x = \delta_0$ the expression for velocity can be written as

$$v_0 = Pr\delta_0/2\eta \cdot l \qquad (22.8)$$

(where it is taken into account that $\delta_0 \ll r$).

During motion of the solution, excess charges are transported which are present in the slip layer. This flux of charges is equivalent to the electrical current in the solution. Taking into account that the perimeter of the slip layer is close to $2\pi r$ we find for the current

$$I = 2\pi r(Q_S)_k \cdot v_0 = (\varepsilon_0 \varepsilon \pi r^2 \zeta/l \cdot \eta) \cdot P. \qquad (22.9)$$

Current flow in a pore of length l and total cross section S produces an ohmic potential drop in the solution, which is the streaming potential:

$$\varphi_{str} = I \cdot R = I \cdot l/\pi r^2 \sigma = (\varepsilon_0 \varepsilon \zeta/\eta \cdot \sigma) \cdot P. \qquad (22.10)$$

Like the velocity of electroosmosis, the value of the streaming potential is independent of geometric parameters of the porous solid through which the liquid is forced.

Electrophoresis. The physical situation of relative motions of a solution and another (insulating) phase during electrophoresis is exactly the same as in electroosmosis. Hence, the linear velocity of a cylindrical particle (which is the equivalent of a cylindrical pore) is given by the value following from Eq. (22.4). With particles of different shape, this velocity can be written as

$$v = \gamma(\varepsilon_0 \varepsilon \mathbf{E}\zeta/\eta), \qquad (22.11)$$

where γ is a coefficient allowing for particle shape. For spherical particles $\gamma = 2/3$.

Sedimentation potential. The equation for the sedimentation potential is the analog of Eq. (22.10). As in the case of electrophoresis, one can allow for the shape of sedimenting particles by using the coefficient γ. The hydrostatic pressure P is replaced by the "pressure" of the column of sedimenting (or floating) particles, which evidently is given by $gM[1 - (\rho_0/\rho)]$, where g is the acceleration of gravity (or other force), M is the mass of particles in a column measuring 1 m² in cross section, ρ_0 and ρ are the densities of the solvent and suspended matter.

Interrelations between the electrokinetic processes. Equation (22.4) for electroosmosis and Eq. (22.10) for the streaming potential, as well as the analogous equations for the other two electrokinetic processes, yield the relation

$$v/\mathbf{E} = \varphi_{\text{str}} \cdot \sigma/P. \tag{22.12}$$

This equation is valid, regardless of solution properties (the values of ε and η), surface properties (the value of ζ) and size of the disperse-phase elements. All parameters of this equation can be determined by independent measurements. The validity of Eq. (22.12) was demonstrated by such measurements. This result is an additional argument for the claim that all four of the electrokinetic processes actually obey the same laws and have the same physical origin.

Range of application of the equations deduced. The equations reported above are not entirely rigorous. A number of assumptions and approximations have been made when deducing them, and hence, the range of application of the equations is somewhat restricted. Motion of the solution has always been regarded as laminar. It was assumed that the second phase is an insulator, and hence will not distort the electrical field existing in the solution. It was assumed that an enhanced surface conductivity is not found close to the interface (this could for instance be caused by the higher concentration of ions in the edl), and that the values of dielectric constant and viscosity of the solution in the layer next to the surface are the same as those in the bulk solution. The assumption was used, finally, that the characteristic size of the disperse phase (e.g., pore radius r) is larger than the thickness δ_0 of the layer of slipping charges.

Departures of the electrokinetic behavior of real systems from that described by the equations reported occurs most often because of breakdown of two of the above assumptions, viz., because of marked surface conductivity (particularly in dilute solutions, where the bulk conductivity is low) and because of a small characteristic size of the disperse-phase elements (e.g., breakdown of the condition of $\delta_0 \ll r$ in extremely fine-porous diaphragms). A number of more complicated equations allowing for these factors have been proposed.

22.3. PRACTICAL USE OF ELECTROKINETIC PROCESSES

Electrokinetic processes are widely used in different fields of science and technology.

We had already mentioned the use of electrokinetic processes for research into the electric properties of surface layers of insulating materials. Such measurements are used, in particular, when studying the surface properties of polymeric materials, their behavior in different media and their interactions with other materials, e.g., with adsorbing surface-active substances. The results of this research are used in textile, cellulose and paper, and other industries.

An important technique for the qualitative and quantitative analysis of different macromolecular materials is based on the electrophoretic separation of particles having different transport velocities (e.g., because they have different zeta potentials). This technique is used for the analysis of proteins, polysaccharides, and other naturally occurring substances having a molecular size approaching that of colloidal particles (for more details see Section 23.3.4). It is an advantage of the electrophoretic method that mild experimental conditions can be used, viz., dilute solutions with pH values around 7, room temperature, etc., which are not destructive to the biological macromolecules.

In manufacturing, electrophoresis is used to apply coatings of different inorganic or organic materials to conducting substrates. The best known example is the protective coating of metal parts, e.g., car bodies, with primers or paint. Another example is the application of mixed coatings of barium and strontium compounds on tungsten wire used in cathode-ray tubes. Both aqueous electrolyte solutions and solutions on the basis of different organic or mixed solvents are used in these applications. The particles in the suspensions have a size between 1 and 50 μm. Voltages between several tens and several hundreds of volts are used to set up the electrical fields. Typical current densities in the (poorly conducting) solutions are 10 to 50 A/m^2. The electrophoretic application of a paint coat of 20 μm takes about 5 min (for comparison: electrolytic plating of a copper layer of same thickness with a current density of 50 A/m^2 would require more than three hours). The electrophoretic coatings obtained are very uniform, and the throwing power of the baths is high, i.e., one can coat parts of complex shape.

Electroosmosis is used to remove liquid (moisture) from different porous solids, e.g., in drying soil for building purposes (which improves the bond between the foundations and the soil). A combination of electrophoresis and electroosmosis is sometimes used to dry peat or clay. In this way, the water content of peat can be reduced from 90 to 55–60%. Unfortunately, the energy required for a further reduction of the water content is very high.

Chapter 23
BIOELECTROCHEMISTRY

Many of the physical and chemical processes and phenomena which are basic to the vital function of all biological systems are electrochemical in nature. It is the primary task of bioelectrochemistry to reveal the mechanisms and basic electrochemical features of such biological processes.

Electrochemistry and bioelectrochemistry have the same sources: they have emerged as sciences from the famous experiments of the Italian physiologist and anatomist Luigi Aloisio Galvani. In 1786, when experimenting with prepared frog legs he discovered a contraction of the muscles which was analogous to that provoked by the discharge of a Leyden jar, when he touched the muscle tissue with two pieces of metal which were dissimilar but in mutual contact. Recognizing the metals as conductors, but not as a source of electricity, he ascribed the effect to a special "animal electricity." The Italian physicist Alessandro Volta in 1794 offered a different interpretation for the effect when he pointed out that the origin of this "galvanic effect" is the contact between the two dissimilar metals and their contact with the muscle tissue. A dispute developed between Galvani and Volta which ended with a temporary victory of Volta. It was in fact discovered that when dissimilar metals are in contact with each other and with an ionic conductor, potential differences arise in the circuit which more specifically will give rise to muscle contraction. This was the starting point for the concepts of electrode potentials, which are a highly important component of modern electrochemistry. On the other hand, a number of observations were made in the first half of the 19th century from which it was concluded that potential gradients exist also in living tissue (in fact, even Galvani when repeating his experiments with identical metal pieces had obtained the same effect as before; but his contemporaries had paid no attention to these studies). Thus, Galvani's ideas were also vindicated, and it is from these ideas that bioelectrochemistry started.

Bioelectrochemistry is a science at the junction of many other sciences: electrochemistry, biophysics, biochemistry, electrophysiology, and others. The biological systems are extremely diverse in their constitution and detailed mechanism of functioning; each system has its own specific morphological and physiological features. In contrast to electrophysiology, bioelectrochemistry is only concerned with the general and basic laws of the electrochemical processes occurring in biological entities, and disregards the particular features of specific systems. For this reason bioelectrochemical studies often are conducted, not on natural objects but on synthetic model systems, e.g., artifical membranes.

From an electrochemical viewpoint, biological systems are highly branched circuits consisting of ionic conductors, viz., of aqueous electrolyte solutions and highly selective membranes. These circuits lack metallic conductors, but it has been found relatively recently that they contain sections which behave like electronic conductors, i.e., sections in which electrons can be transferred over macroscopic distances owing to a peculiar relay-type mechanism.

In the present chapter a brief outline of two major lines of modern bioelectrochemistry is given: studies of transmission of the nervous impulse (Section 23.1) and studies of the bioenergetic processes within cells (Section 23.2). Both are related to the mechanism of cell membrane function and to membrane structure. In Section 23.3 a number of points are discussed which are somewhat outside bioelectrochemistry, but which are relevant for the use of electrochemical methods in biological and medical problems.

23.1. TRANSMISSION OF THE NERVOUS IMPULSE

23.1.1. The Functions and Structure of Cell Membranes

Cells are the working units of all biological systems. All basic vital functions occur in cells: the reproduction of genetic information, chemical metabolism and biosynthesis, the bioenergetic accumulation and transformation of the different kinds of energy, etc. The nerve cells of multicellular organisms transmit the information from various receptors (the organs of sight, hearing, etc.) and internal controlling organs to the central nervous system and brain for analysis of this information, and for the transmission of response instructions to the executive organs (muscle, endocrine, or other system). The nerve cells of the peripheral nervous system have long, thin (dendritic) processes, the axons, which provide possibilities for a direct transmission of needed information over long distances. Muscle cells, too, have the shape of long, thin fibers through which information and commands can be transmitted.

In their vital functions the cells are in constant interaction with their surroundings. Various chemical substances enter the cell, and others leave it; energy effects are also possible (the inflow of photons). Cells are surrounded by thin membranes (surface or plasma membranes) which enclose them and jointly with them regulate all kinds of exchange and interaction with the surroundings. These membranes are 7 to 15 nm thick. Inside the cells again different kinds of intracellular membrane are found which have an extremely strongly extended surface area (folds, numerous bulges, etc.). These membranes surround a nucleus and are the constituents of different cell organelles: the mitochondria, the chloroplasts, etc. Intracellular membranes are 5 to 9 nm thick.

Both the intracellular and the plasma membranes are actively involved in the cell's vital functions. In the surface membranes of axons, processes of information transfer in the form of electrical signals ("nerve impulses") take place. Bioenergy conversion processes occur at the intracellular membranes of the mitochondria and chloroplasts.

The intracellular and plasma membranes have a complex structure. The main components of a membrane are lipids (or phospholipids) and different proteins. Lipids are fatlike substances representing the esters of one di- or trivalent-alcohol and two aliphatic fatty-acid molecules (with 14 to 24 carbon atoms). In phospholipids, phosphoric acid residues $-O-PO(O^-)-O-$ are located close to the ester links $-CO-O-$. The lipid or phospholipid molecules have the form of a compact polar "head" (the ester and phosphate groups) and two parallel, long nonpolar "tails" (the hydrocarbon chains of the fatty acids). The polar head is hydrophilic and readily interacts with water, while the hydrocarbon tails to the contrary are distinctly hydrophobic.

Cell membranes consist of two layers of oriented lipid molecules ("lipid bilayer membranes"). The molecules of these two layers are with their hydrocarbon tails toward each other, while the hydrophilic heads are outside (Fig. 23.1a). The mean distance between lipid heads is 5 to 6 nm. Various protein molecules having a size commensurate with layer thickness "float" in the lipid layer. Part of the protein molecules are located on

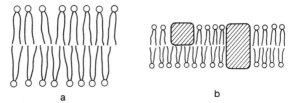

Fig. 23.1. Lipid bilayer membrane (a) and the location of protein molecules in it (b).

the surface of the lipid layer, others thread through the layer (Fig. 23.1b). Thus, the membrane as a whole is heterogeneous and has mosaic structure.

23.1.2. Electrochemical Properties of Cell Membranes

All biological systems contain aqueous electrolyte solutions. These solutions consist of strong electrolytes (inorganic salts) as well as various organic substances with acidic or basic functional groups which usually behave as weak electrolytes. The solutions are often gellike in their consistency because of the polyelectrolytes, proteins, and other macromolecules contained in them. The pH values of biological solutions as a rule are between 6.7 and 7.6.

An outer cell membrane separates the intracellular solution or cytoplasm from the extracellular solution. These two solutions differ in their compositions. The extracellular fluid mainly contains Na^+ and Cl^- ions (0.1 to 0.5 M) as well as minor amounts of K^+, Ca^{2+}, and Mg^{2+} ions, while the cytoplasm has a high concentration of K^+ ions (0.1 to 0.5 M) and low concentrations of Na^+ and Cl^- ions. Principal anions in the cytoplasm are the relatively large anions of different organic acids, including polyanions. As an example we report the major inorganic ions contained in the extra- and intracellular solutions of frog muscle (in mM):

	Na^+	K^+	Cl^-
outside the cell	120	2.5	120
inside the cell	9.2	140	3.5

The specific resistance of the extracellular fluid is 0.2 to 2 $\Omega \cdot m$. The ionic mobilities and specific resistance in the cytoplasm are of the same order of magnitude as those in the extracellular fluid, despite the gellike consistency commonly associated with the cytoplasm.

The membranes also have a certain, though small, ionic conductivity. The electrical resistance of membranes when referred to unit surface area is 10^{-2}–10^{-1} $\Omega \cdot m^2$, which when allowing for the small membrane thickness (about 10 nm) corresponds to the rather high value of specific (volume) resistance of 10^6–10^7 $\Omega \cdot m$.

This high resistance is due to the difficult transfer of hydrated ions through the hydrophobic part of the lipid layer (the permeability of the membrane for nonhydrated organic substances is two to three orders of magnitude higher than that for ions). The transfer can be accomplished in two ways: via special ionic channels formed by the protein molecules, or by ionophores. The latter are large, mobile organic molecules wrapping the ions to be transferred from all sides with a "fur coat" of hydrophobic groups. Such complexes form on one side of the membrane, pass through the lipid layer and decompose on the other side. A distinguishing feature of all these versions of transport is their high selectivity. Hence, the formal values of the mobilities and transport numbers of different ions in the membrane may be highly different.

A very important property of the membranes are the membrane potentials which are established. For biological membranes, the membrane potential is defined as the potential in the fluid within the cell relative to that in the fluid outside the cell: $\varphi_m = \psi^{(i)} - \psi^{(e)}$. Nowadays techniques are available for direct measurements of membrane potentials in various cells; they use microelectrodes or microprobes in the form of Luggin glass capillaries highly drawn out (with diameters of less than 1 μm) which can be inserted into the cell and into the extracellular fluid.

In the cell's nonexcited state (cf. Section 23.1.3) the membrane potential, φ_m, is always negative, i.e., the cytoplasm is negatively charged relative to the extracellular solution. For different cell types the values of φ_m vary between -50 and -100 mV; for frog muscle the value is -90 mV.

Two questions arise in connection with the ionic permeability present in membranes: how can the ionic concentration gradients between the two sides of the membrane be preserved despite of "leakage" of ions, and how is the membrane potential related to these gradients. These questions have been explored in extended studies still continuing today.

One of the founders of experimental neurophysiology, the German Julius Bernstein, in 1902 advanced the concept that biological membranes are permeable only to K^+ ions and not to any other ions. As a result, a Donnan equilibrium is set up (Section 5.4.1) and a certain membrane potential develops the value of which is determined, according to Eq. (5.26), by the K^+ ion concentrations in both solutions. These concentrations in turn depend on the concentrations of other, nonpermeating ions, and particularly on the concentration of organic anions in the cytoplasm. The potential gradient which develops compensates the effect of the concentration gradient (the electrochemical potentials, $\tilde{\mu}_{K^+}$, of the potassium ions in the two solutions become identical), and hence the concentrations of these ions will not become equal. For frog muscle, a value of -103 mV is calculated for the membrane potential from Eq. (5.26), which is close to, though a little more negative than, the experimental value.

In the 1940s a number of phenomena were discovered which do not fit these ideas. It was shown in particular that biological membranes are permeable, not only to K^+ ions but also to Na^+ and Cl^- ions. It is true that these membranes have different permeabilities for different ions. The permeability (the fluxes of ions crossing the membrane under the effect of identical "driving forces," i.e., identical concentration and potential gradients) can be measured with the aid of tracer atoms. Such measurements have shown that the permeability for Na^+ ions is about 75 times lower than that for the K^+ ions. Despite this difference, the total flux of Na^+ ions from the external solution into the cytoplasm (along the potential gradient) is substantially the same as that of K^+ ions in the opposite direction (against the potential gradient). The large organic ions present in the cytoplasm practically do not cross the membrane.

Membrane permeability for the Cl^- ions is not in contrast to the conclusion that a simple membrane equilibrium like that described in Section 5.4.1 is established at the membrane. In fact, the membrane potential calculated for the above example with Eq. (5.26) from the Cl^- ion concentration ratio is exactly -90 mV, i.e., the Cl^- ions in the two solutions are in equilibrium, and there is no unidirectional flux of these ions.

Another situation is found for the Na^+ ions. When the membrane is permeable to these ions, even if only to a minor extent, they will be driven from the external to the internal solution, not only by diffusion but, when the membrane potential is negative, also under the effect of the potential gradient. In the end, the unidirectional flux of these ions should lead to a concentration inside which is substantially higher than that outside. The theoretical value calculated from Eq. (5.26) for the membrane potential of the Na^+ ions is $+66$ mV. Therefore, permeability for Na^+ ions should lead to a less negative value of the membrane potential, and this in turn should lead to a larger flux of potassium ions out of the cytoplasm and to a lower concentration difference of these ions. All these conclusions are at variance with experience.

Thus, the above ideas do not suffice for an interpretation of all experimental results. These ideas include the assumption that the ions move in the membrane, only under the effect of concentration and potential gradients (diffusion and migration), and that transport of one sort of ions is independent of the transport of other sorts of ions (the law of independent migration of ions, see Section 10.4.3). This transport of ions under the effect of external forces has been named passive ionic transport.

In the attempt to overcome the contradictions, the assumption must be made that aside from passive transport in the cell, an active transport of Na^+ ions from the cytoplasm to the external solution is accomplished by the action of peculiar molecular "pumps."

The possibility of active transport of substances across membranes had first been pointed out in the middle of the 19th century by the physiologist Emil Heinrich du Bois-Reymond, a German of Swiss descent. The ability to accomplish active transport of ions and uncharged molecules in the direction of increasing electrochemical potentials is one of the most important features of cell membrane function. The law of independent ionic migration as a rule is violated in active transport.

The ionic composition of the cell is maintained by operation of the sodium–potassium pump, which pumps ions from more dilute to more concentrated solutions, viz., the Na^+ ions from the cytoplasm to the solution outside, and the K^+ in the opposite direction. We shall not discuss the details of the molecular operating mechanisms of this pump (of which many particulars are still not clear), but only point out certain features of its action. The transport of the Na^+ and K^+ ions is coupled; the ions are pumped, strictly in a 2:3 ratio. It follows that transport is the result of a cyclic process; in one half period the Na^+ ions are transferred in one direction, and in the other half period the K^+ ions are transferred in the opposite direction.

Active transport against the action of external forces requires the expenditure of a certain energy. This energy is 0.1–0.2 eV for transport of each Na^+ ion in the direction indicated (for the example reported above: 90 meV for overcoming the potential gradient, and 66 meV for overcoming the concentration gradient, which is a total of 156 meV). For the K^+ ions the transport energy is much lower (10–20 meV), since here the potential gradient aids in overcoming the concentration gradient.

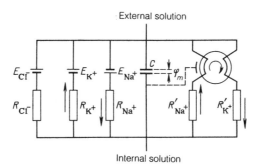

Fig. 23.2. Electrical equivalent circuit of a cell membrane.

The energy needed to transport ions across the membrane is obtained by the cell in chemical reactions occurring in it, viz., the oxidation of organic substances with oxygen (for more details see Section 23.2). Every second about 10^{11}–10^{12} ions are transported across 1 m^2 of membrane area. This process requires 20–30% of all energy generated by the cell. It has been calculated that the total power of the ionic pumps in the cells of the brain is about one watt.

The setting up of a membrane potential φ_m is associated with the formation of an electric double layer which has major part of its charges close to the (outer and inner) membrane surfaces, within the range of the lipid heads. A temporary change in charge density in the edl (e.g., by supply of charge from outside) will produce a change in membrane potential. By operation of the ion pumps and ionic redistribution, the original state of the membrane is reestablished quite rapidly when the external action has ceased. The membrane's specific electrical capacitance can be determined by direct measurements. It is about 10^{-2} F/m^2. According to the electrostatic capacitor relation (Section 2.1), this is exactly the capacitance of a layer 5 nm thick where the relative permittivity ε is about six.

Allowing for all these data one can describe the electrochemical properties of a cell membrane with the electrical equivalent circuit shown in Fig. 23.2. Here the capacitor C_m simulates the membrane's capacitance. Between its plates the potential difference φ_m is established, which is in fact the membrane potential. To the left of the capacitor the branches mediating passive transport (leakage) are shown, and to the right the branches mediating active transport. Each of the branches of passive transport include a power source E_j and a resistance R_j. The power source simulates the work of transfer due to the concentration gradient of the corresponding ion j. The voltage E_j of this source, which is the theoretical value of the membrane potential, is calculated from Eq. (5.26). Resistance R_j is the resistance to ionic transport. It is inversely proportional to the ionic mobilities in the membrane or the membrane's permeability for ions. The resistances for K^+ and Cl^- ions are about 10^{-2}–10^{-1} $\Omega \cdot m^2$; for the Na^+ ions the value is almost two orders of magnitude higher, as had been pointed out above. The resistances for the K^+ and Na^+ ions in the active transport branches, R_j', differ from the corresponding values, R_j, in the passive transport branches, since the transport mechanisms in these branches are different. The ion pump accomplishing coupled transport of the Na^+ and K^+ ions against the electrochemical potential gradients is shown in the right-hand part of the circuit. It has a feedback (shown as broken line in the circuit) which regulates its work depending on the value of φ_m, and maintains the required values of the concentration gradients and membrane potential.

It can be seen from this equivalent circuit that the actual value of membrane potential φ_m should be intermediate between the extreme values of the theoretical membrane potentials E_j for the different ions, i.e., between -103 mV for the K^+ ions and $+66$ mV for the Na^+ ions, in the example mentioned. However, since the resistance is much higher for the latter than for the former ions, the membrane potential actually should be closer to the theoretical value for the K^+ ions, which is confirmed by experiments ($\varphi_m = -90$ mV).

23.1.3. Excitation of Cell Membranes

Another important property of the outer membranes of nerve and muscle cells is their susceptibility to excitation under the effect of electric action. Excitation can be brought about, e.g., by an external electric current pulse. Pulses can be applied to the membrane with the aid of two microelectrodes, one residing in the extracellular fluid and the other introduced through the membrane into the cytoplasm.

When a current pulse is applied, part of the charges introduced will be consumed for changing the charge density in the edl, i.e., for changing the membrane potential. The potential change will depend on the value of current and length of the pulse. When positive charges are supplied to the cytoplasm by the current, the membrane potential will move in the positive direction (its negative value will decrease), which is the so-called membrane depolarization. When the current is in the opposite direction the negative value of φ_m will increase, which is the so-called membrane hyperpolarization.

Ordinarily, when the current pulse is over, the excess charges will be drained through the passive transport channels, and by operation of the sodium–potassium pumps the original values of membrane potential and of the concentration gradients will be reestablished. However, when in the case of depolarization the negative value of φ_m has dropped below a certain threshold value, which is about -50 mV, the picture changes drastically: excitation of the membrane occurs. When the current is turned off, now the membrane potential not only fails to be restored but continues to change in the positive direction. It goes through zero, attains values of 30–50 mV (membrane polarity reversal), and only after this excursion returns rather quickly to the original value (Fig. 23.3). This spontaneous surge or burst of potential has been called the action potential or spike. The burst lasts a few milliseconds. The character of the excitation (the amplitude and duration of the action potential) is independent of the original current pulse parameters; the only condition which must be met to produce this process is attainment of the threshold potential. A higher intensity and (or) longer duration of the original pulse only accelerate attainment of the threshold potential but do not alter the shape of the action potential. Following its return to the original state the membrane can again be excited, only after a certain time (a few milliseconds) during which it is in the refractory (nonexcitable) state.

It has been shown by extremely delicate experiments (including the use of tracer atoms) that membrane excitation occurs because of changes in permeability for the ions. The ionic permeability (or the resistance R_j to the ion flux during passive transport which is inversely proportional to it) is not constant but depends on potential. When the threshold potential is attained the permeability for Na^+ ions and hence the flux of these ions from the extracellular fluid into the cytoplasm increase drastically. This decrease in R_{Na^+} produces an additional shift of the membrane potential in the positive direction, which in turn causes a further increase in permeability and Na^+ ion flux, etc., i.e., the process is self-accelerating. The membrane potential shifts until it comes close to the theoretical value for the Na^+ ions (+66 mV). In this state the permeability for Na^+ ions is two to three orders of magnitude higher than the original value. With a certain time delay (of about 1–2 ms) the permeability for K^+ ions and their flux out of the cell also start to increase. This produces an opposite change of membrane potential, i.e., a shift in the negative direction. At the same

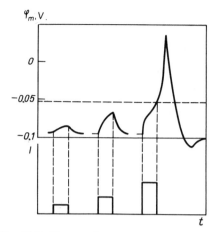

Fig. 23.3. Changes in membrane potential φ_m of a cell membrane occurring upon application of depolarizing current pulses of different amplitude I: (a, b) below-threshold; (c) excitation of the membrane during an above-threshold pulse.

time there is a drastic drop in permeability for Na^+ ions and, little later, for the K^+ ions. In the end the system returns to its original state.

Thus, in biological membranes (in contrast to the ordinary membranes described in Sections 5.4 and 17.2) we encounter a new phenomenon, the ability to regulate the ionic permeability by changes in membrane potential. The reasons for this effect are not completely clear. It may occur, both on account of changes in the properties and geometry of the ionic channels and a partial blocking of these channels by other ions ("gating" of the channels) and on account of changes in the conditions for ionic transport by ionophores.

23.1.4. Propagation of the Excitation

The idea that signals are transmitted along the nerve channels as an electric current had arisen already in the middle of the 19th century. Yet even the first measurements performed by Hermann von Helmholtz showed that the transmission speed is about 10 m/s, i.e., much slower than electric current flow in conductors. It is known nowadays that the propagation of nerve impulses along the axons of nerve cells (which in

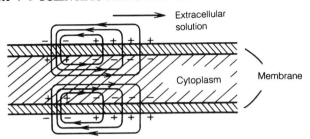

Fig. 23.4. Nerve impulse propagation by longitudinal currents in the extracellular solution and in the cytoplasm.

man are as long as 1.5 m) is associated with an excitation of the axon's outer membrane.

Each nervous impulse is a group of consecutive action potential bursts. It had been pointed out above that these bursts are identical in shape. The information to be transmitted by the impulse is encoded in terms of the number of individual bursts in a group and of the intervals between bursts.

The membrane of a nerve cell at first is excited in a narrow zone close to the point of signal input. The associated local changes in membrane potential and polarity give rise to potential gradients along the membrane as well as to longitudinal electric currents, both in the cytoplasm (in the direction from the excited to neighboring segments) and in the external solution (in the opposite direction). These currents depolarize the neighboring membrane sections (Fig. 23.4), i.e., produce a change in their membrane potential in the positive direction. Owing to the ohmic potential drop and also because of dissipation of the current in the extracellular solution, the amplitude of the induced potential change is lower than that of the original change, i.e., the signal is attenuated during transmission. But as long as the threshold potential is attained, neighboring sections become excited and will in turn relay the signal to the following section, and so on. Hence the excitation propagates along the membrane surface over large distances without any attenuation.

The longitudinal depolarizing effect propagates from an excited section in both directions. But owing to their refractory state, sections which had just been excited previously will not be reexcited, so the impulses continue in only one direction.

The energy needed for transmission of the undamped impulse is taken from the energy expended to maintain concentration gradients, i.e., in the final analysis from the energy that is generated by the cell for ion pump operation.

The speed of impulse propagation is determined by the time required for edl charging in neighboring membrane sections, and depends on edl capacitance and on the amplitude of the longitudinal current. The current in turn depends on the resistance of the extracellular solution and cytoplasm between neighboring sections, at a given potential gradient. The cross section of the cytoplasm is very small, so that its resistance (which is inversely proportional to cross section) provides the main contribution. The propagation speed is 20–30 m/s in squid axon, which has a relatively large cross section (about 1 mm^2). Much lower velocities are found for the usual axons with smaller cross sections.

The axons of the more complex nervous system of vertebrates have a special structure securing higher transmission speed. They are surrounded by a rather compact sheath of secondary cells (the myelin sheath) which completely insulates the membrane from the extracellular solution. It is only at certain points found at intervals of 1–2 mm that this sheath has gaps, the nodes of Ranvier, which are 1–2 μm wide, and where the membrane is in contact with the electrolyte. The insulated axon sections have a low capacitance, hence the longitudinal currents basically are used for the charging of more distant bare sections. The impulse thus jumps as it were along the nodes of Ranvier. The speed of impulse transmission in a myelinated axon is two to three orders of magnitude higher than that in a nonmyelinated axon having the same cross section.

Nerve impulse propagation along axons can be simulated by a rather simple electrochemical device (the model of Ralph Stayner Lillie, 1936). In this device an iron wire is immersed into concentrated nitric acid. Here the iron acquires a rather positive potential and becomes completely passivated. If the wire is activated at some point (e.g., by mechanical disruption of the passive film), the potential at this point immediately becomes more negative. Longitudinal currents have the effect that the passivating layer of a neighboring section begins to be reduced while the original section repassivates. The activation zone propagates at a certain constant speed all along the wire. In a more complex version of this model, a large number of iron electrodes are fused at certain intervals into glass tubing. The inside of the tubing is filled with a salt solution, the outside is in contact with concentrated nitric acid. When one of the electrodes is activated, the activation spreads in jumps to the next, and then to the following electrode, exactly as in myelinated axons.

In 1963 the British scientists Alan L. Hodgkin, Andrew F. Huxley, and Sir John Eccles were awarded the Nobel prize in medicine for their work on the excitation of cell membranes and of nerve impulse propagation.

23.2. BIOENERGETICS

23.2.1. Oxidizing Reactions in Cells

The energy requirements of the cells of living organisms are met by energy set free by chemical reactions involving the oxidation of organic nutrients by air oxygen. These reactions can conditionally be written as

$$(CH_2O) + O_2 \rightarrow CO_2 + H_2O, \tag{23.1}$$

where (CH_2O), a carbohydrate fragment, stands for typical nutrients. The thermodynamic parameters of reaction (23.1) are: $-\Delta H$ = 503 kJ/mol and $-\Delta G$ = 470 kJ/mol. In biological systems, oxidation reactions occur isothermally—"cold combustion"—and instead of yielding thermal energy (which would involve an enthalpy decrease) they immediately produce other forms of Gibbs energy, viz., the energy of chemical bonds in reaction intermediates or the energy of electric fields. Such reactions, which produce Gibbs energy, are called exergonic (as distinguished from exothermal). Analogous reactions proceed in fuel cells operated with organic reducing agents (fuels).

The pathway of the metabolic process converting the original nutrients, which are of rather complex composition, to the simple end products of CO_2 and H_2O is long and complicated, and consists of a large number of intermediate steps. Many of them are associated with electron and proton (or hydrogen-atom) transfer from the reduced species of one redox system to the oxidized species of another redox system (see Section 10.9.3). These steps as a rule occur, not homogeneously (in the cytoplasm or intercellular solution) but at the surfaces of special protein molecules, the enzymes, which are built into the intracellular membranes. Enzymes function as specific catalysts for given steps.

Considering the many steps involved in the overall reaction we may ask about the chemistry (stoichiometry) and energetics of each intermediate step. Through the work of numerous biochemists, the chemistry of most steps has now been determined and the enzymes revealed which catalyze each of these steps.

The energetics of each intermediate step can be characterized by the Gibbs energies of the corresponding redox systems or the equilibrium potentials of a nonconsumable electrode in contact with these systems [which are related to the Gibbs energies through Eq. (3.37)]. Using the scale of the reversible hydrogen electrode in the same solution, i.e., in a solution having a pH of about 7, we find a value of 1.22 V for the electrode potential of air oxygen with a partial pressure of 0.2 atm, and a value of 0.0 V for hydrogen; for major nutrients, the values are between -0.1 and $+0.1$ V (on the scale of the standard hydrogen electrode, all potentials given in the present section would be 0.41 V more negative).

In each oxidation step of the metabolic process the system's oxidation state becomes a little higher, and the electrode potential accordingly moves in the positive direction. On its way to complete oxidation (i.e., during a potential change of about 1.2 V) the system passes through several tens of intermediate steps, hence the energy change in each step is small and corresponds to an increase in electrode potential by 10–20 mV. Thus, each individual redox reaction occurs with very similar electrode potentials (Gibbs energies) of the redox systems involved in this reaction. The high rates of these reactions which are found at extremely low "driving forces" are the amazing special feature of biological systems, and attest to the extreme efficiency of enzyme catalysis.

The high catalytic activity of enzymes has a number of sources. Every enzyme has a particular active site configured so as to secure intimate contact with the substrate molecule (a strictly defined mutual orientation in space, a coordination of the electronic states, etc.). This results in the formation of highly reactive substrate–enzyme complexes. The influence of the individual enzymes also rests on the fact that they act as electron shuttles between adjacent redox systems. In biological systems one often sees multienzyme systems for chains of consecutive steps. These systems are usually built into the membranes, which secures geometric proximity of any two neighboring active sites and transfer of the product of one step to the enzyme catalyzing the next step.

The first steps of the metabolic process are the purely chemical enzymatic breakdown of the original nutrients into simpler, low-molecular-weight compounds. This breakdown occurs, first in the digestive tract and then in the cells. The basic oxidation steps occur at the membranes of the cell mitochondria (the higher the energy demand of a cell, the more mitochondria will be contained in it, which may be tens of thousands). Hydrogen atoms are split off from the low-molecular-weight organic compounds under the effect of the enzyme dehydrogenase. These atoms are transferred to molecules of the organic compound NAD (nicotinamide-adenine dinucleotide), which functions as a hydrogen acceptor and by this reduction process yields the compound $NADH_2$. The Gibbs reaction energy in the $NAD/NADH_2$ system corresponds to an electrode potential of about 0.08 V. The further oxidation (dehydrogenation) of $NADH_2$ then occurs purely electrochemically. The electrons and protons are transmitted through a multienzyme system called the electron transfer chain. The last step is electron and proton transfer from the enzyme cytochrome oxidase to an oxygen molecule existing as a hemoglobin complex. This oxygen reduction step is entirely analogous to electrochemical oxygen reduction without the intermediate formation of hydrogen peroxide (see Section 19.2.2).

A remarkable feature of the bioenergetic oxidation reactions of nutrients in cells is the fact that they are always coupled to another reaction, that of synthesis of the energy-rich chemical substance adenosine triphosphate (ATP) from adenosine diphosphate (ADP) and phosphate (oxidative phosphorylation, Vladimir A. Engelgardt, 1931):

$$A\text{–}PO_3^-\text{–}PO_3^-\text{–}OH + H_2PO_4^- \rightleftarrows A\text{–}PO_3^-\text{–}PO_3^-\text{–}PO_3^-\text{–}OH + H_2O \quad (23.2a)$$

(here A is the complicated heterocyclic compound adenosine) or, in an abbreviated form,

$$ADP + P \rightleftarrows ATP + H_2O. \quad (23.2b)$$

The phosphorylation reaction is also localized in the cell mitochondria. The enzyme ATP synthetase present in the mitochondrial membranes is involved in this reaction. The back reaction, which is the hydrolysis of ATP to ADP, occurs at other points of the cell and involves another enzyme, ATPase. The concentrations of the main reaction components, ADP and ATP, in cytoplasm are about 1 mM.

The synthesis of ATP is endergonic, and associated with an increase in Gibbs energy by about 30.7 kJ/mol. This energy is stored in the ATP molecule as the chemical bond energy between the third (outermost) phosphate group and the remainder of the molecule. These energy-rich chemical bonds are called macroergic. This energy is released in the back reaction, i.e., ATP hydrolysis to ADP.

In the oxidation of organic nutrients in the cell, on an average about six moles of ATP are synthesized for each mole of oxygen consumed. Therefore, the mean efficiency of conversion of the chemical energy in the oxidation of organic substances (470 kJ/mol) to the energy of ATP molecules is about 40%.

The ATP molecule is a universal energy carrier in the cell, and it is the principal one. It will sustain all reactions and processes requiring Gibbs energy: active ion transport (ion pump operation), biosynthesis of proteins and other substances, muscle contraction, etc. It is also employed for temporary energy storage in the cell.

The coupling mechanism of the phosphorylation of ADP molecules and of electron transfer through the membrane during the oxidation of organic substances is not entirely clear, and different viewpoints have been expressed in this respect. According to the chemiosmotic theory of Peter Mitchell (1961), concentration (or electrochemical-potential) gradients of the hydrogen ions develop in the oxidation of organic substances between the two sides of the mitochondrial membrane. As a result, a membrane potential of about 0.2 V is established while the solution on the outside is negatively charged. The electric field in the membrane acts as the driving force of the phosphorylation reaction. Subsequently the existence of a membrane potential has been demonstrated experimentally for the mitochondrial membranes. Thus, even here we have to do with an electrochemical mechanism through which the chemical energy of the oxidation reaction is converted, first to the energy of an electrical field, then to the chemical energy of ATP molecules.

23.2.2. Photosynthesis

Photosynthesis is the reverse of reaction (23.1), viz., the formation of carbohydrates and oxygen from water and carbon dioxide with solar energy.

Photosynthesis occurs in the chloroplasts contained in the cells of green plants. The chloroplasts hold two types of photosynthetic systems, which are called PS I and PS II. These systems operate in synchronization. Each system (containing one reaction site) holds about 200 molecules of chlorophyll (abbreviated Chl) and a set of different enzymes. Several different forms of chlorophyll molecules exist.

Chlorophyll absorbs photons having a wavelength of about 680 nm and an energy of about 1.8 eV. The primary act of the photosynthetic reaction is excitation of the chlorophyll molecules during photon absorption:

$$Chl + h\nu \rightarrow Chl^*. \tag{23.3}$$

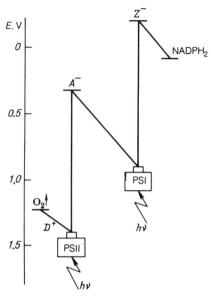

Fig. 23.5. Schematic representation of the energy
levels in photosynthesis.

This raises the electrochemical potential, $\bar{\mu}_j$, of these molecules and alters
the formal value of electrode potential in redox reactions involving the
chlorophyll (see Section 21.4).

In a system containing other redox systems, e.g., the systems
$A + e^- \rightleftarrows A^-$ and $D \rightleftarrows D^+ + e^-$ which in the dark are in equilibrium
with chlorophyll, redox reactions of the type of

$$Chl^* + D \rightarrow Chl^- + D^+ \tag{23.4a}$$

and

$$Chl^- + A \rightarrow Chl + A^- \tag{23.4b}$$

are possible after excitation of the chlorophyll. These reactions result in
a unidirectional electron transfer from donors D to acceptors A, which is
associated with an increase in electrochemical potential of these bound
electrons and transfer of the oxidizing functions to species D^+.

Photosynthesis occurs consecutively, first in PS II and then in PS I
(Fig. 23.5). The excited chlorophyll in PS II is a strong oxidizing agent
and has a redox potential of about 1.4 V. It can set free molecular oxygen
by oxidation of water molecules. According to Eq. (19.17), four electrons

are needed for the formation of one oxygen molecule, i.e., four chlorophyll molecules (each adding one electron upon excitation) and four photons. Hence the reaction proceeds through intermediate enzyme redox systems of the type of D^+/D permitting a temporary "accumulation" of electrons. These intermediate systems are not fully known.

The reducing agent A^- produced in the coupled reaction has a potential of about 0.3 V, i.e., is relatively weak and unable to reduce the CO_2 molecules to CH_2O groups. The latter process requires participation of a further photochemical system PS I involving another variety of chlorophyll molecules. In it the reducing agent A^- is oxidized in an analogous way via the intermediate redox systems Y^+/Y and Z/Z^-, and a strong new reducing agent Z^- is formed which has a redox potential of about -0.2 V. The photochemical reactions as such terminate with the formation of $NADPH_2$ (the reduced form of NAD phosphate; the redox potential is about 0 V). Further steps of CO_2 reduction and carbohydrate formation then occur as enzymatic reactions in the dark.

In the photosynthesis reactions occurring in each of these two consecutive stages, four photons are involved per molecule of oxygen evolved. In the redox steps immediately following excitation of the chlorophyll, the electrochemical potential of electrons in the reducing agents being formed increases by about 1.1 eV at each stage, i.e., by a total of 2.2 eV. It then decreases during the further redox steps owing to inevitable energy losses. This decrease is particularly marked in the transition from PS II to PS I (see Fig. 23.5).

The individual steps of the multistep chemical reduction of CO_2 with the aid of $NADPH_2$ require energy supply. This supply is secured by participation of ATP molecules in these steps. The chloroplasts of plants contain few mitochondria. Hence the ATP molecules are formed in plants, not by oxidative phosphorylation of ADP but by a phosphorylation reaction coupled with the individual steps of the photosynthesis reaction, and particularly with the steps in the transition from PS II to PS I. The mechanism of ATP synthesis evidently is similar to the electrochemical mechanism involved in their formation by oxidative phosphorylation; owing to concentration gradients of the hydrogen ions between the two sides of internal chloroplast membranes, a certain membrane potential develops on account of which the ATP can be synthesized from ADP. Three molecules of ATP are involved in the reaction per molecule of CO_2.

In photosynthesis, the reaction involves eight photons for each oxygen molecule, each with an energy of about 1.8 eV (175 kJ/mol). Hence the efficiency of solar-to-chemical energy conversion in photosynthetic systems with chlorophyll is about 40%.

23.3. ELECTROCHEMICAL METHODS IN BIOLOGY AND MEDICINE

23.3.1. Biological Macropotentials

In nerve impulse propagation, potential gradients develop in the extra-cellular solution, and ionic fluxes and electric currents appear in a direction along the membrane surface. The electrical fields of billions of synchronized cells add up, and in a strongly attenuated form can be measured at the outside of the living organism as biological "macropotentials." For an analysis of individual cell functions, this external potential is not of great interest, but sometimes it will characterize the function of the body or of entire individual body organs.

Biological macropotentials are widely used in medicine for diagnostic purposes: electrocardiograms to characterize heart action, electroencephalograms to characterize brain function, and electromyograms to characterize muscle activity. Inert (absolutely polarizable) metal electrodes are applied to different points of the body surface in order to do such measurements, and potential differences are determined between any two of these electrodes as functions of time (e.g., during contraction of the heart muscle). The potential differences that can be measured at the body surface are very small; in cardiograms they are tenths of a millivolt, and in encephalograms they are hundredths of a millivolt. Hence sensitive measuring devices with electronic amplifiers are used for such analyses.

Cases are known where the external potentials attain high values. In antiquity already, incomprehensible features of certain fishes caught attention. Around 1800 it became clear that these features are associated with electric phenomena, and they were attributed to so-called "animal electricity." It was in 1832, finally, that Michael Faraday could show that the different types of electricity including the animal variety are identical in nature. Studies of the "electric" fishes performed in the first half of the 19th century had a large effect on the development of bioelectrochemistry.

The electricity-producing system of electric fishes is built as follows. A large number of flat cells (about 0.1 mm thick) are stacked like the flat unit cells connected in series in a battery. Each cell has two membranes facing each other. The membrane potentials of the two membranes compensate each other. In a state of rest, no electrostatic potential difference can be noticed between the two sides of any cell or, consequently, between the ends of the stack. The ends of nerve cells come up to one of the membranes of each cell. When a nervous impulse is applied from outside this membrane is excited, its membrane potential changes, and its permeability for ions also changes. Thus the electrical symmetry of the cell is perturbed, and a potential difference of about 0.1 V develops between the two sides. Since nervous impulses are applied simultaneously to one of the membranes in each cell, these small potential differences add up, and an appreciable voltage arises between the ends of the stack. In the electric organ of fishes, a number of such stacks are connected in parallel and in series. The total voltage attains 500 V in the electric eel. A current pulse of about 0.5 A develops when this voltage appears across an external circuit (fresh water or seawater). For the electric ray, these numbers are 60 V and 50 A, respectively. The length of such an electric pulse is comparable with the time of cell membrane excitation, i.e., 1–2 ms, which is quite sufficient to defeat a designated victim. Some species of fish use pulses repeated at certain intervals.

23.3.2. Electrochemical Sensors

The chemical composition of biological objects is extremely complex. They contain the macromolecules of proteins, lipids, and many other substances in addition to low-molecular-weight organic and inorganic compounds. Different external effects can produce both quantitative and qualitative composition changes; some substances disappear, and (or) others appear. Some substances which are essential for the functioning of the cells or of the entire organism are present in very small concentrations, of 10^{-5} M and less.

The qualitative and quantitative analysis of such media is usually performed *in vitro*, i.e., on samples withdrawn from the biological organism. After the sampling, different methods of chemical analysis can be used, including the electrochemical methods described in Chapter 20. However, this method has the important defect that the samples withdrawn no longer interact with the remainder of the organism, which can affect their composition. Hence great efforts have been expended recently in developing *in vivo* methods of analysis, i.e., performing analysis directly in the living organism. The only methods of analysis which are suitable here are the electrochemical methods using different microelectrodes and microsensors introduced into the organism. It is an important special feature of the electrochemical methods that they allow continuous, long-term measurements to be made which could encompass different phases of vital activity of the biological object examined.

Almost all of the methods described in Chapter 20 can be used for *in vivo* analyses, both voltammetric and potentiometric ones. The former are mainly used in the analysis of organic substances which, within certain ranges of potential, can be either oxidized or reduced. Another popular method is the amperometric determination of oxygen in different biological media with the Clark electrode (Section 20.3.1).

Thin wires made of platinum metals or carbon fibers are used as the microelectrodes which are introduced or implanted into the test medium. It must be verified that the electrodes are compatible with the medium, e.g., that there is no clotting effect during contact with blood. Problems often arise owing to the adsorption of substances which are foreign to a given reaction, e.g., of proteins when determining the concentration of low-molecular-weight organic substances. In a number of cases membranes can be used (e.g., membranes glued to the electrode) which will hold back the foreign macromolecules but allow the substances to be examined to penetrate to the electrode surface.

The concentrations of hydrogen ions (solution pH) and of a number of other inorganic ions: Na^+, K^+, Ca^{2+}, Cl^-, PO_4^{3-}, etc., are determined potentiometrically. Glass electrodes are used for the analysis of hydrogen ions; various other types of ion-selective electrodes are used for the other ions. Electrodes with ion-selective solvent membranes have become very popular. These electrodes are made in the form of thin glass capillaries (about 1 μm in diameter) which in the lower part contain a small volume of a liquid that is immiscible with water; the remainder of the capillary is filled with electrolyte solution, e.g., 3 M KCl.

Both in voltammetry and in potentiometry, usually only one electrode (the working electrode) is introduced or implanted into the living organism. The auxiliary or reference electrode is positioned at the most convenient point, for instance at the outside of the organism. Since the working electrodes are extremely small, the currents in the measuring circuit are also small, sometimes hundredths of a microampere only. In certain cases, e.g., when thin capillaries are used, a high resistance arises in the measuring circuit, which could be hundreds or thousands of megohms. Under such conditions one must use very sensitive measuring instruments. Sometimes it may be necessary to provide for protection against extraneous electro-static interference (with a "Faraday cage").

23.3.3. Influence of Electric Currents on the Organism

If, as a result of electrochemical processes, electrostatic potential gradients and electric currents can arise in different sections of a cell or whole organism, then conversely currents or potential gradients applied from outside will produce certain changes in the cells and organisms. It is natural that these changes will depend on the electric field or current parameters.

In the limit a so-called electroshock, possibly with lethal results, can develop when a strong current acts upon the organism. Weak effects are often utilized for medicinal purposes. Methods are known where an electrical stimulation is applied to individual points of the body surface which will have effects on different organs. Such a stimulation is used in particular against pain (electroanaesthetization, electronarcosis), for influencing the psychic state (induce feelings of comfort or discomfort), for the "refreshing" of memory, etc. It is quite possible that the analogous effects of the well-known method of acupuncture rest on local electric pulses which develop under the effect of inserted needles and then spread over the body.

Often an electrical stimulation of cardiac activity is used in medical practice. Under normal conditions the electrical impulses needed for the uninterrupted rhythmic contraction of heart muscles are generated by a special organ, the sinoatrial node. External electrical pulses can be used when this organ's function is disturbed. The so-called pacemakers contain batteries with voltages of 3–5 V and a special electronic device generating electrical pulses with the required periodicity. The entire device is built into a titanium capsule (with a volume of about 50 ml) which is implanted in a convenient place of the body. The capsule also serves as the auxiliary electrode. The platinum or platinum–iridium working electrode is inserted through the vein directly into the muscle of the right ventricle of the heart. A current pulse of about 30 μA is quite sufficient for excitation of the cell membranes in this muscle and subsequent transmission of the impulse over the entire muscle system of the heart. The energy stored in the primary battery secures uninterrupted function of the pacemaker for 5 to 10 years.

Starting in about 1970, research efforts were intensified concerning the use of electrical current for faster union of fractured bones. The mechanism of this acceleration is still not clear. It is assumed that the changes in membrane potential of individual cells and in adsorption properties of the membranes have a substantial effect here. At first constant electric currents had been used, but later it was found that better results are obtained with current pulses. When using current pulses (ac) it is possible to do without the insertion of electrodes into the body (and the attending discomfort), and employ capacitive or inductive coupling to apply the current without contacts.

In recent years work was started in which this sort of phenomena are studied at the cell level, particularly the electrical breakdown of cell membranes and the fusion of neighboring cells under the influence of electric fields. These studies are of great value, in particular, for the further development of cell engineering and gene technology.

23.3.4. The Use of Electrokinetic Methods

It had already been pointed out in Section 22.3 that electrophoresis can be used for the qualitative and quantitative analysis of proteins, enzymes, and other naturally occurring macromolecular substances. According to Eqs. (22.1) and (22.11), the electrophoretic mobility of macromolecules in electric fields depends on their total charge, which differs between different molecules. For the analysis one uses zone electrophoresis (electrochromatography). The method essentially works as follows. The original mixture of macromolecules is introduced at a particular point into a horizontal glass tube filled with dilute electrolyte solution. When an electric field is applied along the tube the different molecules, owing to their different transport rates, will become spatially separated, and after a certain time will have gathered in different segments of the tube. The concentration of each individual fraction can be determined by optical techniques. Thickeners or various other devices which immobilize the solution, e.g., a strip of filter paper to soak up the solution (paper electrochromatography), are used in practice against the remixing caused by convection in the liquid solution. This method is often used in clinical practice in order to diagnose different illnesses from changes in composition of the protein molecules.

The same principle is used for the preparative separation of mixtures of biological materials, the extraction of different individual components from these mixtures, and also their purification. In this case one uses an electrophoretic method with continued introduction of individual portions of the mixture and withdrawal of portions of pure fractions. There have been reports that such processes were accomplished in spacecraft where, since gravitational forces are absent, the liquid solutions can be used without the danger of natural convection.

Electroosmosis is used in medicine for continuously or periodically introducing into the body liquid medications in rigorously defined quantities.

BIBLIOGRAPHY

A. Periodicals and Serials

A-1. *Electrochimica Acta*, Pergamon Press, Oxford.

A-2. *Journal of Electroanalytical Chemistry and Interfacial Electrochemistry*, Elsevier Sequoia, Lausanne.

A-3. *Journal of the Electrochemical Society*, The Electrochemical Society, Inc., Pennington, New Jersey.

A-4. *Elektrokhimiya*, Nauka, Moscow [in Russian; English translation: *Russian Electrochemistry*, Consultants Bureau, New York].

A-5. *Journal of Applied Electrochemistry*, Chapman and Hall, London.

A-6. *Journal of Power Sources*, Elsevier Sequoia, Lausanne.

A-7. J. O'M. Bockris, B. E. Conway, et al., editors, *Modern Aspects of Electrochemistry*, Vols. 1–3, Butterworths, London; Vol. 4 and up, Plenum Press, New York.

A-8. P. Delahay, C. W. Tobias, H. Gerischer, *et al.*, editors, *Advances in Electrochemistry and Electrochemical Engineering*, Wiley-Interscience, New York; continued as *Advances in Electrochemical Science and Engineering*, H. Gerischer and C. W. Tobias, editors, VCH, Weinheim D.

A-9. A. J. Bard, editor, *Electroanalytical Chemistry*, Marcel Dekker, New York.

A-10. *Electrochemistry. Specialist Periodical Reports*, Royal Society of Chemistry, London.

B. Books on Theoretical Electrochemistry

B-1. J. O'M. Bockris and A. K. N. Reddy, *Modern Electrochemistry*, 2 volumes, Plenum Press, New York (1970).

B-2. J. O'M. Bockris, B. E. Conway, E. Yeager, *et al.*, editors, *Comprehensive Treatise of Electrochemistry*, Plenum Press, New York, Vol. 1 (1980) to Vol. 10 (1985).

B-3. G. Kortüm, *Lehrbuch der Elektrochemie, 5. Auflage,* Verlag Chemie, Weinheim/Bergstrasse (1972).

B-4. J. Koryta, J. Dvořák, and V. Boháčková, *Electrochemistry,* Methuen, London (1970).

B-5. A. N. Frumkin, V. S. Bagotzky, Z. A. Iofa, and B. N. Kabanov, *Kinetics of Electrode Processes*, Moscow University, Moscow (1952) [in Russian; English translation available as AD 668917, NTIS, Springfield VA (1967)].

B-6. A. N. Frumkin, *Potentials of Zero Charge*, 2nd edition, Nauka, Moscow (1982) [in Russian].

B-7. P. Delahay, *Double Layer and Electrode Kinetics*, Interscience, New York (1965).

B-8. K. J. Vetter, *Elektrochemische Kinetik*, Springer, Berlin (1961) [English translation: *Electrochemical Kinetics*, Academic Press, New York (1967)].

B-9. L. I. Krishtalik, *Charge Transfer Reactions in Electrochemical and Chemical Processes*, Consultants Bureau, New York (1986).

B-10. R. Landsberg and H. Bartelt, *Elektrochemische Reaktionen und Prozesse*, Deutscher Verlag der Wissenschaften (VEB), Berlin (1977).

B-11. A. J. Bard and L. R. Faulkner, *Electrochemical Methods, Fundamentals and Applications*, Wiley, New York (1980).

B-12. J. S. Newman, *Electrochemical Systems*, Prentice-Hall, Englewood Cliffs, New Jersey (1973).

B-13. J. Koryta, *Ions, Electrodes, and Membranes*, Wiley, Chichester (1982).

B-14. E. Yeager and A. J. Salkind, editors, *Techniques of Electrochemistry*, Wiley-Interscience, New York, Vol. 1 (1972) to Vol. 3 (1978).

B-15. H. Bloom and F. Gutmann, editors, *Electrochemistry, The Past Thirty and the Next Thirty Years*, Plenum Press, New York (1977).

B-16. J. O'M. Bockris, D. A. J. Rand, and B. J. Welch, editors, *Trends in Electrochemistry*, Plenum Press, New York (1977).

B-17. R. Kalvoda and R. Parsons, editors, *Electrochemistry in Research and Development*, Plenum Press, New York (1985).

B-18. R. A. Robinson and R. H. Stokes, *Electrolyte Solutions*, 2nd edition, Butterworths, London (1959).

B-19. B. E. Conway, *Ionic Hydration in Chemistry and Biophysics*, Elsevier, Amsterdam (1981).

B-20. Y. Marcus, *Ion Solvation*, Wiley, Chichester (1986).

B-21. Yu. Ya. Gurevich, Yu. V. Pleskov, and Z. A. Rotenberg, *Photoelectrochemistry*, Consultants Bureau, New York (1980).

B-22. Yu. V. Pleskov and Yu. Ya. Gurevich, *Semiconductor Photoelectrochemistry*, Consultants Bureau, New York (1985).

B-23. S. R. Morrison, *Electrochemistry at Semiconductor and Oxidized Metal Electrodes*, Plenum Press, New York (1980).

B-24. J. O'M. Bockris and S. U. M. Khan, *Surface Electrochemistry: A Molecular Level Approach*, Plenum Press, New York (1993).

C. Books on Applied Electrochemistry

C-1. D. Pletcher, *Industrial Electrochemistry*, Chapman and Hall, London (1982).

C-2. U. Landau, E. Yeager, and D. Kortan, editors, *Electrochemistry in Industry: New Directions*, Plenum Press, New York (1982).

C-3. H. Kaesche, *Die Korrosion der Metalle, 2. Auflage,* Springer, Berlin (1978) [English translation: *Metallic Corrosion,* NACE, Houston, Texas (1987)].

C-4. V. S. Bagotzky and A. M. Skundin, *Chemical Power Sources,* Academic Press, London (1980).

C-5. M. M. Baizer, editor, *Organic Electrochemistry,* Marcel Dekker, New York (1973); M. M. Baizer and H. Lund, editors, *Organic Electrochemistry,* Second edition, Marcel Dekker, New York (1983).

C-6. F. Beck, *Elektroorganische Chemie,* Verlag Chemie, Weinheim/Bergstrasse (1974).

C-7. A. M. Bond, *Modern Polarographic Methods in Analytical Chemistry,* Marcel Dekker, New York (1980).

C-8. B. N. Vassos and G. W. Ewing, *Electroanalytical Chemistry,* Wiley, New York (1983).

C-9. Z. Galus, *Fundamentals of Electrochemical Analysis,* Horwood, Chichester (1976).

C-10. H. W. Nürnberg, editor, *Electroanalytical Chemistry [Advances in Analytical Chemistry and Instrumentation,* Vol. 10], Wiley, London (1975).

C-11. N. Lakshminarayanaiah, *Membrane Electrodes,* Academic Press, New York (1976).

C-12. R. Kalvoda, editor, *Electroanalytical Methods in Chemical and Environmental Analysis,* Plenum Press, New York (1987).

C-13. A. J. Fry, *Synthetic Organic Electrochemistry,* Wiley, New York (1989).

C-14. O. J. Murphy, S. Srinivasan, and B. E. Conway, *Electrochemistry in Transition,* Plenum Press, New York (1992).

D. Reference Books

D-1. A. J. Bard, editor, *Encyclopedia of Electrochemistry of the Elements,* Marcel Dekker, New York, Vol. 1 (1973) to Vol. 15 (1984).

D-2. C. A. Hampel, editor, *The Encyclopedia of Electrochemistry,* Reinhold, New York (1964).

D-3. B. E. Conway, *Electrochemical Data,* Elsevier, Amsterdam (1952).

D-4. R. Parsons, *Handbook of Electrochemical Constants,* Butterworths, London (1959).

D-5. M. Pourbaix, *Atlas d'équilibres électrochimiques,* Gauthier-Villars, Paris (1963) [English translation: *Atlas of Electrochemical Equilibria in Aqueous Solutions,* Pergamon, Oxford (1966)].

AUTHOR INDEX

SUBJECT INDEX

Absolute potential, 335–336, 526
ac measurements, 154–155, 184–192, 315
ac voltammetry, 509–510
Activated state, 362, 366–367
Activation energy, 275, 278, 361–366
Activation polarization, 124–134
    influence of reactant concentration,
        127–129
Activationless reactions, 368
Active ion transport, 559
Activity, 60, 214
    coefficients, 61–62, 240–243, 249–251
    concentration dependence of, 240–242
    of electrolyte solutions, 61–63, 236–240
Admittance, 185
Adsorption, 287–292
    in electrochemical systems, 291–292
    energy of, 289
    of hydrogen, 321, 325
    of ions, 305–310, 316–318, 320–321, 326
    isotherms, 289–291
    of organic substances, 318–320, 326
    of oxygen, 325–327, 432, 442–443,
        479–480
Amperometric titration, 500
Amperometry, 503
"Animal electricity," 553, 572
Anode, 16, 41
Anodic reaction, 18
Aprotic solvents, 270–271
Archie's law, 417
Arrhenius equation, 275, 361
Auxiliary electrodes (AE), 157, 168

Balance equation, 25–27, 85, 89, 91
Barrierless reaction, 368
BET method, 312

Binary electrolyte solution, 12
Biological macropotentials, 572–573
Bipolar electrodes, 410
Born equation, 230–231, 233–234
Born–Haber cycle, 230
Brønsted relation, 363
Brown–Walker reaction, 495
Buffer solutions, 264–265

Calomel reference electrodes, 163–164
Cataphoretic effect, *see* Electrophoretic effect
Cathode, 17, 41
Cathodic protection, 360
Cathodic reaction, 18
Cell membranes, 554–556
    excitation of, 561–566
    potentials of, 557–561
Charge–dipole interaction, 32
Charging current, 148, 180, 314
Charging curves, 314–315, 321, 433–434
Chemical power sources, 41, 398–401
Chemical yield, 413
Chronoamperometry, 176–177
Chronopotentiometry, 180–184
Clark oxygen sensor, 503–504, 574
Concentrations, different units for, 53–54
Conductivity, 5, 8
    of electrolyte solutions, 13
        influence of concentration on, 14,
            216–218
            methods of measurements of, 214–216
            temperature coefficient of, 216
    of ionic melts, 274–275
    of metals, 8
    of semiconductors, 9
Conductometric titration, 500
Conductometry, 499, 500–501

585